BIOTECHNOLOGY IN AGRICULTURE, INDUSTRY AND MEDICINE

BIOREMEDIATION

BIOTECHNOLOGY, ENGINEERING AND ENVIRONMENTAL MANAGEMENT

BIOTECHNOLOGY IN AGRICULTURE, INDUSTRY AND MEDICINE

Additional books in this series can be found on Nova's website under the Series tab.

Additional E-books in this series can be found on Nova's website under the E-books tab.

ENVIRONMENTAL SCIENCE, ENGINEERING AND TECHNOLOGY

Additional books in this series can be found on Nova's website under the Series tab.

Additional E-books in this series can be found on Nova's website under the E-books tab.

BIOTECHNOLOGY IN AGRICULTURE, INDUSTRY AND MEDICINE

BIOREMEDIATION

BIOTECHNOLOGY, ENGINEERING AND ENVIRONMENTAL MANAGEMENT

ALEXANDER C. MASON
EDITOR

Nova Science Publishers, Inc.
New York

Library of Congress Cataloging-in-Publication Data

Bioremediation : biotechnology, engineering and environmental management /
[edited by] Alexander C. Mason.
p. cm.
Includes index.
ISBN 978-1-61122-730-7 (hardcover)
1. Bioremediation. I. Mason, Alexander C.
TD192.5.B55696 2010
628.5--dc22

2010041282

Published by Nova Science Publishers, Inc. † New York

CONTENTS

PREFACE

Chapter 1 - Major reliance on rapid industrialization, urbanization and altered agricultural practices, past and current defense related activities and recent quest for demilitarization of conventional munitions is fraught with inexorable environmental pollution of nitro-aromatic high explosives (HEs), ordinance related compounds (ORCs), munitions and explosives of concern (MECs) and unexploded ordnances (UXO). Due to widespread and indiscriminate use of these nitro-aromatics (mono-, di- and tri-nitro-toluenes) as chemical feedstock for synthesis of explosives, pesticides, herbicides, dyes, pharmaceuticals, etc., they are recognized as recalcitrant to biological treatment, remain in biosphere and hence, categorized as Hazardous Rating-3. The transformation products poised ecological threats and causing growing concern about toxic, carcinogenic, mutagenic, and teratogenic effects to humans, and contaminants to natural receptors (soil, sediment, surface water, and ground water). These increasing threats to the environment provoked the necessity for their elimination. Several physical and chemical environmental cleanup approaches for elimination of recalcitrant nitro-toluenes are available. These processes are prohibitively costly, destructive and responsible for CO_2 and NO_X release. Therefore, attention has recently been focused on alternative biological approaches through varied microbial metabolic processes for complete and cost-effective detoxification of recalcitrant nitro-aromatic pollutants. Microbes, being the pioneer colonizers on earth, underpin all life processes and an essential component in biogeochemical processes and element cycling, maintaining the functioning of the global ecosystem by physiological and functional diversity. Moreover, notwithstanding with the performance assessment and scale-up effectiveness, variations based on approaches viz. (i) biostimulation, (ii) enhanced natural attenuation, (iii) bioaugmentation and (iv) synthetic biology have been emerged out. Of these, enhanced natural attenuation is achieved through exogenous supplementation by indigenous microbes, while, bioaugmentation relies on application of carrier-encapsulated whole cell and gene bioaugmentation to indigenous microbes to confer superior trait(s) for removal of nitro-toluenes. Contrary to these approaches, synthetic biology employs design and construction of new biological parts, devices and systems and the re-design of existing, natural systems for more effective biodegradation.

This article reviews key aspects of current and emerging remediation technologies underlying bioremediation. Subsequently, current knowledge regarding the application of bioremediation in alleviating contaminant toxicity is summarized highlighting the relative merits and demerits of different options. Finally, this review attempts to provide a concise appraisal on recent progresses on biodegradation, concluding with applications of bio-molecular approaches and 'omics' technologies in bioremediation of nitro-toluenes and their derivatives.

Chapter 2 - Perchlorate (ClO_4^-), primarily used in ordnance, explosives, and as a rocket propellant in the defense and aerospace industries, has contaminated the water supplies of more than 20 million people, primarily in the southwest United States (U.S. EPA, 2008). ClO_4^- is manufactured in 44 states, and ClO_4^- contamination is known in 35 states (U.S. EPA, 2008). Because of its high solubility, non reactivity, and poor absorption to soils, ClO_4^- persists and spreads rapidly throughout the environment (Logan, 2001a). Because ammonium perchlorate absorbs water from the air, ammunitions and rockets have a limited shelf life and the ammonium perchlorate must occasionally be replaced. As of 2006, an expected 165 million pounds of ClO_4^- will be disposed and will require treatment (Wallace et al., 1998).

Chapter 3 - In an international context, developed and developing countries must face global challenges such as economic crises and global warming. Latin-America and the Caribbean are highly vulnerable to the effects of global warming and must understand that facing these environmental issues require changes in public policies, green technologies, green jobs, environmental education, and international public affairs. The main challenge of developing countries is to reach a sustainable level of development without compromising said development. Solving conflicts between environment, industry, and society requires multidisciplinary approaches, with a strong focus on the best available technology, science and public policies.

Different approaches based on science and technology could give enough support to decision-makers to solve national and international concerns related to global warming, especially about historic relations among the industry, the environment, and communities. These interactions have been complicated due to environmental impacts, where topics range from exploitation of resources to cultural disturbances. For industries, a tipping point is to reach sustainable production while avoiding increasing the level of pollutants, an area where Bioremediation processes can be exploited as an environmental tool.

The use of innovative bioremediation, based on sustainability concepts, is necessary due to recent changes in waste management regulations around the world, especially in a Post-Kyoto scenario, which aims to build up a low-carbon economy with sustainable development. These approaches must take biotechnologies and sustainable engineering concepts forward incorporating a vision of industrial ecology, which focuses on a more integrated interaction among industries by practical, sustainable, and economic schemes, instead of treatment at the end of the pipe.

The bioremediation of polluted environments is a major environmentally friendly technology which can be employed by several kinds of industries around the world. The advantages of this technology are its feasibility to be applied to a wide range of environments and contaminants. Its disadvantage is the need of preliminaries and feasibility studies at a laboratory scale, using biotechnological tools, and its escalation to industrial processes. Currently, not much literature on sustainable bioremediation technologies to be applied together with industrial ecology exists. This chapter will relate bioremediation and industrial ecology, provide a review of the state of the art related research, provide successful cases of bioremediation applications in several industries, and show how industrial ecology could complement these approaches.

In particular, two cases of successful application will be described: one about bioremediation by means of composting fuel-contaminated desert-mining soils in the Atacama Desert (Chile) and the other in the use of biosolids as a flotation reagent inside the Chilean mining industry. Future needs for sustainable bioremediation and industrial ecology will be identified, and a discussion about technological management, public policies, and regulatory considerations to support environmental tools will be included. The advantages and disadvantages of industrial ecological tools for environmental management will also be discussed.

Chapter 4 - Petroleum is part of our biosphere for much longer than a thousand years. Petroleum is a complex mixture formed by oxygen-, nitrogen-, and sulfur-containing compounds and hydrocarbons. Quantitatively and economically, hydrocarbons are the most important compounds of the petroleum mixture (up to 85%). These compounds lack functional groups, since they are consisted exclusively of carbon and hydrogen which are characterized by apolar C-C and C-H bonds. They are distinguished according to the presence of π-bonds and cyclic structures and are classified as n-alkanes, alkenes or alkynes, and alicyclic or aromatic hydrocarbons, respectively. Hydrocarbon compounds can either have a biological origin, such as waste or metabolic overflow compounds, structural elements, and defense mechanisms, derived from plants, algae, animals, and microbes, or a chemical origin, through the incomplete combustion of organic matter (van Beilen *et al.*, 2003b; Norse, 2005). Actually, hydrocarbons are widespread in low concentrations in the environment; but oil spillages increase their concentration, making them toxic for most living forms.

Chapter 5 - Bioremediation has emerged as a promising treatment technology for restoring contaminated environments, where organisms degrade contaminants to non harmful end products. A growing field of research is investigating the way in which bioremediation can be optimised and applied to widely occurring pollutants in the soil column. As a common by-product of the petroleum industry, Polycyclic Aromatic Hydrocarbons (PAHs) have become a ubiquitous soil pollutant. Bioremediation of PAHs has a number of potential advantages over traditional contaminated site treatment approaches, which are often intrusive, labour intensive, and may themselves produce harmful by-products. A number of commonly occurring bacteria have been identified which can use PAHs as a sole carbon source for metabolic activity. A wide range of methods have been investigated to enhance the bacterial degradation of pollutants in both laboratory trials and broad field scale application. Research indicates that the principal limiting factors to bioremediation involve pollutant/microbe interaction and the ability of an organism to catabolise a pollutant. PAHs are highly hydrophobic and may often be sorbed deep within the soil column; they must be drawn out of soil pores before they may be accessed by microbes. Research has demonstrated that the use of surface acting agents including 'biosurfactant' is effective in increasing hydrocarbon solubility in soil and consequently, bioremediation activity. Recent research is increasingly focusing on the genetics of metabolic pathways by which organisms catabolise pollutants. Certain strains of commonly occurring bacteria contain genes which encode enzymes that degrade PAHs into non harmful end products while attaining carbon for metabolic processes. Enhancing bacterial growth, distribution and the expression of PAH degradation gene pathways may significantly enhance bioremediation activity. In order to characterise and optimise all aspects of bioremediation, novel research technologies are being developed in order to monitor and characterise the bioremediation process in soil. There is rapidly growing potential to monitor bacterial activity and gene expression in-situ with the use of novel genetic techniques. In recent research, 'biosensor' organisms have been created which transcribe genes for fluorescence or bioluminescence such as *gfp* (green fluorescent protein) and *lux* (luciferase) as reporters for activities such as growth and bioremediation. The development of this field is an exciting approach to developing an improved understanding of in-situ bacterial activity and advancing the application bioremediation in the field.

Chapter 6 - Heavy metal pollution is one the most serious environmental problems. Mining and industrial activity are the main sources of heavy metal contamination. Heavy metals, such as copper (Cu) and chromium (Cr), play an important role as trace elements in biochemical reactions but these heavy metal ions are toxic at higher concentrations, producing serious public health problems.

Traditional technologies for the removal of these heavy metals are very expensive and may have several disadvantages. New techniques are required for reducing their concentrations to environmentally acceptable levels at low costs.

Microorganisms are potent bioremediators. To date, studies on heavy metal resistant have mainly concerted in bacteria. Nevertheless, one of the most ubiquitous biomass types utilized on a large scale by man for centuries are yeasts, moreover this microorganism is an inexpensive, readily available source of biomass. Furthermore, yeast cells are able to accumulate a broad range of heavy metals to varying degrees under a wide range of external condition.

Recent researches have made emphasis to the yeasts isolated from contaminated area. This allows isolating indigenous yeasts able to live in these unfavorable environments.

Candida fukuyamaensis and *Rhodotorula mucilaginosa* were isolated from a copper filter at a mine plant. These yeasts show multiple heavy metal tolerance. They are able to remove Cu(II) of the culture medium by a process of bioaccumulation in the cytoplasm. The resistance mechanism in the yeasts is through sequestration in cytoplasmic of copper ions rather than avoidance or compartmentalization. However, the copper accumulation profiles of both strains are different: whereas the metal is later released by *C. fukuyamaensis*. *R. mucilaginosa* has the capacity to keep copper inside the cell. The toxicity of copper is evidenced for the morphological changes and the increased oxidative stress response by both strains. However the copper bioaccumulation by *R. mucilaginosa* is increased by adaptation of the strain under controlled aeration

On the other hand, *Lecythophora* sp. NGV-1, *Candida sp.* NGV-9 and *Aureobasidium pullulans* VR-8 isolated from tannery effluents and from a nickel-copper mine, show Cr(VI) tolerance. In these cases, the yeast Cr(VI) tolerant is due to the reduction of Cr(VI) to Cr(III) and no bioaccumulation. In vivo Cr(VI) toxicity is 1,000-fold more cytotoxic than Cr(III). Apart from its toxicity, Cr(VI) is highly soluble and thus mobile and biologically available in the ecosystems. In contrast, Cr(III) has a high affinity for organic compounds resulting in the formation of complexes that precipitate as amorphous hydroxides. The reduction of Cr(VI) to Cr(III) is an interesting approach for environmental clean up.

The ability of metal processing by yeasts can be used to concentrate, remove and recover metals from streams and could enhance the efficiency of wastewater treatment processes. This capacity depends of the specific interaction of strain and heavy metal. Different mechanisms of heavy metal resistance may affect different bioremediation strategies.

Chapter 7 - Due to rapid developments and increase in mining and industrial activities, excessive levels of heavy metals are being continuously added into the soils leading to accumulation of these toxic metals into the soils. Cadmium, copper, lead, mercury, chromium, nickel and zinc are considered the most hazardous heavy metals and are included on the EPA's list of priority pollutants. The existence and fate of these metals in soil is a matter of concern not only because of their potential impact on microbial communities and plants but also because of chances for groundwater contamination and hence toxicological impact on human health. The mobility of such heavy metals, as well as their bioavailability and related eco-toxicity, strongly depend on their specific chemical forms and hence, speciation of metals is a critical factor in assessing the potential environmental impacts. Unlike organic compounds, toxic metals are not degradable in the environment, and can persist in soils for decades or even centuries. Therefore, it is highly desirable to apply suitable remedial approaches to decontaminate the polluted soil. Though, various approaches have been suggested for remediation of metal-contaminated sites, recent years have witnessed an enormous growth in the field of bioremediation technology for the restoration of contaminated sites. Biosurfactant mediated bioremediation, a new field of endeavor and innovative approach which is getting wide attention due to their low toxicity, biodegradability and better environmental compatibility. Biosurfactants are surface active molecules produced by variety of micro-organisms capable of reducing surface and interfacial tension, thereby facilitating mobilization of the metal ions by forming biosurfactant-metal complex. Metal contaminants in the environment are usually tightly bound to colloidal particles and organic matter of the soil and applications of biosurfactant for removal of available as well as bound forms of these contaminants are well reported. Biosurfactants have diversity in their chemical forms by the virtue of which they can be selected on the basis of their high specificity for different metals and thus metal removal efficiency. The present chapter emphasizes on various classes of biosurfactants, different possible mechanisms of interaction of biosurfactant with (cationic and anionic) metals and their application in bioremediation of metals. As far as eco-restoration of the metal contaminated site is concerned; potential of such bio-molecules should be explored to lower down the concentration of heavy metals below the limits established as a safe or acceptable by regulatory agencies.

Chapter 8 - Natural stones like marbles, granite and dolomite were massively used for building construction during past centuries. High strength and durability of these stones made their uses more pronounced in constructing sculptural buildings and outstanding decorative artworks. The Taj Mahal, European Cathedrals and Roman Catacombs are few amongst them. With the introduction of different tools and their use in cutting, engraving and delicately carving of stones, the valuable and high artistic decoration was facilitated and outstanding beauty was imparted to these monuments and artifacts. Therefore, these monuments and artworks hold their own cultural and symbolic values. In the course of several centuries, from the moments they were built to the day they faced forfeiture they had seen many rise and falls of life. They bear signs of various civilizations, dominations and reign that help educating us about our past life, history, arts and literature. These heritages are our greatest natural treasures. Once decayed or devastated, it will truly become impossible for us to regain or reproduce them. During past few decades, these monuments had experienced extensive deterioration and aging. Persistent pollutants and human activity has posed a big threat to these sculptural monuments and artworks. Due to lack of appropriate knowledge of efficient repairing strategies, the previous traditional methods applied for restoration purposes turned inefficient and incompatible. Modification in the petrophysical properties of stone and altered aesthetic appearance were the characteristic features of these traditional methods. The failure of the tradition methods in restoration of physical strength and conservation of natural appearance has presented necessity of appropriate management tools and conservation strategies. It stimulated the researchers from all over the world to come together to elucidate, unravel and work out the issues. The global concern and consequently global involvement in characterizing the mechanism of deterioration and devising alternative remedial strategies became evident in the past few years. The aim of the studies in this field is to develop a reliable, compatible, and efficient strategies for conservation, preservation and restoration of the natural and cultural heritage of past.

This chapter detailed about the cause and mechanisms of biodeterioration, concisely summarizes the conservation strategies. Consequently, the scope of this chapter is extended to discuss the major research been carried out by some active research groups individually or in the framework of some dedicated research projects. Finally, molecular approaches and related techniques which had made the deterioration mechanisms more comprehensive are also briefly discussed.

Chapter 9 - Surfactant of biological origin has engrossed great deal of expectation as it finds wide applications. Biosurfactant has proven its metal in field of agriculture, bioprocessing, biomedical, cosmetics, environmental, food and petroleum industries. Some specific application like stabilization of nanoparticals has been greatly appreciated in recent past. Most of these applications have been highlighted in several reviews and chapters. However, discussion of biosurfactant in context to agriculture is rare. Microbial surfactant and some microbe producing surfactant have excellent potential to topple petro-chemical based agrochemicals. Microbial surfactant score much higher as compared to their synthetic counterpart as they are biodegradable, less toxic, have specific mode of action and can be produced from cheap substrates. Biosurfactant can serve as bio control agent, agriculture adjuvant and growth enhancers. The ability to beneficially influence phytoremediation and induce systematic resistance in plants has opened new frontiers for their application in agriculture sector. *Bacillus* sp. one of the biosurfactant producers are often considered microbial factories for the production of a vast array of biologically active molecules potentially inhibitory for phytopathogen growth. Their spore-forming ability also makes these bacteria some of the best candidates for developing efficient biopesticide products from a technological point of view. On Commercial gains, use of Novel recombinant varieties of biosurfactant producing microorganisms, which can grow on a wide range of cheap substrates and produce biosurfactant at high yields, along with efficient downstream processing methods, can potentially bring the required breakthrough in the biosurfactant production.

Chapter 10 - Surface active agent of microbial origin has engrossed great deal of expectation as it exhibit wide applications in different fields under environment friendly regime. Microbial surfactants have proven its metal in field of agriculture, bioprocessing, biomedical, cosmetics, environmental, food and petroleum industries. Some specific application like stabilization of nanoparticals has been greatly appreciated in recent past. Most of these applications have been highlighted in several reviews and chapters. However, discussion of biosurfactant in context to agriculture is rare. Microbial surfactant and some microbe producing surfactant have excellent potential to topple petro-chemical based agrochemicals. Microbial surfactant score much higher as compared to their synthetic counterpart as they are biodegradable, less toxic, have specific mode of action and can be produced from cheap substrates. Biosurfactant can serve as bio control agent, agriculture adjuvant and growth enhancers. The ability to beneficially influence phytoremediation and induce systematic resistance in plants has opened new frontiers for their application in agriculture sector. Surfactant producing microbes can be applied as biopesticide and most potent candidate for this is *Bacillus* sp. which is renowned for vast array of biologically active molecules potentially inhibitory for phyto-pathogen growth. Their spore-forming ability also makes these bacteria some of the best candidates for developing efficient microbial surfactant based bio-pesticide products from a technological point of view. On Commercial gains, use of novel recombinant varieties of biosurfactant producing microorganisms, which can grow on a wide range of cheap substrates and produce biosurfactant at high yields, along with efficient downstream processing methods, can potentially bring the required breakthrough in the biosurfactant production.

Chapter 11 - Intensive industrialization generates by-products or residues, which are usually hazardous (toxic and/or persistent) to the environment and to humans. The most hazardous pollutants include heavy metals, polyaromatic hydrocarbons (PAHs), polychlorinated biphenyls (PCBs), pesticides and synthetic dyes. For more than a century industrialized countries contaminate systematically soil, water and air with such materials. Yet, it was only after the 1980s that remedial actions were undertaken and the laisser-faire approach was set aside. With regard to soil and water contamination, the general term "site remediation" is usually used. This term is not limited to the cleaning of contaminants in either a solid matrix (mainly soil) or in liquid media (groundwater or surface water), because in some cases there is no clean-up at all, and in fact the contaminants are contained or immobilized in place. Therefore, the term "remediation" should be seen as having broader meaning than "cleanup".

Biological treatment technologies have played an important role in the degradation of hazardous environmental pollutants. Bioremediation is a diverse clean-up technology, cost-effective and environmentally friendly compared to the traditional commonly used physico-chemical techniques. The term "soft technology" is sometimes used for biotechnology to distinguish from the most drastic and environmentally affecting physico-chemical techniques. Bioremediation involves the use of various microbial organisms for eliminating environmental pollutants. It may be a permanent solution and may even lead to complete mineralization of organic pollutants, leaving the ecosystem intact. The most important factors for the success of this technology are the availability of suitable microbes in the contaminated medium, the type and concentration of pollutants and the environmental conditions.

This chapter reviews the recent bioremediation research on a range of pollutants in wastewaters produced by industries such as alcohol distilleries, olive mills, and paper and pulp mills industries. Treatment and reuse of effluents present significant challenges due to the characteristics of the wastewater such as high chemical oxygen demand (COD), high content of persistent pollutants and dark colour. Many species have been isolated and characterized as pollutant-degrading microorganisms. Biodegradation and biosorption have been extensively studied using pure or mixed cultures of fungi and bacteria. The evaluation of the proper microorganism(s) is normally the first step for a biological treatment of pollutants. An extended review of the microorganisms, emphasized on bacteria and fungi that have been tested for the removal of the specific pollutants in the industrial wastewater is given in this chapter.

Chapter 12 - As strict regulations compelled industries to develop environmental friendly, low-cost and efficient treatment techniques, the biotechnological approaches to clean contaminated environments received more and more attention in the recent years. *Saccharomyces cerevisiae* is a microorganism mainly used as baker's yeast and for some type of fermentation, but its potential as bioremediation effector is becoming more evident. *S. cerevisiae* is an excellent model system to study the biosorption, the uptake and the accumulation of various chemicals, including pollutants. This simple organism can grow on low-cost media, can be manipulated easily, and is not pathogenic. Its genome has been entirely sequenced and many cellular processes have been identified at molecular level. The high biosorption capacity as well as the possibility to combine biosorption with continuous metabolic uptake render *S. cerevisiae* one of the promising bioremediator candidates. This review focuses on the applicabiliy of *S. cerevisiae* cells to remove or harvest from contaminated sites an array of chemicals, from metal ions to complex organics. Bioremediation using dead biomass or growing cells, biosorptive or bioaccumulating strains, strains isolated from contaminated sites or engineered strains is discussed.

Chapter 13 - Biogas production through anaerobic digestion of biodegradable solid and liquid waste appears to be a sustainable approach. It involves hydrolysis of waste by a consortium of aerobes, acidogenesis by facultative microbes, followed by acetogenesis by both facultative or obligate anaerobes and finally methanogenesis through acetotrophic/ hydrogenotrophic/ methylotrophic reactions. For this purpose, habitats of facultative and obligate anaerobes have been reviewed, classification of methanogens undertaken, their substrate-specificity examined and metabolic pathways for methane production from a variety of substrates discussed through major and minor metabolic pathways, including the role of syntrophs. Together, these microbes provide efficiency of digestion and if their integrated processes are not attended in time, it leads to deterioration in methane production and at times digester failure. Finally, eco-friendly applications for biogas before and after up-gradation are considered, concluding intricacies of the process characterized by ecological and economical gains to control pollution.

Chapter 14 - Through evolution, microorganisms have developed effective mechanisms that help them regulate their cellular function in response to changes in its microenvironment. One of these survival strategies is metal resistance, which becomes a useful tool for biotreatment of metal loaded wastewaters. For the development of a successful biotreatment, studies on metal-microbes, microbe-surfaces and metal-environment interactions are required. The correct design of laboratory scale bioreactors for electroplating wastes, containing Cd(II), Zn(II), Ni(II) or Cr(VI), is the first step of the authors research project. In that way, the use of metal resistant indigenous bacteria either in suspension or in packed bed reactors is the chosen methodology to fulfil the authors general aim. Several aspects should be studied in order to implement high-efficient processes: a) bacterial interactions with Cd(II) and Zn(II) in terms of biosorption mechanisms; b) biotransformation of Cr(VI) to Cr(III); c) Ni(II)-microbe interactions; d) metal influence on bacterial motility; e) bacterial capacity of biofilm development and microbial community behaviour and f) complexing capacity of environment components with metals.

Chapter 15 - Nitrogen is ubiquitous in the environment, being present in the atmosphere, lithosphere, hydrosphere and biota. The fast increase in human population and human activities over the course of the 20th century has lead to an increase in the loading of land-derived nitrogen, from diffuse sources with anthropogenic origin, to coastal areas, namely transitional waters. Thus, estuaries are subdue to large discharges of this nutrient, which has different consequences, namely eutrophication. Salt marshes have important ecological functions as bio-stabilizers, i.e. decreasing tidal currents and wave action and enhancing settling of suspended matter. Halophytes sustain an elevate productivity and provide habitats.

Moreover, nitrogen incorporation in salt marsh plants' biomass (N fixation/primary production) gives them the ability to auto-remediate estuarine systems. The availability of the excess of nitrogen, due to eutrophication, to different biological processes is determined by its conversion in different chemical forms through biological and physico-chemical processes. These can be incorporation of inorganic forms (NO_3^-, NH_4^+) and transformation into organic forms (e.g. aminoacids, proteins), biomass mineralization, nitrification and denitrification (NO_3^- conversion to N_2 gas), which return the reactive nitrogen to its main reservoir, the atmosphere. Thus, denitrification is another process that contributes to remove reactive nitrogen from the estuarine environment, contributing to the reduction of eutrophication. In this chapter, nitrogen sequestration, nitrogen uptake and denitrification will be discussed as natural remediation processes provided by salt marshes, contributing to maintain the ecosystem health.

Chapter 16 - In natural water and subsurface soils chromium occurs in two major oxidation states: III and VI. Cr(VI) is approximately 1,000-fold more cytotoxic and mutagenic than Cr(III). As the application of Cr is extensive in several industries, chromium-associated pollution is an increasing problem. Development of suitable methods for cleaning up such contaminated soils is an important topic of environmental recovery and protection, particularly regarding methods for remediation of Cr(VI)-contaminated soils. At present there exist physicochemical technologies to remove Cr(VI) from industrial waste, but these methods are costly and require much energy and specific equipment. Therefore the employment of lower-cost biotechnology could be an important alternative.

Chapter 17 - The present interest in hydrocarbon biodegradation is motivated in large part by their presence as environmental pollutants. The biodegradation of hydrocarbons has a high ecological significance, as it constitutes the major process for remediation of contaminated areas. Micro-organisms with the ability to degrade many hydrocarbons have been described, and their mechanisms of action have been studied. The bioremediation of soils contaminated with hydrophobic pollutants is limited by the poor availability of these contaminants to micro-organisms. Surfactants can help, by solubilization or emulsification, to release hydrocarbons sorbed to soil organic matter and increasing the aqueous concentrations of hydrophobic compounds, resulting in higher mass transfer rates. Microbially-derived surfactants offer several advantages and interest in these biosurfactants continues to grow. Microorganisms, having diverse synthetic capabilities, offer potential in expanding the range of surfactants with novel compounds, thereby providing new possible applications. The main classes of biosurfactant are glycolipids, lipoamino acids and lipopeptides, polymers (lipoproteins, lipopolysaccharides), phospholipids, monoglycerides and diglycerides, and fatty acids. The most important advantage of biosurfactants over chemical surfactants is probably their ecological acceptability. Biosurfactants are biodegradable and thus problems of toxicity and accumulation in natural ecosystems are avoided. In the environmental sector, bio-surfactants have potential applications in bioremediation and waste treatment because of their inherent degradability. Few works have investigated the optimization of yeast biosurfactant production, mainly within the last decade, revealing that the potential of such microorganisms is not well explored in the industrial field. This chapter focuses on the successfully application of *Candida* biosurfactants produced in inexpensive media in the removal of hydrophobic contaminants.

Chapter 18 - Marine algae also known as seaweeds are available in three different colors; Rodophyta (red), Pheophyta (brown) and Chlorophyta (green). They can withstand the extreme polluted conditions and are potential source of phyco-colloids. Agar-agar, alginic acid and carragenan to name a few are the important phyco-colloids extracted from the cellular wall of these algae. Due to economic importance and abundant availability, the seaweeds are important candidate for the *insitu* bioremediation as they can be artificially cultivated in the polluted site. The present chapter aims at the exploration of the different Indian Marine Alga for the potential bioremediation of the soda ash industry effluent both in laboratory condition and in the field. Some of the results from nutrient absorption capacity experiments conducted in the author's laboratory will be discussed. In addition, the interactions of the alga with the effluent and its impact on phyco-colloid availability and diversity will also be presented.

Chapter 19 - Metal (oid) contaminants present in the industrial effluents have toxic effects on the ecosystem. Arsenic is one such toxic contaminant increasingly distributed in the environment. While traditional metal removal processes have one or the other limitation, biosorption could be a potentially cost-effective alternative for treatment of wastewaters. Various biomaterials have been examined for metal removal efficiency. However, facultative marine fungi are comparatively less explored. In this study, As (III) biosorption ability of facultative marine *Aspergillus niger* fungal biomass was examined as a function of initial concentration of arsenic. Heat killed biomass of *A. niger* was used as biosorbent. Biosorption increased with increasing arsenic concentration. The fungus could remove more than 90% As at all test concentrations. The maximum biosorption value recorded was 108.083 mg/g at As concentration 600 mg/L. The data suggested that facultative marine *A. niger* is an efficient biosorbent and can be harnessed for arsenic remediation.

Chapter 20 - Elevated concentrations of heavy metals in soil from anthropogenic sources (fertilizers, amendments) or mining activities pose long-term risk to environmental and sustainable production. The deteriorated physical and biological characteristics of contaminated soils need to be improved to establish a vegetation cover. Heavy metal contaminated soil can be treated by operations based on chemical, physical or biological techniques. An interesting approach applied in in-situ remediation of heavy metal contaminated soil includes a variety of phosphate amendments such as soluble phosphate and P-bearing insoluble sources, mainly rock phosphate (RP). In this chapter, the authors would like to present their results on remediation of heavy metal contaminated soils by microbially-treated RP obtained as a result of fermentation processes using agro-industrial wastes as substrates. This biotechnological product exerts its activities by three components: organic matter, partially solubilized RP, and microbial biomass. All experiments based on microbially-treated agro-industrial wastes and RP in soil-plant systems contaminated with heavy metals showed higher efficacy in the presence of arbuscular mycorrhizal (AM) fungi. The overall effect is enhanced plant growth, improved soil structural characteristics, higher soil enzyme activity, and enhanced diversity of soil microorganisms. Results of these experiments showed that the combination of plant growth enhancement and reduced metal translocation caused by our biotechnological products combined with additional microbial inoculations could be regarded as a promising strategy for remediation of heavy metal contaminated soils. Further strategies and biotechnological schemes are discussed.

In: Bioremediation
Editor: Alexander C. Mason

ISBN: 978-1-61122-730-7

Chapter 1

MICROBIAL DEGRADATION OF NITRO-TOLUENES AND THEIR DERIVATIVES: PROGRESSES, CHALLENGES AND OPPORTUNITIES

Debasree Kundu, Chinmay Hazra and Ambalal Chaudhari*
School of Life Sciences, North Maharashtra University, P.B. No. 80, Jalgaon 425 001, Maharashtra, India

ABSTRACT

Major reliance on rapid industrialization, urbanization and altered agricultural practices, past and current defense related activities and recent quest for demilitarization of conventional munitions is fraught with inexorable environmental pollution of nitro-aromatic high explosives (HEs), ordinance related compounds (ORCs), munitions and explosives of concern (MECs) and unexploded ordnances (UXO). Due to widespread and indiscriminate use of these nitro-aromatics (mono-, di- and tri-nitro-toluenes) as chemical feedstock for synthesis of explosives, pesticides, herbicides, dyes, pharmaceuticals, etc., they are recognized as recalcitrant to biological treatment, remain in biosphere and hence, categorized as Hazardous Rating-3. The transformation products poised ecological threats and causing growing concern about toxic, carcinogenic, mutagenic, and teratogenic effects to humans, and contaminants to natural receptors (soil, sediment, surface water, and ground water). These increasing threats to the environment provoked the necessity for their elimination. Several physical and chemical environmental cleanup approaches for elimination of recalcitrant nitro-toluenes are available. These processes are prohibitively costly, destructive and responsible for CO_2 and NO_X release. Therefore, attention has recently been focused on alternative biological approaches through varied microbial metabolic processes for complete and cost-effective detoxification of recalcitrant nitro-aromatic pollutants. Microbes, being the pioneer colonizers on earth, underpin all life processes and an essential component in biogeochemical processes and element cycling, maintaining the functioning of the global ecosystem by physiological and functional diversity. Moreover, notwithstanding with the performance assessment and scale-up effectiveness, variations based on approaches viz. (i) biostimulation,

* Corresponding author. Tel.: +91-257-2257425. E-mail: ambchasls@gmail.com

(ii) enhanced natural attenuation, (iii) bioaugmentation and (iv) synthetic biology have been emerged out. Of these, enhanced natural attenuation is achieved through exogenous supplementation by indigenous microbes, while, bioaugmentation relies on application of carrier-encapsulated whole cell and gene bioaugmentation to indigenous microbes to confer superior trait(s) for removal of nitro-toluenes. Contrary to these approaches, synthetic biology employs design and construction of new biological parts, devices and systems and the re-design of existing, natural systems for more effective biodegradation.

This article reviews key aspects of current and emerging remediation technologies underlying bioremediation. Subsequently, current knowledge regarding the application of bioremediation in alleviating contaminant toxicity is summarized highlighting the relative merits and demerits of different options. Finally, this review attempts to provide a concise appraisal on recent progresses on biodegradation, concluding with applications of bio-molecular approaches and 'omics' technologies in bioremediation of nitro-toluenes and their derivatives.

Keywords: Nitro-aromatics; munitions and explosives of concern (MECs); Biodegradation; Bioremediation

1.0 Introduction

In tandem with rapid industrial, economic and societal development, human activities have instigated widespread pollution of the natural global environment. Environmental contamination with high explosives (HEs), ordinance related compounds (ORCs), munitions and explosives of concern (MECs) and unexploded ordnances (UXO) are the results of (i) large-scale industrial manufacture, loading, assembling, packing (LAP), storage, testing and deployment of devices containing explosives and the burial of increasing amounts of obsolete munitions in army ammunition plants, (ii) past and current defense related activities, (iii) rapid industrialization coupled with increased urbanization and altered agricultural practices for a more comfortable lifestyle and (iv) recent quest for demilitarization of nuclear and conventional munitions. With the inception of organic chemistry around 1830, chemists have synthesized many new compounds by nitration. By the late nineteenth century, the nitro-aromatic explosives i.e., nitro-toluenes (NTs, 2-, 3-, 4-NT, 2,4-, 2,6-DNT and TNT) had been synthesized and their advantages over the more shock sensitive nitroglycerine were recognized, particularly for military purposes, allowing their incorporation into military ordinances. Modern explosives including poly-nitro-aromatic compound (2,4,6-tri-nitrotoluene; TNT) and di-nitrotoluene (2,4-DNT, 2,6-DNT) are nitrogen-containing organic compounds with the potential for oxidation to N_2, CO_2 and H_2O. NTs had become the major constituent in US military explosive ordnance and by World War I/II were adopted for use by all countries during that war [Yang *et al.*, 2009]. NTs, particularly TNT had been the predominant HE for military purposes before its production was ceased in the mid 1980s in the US.

1.1 Pollution Scenario: Current State-of-the Art

Pollution with NTs of air, water and soil mainly occurred through its enormous production, indiscriminate use and improper disposal (Fig. 1). During World War I, 17,000 poisoning cases including 475 deaths in US munitions factories due to TNT exposure were reported [Darrach *et al.*, 1998]. As per the US Army Corps of Engineers, the US Army has 583 sites at 82 installations and 45 military installations containing explosives in groundwater and 87 additional locations with suspected contamination [Spain, 2000; Esteve-Nunez *et al.*, 2001]. By 1945, TNT production capacity in the US was as high as 65 tons per production line per day [Jenkins *et al.*, 1986], while German production was 2.36×10^4 tons per month [McCormick *et al.*, 1981]. A single munition factory could generate 2×10^6 liters of wastewater containing TNT and its by-products [Hannink *et al.,* 2002]. In the US, the Army alone has estimated the contamination of 1.2 million tons of soils [Daun *et al.*, 1998] and 294,000 field sites [Snellinx *et al.*, 2002]. The US Department of Defense (DoD) has identified more than 2000 sites with explosives contamination, of which >95% with TNT and >87% above permissible limit [Lee *et al.,* 2007]. Besides, 300,000 to 400,000 contaminated civil sites and 294,000 contaminated field sites still exists in the USA alone. Of which, 51% owned by State and private property owners, followed by 43% sites with underground storage tanks. With over 100 sites in the USA earmarked for remediation, volumes of contaminated soils have estimated to the tune of 1,000000 cubic yards [Cunningham *et al.,* 1997].

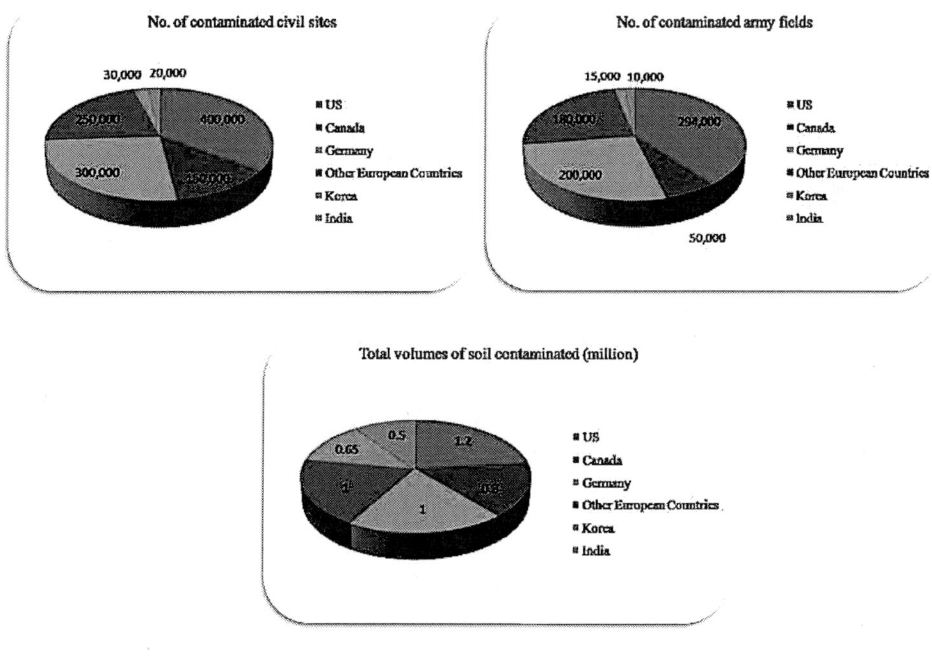

Figure 1. Current global pollution scenario: country wise.

Germany acquired 300,000 tons of old and sub-standard munitions when the former German Democratic Republic (GDR) was reunited with the Federal Republic of Germany (FRG) in 1989 [Conder *et al.,* 2004]. During World War II, Germany produced approximately 800000 tons of TNT [Daun *et al.,* 1998] and the highest amount of TNT was produced at the end of World War II, when the world production was estimated to be 150 000 tons per month [Snellinx *et al*,. 2002]. In 2006, the European Environment Agency reported that approximately 250,000 contaminated field sites are in need of remediation. In Canada, there are 103 training sites (air and land force bases) and three open burning/open detonation sites associated with activities involving TNT [Hawari *et al.,* 2000; Stenuit *et al.,* 2005]. The aqueous wastes (red water) generated from manufacturing of TNT, contained up to 30 nitro-aromatics. While pink water, generated during loading, packing or assembling munitions, often contain high concentrations of other nitro-aromatic explosives. Many sites became contaminated through open detonation and burning of explosives at army depots, evaluation facilities, artillery ranges, and ordnance disposal sites. Before the 1980s, waste was often dumped in unlined pits, and has now contaminated both soil and groundwater. Finally, contamination due to demilitarization has arisen through 'washout operations', which involve removing the fuses from munitions and removing the explosive charge by spraying jets of hot water into the casings. Besides, nitro-toluenes serve as key industrial chemicals, with estimated annual production of 10^8 tons [Kulkarni and Chaudhari, 2007]. In India alone there are about 20,000 abandoned sites covering about 60 different kinds of nitro-aromatics [Prasad *et al.,* 2009]. An environmental watchdog survey revealed that Russia, China and India are among the 'top ten' most polluted places/countries in the world [Anonymous, 2007; Prasad *et al.,* 2009].

1.2 Impacts of Pollution: Problem of Plenty

1.2.1 Chemistry and Properties of Nitro-Toluenes

Only few aromatic compounds, bearing one nitro group on the aromatic ring, are produced by microorganisms. Natural nitro-aromatic compounds described to date are chloramphenicol, aureothin, phidolopin, nitro-pyoluteorin and oxy-pyrrolnitrin. Nitro-aromatics can also be formed abiotically by photochemical reactions in the atmosphere. The majority of organically synthesized nitro-arenes present in the biosphere comes from man-made sources. These compounds are used worldwide as explosives, pesticides and chemical feed-stocks for the synthesis of polyurethane foams, dyes, polymers and pharmaceuticals. Due to the massive industrial use of NTs in the last 75 years, together with the intense military activity of the two World Wars, nitro-aromatics are widely disseminated in the environment and hence, recognized as priority pollutants on the globe by various environmental regulatory agencies.

The recalcitrancy of NTs are mainly attributed to (i) strong electron-withdrawing properties of the nitro substituents [Rieger and Knackmuss, 1995], (ii) high electron deficiency on the entire π-electron system (electophilic nucleus), (iii) steric effects due to the symmetric position of the functional groups and (iv) shielding of π -electrons of the aromatic ring from attack by classical oxidative and electrophilic mechanisms [Heiss and Knackmuss, 2002]. The relatively refractory nature of NTs come from the symmetric location of the nitro ($-NO_2$) groups on the aromatic ring (Fig. 2; Table 1).

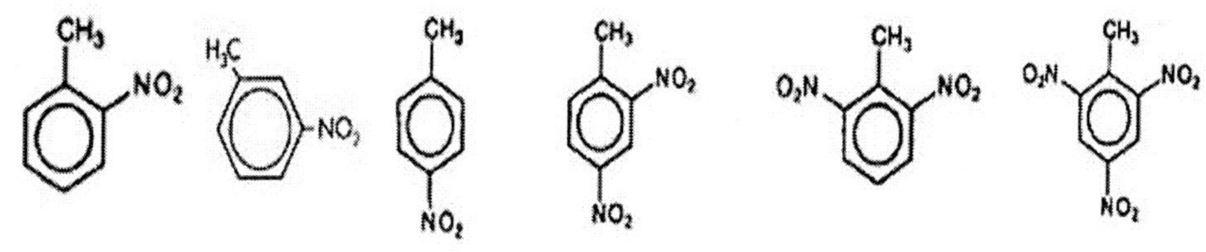

A B C D E F

Figure 2. A: 2-NT; B: 3-NT; C: 4-NT; D: 2,4-DNT; E: 2,6-DNT; and F: 2,4,6-TNT.

The nitro group consists highly electronegative N and O. Of these, oxygen is even more electronegative than the nitrogen atom; hence, the N-O bond is polarized. The partially positive charge of the nitrogen atom, combined with its high electro-negativity, makes the nitro group easily reducible [Esteve-Nunez *et al.,* 2001]. The reduction of the first nitro group is generally much more rapid than that of the rest. The conversion of nitro to amino groups decreases the electron deficiency of the nitro-aromatic ring and consequently a lower redox potential (< -200 mV; anoxic condition) is required to reduce the rest of the nitro group of the molecule. Abiotic reduction involves first proton transfer to the radical anion during the initial one electron reduction.

1.2.2 Environmental Routes and Fate of Nitro-Toluenes: Hazardous Implications

NTs in the eco-system can be subjected to: (i) transfer processes, which re-locate the substance without altering its structure (a) adsorption, (b) retention by crops, (c) run-off movements in dissolved/sorbed state, (d) diffusion and vapor-phase diffusion, and (e) mainly sorption and desorption to soil colloid surfaces. Among these, the interactions at interfaces between organic and inorganic soil colloids and xenobiotics through sorption/desorption mechanisms are significant and can be weakly or strongly associated to inorganic and organic colloids through adsorption. The abiotic soil components involved in the interaction with xenobiotics are pure and dirty clays, humic substances, and humic-clay associations. They may affect the movement of xenobiotics, their availability for plant or microbial uptake, their transformation by abiotic or biotic agents, and their influence on soil processes and (ii) degradation processes alter the chemical structure occur through chemical, biological, and photochemical transformations. The biotic components involved in the biological degradation of xenobiotics are microorganisms, plants, and their enzymatic proteins, as either intra- or extracellular enzymes.

For this purpose, three basic pollutants-based events has been proposed viz. (i) the death of sensitive organisms with the consequent utilization of the organic residues by the surviving populations, (ii) the direct utilization of NTs (as C source for microbial growth) by the organisms with an ability to degrade or to metabolize them and (iii) the development of microbial populations that depend on secondary nutrient sources, e.g., metabolites produced from the decomposition of the pesticide or excreted by the proliferating microflora [Gianfreda and Rao, 2008].

Table 1. Nitro-toluenes and their chemistry

Compound	Chemical properties									
Mono-NTs										
	Molecular formula	**Molar mass ($g.moL^{-1}$)**	**Density ($g.cm^{-3}$)**	**Melting point (°C)**	**Boiling point (°C)**	**Solubility**		**Appearance**	**Shock sensitivity**	**Friction sensitivity**
						Water (g/100g), 30°C	**Solvents**			
***o*-nitrotoluene**	$C_6H_4(CH_3)(NO_2)$	137.1	1.163	-9.3	222	0.065	soluble	pale yellow liquid		
***m*-nitrotoluene**	$C_6H_4(CH_3)(NO_2)$	137.1	1.16	16.1	230	0.05	readily soluble	yellowish-greenish to yellow liquid		
***p*-nitrotoluene**	$C_6H_4(CH_3)(NO_2)$	137.1	1.29	51.7	238.3	0.004	soluble in acetone, di-ethyl ether and benzene	pale yellow rhombic crystals		
Di-NTs										
2,4-DNT	$C_7H_6N_2O_4$	182.13	1.52	70	250-300	0.027	soluble in acetone, di-ethyl ether and ethanol	Pale yellow to orange crystalline solid	No	Very low
2,6-DNT	$C_7H_6N_2O_4$	182.1	1.407			0.0182			No	Very low
Tri-nitrotoluene (TNT)										
2, 4, 6-TNT	$C_7H_5N_3O_6$	227.13	1.654	80.35	295	0.013	soluble in ether, benzene, acetone, pyridine	Pale yellow	No	Insensitive to 353 N

The unsaturated (vadose) zone, reaching from the soil surface down to the saturated zone, can be contaminated by dissolved and volatilized non-aqueous phase liquids (NAPLs) like NTs. In the saturated zone, groundwater flows due to gravity above an impermeable layer can be contaminated by water soluble organics mobilized with the groundwater flow in the saturated zone along-with NAPLs, Light non-aqueous phase liquids (LNAPLs), and dense non-aqueous phase liquids (DNAPLs). DNAPLs may also contaminate the capillary system of fracturated aquitard below the aquifer.

The resultant accumulations in the eco-system can cause (i) toxicity, (ii) potency of bioaccumulation in living systems and (iii) carcinogenicity [Kulkarni and Chaudhari, 2006]. It eventually leads to the formation of methemeglobin on acute exposures and anaemia on chronic exposures. Additionally, they are absorbed through the skin rapidly leading to sensitization dermatitis. Prolonged exposure results in leukocytosis, peripheral neuritis, muscular pains, cardiac irregularities, and renal irritations, liver damage and bladder tumours.

The toxicity of NTs are frequently caused by enzymatic single-electron reduction of the nitro groups [Purohit and Basu, 2000], inducing an oxidative stress by the formation of reactive oxygen species [Kumagai *et al.,* 2004]. This single-electron reduction is mediated by native bacterial oxygen-sensitive nitro-reductases (type II) [Sarlauskas *et al.,* 2004; Ask *et al.,* 2004]. Redox cycling yields a nitrobenzene anion radical that can react with oxygen to form a superoxide radical and the original nitro group. The nitroso- and hydroxyl-amino groups, responsible for toxicity of NTs, react with biomolecules and can elicit chemical mutagenesis and carcinogenesis [Esteve-Nunez *et al.,* 2001]. However, complete reduction of the nitro group to an amino group decreases the mutagenic effect of the compound. On the contrary, hydroxyl-amino-di-nitro-toluenes can cause hemotoxic symptoms in workers exposed to TNT wherein, oxy-hemoglobin is oxidized by molecular oxygen in the presence of hydroxyl-amino-di-nitro-toluenes to yield ferrihemoglobin and nitroso-di-nitro-toluenes [Esteve-Nunez *et al.,* 2001]. The phenomenon is also called as a 'futile cycle' because reduced pyridine nucleotides are oxidized without net reduction of the nitro group. Such reduction products have more affinity to bind covalently to proteins and DNA for additional damage [Sarlauskas *et al.*, 2004]. They are responsible for oxidation of the haemoglobin iron [Peres and Agathos, 2000] as well as uncoupling of oxidative and photosynthetic phosphorylations [Stenuit *et al.,* 2005].

In response to growing concern about the health effects and the ecological threats posed by NTs, several studies have been described on its toxicity, environmental fate, detection methods, size and mass distributions of NTs particles scattered by detonation and, last but not least, (bio)degradation.

2.0 Current Remediation Options for Nitro-Toluenes

2.1 Physical

Increased public awareness about the risks and hazards of toxicity associated with NTs encouraged to develop cost-effective technologies for their remediation. Several conventional (physical) cleanup methods (Table 2) viz. (i) incineration, (ii) filtration, (iii) extraction, (iv) absorption, (v) volatilization, and (vi) hydrolysis are used for the removal of NTs [Kanekar *et al.,* 2003; Kulkarni and Chaudhari, 2006].

2.2 Chemical

Chemical process mainly consists of (i) thermal decomposition or thermolysis, (ii) photo-catalytic degradation, (iii) ozonolysis, (iv) chemical reduction, (v) chemical oxidation, (v) Advanced Oxidation Processes (AOP), (vi) chemically aided electro-kinetics, and (vii) electro-oxidation (Table 2).

2.3 Hybrid Methods

To overcome the drawbacks of physical or, chemical methods combined treatment techniques, viz. (i) physico-chemical and (ii) biologio-chemical approaches that profits from advantages of both methods, while at the same time reduces the disadvantages has come up recently as an alternative remediation technology (Table 2).

UV-H_2O_2, Fenton and photo-Fenton systems are the examples of physico-chemical hybrid methods, while biologio-chemical approaches utilize (i) photochemical treatment followed by microbiological transformation [Schrader and Hess, 2004] or (ii) photochemical mineralization (oxidative AOPs) after a biological reduction step [Kroger and Fels, 2007].

2.4 Bioremediation of Nitro-Toluenes: A Multi-Scale and Multi-Faceted Holistic Approach

Non-compliance of the physical, chemical and combination method due to (i) inefficiency, (ii) cost-intensivity, (iii) non-sustainability, and (iv) lower cash cost impact on sales and revenue shifted attraction towards bioremediation of NTs-polluted sources [Kulkarni and Chaudhari, 2007; Kapley and Purohit, 2009]. With (i) an increasing public awareness and the introduction of stringent environmental regulations with an increased emphasis of the need to move towards a low carbon/carbon neutral economy, (ii) eco-friendly, eco-sustainability and cost-effectiveness and, (iii) the development of sustainable remediation metrics, there is an increased interest from traditional technologies to bioremediation.

The general approaches to bioremediation are basically (i) intrinsic bioremediation/ natural attenuation, (ii) biostimulation, and (iii) bioaugmentation (Table 2). In each approach, microbes endowed with inherent abilities to live, metabolize, thrive, colonize and decontaminate NTs contaminated eco-system (soil/water streams) come to the rescue.

Table 2. Amelioration of NTs: current treatment methods

	Approach	Salient features	Merits	Demerits
I. Physical				
(a)	Incineration	Combustion of organic materials and/or substances into ash, flue gases, particulates, and heat	Reduce the volume of the original waste by 95-96 %	Emits particulate matter, heavy metals, dioxins, furans, sulphur dioxide, and hydrochloric acid Produces ~1 tonne of CO_2/tonne of organic wastes Separates unwanted compounds, but do not totally destroy them
(b)	Extraction	Separation by leaching of metal from ore using acid	Useful in extractive metallurgy since many metals can form soluble salts in aqueous media	Underground water pollution Un-repairable damage to the ecosystem through enhanced solubility of metal pollutants
(c)	Adsorption	Target compounds get adsorbed onto the surface of a properly designed matrix and temporarily holds up the pollutants	Normally used as a platform for preliminary separation	Essentially a surface phenomenon Total separation from the mixture is impossible task

Table 2. (Continued)

II. Chemical				
(a)	Thermal decomposition /thermolysis	Chemical substance breaks up into at least two chemical substances by endothermic process	Normally used as a platform for preliminary separation	Cost-intensive Not suitable for large scale operation.
(b)	Photo-catalysis	Reactive oxidizing agents, viz. H_2O_2 or O_3, semiconductor like TiO_2, UV rays, Fenton's reagent (H_2O_2+Fe^{+2}), peroxone (H_2O_2+O_3) and hydrothermal catalysts accelerates photoreaction	Normally used as a platform for preliminary separation	Energy intensive High cost input Not feasible for industrial hazardous waste treatment High strength contaminants are only slowly degraded (as a percentage ofinitial concentration) Build-up of intermediates does occur
(c)	Ozonolysis	Ozone cleaves NTs to form organic compounds in which the multiple carbon-carbon bonds have been replaced by a double bond to oxygen	Multiple bonds can be oxidized	Non-functional with organo-metallic compounds The outcome of the reaction depends on the workup conditions

Table 2. (Continued)

(d)	Chemical reduction	Mode of action include hydrogenation, noble metal catalysis (using Pd, C, Fe or Al_2O_3), Ni/ zeolite Y, Ni/ SiO_2 and Raney Ni	Normally used as a platform for preliminary separation	Cost-intensive
(e)	Chemical oxidation	Oxidation of NTs using ozone, hydrogen peroxide, Fenton's reagent and modified Fenton's reagents (i.e., iron-catalyzed hydrogen peroxide), permanganate, and persulfate	Multiple bonds can be oxidized Normally used as a platform for preliminary separation	Cost-intensive Effectiveness relies on nature and the level of contamination, the solid matrix characteristics and the particle size distribution
(f)	Advanced Oxidation Processes (AOP)	Uses -OH radicals as the oxidizing species either by (i) UV-irradiation of hydrogen peroxide (H_2O_2), ozone (O_3)and titanium dioxide (TiO_2)	Superior than other chemical methods Relatively faster	Cost-intensive
(g)	Electrokinetics	Under the effect of low-level electric current/voltage, transport/ mobilization of solutes occur	Applied for *in situ* treatment of fine and low-permeability materials De-watering and removal of organic or inorganic species (including salts) can be simultaneously obtained	Chemical amendments need to be added to promote the solubilization and the transport of contaminants

Table 2. (Continued)

(h)	Electro-oxidation	Uses low-voltage direct currents (DC) to degrade organic pollutants by inducing oxidation reactions and water electrolysis	Electrochemical generation of H_2O_2 leads to decomposition into -OH radicals, a strong non-selective oxidizing agents capable of degrading a variety of organic pollutants	Presence of micro-conductors (e.g., iron minerals) allows redox reactions to occur simultaneously
		III. Hybrid methods		
Physico-chemical (i) UV-H_2O_2		-OH radicals are generated by homolytic cleavage of H_2O_2	Effective as compared to either physical/chemical method alone	At high concentrations, H_2O_2 can compete with NTs for hydroxyl radicals, thereby reducing the efficiency of the oxidative degradation
(ii) Fenton		-OH radicals are derived from the reaction of ferrous ion salts with H_2O_2		
(iii) Photo-Fenton systems		Besides -OH radicals as mentioned above, photons can additionally reduce Fe^{3+} back to Fe^{2+}		No extensive mineralization of NTs with reasonable energetic input
Biologio-chemical (i) Photochemical treatment followed by microbiological transformation		Exploits power of strong non-selective oxidizing agents to overcome the rate-limiting step(s); thereafter microbial degradation occur	Effective as compared to either microbial/chemical method alone	Both methods does not essentially ensure complete mineralization Presence of additional organic material (dissolved organic matter, viz. humic acids and other constituents) slower reaction

Table 2. (Continued)

	(ii) Photochemical mineralization after a biological reduction step	Instead of chemical reduction, specific microbial reduction produces intermediates which are then routed to mineralization by photochemical method		
		IV. Bioremediation		
(a)	Natural attenuation/ Intrinsic bioremediation	Uses of degradative microbial strains to degrade/mineralize NTs	Natural process Cost-effective as compared to physical/chemical counterparts	Limited array of effective strains Sometimes relatively slow Very few mineralizes NTs
(b)	Biostimulation	Amendments/supplementation of nutrients and other factors to stimulate natural biodegradation	Substances with similar structure induce the same metabolic pathways Biomass density increases	Often introduced substrates show inhibitory effect to the microbe(s)
(c)	Bioaugmentation	Hastening degradation rate via use of degradative microbial consortia or exogenous addition of catabolic gene(s)	Transfers remediation genes to indigenous microbes Faster than other existing methods	Cost remains a concern Biosafety of engineered strains attracts ethical debates

2.4.1 Microbial Features

The incredible versatility of degradative abilities harbored by microbes lies in the fact that, by the time human beings came on the scene, microbes had already coexisted for billions of years with an immense variety of pollutants [Dua *et al.,* 2002]. The vast diversity of potential substrates for growth led to the evolution of enzyme capable of transforming many unrelated natural pollutants by many different catalytic mechanisms. The resulting giant 'library' of microbial enzymes serves as raw material for further revolution whenever a new chemical becomes available [Ellis, 2000; Dua *et al.,* 2002].

Early work pertaining to microbial remediation of nitro-aromatics has shown that they are recalcitrant compared to non-nitrated compounds [Nishino *et al.,* 2000]. Persistence of such xenobiotics in the environment has led to exposure and recruitment of microbe(s) to degrade an increasing number of NTs. For this purpose, microbes have evolved diverse pathways for their mineralization. Extensive research in the last two decades on this aspect has led to isolation of a number of organisms with potential to degrade nitro-aromatics. However, the number of organisms utilizing NTs as a sole source of C and/or N are very sparse due to (i) generation of additional toxic substances by reduction of the nitro groups, rather than oxidizing the aromatic ring [Rieger and Knackmuss, 1995; Hawari *et al.,* 2000; Esteve-Nunez *et al.,* 2001], (ii) shielding and thereby withdrawing electrons of aromatic ring necessary for oxidative attack by symmetrically located nitro substituents, (iii) failure of oxidative degradation owing to the inaccessibility of aromatic compounds to ring opening reactions [Kroger and Fels, 2007], and (iv) relatively slow pace of aerobic/anaerobic processes [Kulkarni and Chaudhari, 2007].

2.4.2 Biodegradation Modes: Aerobic

In aerobic biodegradation (Fig. 3A&B and Table 3), microbes appear to have evolved four main strategies viz. (a) di-oxygenation of the nitro-aromatic ring, with release of the nitro-group as nitrite and production of di-hydroxy intermediates, (b) monoxygenation to epoxides, (c) formation of a hydride- Meisenheimer complex, and (d) partial reduction of the nitro-group, formation of hydroxyl-amino-benzene derivatives and ammonia release, followed by rearrangement of the hydroxyl-amino-benzene to the corresponding catechol and elimination of another ammonia molecule [Nishino *et al.,* 2000; Ye *et al.,* 2004].

Aerobic degradation of NTs by bacteria involves mainly mono- and di-nitro-toluenes as a source of carbon and/or nitrogen and energy by complete mineralization [Peres and Agathos, 2000; Kulkarni and Chaudhari, 2007] through both oxidative and reductive reactions [Crawford, 1995; Nishino *et al.,* 2000; Peres and Agathos, 2000] and catabolic enzymes [Ye *et al.,* 2004]. Oxygenase catalyzed oxygenolytic metabolism does not occur in TNT due to the electrophilic character of the nitro groups; instead, nitro groups or the aromatic nucleus are reduced [Esteve-Nunez *et al.,* 2001; Heiss and Knackmuss, 2002; Snellinx *et al.,* 2002].

Table 3. Aerobic degradation of nitro-toluenes by various microorganisms

Nitro-toluene (NT)	Organism	Metabolic route	Reference
2-NT	Bacteria		
	Pseudomonas JS42	Degradation via 3-methylcatechol with release of nitrite leading to ring cleavage	Parales *et al.* (1998) Haigler *et al.* (1999)
	Acidovorax sp. strain JS42		Lee *et al.* (2007)
	P. putida F1 and *Pseudomonas* sp. JS 150	Oxidative attack the methyl group of 2-NT, yielding 2- -nitro-benzylalcohol	Lee *et al.* (2007)
	Bacillus sp., *Bacillus flexus* and *Micrococcus* sp.	Degradation via 3-methylcatechol with release of nitrite leading to ring cleavage	Mulla *et al.* (2010)
3-NT	Bacteria		
	P. putida F1 and *Pseudomonas* sp. JS 150	Oxidative attack the methyl group of 3-NT, yielding 3- -nitro-benzylalcohol	Smets *et al.* (2007)
	Pseudomonas putida OU83	70% of the substrate was reduced to 3-aminotoluene, a minor part (30%) was converted, via 3-nitrobenzylalcohol, 3-nitrobenzaldehyde, and 3-nitrobenzoic acid, to 3-nitrophenol.	Smets *et al.* (2007)
4-NT	Bacteria		
	Pseudomonas sp. 4NT	Formation of 4-NBA followed by the partial reductive pathway leading to mineralization	Haigler and Spain (1993)
	Pseudomonas sp. TW3		Rhys-Williams *et al.* (1993)
	Pseudomonas sp.	Formation of 4-hydroxylamino derivative which is converted to protocatechuate with ring cleavage and release of NH_3	Rhys-Williams *et al.* (1993)
	Pseudomonas fluorescens 410PR	Mineralization via 4-NBA	Michan *et al.* (1997)

Table 3. (Continued)

	Mycobacterium sp. strain HL 4-NT-1	Production of 4- hydroxyl-amino-toluene followed by 6-amino-m-cresol before meta-ring cleavage	Spiess *et al.* (1998)
2,4-DNT	Bacteria		
	Pseudomonas sp.	Formation of 4-methyl-5 nitrocatechol	Spanggord *et al.* (1991)
	Burkholderia cepacia JS850, *Hydrogenophaga paleronii* JS863	Mineralization with release of nitrite, formation of methyl-4-nitrocatechol	Spanggord *et al.* (1991)
	Burkholderia sp. DNT		Spanggord *et al.* (1991)
	Burkholderia cepacia R34		Johnson *et al.* (2001)
	Pseudomonas sp. ST53		Snellinx *et al.* (2002)
	Two identified consortia		Snellinx *et al.* (2002)
	Sinorhizobium meliloti DHK1		Van Aken *et al.* (2004)
	P.putida TOD		Van Aken *et al.* (2004)
	Alcaligenes JS867		Van Aken *et al.* (2004)
	Burkholderia sp. DNT		Khleifat (2010)
	Fungi		
	Phanerochaete chrysosporium	Complete mineralization to CO_2	Valli *et al.* (1992)
2,6-DNT	Bacteria		
	Burkholderia cepacia strain JS850	Di-oxygenation of the 2,6-DNT to 3- methyl-4-nitrocatechol (3M4NC), accompanied by the release of nitrite	Nishino *et al.* (2000)
	Burkholderia sp.		Nishino *et al.* (2000)
	Hydrogenophaga palleronii strain JS863		Nishino *et al.* (2000)

Table 3. (Continued)

2,4,6-TNT	Bacteria		
	Pseudomonas sp. strain IIBx	First evidence of [^{14}C]TNT mineralization	Traxler *et al.* (1974)
	Pseudomonas sp. strain 1-2wt	First evidence of [^{14}C]TNT mineralization	Traxler *et al.* (1974)
	Pseudomonas sp. strain FR2	Reduces TNT to DANT	McCormick *et al.* (1976)
	Pseudomonas sp. strain CBS3	Reduces TNT to DANT	Schackmann and Muller (1991)
	Pseudomonas sp.	TNT transformation to 2ADNT and 4ADNT	Jones *et al.* (1995)
	Pseudomonas pseudoalcaligenes	Reduction of nitro groups and nitrite release from 2,4-dihydroxylamino-6- nitrotoluene	Fiorella and Spain (1997)
	Pseudomonas savastanoi	Denitration of TNT to release nitrite and reduction of nitro group to amino	Martin *et al.* (1997)
	Serratia marcensens	TNT as sole carbon and energy source in presence of Tween 80	Montpas *et al.* (1997)
	Enterobacter cloacae PB2	Transformation to 2-amino-4-nitrotoluene	French *et al.* (1998)
	Bacillus sp.	Nitrite release from TNT and transformation to 2-amino-4-nitrotoluene	Kalafut *et al.* (1998)
	Enterobacter sp.	TNT transformation to unknown polar compounds and 3% mineralization	Kalafut *et al.* (1998)
	Pseudomonas aeruginosa	Nitrite release from TNT and transformation to 2-amino-4-nitrotoluene	Kalafut *et al.* (1998)
	Staphylococcus sp.	Nitrite release from TNT and transformation to 2-amino-4-nitrotoluene	Kalafut *et al.* (1998)
	Pseudomonas fluorescens	TNT transformation into ADNT, DANTs, and 4-acetamide-2-amino-6- nitrotoluene	Esteve-Nunez *et al.* (2001)
	Klebsiella sp. strain C1	Generation of hydroxylaminodinitrotoluenes and amino-dinitrotoluenes	Kim *et al.* (2002)

Table 3. (Continued)

Pseudomonas sp. clone A (pWW0-Km)	Partial mineralization	Snellinx *et al*. (2002)
Pseudomonas sp. ST53	Transformation to 2-amino-4-nitrotoluene	Snellinx *et al*. (2002)
Mycobacterium vaccae	50% radioactivity from [^{14}C]TNT incorporated into lipids	Snellinx *et al*. (2002)
Mycobacterium sp. *strain* HL4NT-1	transform TNT to hydride and dihydride Meisenheimer complex	Snellinx *et al*. (2002)
Pseudomonas aeruginosa MA101	Transforms TNT and ADNTs into highly polar compounds	Snellinx *et al*. (2002)
Pseudomonas fluorescens	An NADPH flavoprotein oxidoreductase catalyzes multiple transformations, generating Ar-NHOH, H-TNT and H_2-TNT	Snellinx *et al*. (2002)
Pseudomonas fluorescens B3468	TNT degradation to phloroglucinol and pyrogallol; nitrogen of TNT is released as ammonium	Snellinx *et al*. (2002)
Rhodococcus erythropolis	Cells growing on picric acid catalyze ring hydrogenation, forming hydride and di-hydride TNT-Meisenheimer complexes	Snellinx *et al*. (2002)
Stenotrophomonas maltophilia	TNT as sole nitrogen source	Snellinx *et al*. (2002)
Acinetobacter junii A8	TNT transforms to 2,6-dinitro-4-nitrosotoluene, 4-amino-2,6-dinitrotoluene, 4-amino-2,6- dinitrobenzoic acid	Moslemy *et al*. (2003)
Nocardiodes CB22-2	De-nitration leads to formation of Meisenheirmer complex	Eyers *et al*. (2004)
Yarrowia lipolytica NCIM 3589	Transform TNT to H^--TNT, 2,4 DNT, ADNT	Jain *et al*. (2004)
P. aeruginosa	Formation of 2-amino-4,6-dinitrotoluene and 2,2'-azoxytoluene	Ye *et al*. (2004)

Table 3. (Continued)

Rhizobium meliloti	Partial reduction of TNT to amino-dinitrotoluene	Ye *et al.* (2004)
E. coli	Resting cells transformed TNT to hydroxyl-amino-di-nitro-toluenes (2HADNT, 4HADNT, with 4HADNT as the dominant isomer), aminodinitrotoluenes (4ADNT, with sporadic detection of 2ADNT), 2,4-di (hydroxylamino)-6-nitrotoluene (24D(HA)6NT), 2,4-diamino- 6-nitrotoluene (24DA6NT), and an additional compound which was tentatively identified as a (hydroxylamino) aminonitrotoluene isomer	Thompson *et al.* (2005)
Bacillus subtilis SK1	Biotransformation of TNT to aromatic amines	Galvao *et al.* (2007)
Pseudomonas aeruginosa strain MX	TNT transformation into HADNTs	Smets *et al.* (2007)
Burkholderia cepacia and *Pseudomonas* spp.	Conversion of TNT into 2-Am-4,6-DNT	Muter *et al.* (2008)
Raoultella terrigena strain HB	TNT transforms to aminodinitrotoluenes (2-ADNT, 4-ADNT) and diaminonitrotoluenes (2,4- DANT) as the metabolites	Singh *et al.* (2008)
Pseudomonas putida KT2440	Biodegradation of TNT by nitroreductase	Fernández *et al.* (2009)
Fungi		
Wood-rotting/white rot basidiomycetes		
Phanerochaete chrysosporium	Formation of hydroxylamino dinitrotoluene	Micheles and Gottschalk (1994)
Heterobasidium annosum TM5P2; *Hypholoma fasciculre* TM 5.2; *Pleurotus flabellatus* MWP 127-1; *Pleurotus ostreatus*, *Treametes* sp.	Complete degradation of TNT through conversion of amino-nitrotoluene by fungal manganese peroxidase	Scheibner *et al.* (1997)

Table 3. (Continued)

Basidiomycetous fungi	Converts to hydroxyl amino and aminodinitrotoluene to CO_2 (4-80%)	Hawari *et al.* (1999)
Phlebia radiata	Reduce and mineralize TNT by manganese-dependent peroxidase	van Aken *et al.* (1999)
Phanerochaete chrysosporium	Mineralized under lignolytic condition with formation of 9 different acetylated products	Hawari *et al.* (2000)
P. chrysosporium, *Kuehneromyces mutabilis*, *Hypholoma fasciculare*, and *Phlebia radiata*	Complete degradation under lignolytic and non- lignolytic conditions	Perkins *et al.* (2005)
Litter-decaying basidiomycetes		
Litter decaying fungus	Concentration decreased to 95% in 300 days with no reduced intermediate	Herre *et al.* (1997)
Agaricus eastivalis TMAest1	100% biotransformation of TNT with 9.0% release of CO_2	Hawari *et al.* (2000)
Agrocybe praecox TM70.3.1	100% biotransformation of TNT with 14.5% release of CO_2	
Clitocybe odora TM3	88% biotransformation of TNT with 5.1% release of CO_2	
Coprinus comatus TM6	82% biotransformation of TNT with 0.8% release of CO_2	
Micromycetous fungi		
Nematoloma forwardii	Complete mineralization through conversion of 2- amino-4,6-dinitrotoluene by manganese peroxidase	Scheibner and Hofrichter (1998)
Alternaria sp. TMRZ/WN2	74% biotransformation of TNT	Hawari *et al.* (2000)
Aspergillus terrus MWi458	100% biotransformation of TNT with complete mineralization	
Mucor mucedo DSM810	95% biotransformation of TNT	
Neurospora crassa TM	Complete mineralization	
Penicillium frequentans ATCC96048	Complete mineralization	
Rhizoctonia solani MWi5	90% biotransformation of TNT	
Fusarium oxysporum	Transformation to 2-A-DNT, 4-A-DNT and 2,4-DAT	Smets *et al.* (2007)

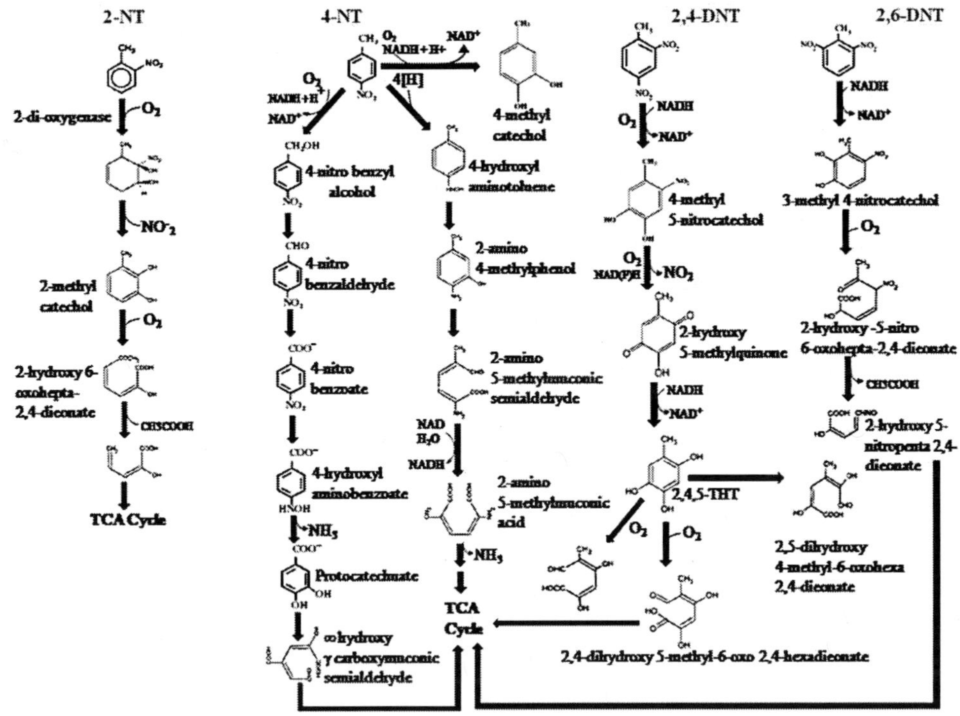

Figure 3A. Aerobic biodegradation pathway of NTs.

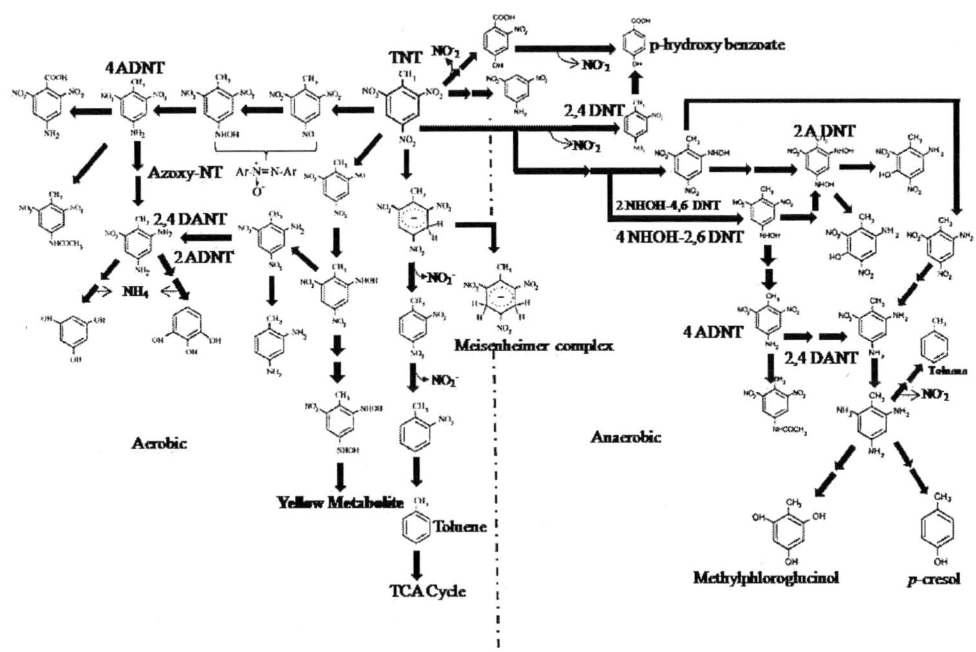

Figure 3B. Aerobic and anaerobic biodegradation pathway of 2,4,6-TNT.

2.4.2.1 Mono-Oxygenase (EC 1.13.14) Catalyzed Reactions

Mono (2,3 and 4-NTs)- and di-nitrotoluenes (2,4 and 2,6-DNTs) can be degraded by an initial attack under aerobic conditions through oxygenases. This class of enzymes has the property of catalyzing the incorporation of molecular oxygen into the aromatic moiety, with NAD(P)H as a co-factor.

Mono-oxygenases generally catalyze the addition of a single hydroxyl group to compounds that already contain such a group [Stenuit *et al.,* 2005]. This initial mono-oxygenase attack on the aromatic ring is followed by the subsequent release of the nitro group as nitrite. Mono-oxygenases often play a role in the metabolism of NTs by increasing either their reactivity and/or the water-solubility through the addition of an oxygen atom. Typical flavin dependent mono-oxygenases contain a tightly bound flavin cofactor, which can be reduced by an NAD(P)H substrate through their own catalytic activities. Well-known example includes two-component flavin diffusible mono-oxygenase family (TC-FDM) with two members Esd and Ese. However, there are no concrete reports of mono-oxygenase catalyzed biotransformation of NTs until now.

2.4.2.2 Di-Oxygenase (EC 1.14.12) Catalyzed Reactions

Aromatic ring di-oxygenases initiate the degradation of a whole range of NTs through addition of two hydroxyl groups, as it is the case for 2-nitrotoluene and 2,6-dinitrotoluene [Nishino *et al.,* 2000]. The addition of two hydroxyl groups results in a hypothetical partially reduced and unstable diol intermediate, that rearranges to a catechol compound (1,2-dihydroxybenzene) with the release of nitrite. Oxidative ring cleavages are mainly catalyzed by a di-oxygenase in the presence of NADPH through either *ortho*-ring cleavage between two hydroxyl groups or *meta*-ring cleavage at the bond proximal to one of two hydroxyl groups [Ye *et al.,* 2004; Stenuit *et al.,* 2005].

Pseudomonas strain JS42 degrades 2- NT through di-oxygenation catalyzed by 2-nitrotoluene-2,3-dioxygenase via 3-methylcatechol with release of nitrite leading to ring cleavage (Fig. 3A). The enzyme is (i) specific towards aromatic compounds and catalyses other di-oxygenation and monoxygenation reactions, (ii) a three-component di-oxygenase, and (iii) similar to 2,4-DNT-dioxygenase [Ye *et al.,* 2004].

Psuedomonas mineralized 2,4-DNT with a nitrite removal pathway involving di-oxygenase and mono-oxygenase enzymes as shown in Fig. 3A [Spanggord *et al.,* 1991; Suen and Spain 1993] to produce 2-hydroxy-5-methylquinone and leads to 2,4,5-trihydroxytoluene via mono-oxygenase and quinine reductase, respectively. Ring fission of 2,4,5-trihydroxytoluene likely occurs at position 5,6 of the aromatic ring to yield 3,5-dihydroxy-2-methyl-6-oxo-hexa-2,4-dienoic acid as ring cleavage product [Ye *et al.,* 2004].

Nishino *et al.* (2000) isolated *Burkholderia cepacia* strain JS850 and *Hydrogenophaga paleronii* strain JS863 to mineralize 2,4-DNT in the same way but degraded 2,6-DNT in a different manner. When 2,4- and 2,6-DNT were used as sole source of carbon and nitrogen together, di-oxygenation of the 2,6-DNT to 3- methyl-4-nitrocatechol (3M4NC) was the initial reaction, accompanied by the release of nitrite via meta-ring cleavage. Although 2,4-DNT-degrading strains convert 2,6-DNT to 3-methyl-4-nitrocatechol, but further catabolism was arrested. The pathway for 2,4-DNT degradation was different from 2,6-DNT degradation. In the latter case, 2-hydroxy-5-nitro-6-oxohepta-2,4-dienoic acid was the first ring fission product. How 3M4NC is converted to 2-hydroxy-5-nitro-6-oxohepta-2,4-dienoic acid is still unclear. It is proposed that position of 3 methyl group appears as a key determinant recognized by initial di-oxygenase to produce high specific 3M4NC in the 2,6-DNT pathway. The gene encoding di-oxygenase showed a nucleotide sequence similar to the ∞ subunit among nitro-arene di-oxygenases. The genes for the initial di-oxygenases involved in 2,4- DNT and 2,6-DNT degradation are all closely related, and the enzymes are produced at low constitutive levels [Spain 1995; Nishino *et al.,* 2000]. After initial di-oxygenation, the two pathways appear to diverge (Fig. 3A).

How DNT degradation is affected by the presence of both isomers is techno-economically relevant, since 2,4-DNT and 2,6-DNT are produced in a 4:1 ratio [Nishino and Spain, 2001], and are therefore, often present together in munitions plant wastewater [Ye *et al.,* 2004]. In an attempt to simultaneously degrade 2,4- and 2,6- DNT, Nishino and Spain (2001) found (i) mineralization rates of 98 and 94%, respectively and (ii) unpredictable behavior until an adapted population is established.

Structurally, TNT is similar to 2,4- and 2,6-DNT, suggesting that di-oxygenases could also hydroxylate TNT. However, the chemical stability and inertness of TNT due to (i) delocalization of the electrons in the aromatic ring, (ii) pronounced electron-withdrawing character of the nitro groups, (iii) high redox potential imposed by three nitro-groups, and (iv) high electron deficiency of the ring, render the electrophilic attack by oxygenases vulnerable to reductive process [Peres and Agathos, 2000; Snellinx *et al.,* 2002; Ye *et al.,* 2004; Roldan *et al.,* 2008]. Aerobic microbial metabolism involves the reduction of one or more nitro groups on the aromatic ring, with accumulation of amino-di-nitro-toluenes (2-A-4,6-DNT, 4-A-2,6- DNT) or di-amino-nitro-toluenes (2,4-DA-6-NT). These intermediate compounds resist further oxidative cleavage by microbe(s) and form a very serious problem due to their higher solubility, toxicity and geno-toxicity as compared to TNT. Besides this, condensation of partially reduced intermediates (hydroxyl-amino- and nitroso-di-nitro-toluenes), forms yet more toxic derivatives viz. tetra-nitro-azoxy-toluenes (2,2'-azoxy, 4,4'- azoxy, 2'4-azoxy) under aerobic conditions [Snellinx *et al.,* 2002].

Although aerobic mode neither removes nitrite nor re-aromatize of TNT to 2,4- or 2,6-DNT under physiological conditions [Lenke *et al.,* 2000], but degradation of TNT to 2,4-DNT is desirable because certain *Pseudomonas* strains can metabolize 2,4-DNT through di-oxygenase to 4-methyl-5-nitrocatechol and release nitrite [Esteve-Nunez *et al.,* 2001].

Studies based on penta-erythritol tetranitrate (PETN) reductase of *Enterobacter cloacae* PB2, French *et al.* (1998) observed slow growth on 2,4,6-TNT without production of DNT under aerobic condition. In another study, oxidative reduction of TNT by carbon monoxide dehydrogenase from *Clostridium thermoaceticum* [Huang *et al.,* 2000], and manganese-dependent peroxidase (MnP) from *Phlebia radiata* [Van Aken *et al.,* 2004] has also reported the same.

Notwithstanding oxidative transformation of reduced metabolites of TNT, attempts have been directed to favour mineralization. Significant mineralization of ^{14}C-TNT (18-20%) by *Phanerachaete chrysosporium* has been achieved in both soil and liquid cultures. While, *Pseudomonas* strain ST53 was identified to utilize 2,4-DNT and TNT as N-source under nitrogen-limiting conditions [Snellinx *et al.,* 2002]. Oxidative de-nitrification of TNT to 3-methyl-4,6-dinitrocatechol has been shown by Tront and Hughes (2005). Similar de-nitration activities were also reported by (i) Fiorella and Spain (1997), Kalafut *et al.* (1998) and Duque *et al.* (1993) in *Pseudomonas*, (ii) Martin *et al.* (1997) in *P. savastanoi*, (iii) Kim *et al.* (2002) in *Klebsiella sp.*, (iv) Bae *et al.* (1995) in *Enterobacter sp.*, (v) Popesku *et al.* (2004) in mixed culture, (vi) Naumova *et al.* (1988) in *P. fluorescens*, and (vii) Vanderberg *et al.* (1995) in *Mycobacterium vaccae*.

2.4.2.3 Meisenheimer Complex Formation

The partial reduction of the aromatic ring of DNT/TNT through addition of hydride ions leads to formation of a hydride-Meisenheimer complex [Spain, 1995; Kulkarni and Chaudhari, 2007]. The deficiency of electrons in the aromatic nucleus of NTs, (especially TNT) makes nucleophilic attack on the ring possible, resulting in a red-orange and negatively charged non-aromatic structure: Meisenheimer complex [Esteve-Nunez *et al.,* 2001]. Such complexes can be formed *in vivo* by hydride attack from a reduced pyridine nucleotide. Essentially, it is a direct $2e^-$ reduction of the aromatic ring catalyzed by aerobic bacteria and consists of adding hydride (negatively charged nucleophile form of atomic hydrogen H^-) to poly-nitro-arenes to form Meisenheimer complexes [Lenke and Knackmuss, 1992; Stenuit *et al.,* 2005]. The Meisenheimer complex can be re-aromatized after nitrite release, with the formation of di-nitro-toluenes [Rieger and Knackmuss, 1995; Kulkarni and Chaudhari, 2007].

2.4.2.4 Partial Reduction of Aromatic Ring

The nitro group is partially reduced to corresponding hydroxyl-amine and upon hydrolysis yields ammonia [Kulkarni and Chaudhari, 2007]. The hydroxyl-amino rearrangement and the release of ammonia are catalyzed either by a lyase or a mutase to provide respectively a di-hydroxy or an aminophenol compound [Stenuit *et al.,* 2005]. This somewhat fortuitously gratuitous and unspecific reduction of the nitro groups to amino groups takes place mainly in 4-NT and TNT.

Pseudomonas strains TW3 and 4-NT degrade 4- nitrotoluene (4-NT) by sequential oxidation of the methyl group to 4-nitrobenzoate and then assimilated by the afore-mentioned reactions [Groenewegen *et al.,* 1992]. Alternatively, the hydroxyl-amino group can undergo a mutase-mediated rearrangement to ortho-aminophenols, by a reaction similar to the non-enzymatic acid-catalysed Bamberger rearrangement [Davis *et al.,* 2000]. The amino-phenols serve as ring fission substrates with production of amino-muconic semi-aldehyde in a di-oxygenase-mediated *meta*-ring cleavage and subsequent release of ammonia. *Mycobacerium* sp. strain HL-4NT-1 also degraded 4-NT, to 4- hydroxyl-amino-toluene followed by 6-amino-*m*-cresol without oxidation of the methyl group before meta-ring cleavage [Ye *et al.,* 2004].

An NADH-dependent flavoprotein oxidoreductase (XenB) from *Pseudomonas fluorescens* catalyzed the reduction of TNT by adding a hydride to the aromatic ring with the concomitant formation of di-hydride Meisenheimer complex or catalyzed the reduction of nitro groups. A *Pseudomonas sp.* clone A used TNT as sole nitrogen source with successive removal of nitro groups to yield 2,4- and 2,6-DNT [Duque *et al.,* 1993]. Non-beneficial catabolism of TNT by direct hydride transfer resulted in a di-hydride-Meisenheimer complex (2H-TNT) and several protonated tautomers of 2H-TNT ($2H\text{-}TNT.H^+$) [Stenuit *et al.,* 2005].

2.4.3 Biodegradation Modes: Anaerobic

Reductive pathway for anaerobic biodegradation of NTs is favoured amongst microbes due to (i) barriers to oxidation imposed by nitro groups, (ii) rapid reduction at low redox potential, thereby minimizing oxidative polymerization of substrates and (ii) requirement of expensive O_2 delivery systems, high maintenance and energy costs in aerobic processes. It involves reduction of nitro groups to aromatic amines through a six-electron transfer mechanism [Razo-Flores *et al.,* 1997; Kulkarni and Chaudhari, 2007]. The reduction of nitro-groups to nitroso derivatives, hydroxyl-amines or amines is catalyzed by nitro-reductases through the successive addition of electron pairs donated by co-substrates [Kulkarni and Chaudhari, 2007]. The resulting amino-aromatic product can be degraded to catechol by an aniline oxygenase. The catechol produced is further mineralized by ring-cleavage reactions and liberation of ammonia [Marvin-Sikkema and de Bont 1994; Stenuit *et al.,* 2005].

Microbes like *Veillonella alkalescens*, *Clostridium* and *Desulfovibrio* sp. have been extensively studied for their pathways transforming DNT and TNT into amino- and hydroxyl-amino-derivatives under anaerobic conditions (Fig. 3B) [Esteve-Nunez *et al.,* 2001; Zhang and Bennett, 2005]. Unlike oxygenolytic mineralization, significant mineralization of TNT and DNT under anaerobic conditions has never been achieved and anaerobic mineralizing bacteria were never isolated [Zhang and Bennett, 2005]. Hence, synergistic participation of a consortium of bacteria is required for partial/complete degradation of NTs [Alexander, 1999].

Table 4 summarizes the performance of various NTs-degrading anaerobic bacteria. Several studies have demonstrated reduction of nitro groups to amino groups as a prelude to degradation of a number of mono-, di- and tri-nitro-toluene compounds [Kulkarni and Chaudhari, 2007]. The rate of reduction of nitro groups is generally governed by (i) chemistry of the entire molecule: number of rings, number and types of external substituents [Kulkarni and Chaudhari, 2007] and (ii) nitro substituents: concomitant conversion to amino groups takes place with decreasing reaction velocity [McCormick *et al.,* 1981; Zhang and Bennett, 2005].

2.4.4 Enzymatic Biotransformation of Nitro-Toluenes

While a comprehensive understanding of the fate of NTs and its transformation products in microbially active systems existed a decade ago, recent upsurge on biotransformation of NTs at the biochemical and genetic level has witnessed a significant expansion of our knowledge [Hawari *et al.,* 2000; Esteve-Nunez *et al.,* 2001; Ramos *et al.,* 2005]. Pragmatic benefit of *in situ* or *ex situ* enzymatic treatment by free/immobilized biocatalysts includes: (i) no involvement of microbial population, (ii) no buildup of biomass or chemicals, (iii) actively sustainable in harsh conditions, and (iv) tailor-made tuning of substrate range (specific/broad) as per needs. Commercialized enzymatic products for bioremediation of NTs consist of (i) freeze-dried enzyme powder with a shelf life of several months, (ii) immobilized enzymes encapsulated in polyurethane pad, and (iii) immobilized enzyme within a self-contained user-friendly device.

Enzymes reported in enzymatic bioremediation are (i) nitro-reductases, (ii) old yellow enzyme family, (iii) respiration-associated nitro-reductase, (iv) ring hydroxylating di-oxygenases, (v) Fe-only hydrogenase, (vi) nitro-reductase I, (vii) carbon monoxide dehydrogenase (CODH), (viii) pyruvate-ferredoxin oxidoreductases, (ix) sulfite reductase and (x) ligninolytic fungal enzymes from white-rot basidiomycetes.

Table 4. Anaerobic degradation of nitro-toluenes by various microorganisms

Nitro-toluene (NT)	Organism	Metabolic route	Reference
2,4-DNT	Bacteria		
	Veillonella alkalescens *Desulfovibrio* sp.	The earliest evidence of anaerobic DNT degradation	Esteve-Nuñez *et al.* (2001)
	Shewanella canadensis *Shewanella sediminis* *Desulfofrigus fragile* *Desulfovibrio* sp.	Biotransformation of 2,4 DNT into 2-amino-4-nitrotoluene (2-A-4-NT), 4-amino-2-nitrotoluene (4-A-2-NT) and 2,4- diaminotoluene (2,4-DAT)	Yang *et al.* (2009)
2,4,6-TNT	Bacteria		
	Veillonella alkalescens	Reduction of TNT to TAT	McCormick *et al.* (1976)
	Clostridium pasterianum	Reduction of TNT to TAT	McCormick *et al.* (1976)
	Methanococcus sp. strain B	Reduction of TNT to DANT	Boopathy and Kulpa (1994)
	Clostridium bifermentans CYS-1	Degrades TNT to alyphatic polar compounds via 4ADNT and 2,4DANT	Shim and Crawford (1995)
	Clostridium bifermentans LJP-1	Transforms TNT into TAT and phenolic compounds	Lewis *et al.* (1996)
	Actinomycete sp.	Reduction of TNT to TAT	Pasti-Grigsby *et al.* (1996)
	Clostridium sordelii	Reduction of TNT to TAT	Ederer *et al.* (1997)
	Escherichia coli	Reduction of TNT to TAT	Ederer *et al.* (1997)
	Lactobacillus sp.	Reduction of TNT to TAT	Ederer *et al.* (1997)
	Salmonella typhimurium	Reduction of TNT to TAT	Ederer *et al.* (1997)
	Mixed microbial population	Transforms TNT to 2-hydroxylamino-4,6-dinitrotoluene, 4-hydroxylamino-2,6- dinitrotoluene, amino-4,6-diinitrotoluene and 4-amino-2,6-dinitrotoluene	Ederer *et al.* (1997)

Table 4. (Continued)

Clitocybula dusenii TMbl 2	Partial mineralization	Scheibner *et al.* (1997)
Desulfovibrio sp.	Transforms TNT into TAT and DANT; 42% of radioactivity from [^{14}C]TNT is associated with cell biomass	Drzyzga *et al.* (1998)
Pseudomonas sp. strain JLR11	TNT as nitrogen source and final electron acceptor	Esteve-Nu´n˜ez and Ramos (1998) Esteve-Nu´n˜ez *et al.* (2000)
Clostridium sp.	Bamberger re-arrangement of di-hydroxyl-amino-di-nitro-toluene	Hughes *et al.* (1998)
Clostridium acetobutylicum	Reduction of TNT to TAT	Esteve-Nuñez *et al.* (2001)
Klebsiella sp. C1	Anaerobic transformation of TNT to HADNTs, ADNTs, DANTs, DNTs, NO_2^-	Kim *et al.* 2002
Desulfovibrio sp.	TNT as the sole nitrogen source; reduction of TNT to TAT	Snellinx *et al.* (2002)
Desulfovibrio sp. strain B	TNT as nitrogen source, toluene as a putative intermediate	Snellinx *et al.* (2002)
Stropharia rugosa-annulata DSM11372	Partial mineralization	Snellinx *et al.* (2002)
Cellulomonas sp. ES6	Anaerobic transformation of TNT to 4-HADNT, 2-HADNT, 2,4-D(HA)6NT, 4-ADNT	Borch *et al.* (2005)
Fusarium oxysporum	2-ADNT, 4-ADNT, 2,4-DA6NT	Hoehamer *et al.* (2006)
P. aeruginosa ESA-5	ADNT and DANT	Eyers *et al.* (2008)
Pseudomonas sp. strain TM15	Reductive metabolism of TNT into 2,4-dihydroxylamino-nitrotoluene (2,4DHANT)	Kubota *et al.* (2008)
Butyrivibrio fibrisolvens, Fibrobacter succinogenes, Lactobacillus vitulinus, Selenomonas ruminantium, Streptococcus caprinus, and *Succinivibrio dextrinosolvens*	Completely degrade TNT, with ~5% of the original TNT recovered as di-amino-nitro-toluene metabolites	Lorme and Craig (2009)

Table 4. (Continued)

Eubacterium ruminantium, Lactobacillus ruminis, Ruminobacter amylophilus, Streptococcus bovis, and *Wolinella succinogenes*	Completely degrade TNT, with 23-60% of the TNT recovered as amino-di-nitro-toluene and/or di-amino-nitro-toluene metabolites	Lorme and Craig (2009)
Clostridium polysaccharolyticum, Megasphaera elsdenii, Prevotella bryantii, Prevotella ruminicola, Ruminococcus albus, and *Ruminococcus flavefaciens*	80-90% degradation of TNT (100 ppm)	Lorme and Craig (2009)
Desulfovibrio desulfuricans subsp. *desulfuricans, Prevotella albensis,* and *Treponema bryantii*	50-80% degradation of TNT (100 ppm)	Lorme and Craig (2009)
Clostridium nitrophenolicum	Amino, Diaminonitrotoluene, and tetranitroazoxytoluene isomers	Moshe *et al.* (2009)
Myriophyllum aquaticum	Reduction of TNT to TAT	Yang *et al.* (2009)

Nitro-Reductases

A major class of enzymes involved in biotransformation of NTs is the nitro-reductases: oxygen insensitive (type I) and oxygen sensitive nitro-reductase (type II). The former catalyze the sequential reduction of the nitro groups through the addition of electron pairs from NAD(P)H to produce the nitroso, hydroxyl-amino and amino derivatives. However, the hydroxyl-amino derivative may be the end product of the reaction and usually, the nitroso intermediate is not detected because the second 2 e^- reaction has a much faster rate than the first two electron transfer [Koder *et al.,* 2002; Race *et al.,* 2005]. On the contrary, type II nitro-reductases catalyze the single-electron reduction of the nitro group to produce a nitro anion radical, which can be re-oxidized aerobically to the original structure with the concomitant production of the superoxide anion in a futile cycle. Nitro-reductases contain flavin mononucleotide (FMN) as the cofactor, use NAD(P)H and/or NADH as the electron donor, and perform sequential 2 e^- reductions of nitro substituents, although the nitroso intermediates are typically not detected [Roldan *et al*., 2008]. Nitro-reductases include NfsA and NfsB in *E. coli* [Rau and Stolz, 2003], SnrA and Cnr in *Salmonella enterica* serovar *typhimurium* [Nokhbeh *et al.* 2002], NfsI in *Enterobacter cloacae* [Roldan *et al*., 2008], RdxA in *Helicobacter pylori* [Roldan *et al*., 2008], flavin reductase P in *Vibrio harveyi* [Roldan *et al*., 2008], Frase I in *Vibrio fisherii,* NADH oxidase in *Thermus thermophilus,* nitrobenzene nitroreductase in *Pseudomonas pseudoalcaligenes,* PnrA and PnrB in *Pseudomonas putida* [Caballero *et al*., 2005], and NitA and NitB in *Clostridium acetobutylicum* [Kutty and Bennett, 2005].

It has become increasingly apparent that nitro-reductases reduce nitro to hydroxyl-amino functional groups and not beyond, even in the presence of excess reducing power [Caballero *et al.,* 2005; Kutty and Bennett, 2005]. On the other hand, in whole-cell assays, reduction typically goes to the level of amino groups. In addition, TNT is reduced faster than di- and mono-NTs or reduced transformation products [Caballero *et al.,* 2005] and can be explained by the one-electron redox potential of these substrates [Roldan *et al*., 2008].

Old Yellow Enzyme (Oye) Family

Another significant family of enzymes that shares the type I nitro-reductase activity (nitro to amino reduction) but has no real homology is called the old yellow enzyme (OYE) family, named after the first enzyme shown to contain a bound flavin (FMN). Some of these enzymes, in addition to being able to transform TNT via nitro-reduction, are able to liberate nitrite from TNT by de-nitration [Williams *et al.,* 2004]. The nitro-reduction pathway is catalyzed by all the enzymes, while de-nitration is only catalyzed by PETN reductase, XenB, and NemA via the formation of TNT hydride-Meisenheimer complex (H-TNT) and TNT dihydride-Meisenheimer complex (2H-TNT; Williams *et al.,* 2004) with subsequent or concomitant nitrite release. The exact physiological role of these OYE enzymes is still unclear, but they are probably part of a general stress response system (e.g., induction upon oxidative stress) and may maintain proper intracellular redox state, by transferring electrons from pyrimidine nucleotides to suitable oxidants such as quinones.

Respiration-Associated Nitro-Reductase

TNT has classically been utilized as a possible nitrogen source by *P. putida* JLR11, and can serve the unexpected role of terminal electron acceptor. *P. putida* JLR11 contains a membrane-associated respiratory nitro-reductase that confers the capacity to use nitro groups from TNT as the sole electron acceptor while oxidizing the electron donor acetate under anaerobic conditions. This novel respiratory process was supported by the measurement of proton translocation and ATP synthesis coupled to the oxidation of acetate or H_2 concomitant with the reduction of TNT to 4-amino-2,6- dinitrotoluene (4ADNT) in whole cells or membrane vesicles, respectively [Smets *et al.,* 2007].

Biologically inert and recalcitrant isomers, 2- amino-4,6-dinitrotoluene (2ADNT) and 4ADNT, formed via aerobic and anaerobic reduction of DNT and TNT, can be attacked oxygenolytically by di-oxygenase. Recombinant *E. coli* expressing the 2,4-DNT di-oxygenase (R34DDO) from *Burkholderia cepacia* R34 or *Burkholderia sp.* strain DNT are capable of (i) oxidizing 2ADNT at the 3,4 position releasing nitrite to form 3-amino-4-methyl-5-nitrocatechol and (ii) hydroxylating the methyl group to form 2-amino-4,6-dinitrobenzyl alcohol [Johnson *et al.,* 2001]. Despite the promises of the oxygenolytic cleavage of the aromatic ring, use of these enzymes are limited due to (i) low activity ($\leq$ 0.5 nmol product min^{-1} mg $protein^{-1}$), especially for the more recalcitrant 4ADNT, and (ii) elusive nature of the formed amino-methyl-nitrocatechols [Smets *et al.,* 2007].

Fe-Only Hydrogenase

The Fe-only hydrogenase from *C. acetobutylicum*, typically associated with H_2 production, catalyzes a H_2-dependent reduction of TNT to 2-hydroxylamino-4,6-dinitrotoluene (2HADNT) and 4-hydroxylamino-2,6-dinitrotoluene (4HADNT) and subsequent reduction to 2,4-dihydroxylamino-6-nitrotoluene (24D(HA)6NT) during acidogenic phase [Watrous *et al.,* 2003]. Inhibition studies using antibodies have revealed core parts of the enzyme necessary for TNT reduction, while [4Fe-4S] centers have been inferred as essential from Cu^{2+} inhibition [Kutty and Bennett, 2006].

Nitroreductase I

Three NAD(P)H-dependent nitro-reductases, without significant sequence homology to the nitro-reductase or OYE family discussed above, have been detected in *Klebsiella sp.* C1 [Kim and Song, 2005]. The enzyme with highest activity, nitro-reductase I, has FMN as its cofactor, catalyzes the two-electron reduction of several nitro-aromatic compounds, and produces 2ADNT and 2,2′,6,6′- tetranitro-4,4′-azoxytoluene from TNT. Hence, unlike other nitro-reductases, it transforms hydroxyl-amino-dinitrotoluenes (HADNTs) into ADNT, with preference for reduction in the ortho position. Even 2,4-DNT was subjected to nitro-reduction in the *ortho* position only to yield 2- hydroxylamino-4-nitrotoluene [Smets *et al.,* 2007].

A growing number of bacterial nitro-reductases have been isolated to transform TNT to HADNTs through direct reduction of the nitro group, and yield de-nitration products. The difference between reaction products observed in pure enzyme and whole-cell transformation experiments suggests the involvement of yet another group of enzymes that can transform NTs and its reduction products [Smets *et al.,* 2007].

Enzymes from White-Rot Basidiomycetes

Several white-rot fungi are able to mineralize DNT to CO_2 and degrade TNT to DNTs: (i) *Phanerochaete chrysosporium*, (ii) *Nematoloma frowardii* and *Stropharia rugosoannulata*,

(iii) *Trametes villosa* and (iv) *Kuehneromyces mutabilis*, *Hypholoma fasciculare*, and *Phlebia radiate* using LiP, MnP, and Lac.

Table 5. Various bioremediation technologies: an overview

Sr. No.	Bioremediation technologies	Process details	Limitations	Average costs (US$)
I.	*In situ* bioremediation: Existing technologies			
(a)	Soil washing	Uses liquids (usually water, occasionally combined with solvents) and mechanical processes to scrub soils Separates fine soil (clay and silt) from coarse soil (sand and gravel) and clean, larger volume of soil is considered to be non-toxic and can be used as backfill	Requirement of pre-treatment Difficulty in removing organics adsorbed onto clay particles Disposal problem of the washed soil	170/t
(b)	Soil vapor extraction	Involves the installation of vertical and/or horizontal wells in the area of soil contamination. Air 'blowers' are often used to aid the evaporation process	Concentration reductions greater than 90% difficult to achieve Not appropriate for sites with a groundwater table located less than 0.9 m below the land surface	20-50/t
(c)	Landfarming	Contaminated soil is mixed with nutrients and periodically aerated Involves spreading of excavated contaminated soils in a thin layer (no more than 1.5 m) on the ground surface of a treatment site and stimulating aerobic microbial activity within the soils through aeration and/or the addition of nutrients, minerals, and water/moisture	Requirement of large land space Pre-treatment of volatilized components	30-60/t

Table 5. (Continued)

(d)	Composting	Soil is mixed with a bulk agent and additional organic substrates such as straw, alfalfa, wood chips, sugar beet stems, leaves, etc. NTs are reduced to amino and di-amino-nitro-toluenes during the anoxic phase, and subsequent aerated phase eliminated most of the transformation products, possibly by covalent binding to the soil	Lack of full-scale information about biological systems Long incubation period	20-30/t
(e)	Biopiles	Piling of contaminated soils into piles or heaps and then simulating aerobic microbial activity by aeration and the addition of minerals, nutrients, and moisture	Difficulty in removing organics adsorbed onto clay particles Concentration reductions greater than 90% difficult to achieve	130-260/cubic yard
(f)	Bioventing	Injects air into the contaminated media at a rate designed to maximize in situ biodegradation and minimize/eliminate the off-gassing of volatilized contaminants to the atmosphere	Lengthy treat times (~2 years) Availability and installment of equipments	30-90/t
(g)	Bioslurping	Combines elements of bioventing and vacuum-enhanced pumping to recover free product from the groundwater and soil while promoting the aerobic bioremediation	Lack of full-scale information about biological systems	30-90/t
II.	*In situ* bioremediation: Newer emerging *in situ* technologies			

Table 5. (Continued)

(a)	Natural attenuation/ Enhanced Natural attenuation	Microbial mineralization to simple, non-hazardous inorganic compounds, microbial transformation to similar or more complex, but innocuous compounds, immobilization due to interactions with the soil or aquifer, and attributed to volatilization, dilution, sorption, and microbial degradation	Require a comprehensive characterization of the site-specific biodegradation processes Failure to demonstrate that decrease in the pollutant concentration actually occur due to mass reduction caused by biodegradation rather than due to dispersion, sorption, or volatilization	20-50/t
(b)	Biostimulation	As substances with similar structure might induce the same metabolic pathways, substrates with similar chemical structure are used to elevate biomass density and degradation rate	Introduction of substrates with toxic effects for organisms is inadvisable Lack of complete chemistry of co-metabolism	50-100/t
(c)	Bioaugmentation	Use of degradative microbial consortia or the augmentation of catabolically-relevant organisms to hasten remediation	May introduce pathogens Biosafety and ethical issues remain to be addressed	100-200/ cubic meter

Table 5. (Continued)

(d)	Phytoremediation	Use of plants and associated microbes to remove, contain, inactivate, stabilize, detoxify or degrade NTs and to revitalize contaminated sites	Long time Relatively new technique Bioavailability, toxicity and fate of metabolized products is not known Risk of contaminants to enter food web	
III.	*Ex situ* bioremediation	Removal of contaminated material from the source (soil/water) and processing it in bioreactors under controlled operating parameters	High operating, capital, equipment, design, and construction cost	130-200/cubic meter

2.4.5.1 *In Situ* Bioremediation

An *in situ* process is applied at the contaminated site, without removal of contaminated material from its original location [Kulkarni and Chaudhari, 2007]. Perceived criteria for effective *in situ* bioremediation include (i) documented loss of contaminants from the site, (ii) laboratory assays showing that microorganisms from site samples have the potential to transform the contaminants, and (iii) evidence that the biodegradation potential is actually realized in the field.

Existing Technologies

Soil Washing

Soil washing uses liquids (usually water, occasionally combined with solvents) and mechanical processes to scrub soils [Chu and Chan, 2003; Urum *et al.,* 2003]. It separates fine soil (clay and silt) from coarse soil (sand and gravel) and clean, larger volume of soil is considered to be non-toxic and can be used as backfill [USEPA, 1996; RAAG, 2000; Chu and Chan, 2003]. Although a cost-effective strategy, but it suffers from several limitations viz. (i) requirement of pre-treatment, (ii) difficulty in removing organics adsorbed onto clay particles, (iii) complex waste mixtures require a combination of solvents, (iv) disposal problem of the washed soil and (v) high average cost ($170 US/t, including excavation).

Soil Vapor Extraction

Soil vapor extraction (SVE), also known as soil venting or vacuum extraction, is an accepted, and cost-effective technology for remediating unsaturated soils contaminated with NTs [Zhan and Park, 2002; Halmemics *et al.,* 2003]. Its applicability is limited since (i) It is difficult, if not impossible, to develop models that permit an accurate prediction of SVE cleanup times from the data collected in short-term pilot studies [Yeung and Hsu, 2002], (ii) concentration reductions greater than 90% difficult to achieve [Zhan and Park, 2002; Khan *et al.,* 2004], (iii) it is generally not appropriate for sites with a groundwater table located less than 0.9 m below the land surface [Khan *et al.,* 2004] and (iv) moderate to high operational costs ($20 to $50 US/t).

Landfarming

In land farming, contaminated soil is mixed with nutrients and periodically aerated [Kulkarni and Chaudhari, 2007]. This technology involves spreading of excavated contaminated soils in a thin layer (no more than 1.5 m) on the ground surface of a treatment site and stimulating aerobic microbial activity within the soils through aeration and/or the addition of nutrients, minerals, and water/moisture [Khan *et al.,* 2004]. Widrig *et al.* (1997) used to ameliorate TNT contaminated soil by application of molasses. Mineralization of (^{14}C) TNT was demonstrated by the presence of ^{14}C in the cell biomass as trichloroacetic acid-precipitable material. The main inherent drawbacks includes: (i) requirement of large land space, (ii) pre-treatment of volatilized components, (iii) possibility for contaminant mobilization from the treatment site to a previously undisturbed site, (iv) concentration reductions greater than 95% and constituent concentrations less than 0.1 ppm are difficult to achieve, (v) capital, installation, operation, and maintenance costs ranges from $30 to $60 US/tonne with a treatment duration of 6 months to 2 years.

Composting

Composting, a most acceptable approach for the remediation of munitions-contaminated soils indicates that NTs can be removed from contaminated soils [Esteve-Nunez *et al.,* 2001]. For this purpose, soil is mixed with a bulk agent and additional organic substrates such as straw, alfalfa, wood chips, sugar beet stems, leaves, etc. Windrow composting and anaerobic-aerobic composting systems are highly efficient for treatment of NT-contaminated soils [Esteve-Nunez *et al.,* 2001]. Under these conditions, the NTs are reduced to amino and di-amino-nitro-toluenes during the anoxic phase, and subsequent aerated phase eliminated most of the transformation products, possibly by covalent binding to the soil [Esteve-Nunez *et al.,* 2001]. Composting of NTs markedly reduced mutagenicity in the contaminated soil [Groenewegen *et al.,* 1992].

The main drawbacks of the composting method are: (i) lack of full-scale information about biological systems, (ii) long incubation period, (iii) lack of effective control on various parameters that may affect the outcome, and (iv) poor chemical characterization of the outcome of the process. The merit of the composting method is its moderate cost by virtue of *in situ* application [Kulkarni and Chaudhari, 2007].

Biopiles

Biopiles (biocells, bioheaps, biomounds, compost cells, heap pile bioremediation, and static-pile composting) typically involves the piling of contaminated soils into piles or heaps and then simulating aerobic microbial activity by aeration and the addition of minerals, nutrients, and moisture [Khan *et al.,* 2004]. Although similar to landfarming, biopile does not aerate the soil through tilling or plowing. However, the typical cost associated with this treatment ($130 to $260 US per cubic yard) hinders its widespread applicability [Khan *et al.,* 2004).

Bioventing

The bioventing process injects air into the contaminated media at a rate designed to maximize *in situ* biodegradation and minimize/eliminate the off-gassing of volatilized contaminants to the atmosphere. Unlike biosparging, which involves pumping air and nutrients into the saturated zone, it pumps the air only into the unsaturated (vadose) zone [Mihopoulos *et al.,* 2001]. Significant advantages of bioventing over other processes include: (i) short treatment times (6 months to 2 years), (ii) readily availability and easy installment of equipments and (iii) cost-competitiveness ($30 US to $90 US/t of contaminated soil).

Bioslurping

Bioslurping, designed to address floating LNAPL layers, is a new in situ remediation technology that combines elements of bioventing and vacuum-enhanced pumping to recover free product from the groundwater and soil while promoting the aerobic bioremediation of hydrocarbon contaminants [Khan *et al.,* 2004]. It is used most often at sites with fine- to medium-grained overburden materials, but it is also used successfully at sites with medium- to coarse-grained materials and in fractured rock. It can be applied at sites with shallow groundwater and sites with the groundwater below 30 m [Yen *et al.,* 2003]. As compared to most other LNAPL remediation systems, bioslurping is considered to be cost-effective. The reduction in the amount of extracted groundwater and the ability to extract soil-gas at concentrations below the regulatory limits minimizes storage, treatment, and disposal costs, thereby, reducing project costs [Khan *et al.,* 2004].

Biosparging

The rationale behind biosparging is to inject air and nutrients into the soil below the water table to enhance the degradation of contaminants by naturally occurring organisms. This *in situ* technology generally uses indigenous microorganisms to treat the contaminated area [Khan *et al.,* 2004]. Inspite of a paucity of field and laboratory data to support design considerations, certain factors which make this method 'cost-competitive' are: (i) availability and easy installation of equipments, (ii) little disturbance to site operations, (iii) requires no storage or discharge of groundwater, and (iv) short treatment times (6 months to 2 years).

Newer Emerging in Situ Technologies

Natural Attenuation (NA) and Enhanced Natural Attenuation (ENA)

The use of naturally selected microbial populations to remove organic contaminants is known as intrinsic remediation/natural attenuation (NA)/monitored natural attenuation. The driving force behind NA lies in the ability of microbes to derive energy and essential growth requirements from the presence of contaminant. NA of explosives may occur through one or more of the following mechanisms: (i) microbial mineralization to very simple, non-hazardous inorganic compounds, (ii) microbial transformation to similar or more complex, but innocuous compounds, (iii) immobilization due to interactions with the soil or aquifer, and (iv) immobilization due to restrictive site geology and/or hydrology. NA processes are attributed to volatilization, dilution, sorption, and microbial degradation. The latter is the most important process since microbial activity decreases the mass of contaminants by breaking down or transforming the pollutants. The increasing number of guidelines for monitoring and enhancing *in situ* biodegradation [US EPA, 1999, 2006; NRC, 2000; UK-EA, 2000; Michels *et al.,* 2008], as well as review articles [Haack and Bekins, 2000; Iwamoto and Nasu, 2001; Roeling and van Verseveld, 2002; Lovley, 2003; Smets and Pritchard, 2003; Scow and Hicks, 2005; Foght, 2008; Weiss and Cozzarelli, 2008; Illman and Alvarez, 2009; Bombach *et al.,* 2010], reflects greater impact of the approach.

The major difficulty of NA as an effective remedial strategy lies in failure to demonstrate that decreases in the pollutant concentration actually occur due to mass reduction caused by biodegradation rather than due to dispersion, sorption, or volatilization. Moreover, application of NA as an innovative cleanup technology at contaminated field sites require a comprehensive characterization of the site-specific biodegradation processes. Key questions pertaining to use of NA strategy are: (i) Does biodegradation occur? (ii) How efficient are the degradation processes? (iii) How will the degradation processes develop over extended time periods? (iv) Is the estimated timeframe for complete decontamination of the field site reasonable? [Bombach *et al.,* 2010]. Performance assessment generally consists of three components: documented contaminant mass loss, geochemical fingerprints associated with biodegradation, and microcosm studies that show direct evidence of biodegradation [Illman and Alvarez, 2009]. Recently, new molecular and isotope fractionation techniques have emerged to complement existing technologies for the forensic analysis and the demonstration of bioremediation and natural attenuation. As there is no single universally reliable method to assess *in situ* biodegradation processes, decisions to employ NA as an innovative cleanup technology at a certain field site require comprehensive studies on various methods suitable for assessing the *in situ* biodegradation of the contaminants of interest.

Biostimulation

Because of high toxicity and resistance to biodegradation, NTs are (i) insufficient source of carbon and energy for microbes and (ii) difficult to be incorporated into central metabolic pathways [Zhao and Ward, 2000; Navratilova *et al.*, 2005; Zhang *et al.*, 2008; Gren *et al.*, 2010]. Thus, biodegradation of these substances requires presence of additional growth substrate to stimulate biomass concentration and degradation [Zhao and Ward, 2000; Karim and Gupta, 2002; Kulkarni and Chaudhari, 2006]. In co-metabolic biodegradation of NTs, growth substrates such as amino acids or carbohydrates are frequently used [Zhao and Ward, 2000; Karim and Gupta, 2002; Qui *et al.*, 2007; Zhang *et al.*, 2008].

In cometabolic systems for degradation of NTs, substrates with similar chemical structure are used to elevate degradation rate. It is based on the assumption that substances with similar structure might induce the same metabolic pathways. Although the use of substances supporting degradation of NTs during re-cultivation of degraded area seems to be justifiable, introduction of substrates with toxic effects for organisms is inadvisable. Good alternative for this problem is the use of aromatic substances naturally present in the environment, as inductors of biodegradation processes. It is known that many species of plants, especially willow (*Salix*), thyme (*Thymus vulgaris*), camomile (*Chamomilla recutita*), bean (*Phaesoli vulgaris*), strawberry (*Fregaria ananasa*), water plants/sweet flag (*Acorus calamus*) and many species of alga produce aromatic compounds as secondary metabolites. These compounds which enter the environment by exudation or are secreted directly into the soil after decay of plants might be used as natural substrates to improve degradation rate of NTs [Singer *et al.*, 2003; Gren *et al.*, 2010]. However, cometabolic transformation of nitro-aromatics in the presence of naturally occurring aromatics in the environment is rare. In a newer approach, Muter *et al.* (2008) used cabbage leaf extract as a biostimulation amendment to enhance the TNT degradation rate. Similarly, with the use of amorphous ferric oxy-hydroxide (ferrihydrite) and TNT as sole nitrogen source, Eyers *et al.* (2008) showed improved TNT de-nitration under anoxic conditions, whereas peat moss combined with crude soybean oil effectively promoted the biodegradation of TNT [Fuller *et al.*, 2004; Hatzinger *et al.*, 2004].

Bioaugmentation

Bioaugmentation entails the use of degradative microbial consortia [Dejonghe *et al.*, 2001] or the augmentation of catabolically-relevant organisms to hasten remediation [Thompson *et al.*, 2005]. Several new approaches of bioaugmentation increase the persistence and activity of exogenous microorganisms and/or genes following introduction into the environment include: (i) bioaugmentation with cells encapsulated in a carrier, (ii) gene bioaugmentation where the goal is for the added inoculants to transfer remediation genes to indigenous microorganisms, (iii) rhizosphere bioaugmentation where the microbial inoculant is added to the site along with a plant that serves as a *niche* for the inoculant's growth, and (iv) phyto-augmentation, where the remediation genes are engineered directly into a plant for use in remediation without a microbial inoculants [Gentry *et al.*, 2004].

Carrier and Encapsulation Technologies

Numerous different carrier materials have been used including biosolids, charcoal-amended soil, clay, lignite, manure, peat and alginate. It has been observed that (i) pre-sterilization of the carrier can increase the inoculant's shelf life [Temprano *et al.,* 2002] and (ii) pre-incubation of the inoculant in a sterile carrier can enhance its ultimate survival in environment [Van Dyke and Prosser, 2000].

The ideal characteristics for a carrier material include: (i) providing an adequate environment for cell survival and growth resulting in a long shelf life and enhanced activity when added to the environment, (ii) being nontoxic to the inoculant microorganisms and the environment, and (iii) allowing targeted introduction of cells and also a means to contain the introduced microorganisms when control is necessary.

Although materials used until now could potentially increase the inoculant's shelf life, but biosolids and manure may introduce pathogens (unless sterilized) or other contaminants into the environment, thereby making it ineffective.

Bioaugmentation with Encapsulated Microbes

Several other materials viz. acrylate copolymers, agarose, alginate, gelatin, gellan gum, kappa-carrageenan, polyurethane, and polyvinyl alcohol gel have been developed to encapsulate or immobilize cells for various purposes. The use of these materials allows the microorganisms to be contained in a relatively non-toxic matrix through which gases and liquids can diffuse. The capsule matrix can buffer the introduced microorganisms against pollutant toxicity in the environment. Additionally, substrates or C compounds can be added to the capsule to confer an advantage to the embedded inoculants; however, Duquenne *et al.* (1999) reported that diffusion of growth substrates from the capsule into the surrounding environment can diminish the desired effect. The capsule may also protect the inoculant from indigenous microorganisms. Another potential benefit of the encapsulation technology is the ability to create microsites with a unique microbial community that works interactively to remediate a given compound.

Activated Soil Bioaugmentation

Cell bioaugmentation can also be achieved by using activated soil directly as both the inoculant and carrier without extracting the degraders from the soil. Activated soil is defined as soil that has been exposed to the contaminant of interest and contains a developed degrader population that can eliminate the contaminant [Gentry *et al.,* 2004]. In spite of being less scientific than other methods, activated soil has several potential advantages: (i) introduction of a naturally developed degrader population(s) that may be composed of several members or even consortia that would be ineffective as pure cultures, (ii) the degraders are not cultured outside of the soil and thus, do not lose their ability to compete in the environment as is often observed for lab-cultured strains, and (iii) potential inclusion of unculturable degraders that would be missed in an attempt to isolate and culture an organism from one site in order to introduce the organism to another site.

Despite of these merits, drawbacks to the use of carriers, encapsulated cells, or activated soils for bioaugmentation still exists. These technologies are more suited to surface applications due to the probability that microbial encapsulation in, or attachment to, larger particles may further impede their movement through soil or sediment [Moslemy *et al.,* 2003]. Depending on the environmental conditions, microorganisms, and encapsulating material used, adverse conditions may develop within the capsule, such as accumulation of toxic compounds or anoxic conditions, which may inhibit or kill the inoculants [Moslemy *et al.*, 2002]. It is, therefore, critical to match the appropriate carrier technology with the specific conditions of the contaminated site.

Gene Bioaugmentation

Since introduced microorganisms often do not survive following bioaugmentation, investigations pertaining to the use of (i) naturally occurring horizontal gene transfer processes, (ii) artificial gene introduction via *mini*-Tn5 transposon system and (iii) gene alteration have gained prominence [Gentry *et al.,* 2004]. Horizontal gene transfer may occur via transformation, transduction, or physical contact and exchange of genetic material like plasmids or conjugative transposons between microorganisms (conjugation). The mini-Tn5 transposon system insert genes into chromosomes of host microbes (mainly gram-negative bacteria) to reduce the potential for transfer of the gene into other microorganisms in the environment. Gene alteration usually results in increased gene expression by modifying (i) transcriptional promoter and terminator sequences, (ii) number of copies of the gene in the host organism, and (iii) stability of the cloned gene protein.

The potential advantages of gene bioaugmentation over the traditional cell bioaugmentation approaches are: (i) introduction of remediation genes into indigenous microorganisms that are already adapted to survive and proliferate in the environment and (ii) no requirement for long-term survival of the introduced host strain. The transfer of plasmids, via conjugation, is the technology most studied with respect to bioaugmentation.

Despite its long-term use in bioremediation, there are no reports on the use of gene bioaugmentation to remediate nitro-toluene contaminated sites [Thompson *et al.,* 2005].

Phytoremediation

The use of plants and associated microbes to remove, contain, inactivate, stabilize, detoxify or degrade NTs and to revitalize contaminated sites is gaining increasing interest [Hannink *et al.*, 2002; Kulkarni and Chaudhari, 2007; Gerhardt *et al.*, 2009; Prasad *et al.*, 2010]. The use of green plants for *in situ* treatment of contaminated soils and groundwater occurs through (i) phytodegradation (or phytotransformation), which transports contaminant into its plant biomass and metabolize it to a less toxic/less bioavailable compound, (ii) phytoextraction, which immobilizes the contaminant into the plant biomass without its metabolism, (iii) rhizodegradation, which breaks down contaminants in the soil through microbial activity, usually enhanced by root exudates, and (iv) phytovolatilization, that transpires contaminant to the atmosphere in a modified form [Hannink *et al.*, 2002].

A variety of plant systems like (i) yellow nutsedge, (ii) bush bean, (iii) switchgrass, (iv) aquatic and wetland species, (v) axenic root cultures, (vi) hybrid poplar [Esteve-Nunez *et al.*, 2001] (vii) barnyard grass (*Echinochloa crusgalli*), (viii) sunflower (*Helianthus annuus*), (ix) Indian mallow (*Abutilon avicennae*), and (x) Indian jointvetch (*Aeschynomene indica*) has been evaluated for phytoremediation of NTs, especially TNT. Although plants do not utilize

NTs as energy source nor direct conjugation to plant macromolecules occur; however, aerobic reduction of NTs is the widely observed transformation reaction by certain plants (Table 6). Little is known about the product profile of TNT biocatalysis of plants beyond mono-aminated metabolites (2-amino-4,6-dinitrotoluene and 4-amino-2,6-dinitrotoluene). While the hydroxyl-amino derivatives are rarely detected, presumably due to their rapid decomposition in aerobic environments, reasons behind rare isolation of di-amino derivatives and di-nitro-toluenes are still unknown [Hannink *et al.*, 2002]. Besides, some plants seem to sequester/integrate TNT metabolites into biomass as conjugates and compartmentalize them into plant vacuoles [Esteve-Nunez *et al.*, 2001]. However, most studies regarding TNT metabolite production are confounded by the parallel production of TNT metabolites by plant-associated microbes. Therefore, in order to exclusively analyze plant transformation of TNT, investigations of the metabolism of TNT in axenic (sterile) experiments shaped a new dimension.

Plant-bacterium combinations to phytoremediate contaminated soil using *Pseudomonas* and meadow bromegrass (*Bromus erectus*) increased the portion of the rhizosphere community involved in nitro-aromatic metabolism and led to a reduction in soil TNT levels [Siciliano *et al.*, 2000]. Similarly, transgenic plants (tobacco) that express the *onr* gene from *Enterobacter cloacae* have been designed for bioremediation of TNT [Kulkarni and Chaudhari, 2007]. Further, Monti *et al.* (2005) fused *dnt* genes from *Burkholderia* sp. to a psychrotolerant strain of *P. fluorescens*, thereby achieving complete degradation of 2,4-DNT as a co-substrate at temperatures as low as 10°C.

2.4.5.2 *Ex Situ* Bioremediation

Ex situ bioremediation refers to removal of contaminated material from the source (soil/water) and processing it in bioreactors under controlled operating parameters (temperature, pH, and aeration). Therefore, this process is more expensive than the *in situ* approach [Kulkarni and Chaudhari, 2007].

In *ex situ* soil bioslurry process, heavily contaminated soil in a soil: water ratio of 1:1 (w/w) is stirred in a reactor to form a slurry and treated under aerobic and anaerobic conditions [Esteve- Nunez *et al.*, 2000] with the purpose of either (i) mineralization of the NTs as the main target or (ii) irreversible binding of NT metabolites to the soil matrix. Initially, the reactor is made anaerobic by addition of co-substrate like starch and metabolizing it by indigenous bacteria to consume the dissolved oxygen. Under these conditions, the TNT concentration was reduced from 3000 mg to 1 $mg.kg^{-1}$ of soil in 5 months. Additive co-substrate(s) served a twofold purpose: (i) oxygen removal by growing aerobes and (ii) availability of electrons for reduction of nitro groups of nitro-aromatics [Kulkarni and Chaudhari, 2007]. The major limitation of this treatment process is its high operating cost (US $130 to 200 per cubic meter) and capital, equipment, design, and construction costs (US $125,000 to 2,000,000) [Khan *et al.*, 2004].

Table 6. Phytoremediation of 2,4,6-TNT

Plant type	Initial TNT concentration	Experimental conditions	Metabolites produced	Reference
Cyperus esculentus (yellow nutsedge)	20 mg L^{-1}	Hydroponic; non-sterile	4-HADNT and ADNTs	Palazzo and Legget (1986)
Phaseolus vulgaris (bush bean)	10 mg L^{-1}	Hydroponic; non-sterile	2 and 4 ADNTs; acid hydrolysable conjugates	Harvey *et al*. (1990)
Allium schoenoprasum (chive)	0.1 -10 mg L^{-1}	Hydroponic; non-sterile	2 and 4 ADNT	Gorge *et al*. (1994)
Medicago sativum (alfalfa)	0.1 -10 mg L^{-1}	Hydroponic; non-sterile	2 ADNTs	Gorge *et al*. (1994)
Phaseolus vulgaris (bush bean)	10 mg g^{-1}	Soil; non-sterile	2 and 4 ADNTs	Scheidemann *et al.* (1998)
Lupinus angustifolius (lupin)	10 mg g^{-1}	Soil; non-sterile	4 ADNTs	Scheidemann *et al.* (1998)
Phacelia sericia (purple fringe)	10 mg g^{-1}	Soil; non-sterile	2 and 4 ADNT	Scheidemann *et al.* (1998)
Triticum aestivum (wheat)	10 mg g^{-1}	Soil; non-sterile	2 ADNTs	Scheidemann *et al.* (1998)
Lolium perenne (rye)	10 mg g^{-1}	Soil; non-sterile	2 ADNT	Scheidemann *et al.* (1998)
Alopecurus pratensis (meadow foxtail)	10 mg g^{-1}	Soil; non-sterile	2 and 4 ADNTs	Scheidemann *et al.* (1998)
Bromus inermus (bromegrass)	10 mg g^{-1}	Soil; non-sterile	2 and 4 ADNTs	Scheidemann *et al.* (1998)
Festuca rubra (turfgrass)	10 mg g^{-1}	Soil; non-sterile	2 and 4 ADNTs	Scheidemann *et al.* (1998)
Medicago sativum (alfalfa)	10 mg g^{-1}	Soil; non-sterile	2 and 4 ADNTs	Scheidemann *et al.* (1998)
Phleum pretense (cat's tail)	10 mg g^{-1}	Soil; non-sterile	2 and 4 ADNTs	Scheidemann *et al.* (1998)

Table 6. (Continued)

Daucus carota (carrot)	1-200 mg kg^{-1}	Contaminated soil; non-sterile	2 and 4 ADNTs; 2,4 and 2,6-DNT	Schneider *et al.* (1996)
Raphamus sativa (radishes)	1-200 mg kg^{-1}	Contaminated soil; non-sterile	2 and 4 ADNTs; 2,4 and 2,6-DNT	Schneider *et al.* (1996)
*Brassica olera*ceae (kale)	1-200 mg kg^{-1}	Contaminated soil; non-sterile	2 and 4 ADNTs; 2,4 and 2,6-DNT	Schneider *et al.* (1996)
*Valerianella locusta (lamb lettuc*e)	1-200 mg kg^{-1}	Contaminated soil; non-sterile	2 and 4 ADNTs; 2,4 and 2,6-DNT	Schneider *et al.* (1996)
Phaseolus vulgaris (bush bean)	1-200 mg kg^{-1}	Contaminated soil; non-sterile	2 and 4 ADNTs; 2,4 and 2,6-DNT	Schneider *et al.* (1996)
*Phaseolus vu*lgaris (bush bean)	10 mg L^{-1}	Hydroponic; non-sterile	2 and 4 ADNTs; 2,4 and 2,6-DNT	Schneider *et al.* (1996)
Catharanthus roseus (Madagascar periwinkle)	100 g L^{-1}	Sterile tissue culture	2 and 4 ADNTs	Hughes *et al.* (1997)
Hybrid poplar	32 mg L^{-1}	Hydroponic; non-sterile	2 and 4 ADNTs; 2,4-DANT	Thompson *et al.* (1998)
Bromus inermus Leyss (smooth bromegrass)	32 mg L^{-1}	Sand culture system; sterile	2 and 4 ADNT	Sun *et al.* (2000)
Myriophyllum aquaticum (parrot feather)	2 mg L^{-1}	Hydroponic; non-sterile	2 and 4 ADNTs	Larson *et al.* (1999)
Myriophyllum aquaticum (parrot feather)	0.681 mg L^{-1}	Hydroponic; non-sterile	2 and 4 ADNTs; 2,4 DNT	Best *et al.* (1997)
Saggitaria latifolia (arrowhead)	0.681 mg L^{-1}	Hydroponic; non-sterile	2 and 4 ADNT	Best *et al.* (1997)
Potamogeton nodosus Poir (pondweed)	0.681 mg L^{-1}	Hydroponic; non-sterile	4 ADNT	Best *et al.* (1997)
Ceratophyllum demersum L. (coontail)	0.681 mg L^{-1}	Hydroponic; non-sterile	4 ADNT	Best *et al.* (1997)

Table 6. (Continued).

Alisma subcordatum Raf (water plaintain)	0.681 mg L^{-1}	Hydroponic; non-sterile	2 ADNT; 2,4 DNT	Best *et al.* (1997)
Corex vulpinoidea Michx (fox sedge)	0.681 mg L^{-1}	Hydroponic; non-sterile	4 ADNT; 2,4 DNT	Best *et al.* (1997)
Scirpus cyperinus (wool grass)	0.681 mg L^{-1}	Hydroponic; non-sterile	2 ADNT	Best *et al.* (1997)
Eleocharis obusta (blunt spikerush)	0.681 mg L^{-1}	Hydroponic; non-sterile	2 ADNT	Best *et al.* (1997)
Phalaris arundinaceae (red canary grass)	0.681 mg L^{-1}	Hydroponic; non-sterile	4 ADNT	Best *et al.* (1997)
Typha angustifolia (narrow leaf cat tail)	0.681 mg L^{-1}	Hydroponic; non-sterile	4 ADNT	Best *et al.* (1997)
Potamogeton nodosus Poir (pondweed)	0.99 mg L^{-1}	Hydroponic; non-sterile	2 and 4 ADNT; polar metabolites	Best *et al.* (1999)
Phalaris arundinaceae (red canary grass)	0.99 mg L^{-1}	Hydroponic; non-sterile	2 and 4 ADNT; polar metabolites	Best *et al.* (1999)
Myriophyllum aquaticum (parrot feather)	0.99 mg L^{-1}	Hydroponic; non-sterile	2 and 4 ADNT; polar metabolites	Best *et al.* (1999)
Myriophyllum aquaticum (parrot feather)	50 g L^{-1}	Hydroponic; sterile	2 and 4 ADNT; polar metabolites	Hughes *et al.* (1997)
Myriophyllum spicatum (Eurasian water milfoil)	50 g L^{-1}	Hydroponic; non-sterile	2 and 4 ADNTs	Hughes *et al.* (1997)
Myriophyllum aquaticum (parrot feather)	1-10 mg L^{-1}	Hydroponic; non-sterile	ADNTs, DANTs, TNB	Rivera *et al.* (1998)
Myriophyllum aquaticum (parrot feather)	22-75mg L^{-1}	Hydroponic; sterile	2 and 4 ADNTs	Bhadra *et al.* (1999)
Myriophyllum aquaticum (parrot feather)	60 mg L^{-1}	Hydroponic; non-sterile	2 and 4 ADNTs; conjugates TNT-1 and TNT-2	Wayment *et al.* (1999)

2.4.5.3 Scale-up Effectiveness for Bioremediation of Nitro-Toluenes

Despite its long-term use, bioremediation of contaminated sites with microbial cells continues to be a source of controversy within environmental microbiology. This largely results from its notoriously unreliable performance record. Some see it as an expensive and ineffective equivalent, while others extol its virtues and even derive a living from its commercialization. Based on often unrealistic extrapolations of data obtained from lab and field experiments, too enthusiastic interpretations and promises have been made [Vangronsveld *et al.*, 2009]. While about 90% reduction in TNT within the first 6 months in ordinance plant at Werk Tanne (Clausthal-Zellerfeld, Harz, Germany) has been achieved using phytoaugmentation with highly active mycorrhiza and white-rot fungi [Koehler *et al.*, 2002], Van Dillewijn *et al.* (2007) reported that neither natural attenuation nor bioaugmentation with *Pseudomonas putida* JLR11 affected TNT levels to a significant degree. Further, too few studies are available at the current time for a comprehensive analysis of laboratory-to-field bioremediation scale-up. Hypotheses pertaining to differences in laboratory- and field scale studies include (i) variability in fates of tested NTs and biodegradation rates in bench-scale, pilot-scale or field trials, (ii) differences in steady-state equilibrium of the microbes involved in field, (iii) mass transport mechanisms and limitations, (iv) the presence of multiple phases, contaminants, and competing microorganisms, (v) spatial geometry heterogeneities, and (vi) sub-surface environmental factors that may inhibit microbial growth (temperature, pH, nutrient, or redox conditions).

At present, the concern that laboratory-scale to field-scale experimental linkage is relatively poor and seems to be justified for many of the reported bioremediations. Where laboratory-to-field linkage is good, more information about the details of successful operation are expected / desired for assurance that scale- up can be achieved robustly in future trials.

2.4.5.4 Bio-Tensioactive Compounds to Alleviate Impediments in Bioremediation of Nitro-Toluenes

Chemical surfactants are basically surface-active agents of amphipathic nature, which alter the properties of fluid interfaces, partition at interface between fluid phases leading to formation of micro-emulsion, impart better wetting, spreading, foaming and detergent traits, rendering them as most versatile process chemicals to be utilized in surfactant-enhanced bioremediation practices [Paria, 2008; Laha *et al.*, 2009]. However, their (i) toxicity, (iii) carcinogenicity, (iii) non-biodegradibility, (iv) bio-accumulation and (v) inconsistent performance with slow desorption kinetics coupled with increased environmental awareness have prompted the use of natural bio-surfactants as an alternative [Hazra *et al.*, 2010]. Several comprehensive reviews have appeared highlighting the environmental applications of biosurfactants in bioremediation arena [Mulligan, 2005; Paria, 2008; Mulligan, 2009; Vardar-Sukan and Kosaric, 2009; Abdel-Mawgoud *et al.*, 2010; Banat *et al.*, 2010].

The application of biosurfactants in the remediation of NTs aims at (i) increasing their bioavailability, (ii) mobilizing and removing the contaminants by pseudo-solubilization and emulsification [Banat *et al.*, 2010], (iii) precipitation-dissolution, (iv) desorption according to Le Chatelier's principle, and (v) direct contact of biosurfactant to sorbed target organic(s) at solid solution interface under conditions of reduced interfacial tension. The afore-mentioned objectives have been achieved either by (i) biosurfactant-enhanced bioremediation or (ii) biosurfactant-enhanced soil washing.

Several adsorption studies on bioremediation of organic pollutants are mainly concentrated on rhamnolipid biosurfactant [Paria, 2008]. Rhamnolipids enhance degradation through several mechanisms by which they (i) have increased hydrocarbon solubility, (ii) interact with degrading cells, to become more hydrophobic for more easy association with hydrophobic substrates for improved bioavailability, and (iii) cause loss of Lipopolysaccharides, leading to increased cell surface hydrophobicity. Popularity of rhamnolipids over other high- and low-molecular biosurfactants are attributed to the following facts: (i) addition of high amount of rhamnolipid would suppress degradation of targets until a substantial portion gets recovered, (ii) lower levels are less appropriate to cause mobilization of organics, which could result in undesirable spreading of a contaminant plume and (iii) lesser it is required, better it is in economic terms. Even though there are no reports on biosurfactant-assisted bioremediation of NTs as of now, modeling of the effect of biosurfactant addition in bioremediation treatment is not predictable, and efficacy has to be evaluated experimentally [Banat *et al.*, 2010; Franzetti *et al.*, 2010]. Although *in situ* production of biosurfactants has been reported as an easier and more cost efficient approach than external addition [Mulligan, 2005], it may lead to numerous secondary effects that have no role in increasing availability of organics. On the other hand, augmentation of organic compounds solubilization is still argued and not necessarily associated with an equivalent increase of its bioavailability to microbes [Tecon and van der Meer, 2009].

2.4.6 Molecular Approaches in Bioremediation of Nitro-Toluenes: New Insights to Old Problem

2.4.6.1 Catabolic Genes and their Regulation

Although microbes at contaminated sites could acquire biodegradative capabilities through an evolutionary or adaptive process, the efficiency of biodegradation of NTs (2-, 3-, 4-NT, 2,4- and 2,6-DNT and TNT) is very low owing to (i) high toxicity, (ii) the electron-withdrawing effect of nitro groups, and (iii) lack of specificity of enzymes involved [Singh *et al.*, 2008; Cao *et al.*, 2009].

Usually, biodegradation at the substrate or the metabolite level is supposed to be coordinately regulated and controlled by a set of structural genes encoding regulatory proteins at the gene level. A number of factors, like structure of the genes, enzymes, substrates, and the metabolites, have been found to influence the expression of catabolic genes. Although these clusters can be encoded in both the chromosomes and the plasmids, but they are usually present in mobile genetic elements, transposons and plasmids. The plasmid-encoded catabolic pathways have the special advantage of facilitating the horizontal transfer of the respective catabolic genes in a microbial population, resulting in a rapid adaptation of microorganisms to the presence of new aromatic pollutants in a particular ecosystem. Some catabolic genes for the biodegradation of NTs are summarized in Table 7.

How evolution of new catabolic pathways govern biodegradation is elegantly elucidated by 2,4-DNT pathway of *Burkholderia*, *Pseudomonas* and *Commamonas*. Molecular characterization of the 27 kb region of DNA encoding 2,4-DNT pathway in these organisms suggest that the genes for pathway are derived from three different sources: (i) initial DNT di-oxygenase, which catalyses the first de-nitration of the substrate, appears to have originated from a naphthalene degradation pathway of *Ralstonia* sp. U2, (ii) second de-nitration step is catalyzed by a 4-methyl-5- nitro-catechol mono-oxygenase that is probably derived from a pathway for degradation of chloro-aromatic compounds, and (iii) the last enzymes of the pathway might originate from a gene cluster for amino acid degradation [Kivisaar, 2009]. The organization of the genes for this pathway indicates progression compaction. However, the presence of (a) open reading frames without any known role in 2,4-DNT degradation, and (b) transposon remnants in the region suggests that the pathway is at an intermediate stage in the evolution of an optimal system from genetic material originating in different bacteria [Johnson *et al.*, 2002]. Thus, both the horizontal transfer of the genes and movement of transposable elements have contributed to the evolution of the 2,4-DNT pathway, at least conclusively in *Burkholderia cepacia* R34. The occurrence of point mutations during this process cannot be excluded when NTs are present at higher concentrations since they are powerful carcinogens [Kulkarni and Chaudhari, 2007; Kivisaar, 2009].

2.4.6.2 Metabolic Engineering

The fast emergence of new enzymes for the degradation of synthetic aromatic compounds is difficult to explain only by horizontal transfer of genes and elevated mutation frequency under stressful conditions. Under the formulated concept of evolution based on the promiscuity of proteins, protein evolution towards a new function involves transitions from a specialized enzyme into a generalized intermediate and thereafter to a new, specialized enzyme. This is further supported by the results of several directed evolution studies on catabolic enzymes. Enzyme promiscuity indicates that (i) the assembly of a catabolic operon and the acquisition of efficient transcriptional control over the system are independent events and (ii) evolution of new effector-responsive regulators is inter-related. On this process, it was demonstrated that the toluene-activated transcriptional activator XylR, encoded by TOL plasmid pWW0, is able to acquire *de novo* responsiveness both to 2,4-DNT and its mono-substituted precursors and to the unrelated isomer of 3-NT [Galvão *et al.*, 2007]. Interestingly, these mutations did not alter amino acids in the effector-binding pocket of the regulator, but amino acid substitutions were located at the protein surface. These changes were believed to affect conformational shifts that follow effector binding and modulate signal transmission between XylR domains [Galvão *et al.*, 2007], thereby indicating an existing but unproductive binding site in XylR for DNT and many others in the toluene catabolic pathway [Kivisaar, 2009]. In contrast, 2-NT degradation genes in *Acidovorax* sp. strain JS42 are controlled by the LysR-type specific regulatory protein NtdR, which is 98% identical to NagR, the activator of the naphthalene degradation operon in *Ralstonia* sp. strain U2 [Jones *et al.*, 2003; Lessner *et al.*, 2003]. Based on how nitrotoluene-responsive regulator NtdR can be generated from a NagR-like ancestor by just a few mutations, Ju *et al.* (2009) showed stepwise broadening of the effector range without loss of the original activity by reconstructing pathway of the evolution of NtdR from NagR.

Table 7. Properties of identified genes involved in degradative pathways of NTs

Target compound	Name	Target metabolite in pathway	Function	Size	Microorganism	Mode of isolation	References
2-NT	*ntdAa*	2-Nitro-toluene	Reductase	4.9 kb	*Pseudomonas* sp. JS42	Genomic library expressed in *E. coli* and screening for metabolic activities	Parales *et al.* (1996)
	ntdAb		Ferrodoxin				
	ntdAc		Iron–sulfur protein α				
	ntdAd		Iron–sulfur protein β				
4-NT	*ntnMA*	4-Nitro-toluene	Mono-oxygenase	14.8 kb	*Pseudomonas* sp. TW3	Genomic library screened with designed probes; activities confirmed by cloning and expression in *E. coli*	James *et al.* (2000)
	ntnD	4-Nitrobenzyl alcohol	Nitrobenzyl alcohol dehydrogenase				
	ntnC	4-Nitrobenzaldehyde	Nitrobenzaldehyde dehydrogenase				
	pnbA	4-Nitrobenzoate	Nitrobenzoate reductase	6 kb	*Pseudomonas* sp. TW3	Genomic library expressed in *P. putida* PaW340 and screening for metabolic activities	Hughes and Williams (2001)
	pnbB	4- Hydroxyl-amino-benzoate	Hydroxylaminobenzoate lyase				
2,4- DNT	*dntA*	2,4-Di-nitro-toluene	Di-oxygenase	27 kb	*Burkholderia cepacia* R34	Genomic library expressed in *E. coli* and screening for metabolic activities	Johnson *et al.* (2002)
	dntB	4-Methyl-5-nitrocatechol	Mono-oxygenase				
	dntD	2,4,5-Trihydroxytoluene	Di-oxygenase				
	dntG	2,4-Dihydroxy-5-methyl-6- oxo-2,4-hexadecanoic acid	Isomerase/hydrolase				
	dntE	Methylmalonic acid semialdehyde	Dehydrogenase				
2,4,6- TNT	*xenB*	2,4,6-Tri-nitro-toluene	Reductase	1.05 kb	*P. fluorescens* I-C	Genomic library expressed in *E. coli* and screening for metabolic activities	Blehert *et al.* (1999)

On this premises, recombinant strains has been constructed for the mineralization of 2,4-DNT by pathway combining from various bacteria via conjugation into a single recombinant host. Such patchwork assembly include: (i) introduction of the genes encoding the 2,4-DNT degradation pathway from *Burkholderia* sp. strain DNT into *Pseudomonas fluorescens* ATCC 17400 for its complete degradation [Monti *et al.*, 2005], (ii) using *E. coli* as a host for cloning and expressing the genes encoding a novel partial reductive pathway for nitro-toluene from *Comamonas* sp. strain CNB-1, (iii) introduction of the TOL plasmid pWWO (*P. putida* mt-2) into *P. putida* F38/D, creating a hybrid strain *P. putida* TB101, and (iv) hybrid *P. putida* TB105, constructed by cloning the genes encoding the TOD pathway on broad host range multicopy vector RSF1010 and introducing into the TOL strain *P. putida* mt-2.

Even though no genes for TNT de-nitration have been discovered up to now, metagenomic libraries are particularly promising for finding de-nitration genes, compared with methods based on the isolation of pure cultures [Eyers *et al.*, 2004].

2.4.6.3 Protein Engineering

In addition to engineering catabolic pathways, modification of certain genes encoding biodegradative enzymes is another alternative approach to improve bioremediation properties. It basically includes (i) rational design, a top-down approach and (ii) site-directed mutation and directed evolution (DNA shuffling), a bottom-up method. Rational design of 2-NT di-oxygenase, responsible for the oxidation of NT to 3-methyl catechol and nitrite, has been carried out by replacing position 258 (N258V) near active sites and significantly changed the enantio-specificity [Singh *et al.*, 2008]. The latter approach has been used to evolve (i) di-oxygenases and mono-oxygenases for 2,3- and 2,5-DNT that previously could not be degraded, (ii) 4-methyl-5-nitrocatechol (second step of DNT biodegradative pathway), (iii) amino-nitro-toluenes, (iv) 2,4-DNT di-oxygenase in *B. cepacia* R34, (v) toluene ortho mono-oxygenase in *B. cepacia* G4, (vi) TOM-Green in *B. cepacia* G4 and (vii) catechol 2,3- di-oxygenase in *P. putida* UCC2. From these insights and by discovering important residues in the active site of toluene mono-oxygenases, it is now possible to control regio-specific oxidation of NTs.

2.4.6.4 Whole-Transcriptome Profiling

For bioremediation, whole-transcriptome profiling using DNA microarrays has been used to determine mutualistic interactions in the rhizosphere for potential strains relevant for bioremediation. The principal approaches toward global protein identification have been categorized into two main categories: (i) the ‘gel based approach’ (2D- MS/MS), use of 2D alone or 2D along-with MS and (ii) ‘gel free approach’, use of multi-dimensional protein identification technology (MudPIT) [Chauhan and Jain, 2010]. In spite of its promise for combining bacteria with strong degradation potential with suitable plant hosts, it has not been explored in bioremediation beyond pesticide lindane and its derivatives.

2.4.6.5 Proteomics in Bioremediation

Since whole-transcriptome is unable to show that changes in transcription lead to changes in protein levels and frequently changes in proteins are necessary for enhanced bioremediation, proteomics is one of the best techniques for gauging changes in metabolism by addressing protein expression, structure, and function. It encompasses three main applications in biodegradation: (i) study of bacterial response to the various environmental contaminants for microbes, (ii) identification of key proteins/ enzymes involved in the metabolism of toxic compounds, and (iii) community structure analysis especially during *in situ* bioremediation. A thorough proteome investigation using 2D- MALDI-TOF in *P. putida* revealed 531 spots: 134 proteins were toluene specific, 35 proteins increased more than two-fold, and 26 proteins repressed under toluene specificity, while a de-nitrifying bacterium expressed about 150 protein spots with increased abundance in presence of toluene [Chauhan and Jain, 2010]. Despite a growing knowledge, the application of proteomics in biodegradation is still in its infancy and certainly provides a ray of hope [Fulekar, 2008; Cao *et al.*, 2009].

2.4.6.6 Biosafety of Engineered Strains

Since the introduction of biomolecular engineering into bioremediation, intensive debates on ethics and risks of the release of engineered strains into the environment remains a major issue due to (i) possibilities of the persistence of naturally undesired genes in the environment and (ii) their undesirable transfer to indigenous species [Diaz, 2004; Cao *et al.*, 2009]. To minimize these potential risks, bio-confinement system, viz., (i) induction of suicide genes and (ii) nucleases-containing killing system has been developed. Surprisingly, research in the development of bio-confinement systems for engineered bacteria appears to have stopped around a decade ago for unknown reasons [Committee on the Biological Confinement of Genetically Engineered Organisms, 2004; Cao *et al.*, 2009]. Apart from this governmental permission and stringent regulatory norms are realized to be an impediment to actualizing field release studies and prevents the integration of state-of-art engineered microbes into field. Nevertheless, a comprehensive and cautious risk assessment is necessary before releasing the engineered strains into the environment to avoid unknown consequences.

3.0 Current R&D Status in Bioremediation of Nitro-Toluenes: Knowledge Explosion for Environmental Solutions

As of May 2010, about 11,684 articles have been published on various aspects of bioremediation starting with only 11 in 1989. Post-genomic era has witnessed several ultra-techniques to (i) characterize partial functional gene specific to the augmented strain(s) or amplification of highly conserved region of a functional gene(s) using DNA Hybridization, Quantitative and Real Time PCR, and FISH [Chauhan and Jain, 2010], (ii) determine community structure by high-throughput fingerprinting methods (SARST, OFRG, rep PCR-genomic fingerprinting, ARDRA, T-RFLP, D/TGGE, SSCP, and ARISA) [Kapley and Purohit, 2009], (iii) analyze global protein data using LC-based shotgun proteomics and (iv) create rational combination of biological properties with central elements of engineering design through synthetic biology.

Much progress has been made in the field of bioremediation in Europe and North America, while developing nations in Africa and Asia remains to be included in this endeavour [Prasad *et al.*, 2010]. There are currently 184 identified sites in the USA and 51 in Europe along with several research centres in Japan and China exploring newer synthetic biology-based bioremediation of DNT and TNT [http://www.synbioproject.org/]. The ever expanding knowledge base about biodegradation pathways, structural and functional knowledge of key enzymes, extensive molecular characterization of naturally detoxifying microorganisms, combined omics analysis of the transcriptome, proteome, metabolome, metabolic flux, or enzyme activities may lead to comprehensive system-level understanding and provide insights into the rational design of microbes for NT bioremediation.

4.0 Bioremediation: Global Market and Costs

At present, cost and performance of various bio-treatment technologies, viz., excavation with on- and off-site incineration, stabilization, soil vapor extraction, thermal desorption for source control, and pump and treat technologies for groundwater are readily available. However, certain issues related to (i) limited number of applications, (ii) inadequate data on cost and performance, (iii) variability in local parameters viz. kind and depth of contamination, soil conditions, disposal of the biomass etc., and (iv) cost:benefit analysis of bio-treatments still remains in nascent stage and needs to be properly addressed. The economic aspect should be evaluated in terms of cost effectiveness, otherwise raises significant uncertainties about (i) the relation between the size and kind of the contamination and the emanating risks and (ii) monetizing health deficiencies [Vangronsveld *et al.*, 2009]. On these premises, even assuming no further planting, existing non-biological remediation technologies seem to be economically non-viable due to (i) impractical cost (US$ 750 billion) and time (about 30 years) of physico-chemical processes, (ii) escalating cost estimates (US$ 30 billion) and hundreds of years of work for explosives, (iii) overburden project estimation of soil excavation in Europe (US$ 270-460/t) and US (US$ 324-552/m^3), and (iv) unrealistic demands of incineration (US$ 1.7 trillion; US $7000/citizen). However, if using bioremediation with the same period of time, the cost would be reduced to US$ 75 billion (US$ 14 billion/year).

5.0. Conclusion

Research in bioremediation has become a societal mandate and integrates various approaches from systems biology (e.g. metabolomics, transcriptomics, proteomics and genome scale metabolic models) to innovative field studies. While microbial mineralization of NTs remains an unrequited aspiration, new biological activity is being generated, important regulatory pathways are being discerned, and the fate of contaminants are being assayed to help us move to a sustainable industrial model. Although there is a need to be cautiously optimistic regarding widespread application of bioaugmentation and phytoremediation, and the limits of this remedial strategy need to be acknowledged, it seems that, from a technological stand-point, we have come very close to acceptable solutions.

Reviewed By

Prof. R. M. Kothari, Director, Rajiv Gandhi Institute of Information Technology and Biotechnology, Bharati Vidyapeeth University, Pune 411 046 (India).

References

Abdel-Mawgoud, A., Lépine, F. & Déziel, E. (2010). *Rhamnolipids: diversity of structures, microbial origins and roles*. Appl Microbiol Biotechnol, 86, 1323-1336.

Alexander, M. (1999). *Biodegradation and Bioremediation: Bioremediation technologies- in situ and solid phase*. San Diego, Academic Press. 159-176.

Anonymous. (2007). *The world's top ten most polluted places*. The New Scientist, 195 (2622), 5.

Ask, K., Decologne, N., Asare, N., Holme, J. A., Artur, Y., Pelczar, H. & Camus, P. (2004). *Distribution of nitroreductive activity toward nilutamide in rat*. Toxicol Appl Pharmacol, 201, 1- 9.

Bae, B; Autenrieth, RL; Bonner, JS. *Aerobic biotransformation and mineralization of 2,4,6-trinitrotoluene*. In: Hinchee, RE, Hoeppel RE, Anderson, DB editors. Bioremediation of recalcitrant organics. Ohio: Battelle Press; 1995; 231-238.

Banat, I. M., Franzetti, A., Gandolfi, I., Bestetti, G., Martinotti, M. G., Fracchia, L., Smyth, T. J. & Marchant, R. (2010). *Microbial biosurfactants production, applications and future potential*. Appl Microbiol Biotechnol, 87, 427-444.

Best, E. P. H., Sprecher, S. L., Larson, S. L., Fredrickson, H. L. & Bader, D. F. (1999). *Environmental behavior of explosives in groundwater from the Milan Army Ammunition Plant in aquatic and wetland plant treatments. Uptake and fate of TNT and RDX in plants*. Chemosphere, 39, 2057-2072.

Best, E. P. H., Zappi, M. E., Fredrickson, H. L., Sprecher, S. L., Larson, S. L. & Ochman, M. (1997). *Screening of aquatic and wetland plant species for phytoremediation of explosives-contaminated groundwater from the Iowa Army Ammunition Plant*. Ann NY Acad Sci, 829, 179-194.

Bhadra, R., Spanggord, R. J., Wayment, D. G., Hughes, J. B. & Shanks, J. V. (1999). *Characterization of oxidation products of TNT metabolism in aquatic phytoremediation systems of Myriophyllum aquaticum*. Env Sci Technol, 33, 3354-3361.

Blehert, D. S., Fox, B. G. & Chambliss, G. H. (1999). *Cloning and sequence analysis of two Pseudomonas flavoprotein xenobiotic reductases*. J Bacteriol, 181, 6254-6263.

Bombach, P., Richnow, H. H., Kästner, M. & Fischer, A. (2010). *Current approaches for the assessment of in situ biodegradation*. Appl Microbiol Biotechnol, DOI: 10.1007/s00253-010-2461-2.

Boopathy, R. & Kulpa, C. F. (1994). *Trinitrotoluene as a sole nitrogen source for a sulfate-reducing bacterium Desulfovibrio sp. (B strain) isolated from an anaerobic digester*. Curr Microbiol, 25, 235-241.

Borch, T., Inskeep, W. P., Harwood, J. A. & Gerlach, R. (2005). *Impact of ferrihydrite and anthraquinone-2,6-disulfonate on the reductive transformation of 2,4,6-trinitrotoluene by a gram-positive fermenting bacterium*. Env Sci Technol, 39, 7126-7133.

Caballero, A., Lazaro, A., Ramos, J. L. & Esteve-Nunez, A. (2005). *PnrA, a new nitroreductase-family enzyme in the TNT degrading strain Pseudomonas putida JLR11*. Env Microbiol, 7, 1211-1219.

Cao, B., Nagarajan, K. & Loh, K.-C. (2009). *Biodegradation of aromatic compounds: current status and opportunities for biomolecular approaches*. Appl Microbiol Biotechnol, 85, 207-228.

Chauhan, A. & Jain, R. K. (2010). *Biodegradation: gaining insight through proteomics.* Biodegradation, DOI: 10.1007/s10532-010-9361-0.

Chu, W. & Chan, K. H. (2003). *The mechanism of the surfactant-aided soil washing system for hydrophobic and partial hydrophobic organics.* Science of the Total Environment, 307(1-3), 83-92.

Committee on the Biological Confinement of Genetically Engineered Organisms (2004) *Biological confinement of genetically engineered organisms*. The National Academies Press, Washington DC.

Conder, J. M., La Point, T. W. & Bowen, A. T. (2004). *Preliminary kinetics and metabolism of 2,4,6-trinitrotoluene and its reduced metabolites in an aquatic oligochaete.* Aquatic Toxicol, 69, 199-213.

Crawford, R. L. (1995). *The microbiology and treatment of nitroaromatic compounds.* Curr Opin Biotechnol, 6, 329-336.

Cunningham, S. D., Shann, J. R., Crowley, D. E., Anderson, T. A. (1997). *Phytoremediation of contaminated water and soil.* ACS Symp Series, 664, 2-17.

Darrach, M. R., Chutjian, A. & Plett, G. A. (1998). *Trace explosive signatures from World War II unexploded undersea ordnance.* Env Sci Technol, 32, 1354-1358.

Daun, G., Lenke, H., Reuss, M. & Knackmuss, H.-J. (1998). *Biological treatment of TNT-contaminated soil. Anaerobic cometabolic reduction and interaction of TNT and metabolites with soil components*. Env Sci Technol, 32, 1956-1963.

Davis, J. K., Paoli, G. C., He, Z., Nadeau, L. J., Somerville, C. C. & Spain, J. C. (2000). *Sequence analysis and initial characterization of two isozymes of hydroxylaminobenzene mutase from Pseudomonas pseudoalcaligenes JS45*. Appl Env Microbiol, 66, 2965-2971.

Dejonghe, W., Boon, N., Seghers, D., Top, E. M. & Verstraete, W. (2001). *Bioaugmentation of soils by increasing microbial richness: missing links*. Env Microbiol, 3(10), 649-657.

Diaz, E. (2004). *Bacterial degradation of aromatic pollutants: a paradigm of metabolic versatility.* Int Microbiol, 7, 173-180.

Drzyzga, O., Bruns-Nagel, D., Gorontzy, T., Blotevogel, K.-H. & Gemsa, D. (1998). *Mass balance studies with 14C-labeled 2,4,6-trinitrotoluene (TNT) mediated by an anerobic Desulfovibrio species and an aerobic Serratia species*. Curr Microbiol, 37, 380-386.

Dua, M., Singh, A., Sethunathan, N. & Johri, A. K. (2002). *Biotechnology and bioremediation: successes and limitations.* Appl Microbiol Biotechnol, 59, 143-152.

Duque, E., Haidour, A., Godoy, F. & Ramos, J. L. (1993). *Construction of a Pseudomonas hybrid strain that mineralizes 2,4,6-trinitrotoluene*. J Bacteriol, 175, 2278-2283.

Duquenne, P., Chenu, C., Richard, G. & Catroux, G. (1999). *Effect of carbon source supply and its location on competition between inoculated and established bacterial strains in sterile soil microcosm.* FEMS Microbiol Ecol, 29, 331.

Ederer, M. M., Lewis, T. A. & Crawford, R. L. (1997). *2,4,6-Trinitrotoluene (TNT) transformation by Clostridia isolated from a munition-fed bioreactor: comparison with non-adapted bacteria*. J Ind Microbiol Biotechnol, 18, 82-88.

Ellis, B. M. L. (2000). *Environmental biotechnology informatics.* Curr Opin Biotechnol, 11, 232-235.

Esteve-Nunez, A. & Ramos, J. L. (1998). *Metabolism of 2,4,6-trinitrotoluene by Pseudomonas* sp. JLR11. Env Sci Technol, 32, 3802-3808.

Esteve-Nunez, A., Caballero, A. & Ramos, J. L. (2001). *Biological degradation of 2,4,6-trinitrotoluene.* Microbiol Mol Biol Rev, 65, 335-352.

Esteve-Nunez, A., Luchessi, G., Philipps, B., Schink, B. & Ramos, J. L. (2000). *Respiration of 2,4,6-trinitrotoluene by Pseudomonas* sp. strain JLR11. J Bacteriol, 182, 1352-1355.

Eyers, L., George, I., Schuler, L., Stenuit, B., Agathos, S. N. &Fantroussi, S. E. (2004). *Environmental genomics: exploring the unmined richness of microbes to degrade xenobiotics.* Appl Microbiol Biotechnol, 66, 123-130.

Eyers, L., Stenuit, B. & Agathos, S. N. (2008). *Denitration of 2,4,6-trinitrotoluene by Pseudomonas aeruginosa ESA-5 in the presence of ferrihydrite.* Appl Microbiol Biotechnol, 79(3), 489-497.

Fernández, M., Duque, E., Pizarro-Tobías, P., Dillewijn, P. V., Wittich, R.-M. & Ramos, J. L. (2009). *Microbial responses to xenobiotic compounds. Identification of genes that allow* Pseudomonas putida *KT2440 to cope with 2,4,6-trinitrotoluene.* Microbial Biotechnol, 2(2), 287-294.

Fiorella, P. D. & Spain, J. C. (1997). *Transformation of 2,4,6-trinitrotoluene by Pseudomonas pseudoalcaligenes JS52.* Appl Env Microbiol, 63, 2007-2015.

Foght, J. (2008). *Anaerobic biodegradation of aromatic hydrocarbons: pathways and prospects.* J Mol Microbiol Biotechnol, 15, 93-120.

Franzetti, A; Tamburini, E; Banat, IM. *Application of biological surface active compounds in remediation technologies.* In: Sen, R. editor. Biosurfactants, 'Advances in experimental medicine and biology'. Berlin: Springer; 2010; 121-134.

French, C. E., Nicklin, S. & Bruce. N. C. (1998). *Aerobic degradation of 2,4,6-trinitrotoluene by Enterobacter cloacae PB2 and by pentaerythritol tetranitrate reductase.* Appl Env Microbiol, 64, 2864-2868.

Fulekar, M. H. & Sharma, J. (2008). *Bioinformatics applied in bioremediation.* Innovative Romanian Food Biotechnol, 2 (25), 28-36.

Fuller, M. E., Lowey, J. M., Schaefer, C. E. & Steffan, R. J. (2004). *A Peat Moss-Based Technology for Mitigating Residues of the Explosives TNT, RDX, and HMX in Soil.* Soil and Sediment Contamination: An International J, 14(4), 373-385.

Galvão, T. C., Mencía, M. & de Lorenzo, V. (2007). *Emergence of novel functions in transcriptional regulators by progression to stem protein types.* Mol Microbiol, 65, 907-919.

Gentry, T. J., Christopher, R. & Pepper, I. L. (2004). *New approaches for bioaugmentation as a remediation technology.* Crit Rev Env Sci Technol, 34(5), 447-494.

Gerhardt, K. E., Huang, X.-D., Glick, B. R. & Greenberg, B. M. (2009). *Phytoremediation and rhizoremediation of organic soil contaminants: Potential and challenges.* Plant Science, 176, 20-30.

Gianfreda, L. & Rao, M. A. (2008). *Interactions between xenobiotics and microbial and enzymatic soil activity.* Crit Rev Env Sci Technol, 38(4), 269-310.

Gorge, E., Brandt, S. & Werner, D. (1994). *Uptake and metabolism of 2,4,6–trinitrotoluene in higher plants.* Env Sci Pollut Res, 4, 229-233.

Gren, I., Wojcieszyńska, D., Guzik, U., Perkosz, M. & Hupert-Kocurek, K. (2010). *Enhanced biotransformation of mononitrophenols by Stenotrophomonas maltophilia KB2 in the presence of aromatic compounds of plant origin.* World J Microbiol Biotechnol, 26, 289-295.

Groenewegen, P. E. J. & De Bont, J. A. M. (1992). *Degradation of 4-nitrobenzoate via 4-hydroxylaminobenzoate and 3,4-hydroxybenzoate in Comamonas acidovorans NBA-10.* Archives Microbiol, 158, 381-386.

Haack, S. K. & Bekins, B. A. (2000). *Microbial populations in contaminant plumes.* Hydrol J, 8, 63-76.

Haigler, B. E. & Spain, J. C. (1993). *Biodegradation of 4-nitrotoluene by Pseudomonas sp. strain 4NT.* Appl Env Microbiol, 59, 2239-2243.

Haigler, B. E., Johnson, G. R., Suen, W. C. & Spain, J. C. (1999). *Biochemical and genetic evidence for meta-ring cleavage of 2,4,5-trihydroxytoluene in Burkholderia sp. strain DNT.* J Bacteriol, 181, 965-972.

Halmemies, S., Grondahl, S., Arffman, M., Nenonen, K. & Tuhkanen, T. (2003). *Vacuum extraction based response equipment for recovery of fresh fuel spills from soil.* J Haz Mat, 97(1-4), 127-143.

Hannink, N. K., Rosser, S. J. & Bruce, N. C. (2002). *Phytoremediation of Explosives.* Crit Rev Plant Sci, 21, 5, 511-538.

Harvey, S. D., Fellows, R. J., Cataldo, D. A. & Bean, R. M. (1990). *Analysis of 2,4,6–Trinitrotoluene and its transformation products in soils and plant-tissues by high-performance liquid chromatography.* Int J Chromatogr, 518, 361-374.

Hatzinger, P. B., Fuller, M. E., Rungkamol, D., Schuster, R. L. & Steffan, R. J. (2004). *Enhancing the attenuation of explosives in surface soils at military facilities: sorption-desorption isotherms.* Env Toxicol Chem, 23, 306-312.

Hawari, J., Beaudet, S., Halasz, A., Thiboutot, S. & Ampleman, G. (2000). *Microbial degradation of explosives: biotransformation versus mineralization.* Appl Microbiol Biotechnol, 54, 605-618.

Hawari, J., Haslaz, A., Beaudet, S., Ampleman, G. & Thioutot, S. (1999). *Biotransformation of 2,4,6-trinitrotoluene (TNT) with Phanerochaete chrysosporium in agitated cultures at pH 4.5.* Appl Env Microbiol, 65, 2977-2986.

Hazra, C., Kundu, D., Ghosh, P., Joshi, S., Dandi, N. & Chaudhari, A. (2010). *Production and characterization of a glycolipid biosurfactant by Pseudomonas aeruginosa using low cost agro-based raw materials.* J Chem Technol Biotechnol, DOI: 10.1002/jctb.2480.

Heiss, G. & Knackmuss, H.-J. (2002). *Bioelimination of nitroaromatic compounds: immobilization versus mineralization.* Curr Opin Biotechnol, 5, 282-287.

Herre, AJ; Micheles, K; Scheibner, K; Fritsche, W. *Bioremediation of 2,4,6-trinitrotoluene (TNT) contaminated soil by a litter decaying fungus.* In: Alleman, BC, Leeson, A editors. In situ and On site Bioremediation. Columbus, OH: Battelle Press; 1997; 493-498.

Hoehamer, C. F., Wolfe, N. L. & Eriksson, K. E. L. (2006). *Biotransformation of 2,4,6-trinitrotoluene (TNT) by the fungus Fusarium oxysporum.* Int J Phytoremed, 8, 95-105.

Huang, S., Lindahl, P. A., Wang, C., Bennett, G. N., Rudolph, F. B. & Hughes, J. B. (2000). *2,4,6-Trinitrotoluene reduction by carbon monoxide dehydrogenase from Clostridium thermoaceticum.* Appl Env Microbiol, 66, 1474-1478.

Hughes, J. B., Shanks, J., Vanderford, M., Lauritzen, J. & Bhadra, R. (1997). *Transformation of TNT by aquatic plants and plant tissue cultures.* Env Sci Technol, 31, 266-271.

Hughes, M. A. & Williams, P. A. (2001). *Cloning and characterization of the pnb genes, encoding enzymes for 4-nitrobenzoate catabolism in Pseudomonas putida* TW3. J Bacteriol, 183, 1225-1232.

Illman, W. A. & Alvarez, P. J. (2009). *Performance assessment of bioremediation and natural attenuation.* Crit Rev Environ Sci Technol, 39, 209-270.

Iwamoto, T. & Nasu, M. (2001). *Current bioremediation practice and perspective.* J Biosci Bioeng, 92, 1-8.

Jain, M. R., Zinjarde, S. S., Deobagkar, D. D. & Deobagkar D. N. (2004). 2,4,6-*Trinitrotoluene transformation b y a tropical marine yeast, Yarrowia lipolytica* NCIM 3589. Mar Pollut Bull, 49,783-788.

James, K. D., Hughes, M. A. & Williams, P. A. (2000). *Cloning and expression of ntnD, encoding a novel NAD(P)$^{+}$-independent 4-nitrobenzyl alcohol dehydrogenase from Pseudomonas sp. strain TW3.* J Bacteriol, 182, 3136-3141.

Jenkins, T. F., Leggett, D. C., Grant, C. L. & Bauer, C. F. (1986). *Reversed-phase high performance chromatographic determination of nitroorganics in munitions wastewater.* Anal Chem, 58, 170-175.

Johnson, G. R., Jain, R. K. & Spain, J. C. (2002). *Origins of the 2,4- dinitrotoluene pathway.* J Bacteriol, 184, 4219-4232.

Johnson, G. R., Smets, B. F. & Spain, J. C. (2001). *Oxidative transformation of aminodinitrotoluene isomers by multicomponent dioxygenases.* Appl Env Microbiol, 67, 5460-5466.

Jones, AM; Greer, CW; Ampleman, G; Thiboutot, S; Lavigne, J; Hawari, J. *Biodegradability of selected highly energetic pollutants under aerobic conditions.* In: Hinchee, E, Hoeppel, RE, Anderson, DB editors. Bioremediation of recalcitrant organics. Columbus, Ohio: Battelle Press; 1995; 251-257.

Jones, R. M., Britt-Compton, B. & Williams, P. A. (2003). *The naphthalene catabolic (nag) genes of Ralstonia sp. strain U2 are operon that is regulated by NagR, a LysR type transcriptional activator.* J Bacteriol, 185, 5847-5853.

Ju, K.-S., Parales, J. V. & Parales, R. E. (2009*). Reconstructing the evolutionary history of nitrotoluene detection in the transcriptional regulator NtdR*. Mol Microbiol, 74, 826-843.

Kalafut, T., Wales, M. E., Rastogi, V. K., Naumova, R. P., Zaripova, S. K. & Wild, J. R. (1998). *Biotransformation patterns of 2,4,6-trinitrotoluene by aerobic bacteria.* Curr Microbiol, 36, 45-54.

Kanekar, P., Doudpure, P. & Sarnaik, S. (2003). *Biodegradation of nitroexplosives.* Ind J Exp Biol, 41, 991-1001.

Kapley, A. & Purohit, H. J. (2009). *Genomic tools in bioremediation.* Ind J Microbiol, 49, 108-113.

Karim, K. & Gupta, S. K. (2002). *Effects of alternative carbon sources on biological transformation of nitrophenols.* Biodegradation, 13, 353-360.

Khan, F. I., Husain, T. & Hejazi, R. (2004). *An overview and analysis of site remediation technologies.* J Env Manage, 71, 95-122.

Khleifat, K. M. (2010). *Characterization of 2,4-Dinitrotoluene dioxygenase from recombinant Escherichia coli* strain PFJS39*: its direct interaction with Vitreoscilla hemoglobin.* Bioremed J, 14, 38 - 53.

Kim, H. Y. & Song, H.-G. (2005). *Purification and characterization of NAD (P)H-dependent nitroreductase I from Klebsiella* sp. C1 *and enzymatic transformation of 2,4,6-trinitrotoluene.* Appl Microbiol Biotechnol, 68, 766-773.

Kim, H., Bennett, G. N. & Song, H. (2002). *Degradation of TNT by Klebsiella sp. isolated form activated sludge.* Biotechnol Lett, 24, 2023-2028.

Kivisaar, M. (2009). *Degradation of nitroaromatic compounds: a model to study evolution of metabolic pathways*. Mol Microbiol, 74 (4), 777-781.

Koder, R. L., Haynes, C. A., Rodgers, M. E., Rodgers, D. W. & Miller, A. F. (2002). *Flavin thermodynamics explain the oxygen insensibility of enteric nitroreductases.* Biochemistry, 41, 14197-14205.

Koehler, J., Warrelmann, J., Frische, T., Behrend, P. & Walter, U. (2002). *In situ phytoremediation of TNT-contaminated soil.* Acta Biotechnol, 22, 67-80.

Kroger, M. & Fels, G. (2007). *Combined biological-chemical procedure for the mineralization of TNT.* Biodegradation, 18, 413-425.

Kubota, A., Maeda, T., Nagafuchi, N., Kadokami, K. & Ogawa, H. I. (2008). *TNT biodegradation and production of dihydroxylaminonitrotoluene by aerobic TNT degrader Pseudomonas sp. strain TM15 in an anoxic environment.* Biodegradation, 19, 795-805.

Kulkarni, M. & Chaudhari, A. (2007). *Microbial remediation of nitro-aromatic compounds: an overview.* J Env Manage, 85, 496-512.

Kulkarni, M. A. & Chaudhari, A. B. (2006). *Biodegradation of p-nitrophenol by P. putida.* Biores Technol, 97, 982-988.

Kumagai, Y., Kikushima, M., Nakai, Y., Shimojo, N. & Kunimoto, M. (2004). *Neuronal nitric oxide synthase (NNOS) catalyzes one-electron reduction of 2,4,6-trinitrotoluene, resulting in decreased nitric oxide production and increased nNOS gene expression: implication for oxidative stress.* Free Radical Biol Med, 37, 350-357.

Kutty, R. & Bennett, G. N. (2005). *Biochemical characterization of trinitrotoluene transforming oxygen-insensitive nitroreductases from Clostridium acetobutylicum* ATCC 824. Arch Microbiol, 184, 158-167.

Kutty, R. & Bennett, G. N. (2006). *Studies on inhibition of transformation of 2,4,6-trinitrotoluene catalyzed by Fe-only hydrogenase from Clostridium acetobutylicum.* J Ind Microbiol Biotechnol, 33, 368-376.

Laha, S., Tansel, B. & Ussawarujikulchai, A. (2009). *Surfactant-soil interactions during surfactant-amended remediation of contaminated soils by hydrophobic organic compounds: a review*. J Env Manage, 90 (1), 95-100.

Larson, S. L., Jones, R. P., Escalon, L. & Parker, D. (1999). *Classification of explosives transformation products in plant tissue.* Env Toxicol Chem, 18, 1270-1276.

Lee, I., Baek, K., Kim, H., Kim, S., Kim, J., Kwon, Y.K, Chang, Y. & Bae, B. (2007). *Phytoremediation of soil co-contaminated with heavy metals and TNT using four plant species*. J Env Sci Health, Part A, 42(13), 2039-2045.

Lenke, H. & Knackmuss, H.-J. (1992). *Initial hydrogenation during catabolism of picric acid by Rhodococcus erythropolis HL 24-2*. Appl Env Microbiol, 58, 2933-2937.

Lenke, H; Achtnich, C; Knackmuss, H.-J. *Perspectives of bioelimination of polynitroaromatic compounds*. In: Spain JC, Hughes JB, Knackmuss H-J editors. Biodegradation of nitroaromatic compounds and explosives. New York; Lewis Publishers; 2000; 91-126.

Lessner, D. J., Parales, R. E., Narayan, S. & Gibson, D. T. (2003). *Expression of the nitroarene dioxygenase genes in Comamonas sp. strain JS765 and Acidovorax sp. strain JS42 is induced by multiple aromatic compounds*. J Bacteriol, 185, 3895-3904.

Lewis, T. A., Goszczynski, S. Crawford, R. L., Korus, R. A. & Admassu, W. (1996). *Products of anaerobic 2,4,6-trinitrotoluene (TNT) transformation by Clostridium bifermentans*. Appl Env Microbiol, 62, 4669-4674.

Lorme, M. D. & Craig, M. (2009). *Biotransformation of 2,4,6-Trinitrotoluene by pure culture ruminal bacteria*. Curr Microbiol, 58, 81-86.

Lovley, D. R. (2003). *Cleaning up with genomics: applying molecular biology to bioremediation*. Nat Rev Microbiol, 1, 35-44.

Martin, J. L., Comfort, S. D., Shea, P. J., Kokjohn, T. A. & Drijber, R. A. (1997). *Denitration of 2,4,6-trinitrotoluene by Pseudomonas savastanoi*. Can J Microbiol, 43, 447-455.

Marvin-Sikkema, F. D. & de Bont, J. A. M. (1994). *Degradation of nitroaromatic compounds by microorganisms*. Appl Microbiol Biotechnol, 42, 499-507.

McCormick, N. G., Cornell, J. H. & Kaplan, A. M. (1981). *Biodegradation of hexahydro-1,3,5-trinitro-1,3,5-triazine*. Appl Env Microbiol, 42, 817-823.

McCormick, N. G., Feeherry, F. E. & Levinson, H. S. (1976). *Microbial transformation of 2,4,6-TNT and other nitroaromatic compounds*. Appl Envl Microbiol, 31, 949-958.

Michan, C., Delgado, A., Haidour, A., Lucchesi, G. & Ramos, J. L. (1997). *In vivo construction of a hybrid pathway for metabolism of 4-nitrotoluene in Pseudomonas fluorescens*. J Bacteriol, 179, 3036-3038.

Micheles, J. & Gottschalk, G. (1994). *Inhibition of lignin peroxidase of Phanerochaete chrysosporium by hydroxylamino dinitrotoluene, an early intermediate in the degradation of 2,4,6-trinitrotoluene*. Appl Env Microbiol, 60, 187-194.

Michels, J., Stuhrmann, M., Frey, C., Koschitzky, H.-P. (2008). *KORA-Handlungsempfehlung mit Methodensammlung, Natürliche Schadstoffminderung bei der Sanierung von Altlasten*.

Mihopoulos, P. G., Suidan, M. T. & Sayles, G. D. (2001). *Complete remediation of PCE contaminated unsaturated soils by sequential anaerobic- aerobic bioventing*. Water Sci Technol, 43(5), 365-372.

Monti, M. R., Smania, A. M., Fabro, G., Alvarez, M. E. & Argarana C. E. (2005). *Engineering Pseudomonas fluorescens for biodegradation of 2,4-dinitrotoluene*. Appl Env Microbiol, 71, 8864-8872.

Montpas, S., Samson, J., Langlois, E., Lei, J., Piche, Y. & Chevenert, R. (1997). *Degradation of 2,4,6-trinitrotoluene by Serratia marcescens*. Biotechnol Lett, 19, 291-294.

Moshe, S. S., Ronen, Z., Dahan, O., Weisbrod, N., Groisman, L., Adar, E. & Nativ, R. (2009). *Sequential biodegradation of TNT, RDX and HMX in a mixture*. Env Pollut, 157, 2231-2238.

Moslemy, P., Neufeld, R. J., Millette, D. & Guiot, S. R. (2003). *Transport of gellan gum microbeads through sand: an experimental evaluation for encapsulated cell bioaugmentation*. J Env Manage, 69, 249.

Mulla, S. I., Hoskeri, R. S., Shouche, Y. S. & Ninnekar, H. Z. (2010). *Biodegradation of 2-Nitrotoluene by Micrococcus* sp. *strain SMN-1*. Biodegradation, DOI: 10.1007/s10532-010-9379-3.

Mulligan, C. N. (2005). *Environmental applications of biosurfactants*. Environ Pollut, 133, 183-198.

Mulligan, M. N. (2009). *Recent advances in the environmental applications of biosurfactants.* Curr Opin Colloid Interface Sci, 14, 372-378.

Muter, O., Versilovskis, A., Scherbaka, R., Grube, M. & Zarina, D. (2008). *Effect of plant extract on the degradation of nitroaromatic compounds by soil microorganisms*. J Ind Microbiol Biotechnol, 35, 1539-1543.

Naumova, R. P., Selivanovskaya, S. L. U. & Mingatina, F. A. (1988). *Possibilities for the deep bacterial destruction of 2,4,6- trinitrotoluene*. Mikrobiologiya, 57, 218-222.

Navratilova, J., Tvrzova, L., Durnova, E., Sproer, C., Sedlacek, I., Neca, J. & Nemec, M. (2005). *Characterization of Rhodococcus wratislaviensis strain J3 that degrades 4-nitrocatechol and other nitroaromatic compounds.* Anton van Leeuwenhoek, 87, 149-153.

Nishino, S. F. & Spain, J. C. (2001). *Technology status review: Bioremediation of dinitrotoluene (DNT).* http://www.estcp.org/documents/techdocs/DNT_Report.pdf, accessed on: July 22, 2010.

Nishino, SF; Spain, JC; & He, Z. *Strategies for aerobic degradation of nitroaromatic compounds by bacteria: process discovery to field application.* In: Spain JC, Hughes JB, Knackmuss H-J editors. Biodegradation of nitronomatic compounds and explosives. New York: Lewis Publishers; 2000; 7-62.

Nokhbeh, M. R., Boroumandi, S., Pokorny, N., Koziarz, P., Paterson, E. S. & Lambert, I. B. (2002). *Identification and characterization of SnrA, an inducible oxygen-insensitive nitroreductase in Salmonella enterica* serovar *Typhimurium* TA1535. Mut Res, 508, 59-70.

NRC. (2000). *Natural attenuation for groundwater remediation.* National Academy Press, Washington DC.

Palazzo, A. J. & Leggett, D. C. (1986). *Effect and disposition of TNT in a terrestrial plant.* J Env Qual, 15, 49-52.

Parales, J. V., Kumar, A., Parales, R. E. & Gibson, D. T. (1996). *Cloning and sequencing of the genes encoding 2-nitrotoluene dioxygenase from Pseudomonas* sp. JS42. Gene, 181, 57-61.

Parales, J. V., Parales, R. E., Resnick, S. M. & Gibson, D. T. (1998). *Enzyme specificity of 2-nitrotoluene 2, 3-dioxygenase from Pseudomonas sp. strain JS42 is determined by the C-terminal region of the alpha subunit of the oxygenase component.* J Bacteriol, 180, 1194-1199.

Paria, S. (2008). *Surfactant-enhanced remediation of organic contaminated soil and water.* Adv Colloid Interface Sci, 138(1), 24-58.

Pasti-Grigsby, M. B., Lewis, T. A., Crawford, D. L. & Crawford, R. L. (1996). *Transformation of 2,4,6-trinitrotoluene (TNT) by Actinomycetes isolated from TNT-contaminated and uncontaminated environments*. Appl Env Microbiol, 62, 1120-1123.

Peres, C. M. & Agathos, S. N. (2000). *Biodegradation of nitroaromatic pollutants: from pathways to remediation.* Biotechnol Ann Rev, 6, 197-220.

Perkins, M. W., De, S., Frederick, L. & Dutta, S. K. (2005). *Ligninolytic and nonligninolytic mineralization of tri-nitrotoluene by several white rot basidiomycetes*. Bioremed J, 9(2), 77-85.

Popesku, J. T., Singh, A., Zhao, J.-S., Hawari, J. & Ward, O. P. (2004). *Metabolite production during transformation of 2,4,6-trinitrotoluene (TNT) by a mixed culture acclimated and maintained on crude oil-containing media.* Appl Microbiol Biotechnol, 65, 739-746.

Prasad, M. N. V., Freitas, H., Fraenzle, S., Wuenschmann, S. & Markert, B. (2010). *Knowledge explosion in phytotechnologies for environmental solutions.* Env Pollut, 158, 18-23.

Purohit, V. & Basu, A. K. (2000). *Mutagenicity of nitroaromatic compounds.* Chem Res Toxicol, 13, 673-692.

Qui, X., Zhong, Q., Li, M., Bai, W. & Li, B. (2007). *Biodegradation of p-nitrophenol by methyl parathion-degrading Ochrobactrum sp.* B2. Int Biodeterior Biodegrad, 59, 297-301.

RAAG. (2000). *Evaluation of risk based corrective action model, Remediation Alternative Assessment Group,* Memorial University of Newfoundland, St John's, NF,Canada.

Race, P. R., Lovering, A. L., Green, R. M., Ossor, A., White, S. A., Searle, P. F., Wrighton, C. W. & Hyde, E. I. (2005). *Structural and mechanistic studies of Escherichia coli nitroreductase with the antibiotic nitrofurazone.* J Biol Chem, 280, 13256-13264.

Ramos, J. L., Mar Gonzalez-Perez, M., Caballero, A. & Van Dillewijn, P. (2005). *Bioremediation of polynitrated aromatic compounds: Plants and microbes put fight.* Curr Opin Biotechnol, 16, 275-281.

Rau, J. & Stolz, A. (2003). *Oxygen-insensitive nitroreductases* NfsA *and* NfsB *of Escherichia coli function under anaerobic conditions as lawsone-dependent azo reductases.* Appl Env Microbiol, 69, 3448-3455.

Razo-Flores, E., Donlon, B., Lettinga, G. & Field, J. A. (1997). *Biotransformation and biodegradation of N-substituted aromatics in methanogenic granular sludge.* FEMS Microbiol Rev, 20, 525-538.

Rhys-Williams, W., Taylor, S. C. & Williams, P. A. (1993). *A novel pathway for catabolism of 4-nitrotoluene by Pseudomonas.* J Gen Microbiol, 139, 1967-1972.

Rieger, PG; Knackmuss HJ. *Basic knowledge and perspectives on biodegradation of 2,4,6-trinitrotoluene and related nitroaromatic compounds in contaminated soil.* In: Spain JC, editor. Biodegradation of nitroaromatic compounds. New York: Plenum Press; 1995; 1-18.

Rivera, R., Medina, V. F., Larson, S. L. & McCutcheon, S. C. (1998). *Phytotreatment of TNT-contaminated groundwater.* J Soil Contam, 7, 511-529.

Roeling, W. F. M. & van Verseveld, H. W. (2002). *Natural attenuation: what does the subsurface have in store?* Biodegradation, 13, 53-64.

Roldan, M. D., Perez-Reinado, E., Castillo, F. & Moreno-Vivian, C. (2008). *Reduction of polynitroaromatic compounds: the bacterial nitroreductases.* FEMS Microbiol Rev, 32, 474-500.

Sarlauskas, J., Nemcikaite-Ceniene, A., Anusevicius, Z., Miseviciene, L., Julvez, M. M., Medina, M., Gomez-Moreno, C. & Cenas, N. (2004). *Flavoenzyme-catalyzed redox cycling of hydroxylamino- and amino metabolites of 2,4,6-trinitrotoluene: implications for their cytotoxicity.* Arch Biochem Biophys, 425, 184-192.

Schackmann, A. & Muller, R. (1991). *Reduction of nitroaromatic compounds by different Pseudomonas species under aerobic conditions.* Appl Envl Microbiol, 34, 809-813.

Scheibner, K. & Hofrichter, M. (1998). *Conversion of aminonitrotoluenes by fungal manganese peroxidase.* J Basic Microbiol, 38, 51-59.

Scheibner, K., Hofrichter, M., Here, A., Michels, J. & Fritsche, W. (1997). *Screening of fungi intensively mineralizing 2,4,6-TNT.* Appl Microbiol Biotechnol, 47, 452-457.

Scheidemann, P., Klunk, A., Sens, C. & Werner, D. (1998). *Species dependent uptake and tolerance of nitroaromatic compounds by higher plants.* J Plant Physiol, 152, 242-247.

Schneider, K., Oltmanns, J., Radenberg, T., Schneider, T. & PaulyMundegar, D. (1996). *Uptake of nitroaromatic compounds in plants-implications for risk assessment of ammunition sites.* Env Sci Pollut Res, 3, 135-138.

Schrader, P. S. & Hess, T. F. (2004). *Coupled abiotic-biotic mineralization of 2,4,6-trinitrotoluene (TNT) in soil slurry.* J Environ Qual, 33, 1202-1209.

Scow, K. M. & Hicks, K. A (2005) *Natural attenuation and enhanced bioremediation of organic contaminants in groundwater.* Curr Opin Biotechnol, 16, 246-253.

Shim, CY; Crawford, DL. *Biodegradation of trinitrotoluene (TNT) by a strain of Clostridium bifermentans.* In: Hinchee, RE, Fredrickson, J, Alleman, BC editors. Bioaugmentation for site remediation. Columbus, Ohio: Battelle Press; 1995; 57-69.

Siciliano, S. D. & Greer, C. W. (2000). *Plant-bacterial combinations to phytoremediate soil contaminated with high concentrations of 2,4,6-trinitrotoluene.* J Env Qual, 29, 311.

Singer, A. C., Crowley, D. E. & Thompson, I. P. (2003). *Secondary plant metabolites in phytoremediation and biotransformation.* Trends Biotechnol, 21, 123-130.

Singh, S., Kang, S., Mulchandani, A. & Chen, W. (2008). *Bioremediation: environmental clean-up through pathway engineering.* Curr Opin Biotechnol, 19, 437-444.

Smets, B. F. & Pritchard, P. H. (2003). *Elucidating the microbial component of natural attenuation.* Curr Opin Biotechnol, 14, 283-288.

Smets, B. F., Yin, H. & Esteve-Nunez, A. (2007). *TNT biotransformation: when chemistry confronts mineralization.* Appl Microbiol Biotechnol, 76, 267-277.

Snellinx, Z., Nepovim, A., Taghavi, S., Vangronsveld, J., Vanek, T. & Vander Lelie, D. (2002). *Biological remediation of explosives and related nitroaromatic compounds.* Env Sci Pollut Res, 9, 48-61.

Spain, JC. *Bacterial degradation of nitroaromatic compound under aerobic conditions.* In: Spain, JC, editor. Biodegradation of nitroaromatic compounds. New York: Plenum Press; 1995; 19-35.

Spain, JC. *Introduction.* In: Spain, JC, Hughes, JB, Knackmuss, H-S editors. Biodegradation of nitroaromatic compounds and explosives. Boca Raton: Lewis Publishers; 2000; 1-5.

Spanggord, R. J., Spain, J. C., Nishino, S. F. & Mortelmans, K. E. (1991). *Biodegradation of 2,4,-dinitrotoluene by Pseudomonas* sp. Appl Env Microbiol, 57, 3200-3205.

Spiess, T., Desiere, P., Fischer, P., Spain, J. C., Kanackmuss, H. J. & Lenke, H. (1998). *A new nitrotoluene degradation pathway in a Mycobacterium strain.* Appl Env Microbiol, 64, 446-452.

Stenuit, B., Eyers, L., Fantroussi, S. E. & Agathos, S. N. (2005). *Promising strategies for the mineralization of 2,4,6-trinitrotoluene.* Rev Env Sci Bio/Technol, 4, 39-60.

Suen, W. C. & Spain, J. C. (1993). *Cloning and characterization of Pseudomonas* sp. *strain dinitrotoluene genes for 2,4-dinitrotoluene degradation.* J Bacteriol, 175, 1831-1837.

Sun, W. H., Horst, G. L., Drijber, R. A. & Elthon, T. E. (2000). *Fate of 2,4,6-trinitrotoluene in axenic sand culture systems containing smooth bromegrass.* Env Toxicol Chem, 19, 2038-2046.

Tecon, R. & van der Meer, J. R. (2009). *Effect of two types of biosurfactants on phenanthrene availability to the bacterial bioreporter Burkholderia sartisoli strain RP037.* Appl Microbiol Biotechnol, DOI: 10.1007/s00253-009-2216-0.

Temprano, F. J., Albareda, M., Camacho, M., Daza, A., Santamaria, C. & Rodriguez-Navarro, D.N. (2002). *Survival of several Rhizobium/Bradyrhizobium strains on different inoculant formulations and inoculated seeds.* Int Microbiol, 5, 81.

Thompson, I. P., van der Gast, C. J., Ciric, L. & Singer, A. C. (2005). *Bioaugmentation for bioremediation: the challenge of strain selection.* Env Microbiol, 7(7), 909-915.

Thompson, P. L., Ramer, L. A. & Schnoor, J. L. (1998). *Uptake and transformation of TNT by hybrid poplar trees.* Env Sci Technol, 32, 975-980.

Traxler, R. W., Wood, E. & Delaney, J. M. (1974). *Bacterial degradation of alpha-TNT.* Dev Ind Microbiol, 16, 71-76.

Tront, J. M. & Hughes, J. B. (2005). *Oxidative microbial degradation of 2,4,6-trinitrotoluene via 3-methyl-4,6-dinitrocatechol.* Env Sci Technol, 39, 4540-4549.

UK-EA. (2000). *Guidance on the assessment and monitoring of natural attenuation of contaminants in groundwater.* Environment Agency, Bristol, UK.

Urum, K., Pekdemir, T. & Gopur, M. (2003). *Optimum conditions for washing of crude oil-contaminated soil with biosurfactant solutions.* Process Safety and Environmental Protection: Transactions of the Institution of Chemical Engineers, Part B 81(3), 203-209.

US EPA. (1999). *Use of monitored natural attenuation at superfund, RCRA corrective action and underground storage tank sites.* Office of Solid Waste and Emergency Response, Washington DC.

US EPA. (2006). *Evaluation of the role of Dehalococcoides in the natural attenuation of chlorinated ethylenes in ground water.* Office of Research and Development, Ada, Oklahoma.

US EPA. (1996). *A citizen's guide to soil washing. Office of Solid Waste and Emergency Response,* US Environmental Protection Agency Publication#EPA542-F-96-002,Washington, DC.

Valli, K., Brock, B. J., Joshi, D. K. & Gold, M. H. (1992). *Degradation of 2,4- dinitrotoluene by lignin degrading fungus Phanerochaete chrysosporium.* Appl Envl Microbiol, 58, 221-228.

Van Aken, B., Hofrichter, M., Scheibner, K., Hatakka, A.I., Naveau, H. & Agathos, S. N. (1999). *Transformation and mineralization of 2,4,6- trinitrotoluene (TNT) by manganese peroxidase from the white-rot Basidiomycete Phlebia radiata.* Biodegradation, 10, 83-91.

Van Aken, B., Yoon, J. M. & Schnoor, J. L. (2004). *Biodegradation of nitro-substituted explosives 2,4,6-trinitrotoluene, hexahydro- 1,3,5-trinitro-1,3,5-triazine, and octahydro-1,3,5,7-tetranitro- 1,3,5-tetrazocine by a phytosymbiotic Methylobacterium sp. associated with poplar tissues (Populus deltoides × nigra DN34).* Appl Env Microbiol, 70, 508-517.

Van Dillewijn, P., Caballero, A., Paz, J. A., Gonzalez-Perez, M. M., Oliva, J. M. & Ramos, J. L. (2007). *Bioremediation of 2, 4, 6-trinitrotoluene under field conditions.* Env Sci Technol, 41, 1378-1383.

Van Dyke, M. I. & Prosser, J. I. (2000). *Enhanced survival of Pseudomonas fluorescens in soil following establishment of inoculum in a sterile soil carrier.* Soil Biol Biochem, 32, 1377.

Vanderberg, L. A., Perry, J. J. & Unkefer, P. J. (1995). *Catabolism of 2,4,6-trinitrotoluene by Mycobacterium vaccae.* Appl Microbiol Biotechnol, 43, 937-945.

Vangronsveld, J., Herzig, R., Weyens, N., Boulet, J., Adriaensen, K., Ruttens, A. Thewys, T., Vassilev, A., Meers, E., Nehnevajova, E., van der Lelie, D. & Mench, M. (2009). *Phytoremediation of contaminated soils and groundwater: lessons from the field.* Env Sci Pollut Res, 16, 765-794.

Vardar-Sukan, F; Kosaric, N. Biosurfactant. In: Lederberg, J editor. *Encyclopedia of Microbiology.* New York: Academic Press; 2009; 618-635.

Wayment, D. G., Bhadra, R., Lauritzen, J., Hughes, J.B. & Shanks, J. V. (1999). *A transient study of formation of conjugates during TNT metabolism by plant tissues.* Int J Phytoremed, 1, 227-239.

Weiss, J. V. & Cozzarelli, I. M. (2008). *Biodegradation in contaminated aquifers: incorporating microbial/molecular methods.* Ground Water, 46, 305-322.

Widrig, D. L., Boopathy, R. & Manning Jr. J.F. (1997). *Bioremediation of TNT-contaminated soil: a laboratory study.* Env Toxicol Chem, 16, 1141-1148.

Williams, R. E., Rathbone, D. A., Scrutton, N. S. & Bruce, N. C. (2004). *Biotransformation of explosives by the old yellow enzyme family of flavoproteins.* Appl Env Microbiol, 70, 3566- 3574.

Yang, H., Zhao J.-S. & Hawari, J. (2009). *Effect of 2,4-dinitrotoluene on the anaerobic bacterial community in marine sediment.* J Appl Microbiol, 107(6), 1799 - 1808.

Ye, J., Singh, A. & Ward, O. P (2004). *Biodegradation of nitroaromatics and other nitrogen-containing xenobiotics.* World J Microbiol Biotechnol, 20, 117-135.

Yen, H. K., Chang, N. B. & Lin, T.F. (2003). *Bioslurping model for assessing light hydrocarbon recovery in contaminated unconfined aquifer.* Practice Periodical of Hazardous, Toxic, and Radioactive Waste Manage, 7(2), 114-130.

Yeung, A. T. & Hsu, H. T. (2002). *Semi-analytical simulation of soil vapor extraction.* Practice Periodical of Hazardous, Toxic, and Radioactive Waste Manage, 6(1), 14-22.

Zhan, H. & Park, E. (2002). *Vapor flow to horizontal wells in unsaturated zones.* Soil Sci Soc America J, 66 (3), 710-721.

Zhang, C. & Bennett G. N. (2005). *Biodegradation of xenobiotics by anaerobic bacteria.* Appl Microbiol Biotechnol, 67, 600-618.

Zhang, J., Sun, Z., Li, Y., Peng, X., Li, W. & Yan, Y. (2008). *Biodegradation of p-nitrophenol by Rhodococcus* sp CN6 *with high cell surface hydrophobicity.* J Hazard Mat, DOI:10.1016/j.jhazmat.2008.07.018.

Zhao, J. S. & Ward, O. P. (2000). *Cometabolic biotransformation of nitrobenzene by 3-nitrophenol degrading Pseudomonas putida* 2NP8. Can J Microbiol, 46, 643-652.

In: Bioremediation
Editor: Alexander C. Mason

ISBN: 978-1-61122-730-7

Chapter 2

ADVANCES IN THE BIOREMEDIATION OF PERCHLORATES

***Yanguang Song*[1] *and Steven W. Van Ginkel*[2]**
[1]Song Pure Water, Fremont, CA, US
[2]Center for Environmental Biotechnology, Arizona State University, Tempe, Arizona, US

Perchlorate (ClO_4^-), primarily used in ordnance, explosives, and as a rocket propellant in the defense and aerospace industries, has contaminated the water supplies of more than 20 million people, primarily in the southwest United States (U.S. EPA, 2008). ClO_4^- is manufactured in 44 states, and ClO_4^- contamination is known in 35 states (U.S. EPA, 2008). Because of its high solubility, non reactivity, and poor absorption to soils, ClO_4^- persists and spreads rapidly throughout the environment (Logan, 2001a). Because ammonium perchlorate absorbs water from the air, ammunitions and rockets have a limited shelf life and the ammonium perchlorate must occasionally be replaced. As of 2006, an expected 165 million pounds of ClO_4^- will be disposed and will require treatment (Wallace et al., 1998).

ClO_4^-, a known competitive inhibitor of the sodium/iodide symporter (NIS) (De Groef et al. 2006), interferes with iodide uptake into the thyroid gland. In fetuses and children, ClO_4^- may interfere with proper development (U.S. EPA, 2008). Risks posed by ClO_4^- must be considered in conjunction with iodine intake, since a healthy diet consisting of iodine counteracts any negative effects induced by ClO_4^- (Kirk, 2006). The U.S. Environmental Protection Agency (EPA) set a preliminary remediation goal (PRG) for ClO_4^- of 24.5 μg/L to prevent adverse effects of exposure to unborn children. This goal assumes that exposure comes solely from drinking water and does not include exposures from food (i.e., mother's milk, lettuce, etc.). However, Ginsberg et al. (2007) states that the PRG can lead to a 7-fold increase in breast milk concentration, causing 90% of nursing infants to exceed the reference dose (RfD) of 0.0007 milligrams of ClO_4^- per kilogram of body weight per day. Ting et al. (2006) state that 6 μg/L is adequately protective of sensitive subpopulations, including pregnant women, their fetuses, infants, and people with hypothyroidism, and the California Department of Public Health as of October 18, 2007 set 6 μg/L as a maximum contaminant level. Massachusetts promulgated a ClO_4^- MCL of 2 μg/L in 2006 (MADEP, 2006).

Nitrate (NO_3^-) is often a co-contaminant with ClO_4^-, since nitrogen is a major element in the manufacture of explosives, which primarily utilizes ammonium nitrate and diesel fuel (ITRC, 2000). NO_3^- is also indirectly introduced into the environment through the oxidation of ammonia-based fertilizer. In affected groundwater, the NO_3^- concentration is usually 2 to 5 orders of magnitude higher than the ClO_4^- concentration (Gu et al. 2003). NO_3^- also has been shown to inhibit iodine uptake in the thyroid gland and accounts for a much larger proportion of iodine uptake inhibition than ClO_4^-, since at the legally accepted maximal NO_3^- levels in drinking water, NO_3^- far exceeds the potential effect of the proposed RfD for ClO_4^- (De Groef et al. 2006). This chapter focuses on both NO_3^- and ClO_4^- remediation since NO_3^- may have to be removed before ClO_4^- is removed. NO_3^- and ClO_4^- can be removed from contaminated water by reduction catalyzed by bacteria. (For other reviews regarding the remediation of ClO_4^-, please refer to NASA 2006, Srinivasana and Sorial 2009, and Sutton, 2006). Bacterial reduction totally detoxifies the contaminants, since they are reduced to harmless N_2 gas for NO_3^- and Cl^- and H_2O for ClO_4^-.

PERCHLORATE RESPIRING BACTERIA

Ubiquity and Diversity of (Per) Chlorate Reducing Bacteria

It has been known for several years that the ability to reduce ClO_4^- is not limited to a single bacterial species, although some of this earlier evidence of ClO_4^- degradation was inferred from information on biological chlorate (ClO_3^-) degradation (Logan, 1998). Bacteria capable of ClO_3^- reduction were found to inhabit a variety of environments including rivers, sediments, soils, and wastewater treatment plants (Van Ginkel et al. 1995). Utilizing media with ClO_3^- as the only electron acceptor, Coates *et al.* (1999) similarly found that the acetate-oxidizing ClO_3^- reducing bacteria (CRB) represented a significant population whose abundance ranged from 2.31×10^3 to 2.4×10^6 cells per gram of samples obtained from a variety of sources, including pristine and hydrocarbon-contaminated soils, aquatic sediments, paper mill waste sludges, and farm animal waste lagoons. All thirteen isolates obtained in this study were also capable of growth on acetate using ClO_4^-, leading to early speculation, via the abbreviation of (per) chlorate, that all CRB were also capable of ClO_4^- reduction (Coates *et al.*, 1999).

More recent studies have provided evidence that not all CRB are PRB, although the converse is true, that all PRB are CRB. Wu *et al.* (2001) used ClO_4^- or ClO_3^- in anaerobic growth medium to compare the abundance of PRB and CRB in different environmental samples. They found that when ClO_4^- was the sole terminal electron acceptor in the medium, the number of PRB in a pristine soil was up to 1000-fold lower than that found using ClO_3^- in the medium to enumerate CRB for the same samples (Wu *et al.*, 2001). This provided indirect evidence that not all CRB were PRB, and that PRB were less abundant than CRB in these environments. Confirmation that not all CRB were PRB was provided by three different studies. Only 8 of 10 CRB isolated from wastewater were found to be capable of respiring ClO_4^- (Logan *et al.*, 2001b). In another study, it was mentioned that a microorganism was isolated that was capable of using ClO_3^-, but not ClO_4^- (Coates *et al.*, 2000). Wolterink *et al.* (2002) isolated a CRB strain *Pseudomonas chloritidismutans* AW-1 from an anaerobic bioreactor treating wastewater containing ClO_3^-, and found that this strain did not reduce ClO_4^-. These studies indicate that while most CRB are able to degrade ClO_4^-, there is a subset of CRB that cannot use ClO_4^- as an electron acceptor for cell respiration.

Perchlorate Reducing Isolates

Many PRB have now been isolated. These PRB are all Gram-negative, facultative anaerobes. Analysis of the 16S rRNA sequences of tested strains indicated that all isolates were members of the class Proteobacteria. The majority of these PRB, including GR-1, perc1ace, KJ, PDX, HZ, JDS5, and 15 other strains (Achenbach *et al.*, 2001), are located in the β-subclass of Proteobacteria. Achenbach *et al.* (2001) proposed two new genera, *Dechloromonas* and *Azospira* for these β-subclass lineages, which represent the predominant PRB in the environment.

Electron Donors and Acceptors used by PRB for Growth

All PRB are capable of dissimilatory reduction of ClO_3^- to Cl^- for energy and growth, and many PRB can also reduce NO_3^-. Although many PRB are capable of complete denitrification, isolate CKB does not grow on NO_3^- (Bruce *et al.*, 1999), and Cuznesove B-1168 (Korenkov *et al.*, 1976) and HAP-1 (Wallace *et al.*, 1996) reduce NO_3^- only to nitrite (NO_2^-) and do not produce ammonia or N_2.

For both ClO_4^- and ClO_3^-, reduction does not occur in the presence of a high concentration of dissolved O_2 (DO) though most isolates can use O_2 as a terminal electron acceptor. HAP-1 was initially reported to be an obligate anaerobe (Wallace *et al.*, 1996), but in a later study it was reported that it was a microaerobic organism (Wallace *et al.*, 1998) although no further details on O_2 tolerance were provided. It seems reasonable that PRB should be able to use, or at least tolerate, dissolved O_2 as it is produced during the decomposition of ClO_3^- and ClO_4^-. It has been suggested that PRB contain a complete high-O_2 aerobic respiratory chain including at least cytochrome *c* oxidase, and that an additional low-O_2 enzyme is involved in ClO_4^- and O_2 reduction (Song and Logan, 2004a). Song (dissertation, 2004) also demonstrated that O_2 increasingly inhibited ClO_4^- reduction of *Azospira* sp. KJ over an O_2 saturation range of 0.5% to 5% (0.04 to 0.39 mg O_2/L). At a DO concentration of 0.5% saturation, only 4.1 % of ClO_4^- was reduced, indicating that the threshold DO concentration to inhibit ClO_4^- reduction is < 0.5%. Although DO concentrations could not be directly measured below 0.5%, a redox potential of 150 mV obtained under anoxic conditions suggests that a DO level of 0.2% (0.02 mg O_2/L) did not affect ClO_4^- degradation. Therefore, the threshold DO level to inhibit ClO_4^- reduction is in a small range of 0.2 to 0.5% of saturation (0.02 to 0.04 mg O_2/L).

In some cases PRB have been tested for their ability to use other electron acceptors such as metals and SO_4^{2-}. GR-1 can utilize Mn (IV) as an electron acceptor. Most PRB and CRB cannot reduce SO_4^{2-} (Rikken *et al.* 1996; Wallace *et al.* 1996; Bruce *et al.* 1999; Coates *et al.* 1999; Herman and Frankenberger, 1999; Michaelidou *et al.* 2000; Wolterink *et al.* (2002). The only report of CRB (*Acinetobacter* spp.) capable of growth using SO_4^{2-} was provided by Stepanyuk *et al.* (1992), but the ability to respire ClO_4^- was not tested and these bacteria could not use NO_3^-. PRB have not been found to be capable of using other electron acceptors, including: Fe (III), selenate, malate, and fumarate (Coates *et al.* 1999).

Both heterotrophic and autotrophic PRB have been isolated. Acetate has been most frequently used as a single substrate for heterotrophic ClO_4^- reduction (Korenkov *et al.*, 1976; Wallace *et al.*, 1996; Bruce *et al.*, 1999; Coates *et al.*, 1999; Herman and Frankenberger, 1999; Logan *et al.*, 2001b), but H_2 or formate was required as an electron donor for the growth of HAP-1 on acetate (Wallace *et al.*, 1996). ClO_4^- reduction by *Citrobacter* sp. IsoCock1 (Okeke *et al.*, 2002) was sustained on acetate, but yeast extract was found to improve growth. A wide variety of organic substrates, including alcohols and carboxylic acids, can be used as growth substrates by PRB although the use of these substrates is strain-dependent. *Dechloromonas* sp. JM (Miller and Logan, 2000), a strain isolated from the bacterial consortium in an autotrophic packed-bed biofilm reactor, reduced ClO_4^- using H_2, but could not grow using H_2 as the sole electron donor. Two autotrophic isolates, *Dechloromonas* sp. HZ (Zhang *et al.*, 2002) and *Dechloromonas* sp. JDS5 (Shrout and Parkin, 2002), have been isolated that can grow in a minimal inorganic medium using ClO_4^-, H_2, CO_2 and nutrients.

Nutritional Requirements for PRB

There is no detailed information on the best medium to use for PRB or what trace nutrients or metals are needed for growth. Several research groups have used a phosphate buffer system (Attaway and Smith, 1993; Rikken *et al.*, 1996; Wallace *et al.*, 1996; Herman and Frankenberger, 1999; Logan *et al.*, 2001b; Wu *et al.*, 2001), while others (Bruce *et al.*, 1999; Coates *et al.*, 1999) have used a bicarbonate-buffered freshwater medium amended with a complex vitamin solution. There is no evidence that these additional vitamins are necessary for PRB growth, but in one study it was shown that several PRB strains could not grow without the trace metal solution (Michaelidou *et al.*, 2000). Iron, molybdenum and selenium appear to be important for PRB growth and ClO_4^- degradation. ClO_4^- reductase purified from GR-1 was found to contain 11 mol of iron, 1 mol of molybdenum and 1 mol of selenium per mol of heterodimer (Kengen *et al.*, 1999). Bender *et al.* (2002) also reported that molybdenum was important for PRB growth. Wallace *et al.* (1996) used yeast extract and peptone in their medium for HAP-1. Zhang *et al.* (2002) found that yeast extract improved the growth of the autotrophic PRB strain HZ, but that yeast extract was not needed for growth.

It is likely that the different trace nutrients used for laboratory media are not needed for bioremediation of low ClO_4^- concentrations in natural systems. As explained below, several field studies have achieved ClO_4^- degradation only through the addition of an oxidizable substrate (acetate, ethanol, etc.), nitrogen, and phosphorus (Green and Pitre, 1999; Hatzinger *et al.*, 2000; Logan *et al.*, 2001b; Evan *et al.*, 2002; Min et al. 2004).

The Perchlorate Degradation Pathway

Much of what is known about bacterial degradation of ClO_4^- has resulted from earlier studies on ClO_3^- reduction. Quastel *et al.* (1925) found that ClO_2^- was produced from ClO_3^-

by one strain of *E. coli* without further reduction of ClO_2^-. The failure of ClO_3^- respiration in this strain was likely due to the toxic effects of ClO_2^- and an absence of the enzyme ClO_2^- dismutase. Later, it was found that some bacteria could reduce ClO_3^- to Cl^- (Aslander, 1928; Bryan and Rohlich, 1954). The reduction of ClO_4^- to Cl^- by several species of heterotrophic bacteria was first demonstrated with the use of ^{36}Cl-labeled ClO_4^- (Hackenthal *et al.*, 1964). Hackenthal (1965) concluded that ClO_3^- was the product of ClO_4^- reduction by cell free extracts obtained from NO_3^--adapted cells of *Bacillus cererus*. They also found that ClO_3^- could be reduced by the same cell free preparation and that it competitively inhibited ClO_4^- reduction. The first proposed ClO_4^- reduction pathway was $ClO_4^- \rightarrow ClO_3^- \rightarrow ClO_2^- \rightarrow Cl_2O^- \rightarrow Cl^-$ (Hackenthal *et al.*, 1964; Hackenthal, 1965). The reduction of ClO_4^- or ClO_3^- to Cl^- by bacteria was subsequently confirmed by other researchers (Korenkov *et al.*, 1976; Malmqvist *et al.*, 1991a; Rikken *et al.*, 1996; Coates *et al.*, 1999; Wu *et al.*, 2001).

Research on the ClO_4^- reduction pathway did not make further progress until a new enzyme, ClO_2^- dismutase, was purified from the PRB strain GR-1 and found to produce O_2 from ClO_2^- (Rikken *et al.*, 1996; van Ginkel *et al.*, 1996). Rikken *et al.* (1996) proposed a three-step mechanism of ClO_4^- reduction (Fig. 1) in which ClO_3^-, ClO_2^-, and O_2 were sequentially produced. This pathway, which is now widely accepted for bacterial respiration using ClO_4^- and ClO_3^-, is: $ClO_4^- \rightarrow ClO_3^- \rightarrow ClO_2^- \rightarrow O_2 + Cl^-$. ClO_2^- dismutase has been isolated from two other bacteria, *Ideonella dechloratans* and strain CKB. Stenklo *et al.* (2001) purified ClO_2^- dismutase from *I. dechloratans* that had catalytic properties that were similar, but not identical, to those found for the ClO_2^- dismutase enzyme obtained from GR-1. Stenklo *et al.* (2001) determined the 22-residue N-terminal amino acid sequence for this enzyme and found no homologue in the protein sequence of the purified enzyme from *I. dechloratans*. Coates *et al.* (1999) also purified a ClO_2^- dismutase from strain CKB, which had characteristics similar to the enzyme purified from strain GR-1. The finding by Coates *et al.* (1999) that all thirteen CRB isolates obtained in their laboratory could disproportionate ClO_2^- into Cl^- and O_2 makes it likely that ClO_2^- dismutase is the central enzyme for the dissimilatory reduction of ClO_2^- to Cl^- in PRB.

Effects of NO_3^- on ClO_4^- Removal

Since the NO_3^- concentration in water can be orders of magnitude higher than the ClO_4^- concentration, NO_3^- may have an effect on the removal of ClO_4^-. The effect of NO_3^- may depend on the bacterial species, NO_3^- and ClO_4^- concentrations, and electron donor – the H_2 pressure in the case of the membrane biofilm reactor (MBfR). For example, NO_3^- has been shown to regulate ClO_2^- dismutase expression and repress ClO_4^- reduction by *D. suillum* (Chaudhuri et al. 2002). This species preferred NO_3^- over ClO_4^- when both compounds were present at equimolar concentrations. ClO_4^- reduction was shown to decrease in the presence of NO_3^- in some studies (Herman and Frankenberger, 1999; Brown *et al.*, 2002; van Ginkel et al. (1995), but not in others (Burns *et al.*, 2001; Logan and LaPoint, 2002).

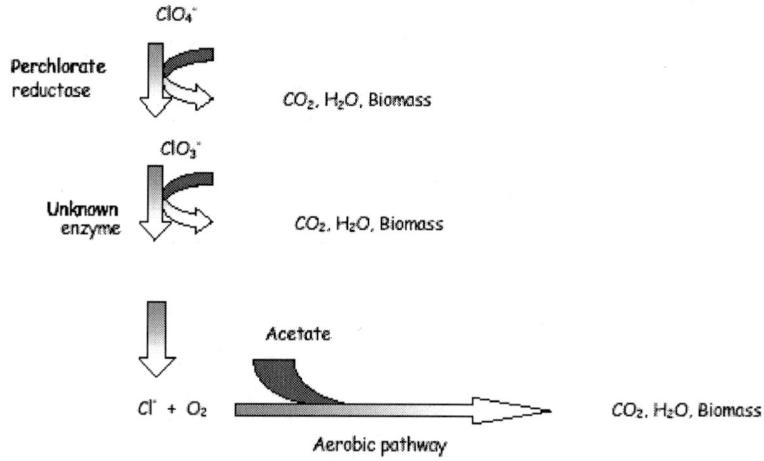

Figure 1. The perchlorate reduction pathway (adapted from Rikken *et al.*, 1996).

A PRB isolate, perclace, was able to completely remove ClO_4^- (130 μg/L) and to simultaneously remove more than 95% of the NO_3^- (20 mg/L) in a sand-packed bioreactor (Herman and Frankenberger, 1999). Achenbach et al. (2001) hypothesized that NO_3^- levels must be depleted before ClO_4^- reduction is initiated when *Dechloromonas* species are grown in media containing ClO_4^- and NO_3^-. Ziv-El and Rittmann (2009) showed that reduction happens in the order – NO_3^- - NO_2^- - ClO_4^- in the MBfR. However, in contrast to *D. suillum*, NO_3^- had no inhibitory effects on ClO_4^- reduction by the ClO_4^- reducer *D. agitata* strain CKB. Furthermore, using ClO_4^--grown *Citrobacter*, Bardiya and Bae (2004) observed preferred usage of ClO_4^- to NO_3^- as an electron acceptor. Another ClO_4^--reducing bacterium not subject to NO_3^- inhibition is the isolate HAP-1 (Wallace et al. 1996). Coppola and McDonald (2000) reported that a mixed culture of *Wollinella succinogenes* could completely degrade ClO_4^- (1.2-1.5 g/L), ClO_3^- (3-3.5 g/L), and NO_3^- (0.2 g/L) in a 7-L continuously stirred tank reactor at a HRT of 16 hr. In pilot-scale tests, Min *et al.* (2004) found that ClO_4^- and NO_3^- were both completely removed in a fixed bed bioreactor. Thus, it appears likely that ClO_4^- and NO_3^- can be simultaneously removed in bioreactors. However, the concentration of NO_3^- will increase the total amount of electron donor needed by the system for complete ClO_4^- removal and with some bacterial species; NO_3^- may have to be completely removed before ClO_4^- is reduced to low levels (Xu et al, 2003).

ClO_4^- may be removed through secondary utilization when NO_3^- is the primary electron acceptor. Nerenberg and Rittmann (2002) and Nerenberg *et al.* (2008) showed that ClO_4^- can be reduced secondarily when NO_3^- is the primary electron acceptor and that NO_3^- reduction is helpful to lower ClO_4^- concentrations to the maximum contaminant or advisory levels (6 - 25 ppb). Steinberg *et al.* (2005) observed that ClO_4^- respiratory enzymes were inducible by ClO_3^- or ClO_4^- and were separate from those used for denitrification when using *Dechlorosoma* sp. KJ. These authors stated that the presence of small amounts of NO_3^- in contaminated groundwater may actually help to increase rates of ClO_4^- reduction once the NO_3^- is completely removed. For example, when strain KJ was grown on ClO_4^- and NO_3^- together and then switched to a ClO_4^--only media, ClO_4^- degradation was higher than when strain KJ was initially grown in a NO_3^--only media.

The impact of NO_3^- on ClO_4^- reduction is important because NO_3^- is a common co-contaminant of ClO_4^--contaminated waters. Many PRB are capable of partial or complete denitrification, and the presence of NO_3^- usually decreases the rate of ClO_4^- reduction. The similarities between the ClO_4^- reduction and denitrification pathways led to an early suggestion that the ClO_4^- and NO_3^- pathways were catalyzed by the same NO_3^- reductase (Hackenthal *et al.*, 1964; Stouthamer, 1967). In addition, it was found that ClO_4^- reductase obtained from GR-1 reduced NO_3^- as well as ClO_3^- and ClO_4^- (Kengen *et al.*, 1999).

The concept of a shared enzyme for NO_3^- and ClO_4^- reduction is unlikely given more recent findings. It has been found that some PRB, for example *Dechloromonas agitata* CKB (Bruce *et al.*, 1999), can respire using ClO_4^- but not NO_3^-. In a PRB capable of denitrification (strain perc1ace), it was found that NO_3^- and ClO_4^- reductases were located in different cell fractions (membrane and periplasmic fractions, respectively), indicating that these two enzymes were separate (Giblin and Frankenberger, 2001). Further evidence of completely separate denitrifying and ClO_4^- pathways was provided by Xu *et al.* (2002). These researchers found that ClO_4^- degradation and denitrification in *Azospira* sp. KJ and PDX were separately induced. When cells were grown on only ClO_4^-, there was minimal NO_3^- reduction. Also, there was no ClO_4^- or ClO_3^- reduction when cells were initially grown on only NO_3^-. However, the presence of both ClO_4^- and NO_3^- in the medium stimulated the degradation of both electron acceptors. Their kinetic results were supported by whole cell protein profiles using SDS-PAGE that showed the appearance of bands in the gels at locations expected for ClO_4^- reductase, ClO_2^- dismutase, and NO_3^- reductase under conditions that agreed with the kinetic studies (Xu *et al.*, 2002).

It is still not clear if only a single enzyme is used by PRB for ClO_3^- and ClO_4^- reduction, or if there are separate enzymes used for ClO_4^- and ClO_3^- reduction. Research by Kengen *et al.* (1999) suggests only one enzyme is needed for ClO_4^- and ClO_3^- respiration. They found that a single enzyme could catalyze both ClO_4^-- and ClO_3^--reduction, and that this O_2-sensitive enzyme was located in the periplasm. However, the maximum reaction rates (measured with methyl viologen) for ClO_4^- (3.8 U/mg) were actually less than those with either NO_3^- (6.2 U/mg) or ClO_3^- (11.3 U/mg) (Kengen *et al.* 1999). Because it appears likely that separate enzymes are used for NO_3^- and ClO_4^- (as explained above), it may also be that there are separate ClO_3^- and ClO_4^- enzymes. Additional evidence for separate enzymes is provided indirectly by the fact that not all CRB are capable of respiration with ClO_4^-, although this question will require further research to resolve.

Effects of O_2 on ClO_4^- Removal

Exposure to DO may pose an adverse effect on perchlorate reduction. A temporary reactor shut-down due to a power outrage at the Texas Street Well Facility exposed the biofilm to high concentrations of DO for more than 48 hours, which led to the system malfunction for two weeks. Further research (Song and Logan, 2004b) found that exposure of cultures (*Azospira sp. KJ*) growing anaerobically, and using ClO_4^-, to DO for 12 hours or more destroyed the capability of these bacteria to reduce ClO_4^- even when they were returned to an anaerobic environment, and the remaining DO was scavenged using Oxyrase™. This exposure time is consistent with that found by others. Attaway et al. (1993) found that exposure of ClO_4^- respiring culture containing primarily the aero-tolerant bacterium *W. succinogenes* HAP1 (Wallace et al. 1996) to DO for 12 hours terminated the ability of the cell to reduce ClO_4^-. However, the short time needed to inhibit ClO_4^- reduction is inconsistent with an explanation that O_2 affected key enzymes known to reduce ClO_4^-. ClO_4^- reductase activity is O_2 sensitive, but it has a half time of inactivation of 2 to 3 days (Kengen et al, 1999), which is much larger than the 12 hours observed here. ClO_2^- dismutase activity is not affected by DO (van Ginkel et al, 1996; Bruce et al, 1999; Stenklo et al, 2001), consistent with other tests in our laboratory showing that cells retained ClO_2^- dismutase activity when the cells were exposed to DO for 12 hours (data not shown). Thus, the lack of ClO_4^- reduction by the cells exposed to high DO for 12 or more hours does not appear to be due to the effect of O_2 on either ClO_4^- reductase or ClO_2^- dismutase.

Cells were capable of resuming ClO_4^- reduction only after exposure to DO when they were able to rapidly reduce the redox potential to negative values. Thus, there must be a link between the inability of the cells to reduce ClO_4^- and the ability to create this negative redox potential. When anaerobically growing cells were exposed to DO, ClO_4^- reduction resumed only when cells reduced the redox potential to less than –127 mV in the presence of Oxyrase™. This large initial decrease in redox potential for the cells able to degrade ClO_4^- cannot be attributed solely to Oxyrase™, as this chemical reduced the redox potential to only 277 mV in an abiotic medium. The ability to produce a redox potential of less than –127 mV must therefore due to microbial activity (Song and Logan, 2004b).

The lack of an effect of DO on the two enzymes known to be involved with ClO_4^- reduction (ClO_4^- reductase and ClO_2^- dismutase), suggest that there must be at least one O_2-sensitive enzyme involved in ClO_4^- reduction. The activity of this enzyme would function to sharply reduce the redox potential of the medium when cells were first returned to anaerobic conditions. A logical candidate for this unknown O_2^- sensitive enzyme would be a high-affinity terminal oxidase (Rice and Hempfling, 1978; Arnold et al. 1986). The ability of cells to reduce the redox potential to negative values demonstrated the presence and activity of a low O_2 tension enzyme. When ClO_4^- reduction was lost (between 8 and 12 hours), this high-affinity terminal oxidase was either blocked or became damaged by exposure to high O_2 concentrations.

These results suggest that the ability of bacteria to reduce ClO_4^-, which produces O_2 from ClO_2^-, can occur as long as low O_2 tensions are maintained. However, exposure of cells to high O_2 concentrations for 12 or more hours will inhibit ClO_4^- reduction likely due to the loss of some key respiratory enzyme that has not yet been isolated from PRB (Song and Logan, 2004b).

Biotreatment Processes

ClO_4^- is highly soluble and chemically stable in water. As a groundwater contaminant, its properties most closely resemble NO_3^-, which is nonadsorptive and nonvolatile. Therefore, many traditional water treatment approaches, such as carbon adsorption, chemical oxidation, and air-stripping are ineffective for ClO_4^- removal from groundwater or drinking water (Logan, 2001a). The two primary effective technologies are ion-exchange (IX) and biological treatment. In addition, IX and biological treatment can be combined such that bacteria are used to treat spent IX brine and make it reusable.

Biological reduction can be carried out with heterotrophic and autotrophic processes. While heterotrophic processes may use many organic electron donors, autotrophic processes are mainly limited to hydrogen gas (H_2) as the electron donor. The advantages of autohydrogenotrophic denitrification and ClO_4^- reduction over heterotrophic processes include no carryover of electron donor, little organic carbon in the effluent, and the relatively low cost of H_2 (Lee and Rittmann, 2002). Another advantage of autohydrogenotrophic denitrification and ClO_4^- reduction is the low cell yield, which results in less excess biomass. Typical stoichiometry of autohydrogenotrophic denitrification is (McCarty, 1975):

$$H_2 + 0.32\ NO_3^- + 0.32\ H^+ + 0.09\ CO_2 \rightarrow 0.15\ N_2 + 1.1\ H_2O + 0.017\ C_5H_7O_2N$$

This reaction shows that NO_3^- is reduced to harmless N_2 gas, and the consumption of H_2 is 3.1 mol H_2/mol NO_3^-, or 0.44 g H_2/g N. The cell yield is approximately 0.43 g cells/g NO_3-N which is approximately 1/4th the yield of heterotrophic denitrifiers utilizing acetate (2.0 g cells/g N):

$$C_2H_3O_2 + 0.8\ NO_3 + 0.8\ H^+ \rightarrow 0.3\ N_2 + 0.8\ H_2O + 0.1\ CO_2 + 1\ HCO_3 + 0.2\ C_5H_7O_2N$$

While the low cell yield is an advantage in terms of excess biomass, it can in some case be a disadvantage, since the autotrophs have lower specific growth rates. Typical stoichiometry of autohydrogenotrophic ClO_4^- reduction using NO_3^- as the nitrogen source is (McCarty, 1975):

$$H_2 + 0.18\ ClO_4^- + 0.02\ NO_3^- + 0.02\ H^+ + 0.10\ CO_2 \rightarrow 0.18\ Cl^- + 0.94\ H_2O + 0.02\ C_5H_7O_2N$$

This reaction shows that ClO_4^- is reduced to the harmless Cl^-, and the consumption of H_2 is 5.6 mol H_2/mol ClO_4^-. The biomass yield with ClO_4^- as an electron acceptor is 12.4 g cells/mol ClO_4^-, which is more than double the biomass yield with NO_3^- as the electron acceptor (6.4 g cells/mol NO_3^-).

BIOREDUCTION OF ION-EXCHANGE BRINE

The most commonly used process today for removing NO_3^- and ClO_4^- from groundwater is IX, in which NO_3^- and ClO_4^- are exchanged for harmless anions as the water passes through a bed of IX resin (Aldridge et al. 2004, McAdam and Judd 2008). There have been many recent advances in the cost and selectivity of resins for ClO_4^- (please refer to the work of Gu et al. (1999) and similar studies for information about ClO_4^-selective anion exchange resins - Siemens 2010, Calgon 2010, Purolite 2010). Selective IX resins are very effective in removing ClO_4^- from contaminated waters. However, these resins are currently incinerated after one time use, making the IX process incomplete and unsustainable for ClO_4^- removal. Although IX is efficient at removing NO_3^- and ClO_4^- from the water, it does not detoxify them. Instead, they accumulate on the IX resin or in the salt-brine used to regenerate the resin. Treatment of the IX brine so that it can be reused or disposed of safely is a major unresolved challenge. One approach for treating the brine is bioreduction, the speed of which depends on the composition of the IX brine.

The composition of spent brine depends on resin type, removal efficiency, and influent characteristics. NaCl is widely used for IX resin regeneration. Typical NaCl concentrations for regeneration are 6–12% for ClO_4^- and 3–12% for NO_3^- (McAdam and Judd, 2008). Of all the cations, sodium (Na^+) causes the most inhibition to bacteria. While potassium (K^+) is the most common cation in the cell, Na^+ is most common in the environment. At the high concentration gradients when a cell is in IX brine, Na^+ can leak into the cell or be brought in via substrate symport mechanisms. Bacteria either have to use ATP to remove Na^+ or dissipate the proton motive force to remove sodium via Na^+/H^+ antiports. All of this costs the bacteria energy and reduces growth. Furthermore, Na^+ interferes with the tertiary structure of enzymes and reduces substrate degradation and subsequently energy generation and growth. Also, Na^+ can disintegrate flocs by displacing multivalent ion bridges that hold extrapolymeric substances (EPS) together and to the bacterial cell. This causes a release of EPS (carbohydrates, proteins, etc.) which can foul the resin if not removed by membrane filtration or foul membranes used for the retention of biomass during IX brine treatment. The fouling tendency of EPS from non-halophilic bacteria is greater than that of halophilic bacteria likely due to the tighter cell membranes of halophilic bacteria. Thus, biotreatment of IX brine should use halophilic bacteria.

Typically, freshwater bacteria undergoing NO_3^- and ClO_4^- reduction can tolerate NaCl concentrations up to ~20 g/L (McAdam and Judd 2008). Above this concentration, degradation rates decrease and halophilic bacteria are needed. Since activated sludge (AS) is a great source of freshwater bacteria, many researchers have tried adapting AS to salinities typical of IX brine. Slow adaptation is recommended as if bacteria can change their cell membrane permeability to restrict Na^+ intake. Park and Marchand (2006) state that ClO_4^- reduction using AS as inoculum was feasible up to 4% salinity. Yoshie et al. (2006) observed higher denitrification rates at 10% than at 2% salinity and attributed the difference to the halophilic genera *Halomonas* and *Marinobacter*. Similar results were found in Van Ginkel et al. (2010a, b). Okeke et al. (2002) obtained sediments from salt evaporation ponds and the Salton Sea, and successfully isolated a *Citrobacter sp.* able to reduce ClO_4^- up to 50 g/L TDS. Cang et al. (2004), using marine sediment as inoculum, observed NO_3^- and ClO_4^- reduction at salinities from 3-6%.

Halophiles are distinguished by their requirement for and tolerance of salt. Slight halophiles grow optimally at 2-5% NaCl, moderate halophiles grow optimally at 5-20% NaCl, and extreme halophiles grow optimally above 20% NaCl. In order to prevent loss of cellular water to the external environment, halophiles accumulate high solute concentrations within the cell. Halophiles may accumulate inorganic ions such as Na^+, K^+, Mg^{2+}, or Ca^{2+} (the "salt-in" strategy), or they may accumulate organic compounds, also called compatible solutes, including glycine betaine, ectoine, choline, sucrose, and glycerol, which do not disrupt cellular processes and have no net charge at physiological pH (DaSarma and Arora, 2001). In many species, Na^+ and K^+ are accumulated to molar concentrations, but never at the same concentration as the external environment. The salt-in strategy requires that intracellular systems are adapted to high salt concentrations, while no special adaptation is required for the compatible solutes strategy (Oren, 1999a). Proteins of halophiles are either resistant to high salt concentrations or require salts for activity. Halophiles contain an excess ratio of acidic to basic, hydrophobic amino acids; this feature is likely to be required for activity at high salinities (DaSarma and Arora, 2001). Several trends are clear in regard to Na^+ and K^+ concentrations (Ventosa et al. 1998):

- The intracellular K^+ concentration is generally higher than that in the medium.
- The Na^+ concentration inside the cells is generally lower than that in the medium.
- The apparent intracellular Na^+ and K^+ concentrations increase with increasing external NaCl concentration in a nonlinear fashion.
- The sum of the internal concentrations is generally insufficient to balance the osmotic pressure of the medium. A balance may be achieved if compatible solutes are taken into account.

The review by Oren (1999a) compares the energetic cost between the salt-in and the compatible solute strategies. Oren concluded that the synthesis of compatible solutes is much more energetically costly than maintaining high KCl concentrations in the cytoplasm. In a theoretical study by Stouthamer (1977), a maximum value of 6.5 g (dry weight) of autotrophic microbial cells can be formed from CO_2 per mole of ATP. Considering that nonhalophilic microorganisms are 80% water (by weight) and that the density of the dry cell material is around 2 g/cm^3, the production of 1 cm^3 of biovolume of nonhalophilic bacteria, containing 0.2 g of dry material of structural components and 0.9 mL water, would cost a minimum of 30 mmol of ATP. The energetic cost of producing compatible solutes is much more than this. For example, organisms living in 4 M of salt may contain 8 M of compatible solutes to maintain osmotic balance. When using glycerol, glycine betaine, or ectoine as the compatible solute, the theoretically added dry weight per 0.2 g of structural cell material not directly involved in osmotic regulation would be 0.66, 0.85, and 1.03 g, respectively. The synthesis of these compounds from CO_2 would cost 216, 389 to 421, and 389 mmol of ATP, respectively. In conclusion, the amount of energy needed to produce compatible solutes greatly exceeds the requirement for basic cellular components. In regard to growth on glucose at a high salinity, the synthesis of glycerol and ectoine may require 4.4 to 6.9 times more ATP than in a low salinity condition. Thus, it should be expected that the cell yield of organisms will decrease as the salinity increases. However, if expressed as dry weight, the difference in yields may be low or absent due to the accumulation of the solutes. In contrast, considering the transport of K^+, Na^+, and H^+ across cell membranes and ATP utilization coupled to this transport in the salt-in strategy, 0.5-0.67 mol of ATP is required per KCl accumulated. Oren (1999a) concluded that this is a small fraction of the energy required for the compatible solute strategy and stated, “For osmolarity generation, the computations show that compatible solutes costs at least an order of magnitude more than inorganic solutes on either an energy or H_2O basis.”

The example above is for extremely halophilic bacteria and may not apply to the slight to moderately halophilic bacteria that would be present in the cultures used in the treatment of IX brine. Moderately halophilic bacteria tend to maintain low intracellular ionic concentrations. These concentrations are achieved by tightening of the permeability barrier of the membrane and not by constant pumping of ions by active, energy-dependent mechanisms. The maintenance energy of a continuous culture of *Halomonas elongata* did not change with salinity providing evidence in favor of an ion-tight membrane (Galinski, 1995).

Many organisms are able to grow better at higher salinities in their natural environment than in culture. This may be due to uptake of secreted compatible solutes by organisms that don’t ordinarily produce them. Glycine betaine and ectoine are excreted from cells during salt down shock. Many halophiles use a variety of compatible solutes and the synthesis of each of these depends on salinity and the nutrient availability. Many bacteria contain transport mechanisms that enable the uptake of compatible solutes, which is more energetically favorable than synthesis. For example, the use of glycine betaine partially restored both biofilm formation and benzoate degradation by *Pseudomonas aeruginosa* under an osmotic shock of 30 g NaCl/L (Bazire et al. 2006). Robert et al. (2000) used glycine betaine, carnitine, and choline to confer an increased salt tolerance of *T. halophila*. They stated that the solute uptake mechanisms of *T. halophila* are constitutive and the intracellular concentration of NaCl is less in the presence of these compatible solutes. Nagata et al. (2005) used proline (1 mM) to recover a culture of *Escherichia coli* under high salinity (60 g NaCl). They also concluded that the co-existence of K^+ with proline enhanced the recovery of growth under high salinity more than that in the presence of proline alone. Lin et al. (2007) found that a Mg^{2+}:Na^+ molar ratio of 0.11 can generate osmotolerance and maintain long-term ClO_4^- reduction. This ratio was used in pilot plant tests described in Lehman et al. (2008). However, Mg^{2+} can precipitate out via carbonate or phosphate precipitates if the pH increases too high due to denitrification.

Oren (1999b) stated that denitrification can occur at salinities up to 30%, possibly due to thermodynamic favorability of the reaction. If this is the case, ClO_4^- reduction should happen at 30% salinity as well. Other less energetic reactions may preclude the existence of microorganism at high salinities. In particular are autotrophic reactions, which require the energetically costly fixation of carbon. Microorganisms that obtain only a little energy in the course of their metabolism may be unable to cope with high salt concentrations.

Biotreatment of IX Brine – Bench-Scale Studies

Most of the work regarding the biotreatment of IX brine has been conducted at the bench-scale. Giblin et al. (2002) treated a twice-concentrated RO stream using a PBR and observed 98% reduction of ClO_4^- in 2 hours with an influent ClO_4^- concentration of 8 mg/L. NO_3^- was simultaneously removed from as high as 900 mg/L to 4 mg/L. The TDS concentration was 4 g/L, the bulk of which was NaCl. They used a pure culture of perl ase – a known ClO_4^--reducer. Acetate was the electron donor and carbon source. The reactor suffered from clogging due to the high TDS of the twice-concentrated rejectate.

Okeke et al. (2002) used a variety of cultures to remove ClO_4^- and NO_3^- in simulated IX brine. Spent regenerate brines typically contain 60-70 mg /L ClO_4^- and 400-4000 mg/L NO_3^-. Acetate and yeast extract were used. Sources of inocula included a Cargill salt evaporation facility and the Salton Sea, both in California. They also used percl ase (mentioned above) and two known halophilic denitrifiers. One microbe they isolated was a *Citrobacter* species. At 7.5% total solids (includes NaCl, NO_3^-, and SO_4^{2-}), this isolate was able to reduce ClO_4^- and NO_3^- by 34.9 and 15.6%, respectively, in one week. Initial ClO_4^- and NO_3^- concentrations were 100 mg/L and 0.25 – 12 g/L, respectively. A co-culture of percl ase and *Citrobacter sp.* was most effective at ClO_4^- removal (46.4% removal in one week in the 7.5% TDS). *Citrobacter sp.* was able to remove 84, 105, 88, mg/L ClO_4^- from an initial 500 mg/L at 0, 2.5, and 5% NaCl, respectively and proved to be more salt tolerant than perl ase. NO_3^- may have inhibited ClO_4^- removal in these experiments.

Some ClO_4^--laden resins can also be regenerated using 0.5% ammonium (Gingras and Batista, 2002). Using AS as the inoculum and lactate as the electron donor, Gingras and Batista (2002) performed growth experiments (based on absorbance) on ClO_4^- in simulated IX brine. NaCl and NH_4OH concentrations were varied from 1-1.5% and 0.4-1.0%, respectively. NaCl solutions of 0.5 and 1.0% reduced ClO_4^- reduction by 30 and 40%, respectively. Ammonium levels of 0.4 – 1.0% reduced ClO_4^- reduction rates by 90%. They mentioned that typical NaCl regenerate brines vary between 6-12% salinity. Generally, 3-5 bed volumes of regenerant are needed for regeneration and 2-3 bed volumes are needed for rinsing. The salinity of the resulting brine is 40-50% of the original brine stream.

There are two ranges of ClO_4^- concentrations in groundwater, depending on how far the groundwater is collected from the source of contamination. At the source of contamination, ClO_4^- concentrations are typically 80 mg/L while at drinking water wells distant from the source, ClO_4^- concentrations typically are 18-75 μg/L. The resulting brines from these two waters are 500 mg/L and 2-3 mg/L, respectively.

Salt toxicity (0-3% NaCl) was studied using three different cultures in batch experiments by Park and Marchand (2006). They included return activated sludge (RAS), anaerobic return activated sludge (ARAS), and anaerobic digester sludge (AD). The initial ClO_4^- concentration was 650 mg/L. Acetate was used as the electron donor. At 3% salinity, the RAS and ARAS cultures consumed the ClO_4^- in 100 and 120 hours with specific ClO_4^- removal rates of 4.4 and 3.5 mg ClO_4^-/g MLVSS-hr, respectively. Above 1.5% salinity, the AD culture failed to reduce ClO_4^-. However, after gradual adaptation, the AD culture was able to reduce ClO_4^- up to 4% salinity with a specific ClO_4^- removal rate of 16.2 mg ClO_4^-/g MLVSS-hr at a MLVSS concentration of 1.82 g/L. They stated that the key factor involved in overcoming inhibition at high salinities relates to maintaining ClO_4^--reducing populations at high cell densities, which could act as a buffer against changes to elevated salinity. Another important aspect could be the aggregation of biomass. Above 1% salinity, they observed the aggregation of biomass. The outer layers of the floc could shield inner microbes from the high salinity (which is contrary to the point made above about Na^+ and EPS). Another important aspect is the use of mixed cultures to remove NO_3^- and oxygen (O_2) down to low levels so ClO_4^- reduction can occur.

Generally, microbes harvested from non-saline environments do not reduce ClO_4^- well at salinities above 2%. The growth of *Dechlorisoma suillus* strain PS and *Dechlorospirillum anomalous* strain WD, both isolated from sediment samples in swine waste primary lagoons, were completely inhibited at above 1% NaCl (Michaelidou et al., 2000). Another ClO_4^--reducing microbe isolated from paper mill waste, *Dechlorimonas agitata* strain CKB, grew well in environments with less than 2% NaCl (optimal growth at 1% salinity), but no growth occurred at 4% NaCl (Coates et al., 2000). Furthermore, *Ideonella dechlorotans* isolated from AS failed to grow at 3% NaCl (Malmqvist et al. 1991b).

Logan et al. (2001c) performed growth experiments (based on absorbance) on various cultures (marine, salt marshes, seawater filter sludge) at salinities of 1-15% using ClO_4^- as the terminal electron acceptor. The cultures taken from the Great Salt Lake had the highest growth rate at 0.06 d-1 (doubling time of 11.6 d) at a salinity of 5%. Growth rates decreased to 0.037 d-1 at a salinity of 11%, and no growth occurred at 13 - 15% salinity.

Using acetate as the electron donor and marine sediment as inoculum, cultures were developed at 3% and 6% salinity by Cang et al. (2004). The culture at 3% salinity was able to simultaneously reduce 100 mg/L ClO_4^- and 500 mg/L NO_3^- in synthetic medium within 5 hours. However, ClO_4^- reduction was slow when the culture was fed actual IX brine. In a later study (mentioned above), Lin et al. (2007) found that adding Mg^{2+} to the medium helped stabilize the a culture in brine solution. When the mole ratio of Mg^{2+} to Na^+ was 0.11, the culture could remove ClO_4^- at the same rate when the salt concentration ranged from 3-6% NaCl. It was also demonstrated that the culture was able to rapidly reduce ClO_4^- and NO_3^- in the presence of varying concentrations of brine constituents such as salt (3 to 10% NaCl), NO_3^- (400 to 4000 mg/L), SO_4^{2-} (600 to 6000 mg/L) and pH (5 to 9) (Hiremath et al., 2006). Inoculum from this and derivative cultures has been used to seed a 200 L pilot plant sequencing batch reactor (SBR) that was used to remove ClO_4^- and NO_3^- from an IX brine pilot-scale demonstration in southern California (Aldridge et al., 2004). A separate 76 L pilot plant biological treatment unit was implemented at a different Southern California site (Lehman et al. 2008). In both cases, the treated brine was recycled back into the IX process as regenerant for 20 cycles with no effect on the ability of the resin to remove ClO_4^- or NO_3^- from the source water.

Of the other anions removed by IX, sulfate (SO_4^{2-}) is selectively removed in conventional resins and will consume electron donor when treated and produce H_2S which is toxic to bacteria at high concentrations. New resins are available which selectively remove NO_3^- and ClO_4^-. Bicarbonate is also removed and is relatively harmless unless precipitated out at high pH.

Biotreatment of IX Brine – the Ion-Exchange Membrane Bioreactor (IEMB)

In a process different than traditional IX, Matos et al. (2006) simultaneously removed NO_3^- and ClO_4^- from drinking water using an IX membrane bioreactor (IEMB) involving Donnan dialysis. In this process, NO_3^- and ClO_4^- contaminated water would enter the 'water compartment' side of the IEMB and would diffuse through an anion-exchange membrane to the 'biocompartment' side of the membrane where they would be bioreduced to N_2 and Cl^-, respectively. Membrane transport studies showed that Donnan dialysis is suitable for obtaining water with concentrations of NO_3^- and ClO_4^- below recommended levels. The IEMB process operated with hydraulic retention times (HRT) ranging from 1.4 to 8.3 h in the water compartment, proved to remove NO_3^- and ClO_4^- while preserving the water composition with respect to other ions, thus avoiding secondary contamination of the treated water. For a polluted water stream containing 100 ppb ClO_4^- and 60 ppm NO_3^-, the concentrations of both ions in the treated stream were kept below the recommended levels of 4 ppb for ClO_4^- and 25 ppm for NO_3^-.

Biotreatment of IX Brine – Fluidized Bed Reactor Studies

The biological fluidized bed reactor (FBR) has been commercially implemented for NO_3^- reduction from fresh water. However, the application of biological FBR system for NO_3^- and ClO_4^- treatment from IX brine is a new concept. Previous work at the University of Houston and California pilot plants has utilized the culture in SBRs (Lehman et al. 2008). Due to the high volume of brine to be treated, the continuous nature of many of the newer ion exchange processes (e.g. ISEP®, Calgon Co.), the reaction time required, and the number of reactors required for an FBR system, conversion of the batch reactor system to a flow-through such as fluidized bed reactor was deemed appropriate. A fluidized bed was chosen over a packed bed design because the latter have shown a need for frequent backwashing due to the column blockage and short circuiting (Kim and Logan 2000; Kim and Logan 2001; Logan 2001a; Logan 2001b; Min et al. 2004). The fluidized bed should allow consistent biomass sloughing and removal of trapped gasses decreasing the need for backwashes. Previous research showed success using granular activated carbon (GAC) and sand as a support medium in ClO_4^- reducing packed and fluidized beds treating contaminated ground water (Kim and Logan 2000). Another study comparing the performance of sand and GAC filled fluidized bed reactors found both beds removed similar amounts of ClO_4^- from ground water. When the beds were fed groundwater, the GAC filled FBR performed slightly better (Greene and Pitre 2000). Patel et al. (2008) described bench-scale research on operation of acetate-fed granular activated carbon (GAC) based FBR for ClO_4^--only, and combined NO_3^- and ClO_4^- removal from synthetic brine (6% NaCl). The FBR was shown to be an effective design for ClO_4^- reduction and exhibited first-order degradation kinetics with respect to ClO_4^- concentrations. NO_3^- was also removed by the organisms in the column and had no negative effects on the removal of ClO_4^- using the FBR design. However, at higher concentrations of NO_3^- the FBR was more difficult to operate due to loss of carbon and biomass from the formation of nitrogen bubbles and the high recycle flow rates needed. Xiao et al. (2010) also demonstrated NO_3^- and ClO_4^- reduction in 6% IX brine in a two-stage pilot-scale FBR. They stated denitrifying *Halomonas* reduced NO_3^- in the first stage which allowed for ClO_4^- reducing bacteria - *Azoarcus/Denitromonas* to establish themselves in the second reactor.

Biotreatment of IX Brine – Direct Treatment of IX Resins

Using an FBR, Venkatesan et al. (2010) developed a novel concept by directly contacting ClO_4^--laden gel-type anion-exchange resins with a ClO_4^--reducing bacterial culture. The study showed that direct bioregeneration of gel-type resin was effective in a fluidized-bed reactor, and that the resin could be defouled, reused, and repeatedly regenerated. Using batch tests, Wang et al. (2008) also showed the direct bioregeneration of ClO_4^--laden resins. In a companion study, Wang et al. (2009) demonstrated that ClO_4^- desorption from the loaded resin could be the rate-limiting step. The addition of a proper amount of counter anions such as Cl^- and SO_4^- improved the regeneration efficiency because these anions could promote both the extent and rate of ClO_4^- desorption from the loaded resin. They also suggested that the application of bacteria that could efficiently reduce ClO_4^- in highly saline solution would make the method more promising for the regeneration of high perchlorate affinity resins.

Biotreatment of IX Brine using the Autotrophic H_2-based MBfR

In order to assure complete NO_3^- and ClO_4^- removal, substrates and nutrients added to anaerobic reactors must not be reduced to limiting levels within the system. Therefore, they will exit the reactor at concentrations that will require subsequent treatment. Failure to remove these constituents will most certainly result in distribution system regrowth issues. Other issues of concern could include health effects from the additives, or rejection of their presence in tap water by consumers. Certainly, any excess of these carbon compounds entering a distribution system will also contribute to disinfectant demand and byproduct formation.

While the basic premise of heterotrophic biological treatment of ClO_4^- is positive, the use of carbon compounds in the treatment scheme raises questions about the cost of post-treatment. Autotrophic biological treatment process, on the other hand, can be a feasible, effective, and sustainable process for removing ClO_4^- from drinking water. One such system is autohydrogenotrophic bioreduction using MBfR, because the bacteria are autotrophs that utilize inorganic carbon as the carbon source and use H_2 as the electron donor.

The degradation of NO_3^- and ClO_4^- in groundwater and IX brine has been demonstrated using MBfR. MBfRs consist of a bundle of fibers that are encased in a continuous-flow reactor vessel. H_2 gas is supplied to the inside of the fibers and diffuses to an autotrophic biofilm that grows on the outside of the fiber wall. The biofilm oxidizes H_2 while reducing NO_3^- and ClO_4^- in the contaminated water. The use of the MBfR provides efficient means to deliver H_2 to the bacteria, because it does not rely on the solubility of H_2 in water, which is low: e.g., 1.3 mg H_2/L in 4.5% brine in equilibrium with 1 atm of H_2 (Weisenberger and Schumpe, 1996). Since this technology uses a biofilm, bacteria largely remain attached to the membrane rather than living in a planktonic state and transported with treated brine back to the IX column. Besides, any bacteria that would be carried over in the treated brine would be subject to severe osmotic down shock in the IX column. Should the bacteria be released in the IX effluent, they are easily removed by usual post treatment systems, such as deep-bed or membrane filtration. The H_2, being insoluble, will not remain in the treated brine and will not cause bacterial growth on the IX resin or lead to the production of disinfection by-products.

MBfR provides an efficient and simple means to deliver H_2 to a contaminant-reducing biofilm. Directly supplying H_2 via membranes is becoming a proven technology and is more efficient than using an organic donor, since the biofilm uses only the stoichiometric amount of H_2 it needs to reduce the contaminants in the feed stream; this leads to reduced costs and precludes leakage of O_2 demand to the effluent. For these reasons, MBfR is being researched and field tested for a range of bioreduction applications when the water can be routed through an *ex situ* bioreactor, as is the case with IX brines.

NO_3^- reducers may out-compete ClO_4^- reducers for the available H_2, especially when the NO_3^- concentration is much higher than the ClO_4^- concentration. Previous studies have shown that limiting the H_2 pressure to MBfR fibers can limit reduction of secondary electron acceptors (Celmer et al. 2006; Chung et al. 2007a; Ziv-El and Rittmann 2009). The competition between NO_3^- and ClO_4^- reducers for the available H_2 may be competition for space on MBfR fibers, which are the source of the H_2, or the available H_2 may be reduced to a very low level by denitrifiers, leaving very little H_2 for ClO_4^- reduction.

Van Ginkel et al. (2008, 2010), Chung et al. (2007b), and Ahn et al. (2009) investigated the bioreduction of NO_3^- and ClO_4^- in IX brine (30 to 60 g/L NaCl) using MBfR. MBfR technology for brine treatment offers a potentially effective brine-treatment method, because it can remove the anions from the brine, thereby making it possible to reuse the brine or avoid problems of brine disposal. Since the MBfR technology treats the brine, a side stream, and not the water directly, the chance of bacteria being released into the distribution system is very low. Since this technology uses a biofilm, the bacteria largely remain attached to the membrane rather than living in a planktonic state and being carried with the treated brine back to the IX column. Besides, any bacteria that would be carried over in the treated brine would be subject to severe osmotic down shock in the IX column. Should the bacteria be released in the IX effluent, they are easily removed by usual post treatment systems, such as deep-bed or membrane filtration. The H_2, being insoluble especially in brine (1.3 mg H_2/L at 1 atm of H_2, 45 g/L NaCl), will not remain in the treated brine and will not cause bacterial growth on the IX resin or lead to the production of disinfection by-products (Weisenberger and Schumpe, 1996).

Chung et al. (2007b) demonstrated autotrophic bioreductions for the entire range of salinities of IX brine using MBfR, but the kinetics slowed noticeably as the salinity was increased from 20 to 40 g/L NaCl. Using inoculum from a salt pond, low reduction rates were observed using Purolite brine (150 g/L NaCl). Greater reductions were observed at salinities from 2-4%. The focus of Van Ginkel et al. (2008, 2010) and Ahn et al. (2009) was to expand upon the work of Chung et al. (2007b) by performing a much more comprehensive investigation of the factors that control NO_3^- and ClO_4^- bioreduction in IX brine. These factors include acclimation period, salinity, H_2 availability, NO_3^- concentration, and inoculum source. The sources of inocula included sediments from three saline sources - Freeport, TX, the Great Salt Lake, and the Salton Sea; and one freshwater source – backwash sludge from a ClO_4^- degrading PBR. Van Ginkel et al. (2008) concluded that the fluxes of NO_3^- and ClO_4^- were high compared to Chung et al. (2007b) and even to freshwater in some cases. Compared to Chung et al. (2007b), NO_3^- reduction rates were increased by ten times, likely due to the acclimation period. NO_3^- and ClO_4^- removal fluxes decreased with increasing salinity. The NO_3^- fluxes were roughly first order in H_2 pressure, but roughly zero-order with NO_3^- concentration.ClO_4^- reduction was higher at lower NO_3^- loadings, compared to high NO_3^- loadings; this is a sign of competition for H_2. The choice of inoculum and the length of the acclimation period made a major difference in regard to NO_3^- and ClO_4^- reduction rates.

Detoxification of ClO_4^- by microbial communities under denitrifying conditions has been recently reported, although the identity of the mixed populations involved in ClO_4^- reduction is not well understood, especially in IX brines. In order to address this, the bacterial diversity of MBfRs set up under high salinity, autotrophic, denitrifying and ClO_4^--reducing conditions was examined by analysis of 16S rRNA gene sequence of clone libraries (Van Ginkel et al. 2010; Ahn et al. 2009). Inocula from diverse locations were tested for their ability to reduce NO_3^- and ClO_4^- in synthetic IX brine (30-45 g/L NaCl) using H_2-based MBfRs. Phylogenetic analysis of 16S rRNA gene sequences showed that proteobacterial species dominated the biofilm communities, particularly NO_3^--reducing γ-proteobacteria. Even though the inocula to the MBfRs came from different sources, clones closely related to *Marinobacter hydrocarbonoclasticus* represented 53% of all clones in the MBfR biofilms at 45 g/L NaCl. The clone libraries contained no known ClO_4^--reducing bacteria, which suggest that denitrifiers carried out ClO_4^- reduction, probably by secondary utilization.

Sahu et al. (2010) developed halophilic, hydrogenotrophic, denitrifying bacteria for the treatment of NO_3^- and ClO_4^- in IX brine. The cultures were inoculated into bench-scale MBfRs with an ''outside in'' configuration where the contaminated water was supplied to the lumen of the membranes and H_2 supplied to the outside. Abiotic mass transfer tests showed that H_2 mass transfer coefficients were lower in brines than in tap water, possibly due to increased transport of salts and decreased H_2 solubility at the membrane/ liquid interface. An average NO_3^- removal efficiency of 93% was observed for the MBfR operated in continuous flow mode with synthetic contaminated groundwater. Removal efficiencies of 30% for NO_3^- and 42% for ClO_4^- were observed for the MBfR operated with synthetic IX brine in batch operating mode with a reaction time of 53 h. Phylogenetic analysis revealed that halotolerant, NO_3^--reducing bacteria of the bacterial classes *GammaProteobacteria* and *Sphingobacteria* were the metabolically dominant members. This study shows that, despite decreased H_2 transfer under high salt conditions, hydrogenotrophic biological reduction may be successfully used for the treatment of NO_3^- and ClO_4^- in the MBfR.

DIRECT BIOREDUCTION OF NO_3- AND ClO_4-

Direct Bioreduction of NO_3^- and ClO_4^- – Bench-Scale Reactor Studies

Table 1. Bench-scale bioreactors for treating perchlorate in water

Reactor type (media)	Inoculum	Substrates	Perchlorate (mg/L)	References
Suspended growth	Wollinella succinogenes HAP-1 in mixed culture	Protein nutrients*	7750	Attaway et al., 1993
Fixed bed (diatomaceous earth pellets)	Wollinella succinogenes HAP-1 in mixed culture	BYF-100**	500, 1500	Wallace et al., 1998
Fixed bed (sand or Celite)	Perclace	Acetate	0.13, 0.738	Herman and Frankenberger, 1999; Giblin et al., 2000a
Fixed bed (sand or GAC)	Mixed culture	Acetate	20	Kim and Logan, 2000
Fixed bed (GAC)	Tapwater	A mixture of acetate, lactate, and pyruvate	0.051-0.055	Brown et al., 2000
Fixed bed (sand)	Azospira sp. KJ	Acetate	20	Kim and Logan, 2001
Fixed bed (cylindrical pall rings)	Primary sludge and effluent from wastewater treatment plant	Acetate	0.1, 1	Burns et al., 2001
Fixed bed (Celite pellets)	Perclace	Acetate	0.8	Losi et al., 2002
Fixed bed	Mixed culture	Hydrogen	0.740	Miller and Logan (2000)
Fixed bed (Celite R-635)	Biosolids from the municipal water treatment plant	Hydrogen	0.740	Giblin et al., 2000b
Fixed bed (glass beads)	Mixed culture	Hydrogen	0.073	Logan and LaPoint, 2002
Fluidized bed (sand or GAC)	Biological solids from an anaerobic digester	Ethanol, methanol, or a mixture of the two alcohols	25	Greene and Pitre, 1999; Hatzinger et al., 2000
Fludized bed (GAC)	Mixed culture	Acetic acid and ethanol	11-23	Togna et al., 2001
Hollow-fiber membrane bioreactor	Ralstonia eutropha	Hydrogen	0.006-0.100	Nerenberg et al., 2002

*Aged brewer's yeast, cottonseed protein, or whey power.

**Naturally occurring protein (54%), peptides, free amino nitrogen, vitamins, and trace elements.

The first bench scale treatment system reported to degrade ClO_4^- was a suspended growth system developed to treat high concentrations of ClO_4^- (Attaway and Smith, 1993). More relevant for drinking water treatment and bioremediation studies are reports on fixed and fluidized bed bioreactors primarily designed to treat low concentrations of ClO_4^-, and laboratory batch tests to evaluate the feasibility of *in situ* bioremediation. These different bench scale studies are summarized in Table 1 (Xu et al, 2003).

1) Heterotrophic fixed bed (FXB) reactor studies

The first reported fixed bed bioreactor was inoculated with a mixed culture containing *Wollinella succinogenes* HAP-1 (Wallace *et al.*, 1998). This system was designed to treat high ClO_4^- concentrations (~ 10%) found in rocket washout wastewaters. The reactor was filled with diatomaceous earth pellets as media. The substrate for the mixed culture (BYF-100) consisted of 54% naturally occurring protein, peptides, free amino nitrogen, vitamins, and trace elements. ClO_4^- was reduced to <100 mg/L (the detect limit of the system) for 95% of the samples taken over the entire 43 day period of operation at a HRTs and influent ClO_4^- concentrations of 1.17 hours and 1500 mg/L, and 0.46 hours and 500 mg/L, respectively.

A much lower effluent ClO_4^- concentration was reached by a sand media bioreactor inoculated with perc1ace, which was isolated from wastewater biosolids (Herman and Frankenberger, 1999). The reactor was packed with sterilized, oven-dried sand (40 to 70 mesh size). Following a start up period, it was found that ClO_4^- could be degraded from a concentration of 130 μg/L to below the detection limit (4 μg/L) at a HRT of 3 hours. It was also reported that a celite packed bioreactor inoculated with the same isolate completely removed ClO_4^- from 738 μg/L at a flow rate of 1 mL/min (Giblin *et al.*, 2000a).

Complete ClO_4^- removal was found by a sand packed bioreactor inoculated with a ClO_4^--degrading enrichment as well (Kim and Logan, 2000). The enrichment was developed with acetate and ClO_4^- using primary digester solids from a wastewater treatment plant. ClO_4^- was removed to non-detectable levels from 20 mg/L to < 4 μg/L over 35 days, after a 10-day start up period, at HRTs ranging 18 to 51 min. The same sand column system was also used to examine ClO_4^- degradation rates when the reactor was inoculated with a pure culture of *Azospira* sp. KJ versus that obtained using a ClO_4^--degrading enrichment (Kim and Logan, 2001). The reactor inoculated with the pure culture removed ClO_4^- from 20 mg/L to less than 4 μg/L at a much shorter HRT of ~ 1 min than the mixed culture reactor (12 min). Acetate was used as the substrate for the culture, and the molar ratio (6.6, n = 156) of acetate-to-ClO_4^- for the pure culture was more than twice the ratio (2.9, n=6) of the mixed culture.

A fixed-bed bioreactor using granular-activated carbon (GAC) didn't perform as well as the same reactor filled with sand media (Kim and Logan, 2000). The GAC media (12 × 40 mesh-sieved) had a bulk density of $0.44 g/cm^3$ (NORIT Americas Inc., Atlanta, GA), while the sand had an average size of 0.425 mm (density of $2.67 g/cm^3$). Both reactors were fed with the same ClO_4^--degrading enrichment. The GAC-packed reactor completely removed ClO_4^- for the first 4 days of operation, but after backwashing the media on day 6, the reactor performed erratically with complete ClO_4^- removal on some days and detectable ClO_4^- concentrations on others. The poor operation of the GAC reactor was attributed to desorption of ClO_4^- from the GAC in the lower parts of the reactor where biological activity was insufficient to degrade the ClO_4^-. Based on these results, the researchers recommended that

GAC would not be suitable for fixed bed bioreactors that would need backwashing to control biofilm formation and prevent clogging.

A slightly different GAC system was examined by Brown *et al.* (2000) for ClO_4^- degradation. Biologically active carbon (BAC) filters were tested for ClO_4^- removal at low influent levels (Brown *et al.*, 2000). The BAC filters were rendered biologically active by feeding 600 bed volumes (BV) of dechlorinated tap water containing 50 μg/L ClO_4^- and 20 μg/L bromate. Deionized water containing 50-55 μg/L of ClO_4^- was then fed to the column for 5800 empty bed volumes (BV), with ClO_4^- removal occurring mainly by sorption to the carbon. ClO_4^- removal was then stimulated by adding a mixture of acetate, lactate, and pyruvate (2 mg/l as carbon) to the water that also contained 2.5 mg/L of DO and NO_3^-. ClO_4^- removal was found to be a function of the concentration of NO_3^- in the water. ClO_4^- was removed from 52 μg/L to below detection limit as 4 μg/L with an effluent NO_3^- concentration of 0.07 mg/L. However, ClO_4^- was not appreciably removed (from 51 to 50 μg/L) at a higher effluent NO_3^- concentration 2.0 mg/L. The effects of NO_3^- on ClO_4^- reduction are discussed above.

Burns *et al.* (2001) reported that two fixed bed bioreactors placed in series could remove ClO_4^- from an influent concentration in the range of 0.1 to 1 mg/L, to below the detection limit even in the presence of high concentrations of NO_3^- (50 mg/L). Both reactors (70 mm diameter; 460 mm in a height) initially inoculated with primary sludge were fed a synthetic solution containing acetate (250 mg/L) and sodium sulfite (to remove DO in the reactor influent). HRTs of 1 h (0.1 mg/L) and 10 h (1 mg/L) were needed for complete removal of ClO_4^-.

Another bench-scale PBR reactor was examined in a 30-day study for its efficiency at removing low concentrations of ClO_4^- (<1 mg/L) from groundwater (Losi *et al.*, 2002). The plexiglass column reactor (total volume of 3.1 L) was filled with Celite pellets and inoculated with the ClO_4^--reducing bacterium perclace (Herman and Frankenberger, 1999). Groundwater from a ClO_4^--contaminated site in California was pumped (upflow mode) into the reactor along with nutrients (N and P) and acetate (approximately 500 mg/L). Influent ClO_4^- concentrations of about 800 μg/L were degraded to nondetectable levels (<4 μg/L) at residence times as low as 0.3 hr. Acetate concentrations were less than 50 mg/L in the effluent. NO_3^- (20 mg/L NO_3^--N) was also completely removed while SO_4^- reduction was not observed.

2) Heterotrophic fluidized bed reactor (FBR) studies

Fluidized bed reactors (FBRs) have also been examined for ClO_4^- removal in bench scale studies (Greene and Pitre, 1999; Hatzinger *et al.*, 2000). The FBRs contained either sand or GAC with particle sizes of 0.2 to 0.6 mm and 0.9 to 1.4 mm, respectively. The reactors were fed ethanol, methanol, or a mixture of the two alcohols as substrates. Ethanol or the ethanol-methanol mixture produced better performance than just methanol. The FBR packed with GAC showed better performance than the sand reactor, and achieved consistent ClO_4^- removal from 25 mg/L to less than the detection limit of 4 μg/L (Greene and Pitre, 1999; Hatzinger *et al.*, 2000).

A laboratory-scale FBR was used to treat ClO_4^--contaminated groundwater from the Longhorn Army Ammunition Plant (LHAAP) (Togna *et al.*, 2001). The glass column reactor containing GAC as fluidization medium was fed both acetic acid and ethanol as electron donors. Effluent ClO_4^- concentrations were less than detection limit except during a low substrate study conducted to determine the point of treatment failure.

3) Autohydrogenotrophic reactor studies – packed bed reactors

NO_3^- has been degraded in H_2-fed bioreactors for many years (Gayle *et al.*, 1989; Kapoor and Viraraghavan 1997), but only very recently there have been reports of using autotrophic bioreactors for the degradation of ClO_4^-. Miller and Logan (2000) developed an unsaturated-flow PBR for the degradation of ClO_4^- in drinking water. The bioreactor was inoculated with a H_2-oxidizing, autotrophic, and ClO_4^- degrading enrichment developed in a chemostat. The PBR was fed with a gas mixture of H_2 (5%) and carbon dioxide (CO_2) and a feed stream containing only ClO_4^- and inorganic nutrients. The reactor was initially operated for a 10-day start up period using a high concentration of ClO_4^- (50 mg/L) in order to build up the biofilm, and was then switched to a lower level of ClO_4^- (740 ± 110 μg/L) for the next 145 days of operation. During this operational period, ClO_4^- was reduced to 460 ± 80 μg/L at a constant hydraulic loading rate of 0.45 cm/min (1.1-1.3 min detention time).

A slightly modified version of the reactor used by Miller and Logan (2000) reactor was tested to examine the removal of ClO_4^- from a ClO_4^- contaminated groundwater (Logan and LaPoint, 2002). The reactor was also fed a gas mixture of H_2 (5%) and CO_2 as the electron donor and carbon source, respectively. An average of 25 ± 5% of ClO_4^- was removed at a HRT of 1.5 min from a groundwater containing 73 ± 2 μg/L of ClO_4^- and 21 ± 5 mg/L of NO_3^-. It was estimated that a reactor with a detention time of 8 to 42 min (95% C.I.) a length of 0.65 to 3.4 m would be necessary to completely remove ClO_4^- in this type of unsaturated-flow system. Zhang et al. (2002) isolated *Dechloromonas* sp. strain HZ from this PBR which was found to grow autotrophically with CO_2 as the sole carbon source and H_2 oxidation was coupled to the complete reduction of ClO_4^-. Similarly, Shrout et al. (2005) also isolated two strains, *Dechloromonas* sp. JDS5 and JDS6, which were incapable of growth at atmospheric O_2 concentrations.

A 120-mL bioreactor packed with Celite R-635 (Celite Corporation, Lompoc, CA) was used to degrade ClO_4^- with H_2 (Giblin *et al.*, 2000b). An autotrophic consortium of bacteria was fed a pressurized water stream containing H_2 and bicarbonate. ClO_4^- (740 μg/L) was removed to below the detection limit at a 2-hr detention time. ClO_4^- removal was incomplete at shorter detention times, with ClO_4^- concentrations of 100 and 200 μg/L and detention times of 1 and 0.67 h, respectively. It was speculated that the ClO_4^- breakthroughs at the shorter detention times resulted from a non-uniform distribution of biomass, pH variations of the groundwater, and limited H_2 transport to the bacteria.

4) Autohydrogenotrophic reactor studies using MBfR

The study by Nerenberg *et al.* (2002) used a MBfR containing 98 hollow fibers (93-cm long; 280-μm outside diameter) in a PVC pipe shell to remove NO_3^- and ClO_4^- from groundwater. The groundwater contained both ClO_4^- (6-100 μg/l) and NO_3^- (2.5-3 mg/L). It was found that ClO_4^- could be removed to below the detection limit when the NO_3^- concentration in the effluent was less than 30 μg/L in the presence of at least 300 μg/L of H_2 in the bulk solution. The optimal pH was 8, and the accumulation of intermediates was unlikely. Nerenberg and Rittmann (2004) and Chung et al. (2007a) show that ClO_4^- can be reduced simultaneously with other contaminants including selenate, selenite, bromate, dichloromethane, arsenate, chromate, and dibromochloropropane. As mentioned earlier, Ziv-El and Rittmann (2009) showed that the order of reduction was NO_3^- - NO_2^- - ClO_4^-.

Direct Bioreduction of NO_3^- and ClO_4^- – Pilot-Scale Reactor Studies

1) Fixed bed reactor studies

A pilot-scale bioreactor was operated as an *ex-situ* treatment process in order to treat ClO_4^--contaminated groundwater at the Naval Weapons Industrial Reserve Plant in McGregor, TX (Perlmutter *et al.*, 2000). The reactor was constructed of steel (5ft diameter, 18 ft in height) and fed water containing 7 to 20 mg/L of ClO_4^-. It was found that adding only acetate (acetate:ClO_4^- ratio of 5:1) and nutrients (nitrogen and phosphorus) at a flow rate of 20 gpm, the ClO_4^- concentration could be completely removed (<20 μg/L, the detection limit at that site).

Two pilot-scale bioreactors were operated for about 7 months to treat ClO_4^--contaminated groundwater (average 76 μg/L) at Texas Street Well Facility in Redlands, California (Evans *et al.*, 2002; Logan *et al.*, 2001a; Min *et al.*, 2004). The 7-ft tall reactors contained two rectangular beds each 2-ft by 1-ft. One reactor contained sand media (1 mm diameter) and the other one plastic media (3.76 cm diameter). The reactors were inoculated with a pure culture of *Azospira* sp. KJ. Acetic acid (47 ± 9 mg/L for the sand media reactor; 53 ± 10 mg/L for the plastic media reactor) and an ammonium phosphate solution (C:N ratio of 5:1) were added to the groundwater. The groundwater contained NO_3^- (4 mg/L), O_2 (7 mg/L), and SO_4^- (33 mg/L) as potential terminal electron acceptors in addition to ClO_4^-. ClO_4^- was completely and consistently removed to nondetectable levels (4 μg/L) at flow rates of 1 gpm in the plastic media reactor, and 2 gpm in the sand reactor. Backwashing (weekly) was critical to maintain consistent ClO_4^- removal rates in the sand reactor. In the absence of regular backwashing, there was short circuiting in the sand reactor resulting in insufficient groundwater detention time for consistent ClO_4^- removal.

A 6 month pilot study using fixed-bed (FXB) bioreactors was also conducted to evaluate the use of FXB bioreactors and to address some of the concerns associated with the full-scale implementation of such a process (Brown et al, 2005). The pilot data illustrated that ClO_4^--reducing FXB bioreactors can be acclimated using organisms indigenous to the local aquifer; can achieve sustained ClO_4^- removal to below the analytical detection limit using reasonable contact times; can produce biological stable effluent; do not foster the growth of pathogenic bacteria; and are robust with respect to system upsets. Based on this project, the California Department of Health Services conditionally approved FXB bioreactors for treating ClO_4^--contaminated drinking water, making it possible to consider the full-scale design and implementation of FXB biological ClO_4^- treatment in California (Brown et al, 2005).

2) Fluidized bed bioreactor studies

A pilot scale FBR was tested in 1996 at the Aeroject facility in Sacramento, CA, using water containing 7,000 to 8,000 μg/L of ClO_4^- and 1.5 mgN/L of NO_3^- (Catts and McCullough, 1998). There are few details available on this system, but it was reported that effluent concentrations of <400 μg/L were achieved for ClO_4^-, and <50 μgN/L for NO_3^-. Ethanol was added in proportion to ClO_4^- at a molar ratio of 4:1.

In 1998, a second FBR bioreactor was tested at the Baldwin Park site in the San Gabriel Basin in California. The reactor (20 in high, 15 ft long) contained activated carbon (10 x 30 mesh) and was similar to the one used at the Aerojet site, except this reactor was inoculated with sludge from a food processing industry (versus a wastewater treatment plant). The reactor was fed groundwater (30 gpm, 114 L/min) amended with ethanol (40 to 70 mg/L) and nutrients. Effluent ethanol concentrations ranged from <10 mg/L to non-detectable concentrations. ClO_4^- (50 to 100 μg/L) removal was generally greater than 90% when DO was low (near 1 mg/L). At other times, removal varied from 23 to 45%. NO_3^- (5 to 6 mg/L NO_3^--N) removal was generally greater than 99%. The denatured ethanol used for the study contained low concentrations of methanol, methyl isobutyl ketone (MIBK), and isopropyl alcohol (IPA) (Catts and McCullough, 1998). These chemicals can be toxic to humans, and therefore the addition of these chemicals to drinking water sources should be avoided. Nitrosodimethylamine (NDMA) present in the water at 70 to 80 μg/L was not affected by the bioreactor treatment (Xu et al, 2003).

Direct Bioreduction of NO_3^- and ClO_4^- – Full-Scale Reactor Studies

Fixed bed reactors had not yet been applied at full scale for ClO_4^- treatment, although all bench and pilot tests to date suggest that fixed bed reactors can effectively remove ClO_4^- to a non-detectable level. For this technology to be practical at full sale, a reliable post-treatment system is needed to remove the biomass and extra substrate and nutrients released from the bioreactor over long periods while maintaining ClO_4^- levels below applicable standards. Also, at full scale treatment of groundwater, it remains to be seen whether backwashing cycles will work.

Four full-scale FBR systems were in operation for ClO_4^- treatment till 2005 (Hatzinger, 2005). The first full-scale FBR was built at a site in Rancho Cordova, CA (www.envirogen.com/perchlorate.htm; Green and Pitre, 2000; Hatzinger *et al.*, 2000). The system consisted of 4 FBRs each 14-ft in diameter treating 4000 gpm of groundwater. The reactors were filled with GAC, inoculated with a ClO_4^--degrading enrichment, and fed ethanol as the electron donor. Following a start up period of 3 weeks, the effluent concentration of ClO_4^- was removed from 6 to 8 mg/L to <4 μg/L for 8 weeks of operation during periods of excess ethanol addition. Periods of incomplete ClO_4^- removal occurred at lower ethanol doses (Green and Pitre, 2000; Hatzinger *et al.*, 2000). Three additional full-scale FBR systems have been constructed since 1998: the Longhorn Army Ammunition Plant, the McGregor Naval Weapons Industrial Reserve Plant, and the Kerr-McGee production facility. Longhorn and McGregor have single reactors with influent flows of 190 and 570 L/min, respectively. Each has consistently produced effluent water with ClO_4^- levels less than 4 μg/L from 2.3 mg/L for the McGregor system and 25 mg/L for the Longhorn reactor, respectively. The system in Henderson consists of 8 FBRs; 4 with sand media and 4 with GAC as supporting media. These reactors are operated in sequence such that the influent water flows through the four sand-based FBRs followed by the four GAC-based reactors. This system is designed to treat 3800 L/min of groundwater with ClO_4^- levels up to 400 mg/L along with ClO_3^- and NO_3^- in the influent (Hatzinger, 2005).

FBRs were conditionally approved for treatment of ClO_4^- contaminated drinking water by the California Department of Health Service (DHS, 2002). However, the presence of methanol and other impurities in ethanol may prohibit the use of ethanol as a substrate in these bioreactors if they are to be used for drinking water treatment.

Soil Bioremediation Feasibility Tests

Important questions for soil bioremediation are whether the added substrate will be used for ClO_4^- degradation or lost to other biochemical routes (such as methanogenesis), and whether there are sufficient microorganisms in the soil to achieve ClO_4^- degradation using added substrate. The first indication that ClO_4^- bioremediation could be easily accomplished were studies that demonstrated ClO_4^- reduction could be achieved solely by adding substrate and nutrients to wastewater and soil samples. For example, it was shown that by adding acetate to headspace-free BOD bottles 50% of 20 mg/L ClO_4^- was degraded within 4 days, and 74% was removed within 9 days (Kim and Logan, 2000). The cultures obtained after 9-days of acclimation were transferred (5% by volume) to freshly prepared media containing acetate (200 mg/L) and either ClO_4^- (50 mg/L) or ClO_3^- (50 mg/L), or same concentrations of both anions (100 mg/L). The ClO_4^- and chlorate degradation rates were similar (75 and 78%, respectively) within 5 days using the acclimated cultures (Kim and Logan, 2000).

Other researchers have shown that CRB are ubiquitous in soil samples, providing indirect evidence that bioremediation to remove ClO_4^- would not require bioaugmentation. For example, van Ginkel et al. (1995) found CRB in rivers, sediments, soils, and wastewater treatment plants (van Ginkel *et al.*, 1995) and Coates *et al.* (1999) found CRB at concentrations of 2.31×10^3 to 2.4×10^6 cells per gram in pristine and hydrocarbon-contaminated soils, aquatic sediments, paper mill waste sludges, and farm animal waste lagoons.

Direct evidence for the feasibility of ClO_4^- bioremediation with different substrates was provided by Wu *et al.* (2001). They examined ClO_4^- degradation using samples from soils, natural waters, and wastewaters. Complete degradation of 210 mg/L ClO_4^- was achieved within 4 to 7 days and 8 to 29 days for raw wastewater and creek water, respectively, using four different substrates (acetate, lactate, citric acid, or molasses). PRB were non-measurable using a five-tube MPN method in some soil samples, but were present at high concentrations at ClO_4^--contaminated sites. For example, ClO_4^- reduction was not achieved with 10 g of "pristine" soil, but it was observed with 100 g of soil. Microorganisms in 10 g of a ClO_4^--contaminated soil (from a site in Texas) completely removed ClO_4^- from 100 mg/L within only 7 days using polylactate or lactate.

Hunter (2001) demonstrated that injecting sand columns with a slowly degradable substrate (vegetable oil) could remove ClO_4^- without the need for bioaugmentation with ClO_4^--acclimated bacteria. Columns were seeded with bacteria from soil, and ClO_4^- (20 mg/L) was added to water and pumped into the columns at a rate of ~25 mL/day. After 14 days of operation, 0.47 mg of soybean oil was injected onto the treatment columns. While ClO_4^- levels remained high in the control columns (no soybean oil), ClO_4^- in the effluent of the treatment columns decreased by ~99% over a 17 week operation in the columns containing the vegetable oil.

In-Situ Bioremediation Studies

Pilot scale tests for ClO_4^- remediation have shown that ClO_4^- reduction can be achieved by adding various electron donors to create anoxic conditions in the subsurface. An excess of an electron donor can be used for the degradation of ClO_4^-, ClO_3^-, and other pollutants. *In situ* bioremediation field tests of ClO_4^- were conducted at the Naval Weapons Industrial Reserve Plant in McGregor, TX (Perlmutter *et al.*, 2000). Trenches were dug at the site to collect groundwater and direct the flow of this water through mounds of material containing various organic sources such as compost (from an edible mushroom production facility) and cottonseed meal. ClO_4^- in the water was removed from 16-27 mg/L to <100 μg/L (detection limit in this study) after only two weeks of operation. NO_3^- was also reduced from 15 mg/L to non-detectable concentrations.

Cox *et al.* (2001) reported that the *in-situ* biodegradation of ClO_4^- was successfully achieved in a deep aquifer at the Aerojet Superfund site in California. Acetate was added into the aquifer to support ClO_4^- reduction. ClO_4^- concentrations in the groundwater decreased from 12 mg/L to less than the detection limit (4 μg/L) within 15 feet of the electron donor delivery well. In addition to ClO_4^-, trichloroethylene (TCE), another common contaminant of groundwater, was completely dechlorinated to ethane.

Direct bioreduction of NO_3^- and ClO_4^- is a promising method to remove these contaminants from groundwater and other drinking water sources. However, at present, the public fears treatment of drinking water using bacteria since the common belief is that we should keep bacteria out of our drinking water. Furthermore, in heterotrophic bacterial treatment of drinking water, there is the risk of substrate carryover which would cause bacterial growth in the distribution system. Thus, the belief is that the biological treatment of drinking water will require significant downstream treatment. It is the authors' belief and this belief has been a reality for more than 30 years in Europe that bacteria can pre-treat drinking water and conventional treatment technologies can be used to remove bacteria and substrate, all without post-treatment chlorination which has been shown to produce disinfection byproducts in the distribution system. In addition, as explained above, MBfR uses H_2 as the electron donor which is volatile and won't stay in the water and cause bacterial growth. MBfR also uses autotrophic bacteria which inherently have low cell yields, making post treatment not needed or less expensive.

References

Achenbach, L.A., Michaelidou, U., Bruce, R.A., Fryman, J., and Coates, J.D. 2001. *Dechloromonas agitata gen. nov., sp nov and Dechlorosoma suillum gen. nov., sp nov., two novel environmentally dominant (per)chlorate-reducing bacteria and their phylogenetic position.* Int. J. Syst. Evol. Microbiol. 51, 527-533.

Ahn, C.H., Oh, H., Ki, D., Van Ginkel, S.W., Rittmann, B.E., Park, J. 2009. *Bacterial biofilm-community selection during autohydrogenotrophic reduction of nitrate and perchlorate in ion-exchange brine.* Appl Microbiol Biotechnol 81,1169-1177.

Aldridge, L., Gillogly, T., Lehman, G.,Clifford, D., Roberts, D., and Lin, X. 2004. *Treatability of Perchlorate in Ground Water Using Ion Exchange Technology,* AWWA Research Foundation.

Arnold, R.G., DiChristina, T.J. and Hoffmann, M.R. 1986. *Inhibitor studies of dissimilative Fe(III) reduction by Pseudomonas sp. Strain 200 ("Pseudomonas ferrireductans").* Appl. Environ. Microbiol. 52(2), 281-289.

Aslander, A. 1928. *Experiments on the eradication of Canada Thistle, Cirsium arvense, with chlorates and other herbicides*. J. Agric. Res. 36, 915-928.

Attaway, H., and Smith, M. 1993. *Reduction of Perchlorate by an Anaerobic Enrichment Culture.* J. Indust. Microbiol. 12, 408-412.

Bardiya, N. and Bae, J.H. 2004. *Role of Citrobacter amalonaticus and Citrobacter farmeri in dissimilatory perchlorate reduction.* J. Basic Microbiol. 44(2),88 – 97.

Bazire, A., Diab, F., Jebbar, M., and Haras, D. 2006. *Influence of high salinity on biofilm formation and benzoate assimilation by Pseudomonas aeruginosa.* J. Ind. Microbiol. Biotechnol. 34(1), 5-8.

Bender, K.S., Chakraborty, R., Lack, J.G., Coates, J.D., and Achenbach, L.A. 2002. *Isolation and characterization of genes involved in (per)chlorate reduction and the utility of these genes as metabolic probes*. In The 102nd general meeting of American Society for Microbiology, Salt Lake City, Utah, USA.

Brown, J.C., Snoeyink, V.L., Kirisits, M.J. 2002. *Abiotic and biotic perchlorate removal in an activated filter.* J. Am. Wat. Work. Assoc. 94, 70-79.

Brown J.C., Snoeyink V.L., Liang S., Raskin L.M., Chee-Sanford J.C., and Lin R. 2000. *Removal of perchlorate in biologically active carbon adsorption systems*. The AWWA Inorganic Contaminants Workshop, February 27-29, Albuquerque, New Mexico, USA.

Brown, J.C., Anderson, R.D., Min, J.H., Boulos, L. Prasifka, D. Juby, G. J.G. 2005. *Fixed-bed biological treatment of perchlorate-contaminated drinking water.* J. Am. Wat. Work. Assoc. 97, 70-81.

Bruce, R.A., Achenbach, L.A., and Coates, J.D. 1999. *Reduction of (per)chlorate by a novel organism isolated from paper mill waste. Environ.* Microbiol. 1, 319-329.

Bryan, E.H., and Rohlich, G.A. 1954. *Biological Reduction of sodium chlorite as applied to measurement of sewage B.O.D.* Sewage Ind. Wastes. 26, 1315-1324.

Burns, N.L., Schulte, D.D., and Dahab, M.F. 2001. *Dual removal of perchlorate and nitrate from groundwater through biological reduction.* In WQTC 2001 proceedings, American Water Works Association.

Calgon. 2010 Accessed 8-26-10 *http://www.calgoncarbon.com/ion_exchange/documents/PerchlorateWhitePaper_CANV.pdf*

Cang, Y., Roberts, D.J., Clifford, D.A. 2004. *Development of cultures capable of reducing perchlorate and nitrate in high salt solutions.* Wat. Res. 38, 3322-3330.

Catts, J.G. and McCullough, P.E. 1998. Draft final: Phase 1 *Treatment study report perchlorate in groundwater Baldwin Park Operable Unit San Gabriel Basin*. Technical Report. Harding Lawson Associates Engineering and Environmental Services, Irvine, CA.

Celmer, D., Oleszkiewicz, J., Cicek, N., Husain, H. 2006. *Hydrogen limitation-a method for controlling the performance of membrane biofilm reactor for autotrophic denitrification of wastewater.* Wat. Sci. Tech. 54(9),165-172.

Chaudhuri, S.K., O'Connor, S.M., Gustavson, R.L., Achenbach, L.A. Coates, J.D. 2002. *Environmental factors that control microbial perchlorate reduction.* Appl. Environ. Microbiol. 68(9), 4425-4430.

Chung, J., Rittmann, B.E., Wright, W.F., Bowman, R.H. 2007a. *Simultaneous bio-reduction of nitrate, perchlorate, selenate, chromate, arsenate, and dibromochloropropante using a hydrogen-based membrane biofilm reactor.* Biodegradation. 18, 199-209.

Chung, J., Nerenberg, R., Rittmann, B.E. 2007b. *Evaluation for biological reduction of nitrate and perchlorate in brine water using the hydrogen-based membrane biofilm reactor.* J Environ. Eng. 133, 157–164.

Coates, J.D., Michaelidou, U., Bruce, R.A., O'Connor, S.M., Crespi, J.N., and Achenbach, L.A. 1999. *Ubiquity and diversity of dissimilatory (per)chlorate-reducing bacteria.* Appl. Environ. Microbiol. 65, 5234-5241.

Coates, J.D., Michaelidou, U., O'Connor, S.M., Bruce, R.A., and Achenbach, L.A. 2000. *The diverse microbiology of (per)chlorate reduction.* In E.T. Urbansky, ed., Perchlorate in the Environment, New York, Kluwer Academic/Plenum Publishers. pp. 257-270.

Coppola, E.N. and McDonaold, G.R. 2000. *Bio-degradation of amonium perchlorate, nitrate, hydrolysates and other energenic materials.* U.S. patent 6,077,432.

Cox, E.E., McMaster, M.L., Neville, S.L., Bonsack, L.T. 2001. *Successful field demonstration of in situ bioremediation of perchlorate in groundwater.* Abstract for In situ and on-situ bioremediation in the sixth international symposium, San Diego, June 4-7, 2001.

DaSarma, S. and Arora, P. 2001. *Halophiles. Encyclopedia of Life Sciences, 1- 9.*

De Groef, B., Decallonne, B.R., Van der Geyten, S., Darras, V.M., Bouillon, R., 2006. *Perchlorate versus other environmental sodium/iodide symporter inhibitors: potential thyroid-related health effects.* Eur. J. Endrocrinol. 155(1), 17–25.

DHS. 2002. *Biological treatment to remove perchlorate given conditional California DHS approval.* Department of Health Services, Sacramento, California. http://www.safedrinkingwater.com/archive/sdwn051502.htm.

Evans, P., Chu, A., Liao, S., Price, S., Min, B., Logan, B.E. 2002. *Pilot testing of a reactor for perchlorate-contaminated groundwater treatment.* In Proceedings Third International Conference on Remediation of Chlorinated and Recalcitrant Compounds, May 20-23, 2002.

Galinski, E.A. 1995. *Osmoadaptation in bacteria.* Adv. Microbiol. Physiol. 37, 272–328.

Gayle, B.P., Boardman, G.D., Sherrad, J.H., Benoit, R.E. 1989. *Biological denitrification of water.* J. Environ. Eng. 115, 930-943.

Giblin, T., Herman, D., Deshusses, M. A. Frankenberger, W.T. 2000a. *Removal of perchlorate in ground water with a flow-through bioreactor.* J. Environ. Qual. 29, 578-583.

Giblin, T.L., Herman, D.C., Frankenberger, W.T. 2000b. *Removal of perchlorate from ground water by hydrogen-utilizing bacteria.* J. Environ. Qual. 29, 1057-1062.

Giblin, T. and Frankenberger, W.T. 2001. *Perchlorate and nitrate reductase activity in the perchlorate- respiring bacterium perclace.* Microbiol. Res. 156, 311-315.

Giblin, T., Losi, M.E., Hosangadi, V., Frankenberger, W.T. 2002. *Bacterial Perchlorate reduction in simulated reserve osmosis rejectate.* Bioremed. J. 6,105-111.

Gingras, T.M., and Batista, J.R. 2002. *Biological reduction of perchlorate in ion exchange regenerant solutions containing high salinity and ammonium levels*. J. Environ. Monit. 4, 96-101.

Ginsberg, G.L., Hattis, D.B., Zoeller, R.T., Rice D.C. 2007. *Evaluation of the U.S. EPA/OSWER preliminary remediation goal for perchlorate in groundwater: focus on exposure to nursing infants.* Environ. Health Perspect. 115, 361-369.

Greene M.R. and Pitre M.P. 2000. *Treatment of groundwater containing perchlorate using biological fludized bed reactors with GAC or sand media.* In E.T. Urbansky, ed., Perchlorate in the Environment, New York, Kluwer Academic/Plenum Publishers. pp. 241-256.

Gu, B., Brown, G.M., Alexandratos, S.D., Ober, R., Patel, V. 1999. *Selective anion exchange resins for the removal of perchlorate ClO4- from groundwater*. Oak Ridge National Laboratory. Environmental Science Division. Publication No. 4863. Available at http://www.clu-in.org/download/contaminantfocus/perchlorate/ORNL1.pdf. Accessed August 2010.

Gu, B., Ku, Y.K., and Brown, G.M., 2003. *Treatment of perchlorate contaminated grounwater using highly selective, regenerable ion-exchange technology: A pilot-scale demonstration.* Federal Facilities Environmental Journal, Spring. 75-94.

Hackenthal, E., Mannheim, W., Hackenthal, R., Becher, R. 1964. *Die reduktion von perchlorat durch bakterien 1. Untersuchungen an intakten zellen.* Biochem. Parmacol. 14, 195-206.

Hackenthal, E. 1965. *Die reduktion von perchlorat durch bakterien 2. Die identitat der nitratreduktase und des perchlorat reduzierenden enzyms aus B. cereus.* Biochem. Parmacol. 14, 1313-1324.

Hatzinger, P.B., Greene, M.R., Frisch, S., Togna, A.P., Manning, J., Guarini, W.J. 2000. *Biological treatment of perchlorate-contaminated groundwater using fluidized bed reactors.* In The Second International Conference on Remediation of Chlorinated and Recalcitrant Compounds. Monterey, California, USA.

Hatzinger, P.B. 2005. *Perchlorate biodegradation for water treatment*. Environ. Sci. Technol. 39, 239A-247A.

Herman, D.C., and Frankenberger, W.T. 1999. *Bacterial reduction of perchlorate and nitrate in water.* J. Environ. Qual. 28, 1018-1024.

Hiremath, T., Roberts, D. J., Clifford, D.A., Gillogly, T.E., Lehman, S.G. 2006. *Biological Treatment of Perchlorate in Spent ISEP Ion-Exchange Brine*. J. Environ. Eng. Sci. 23, 1009-1016.

Hunter, W.J. 2001. *In situ removal of perchlorate from groundwater.* In The Sixth International Symposium on In Situ and On Situ Bioremediation. San Diego, California, USA.

Interstate Technology Regulatory Cooperation, June 2000. *Technology Overview. Emerging Technologies for Enhanced* In Situ Biodenitrification (EISBD) of Nitrate-Contaminated Ground Water Available from: http://www.itrcweb.org/ Documents/EISBD-1.pdf.

Kapoor, A. and Viraraghavan, T. 1997. *Nitrate removal from drinking water-Review.* J. Environ. Eng. 123, 371-380.

Kengen, S.W.M., Rikken, G.B., Hagen, W.R., Van Ginkel, C.G., Stams, A.J.M. 1999. *Purification and characterization of (per)chlorate reductase from the chlorate-respiring strain GR-1.* J. Bacteriol. 181, 6706-6711.

Kim, K. and Logan, B.E. 2000. *Fixed-bed bioreactor treating perchlorate-contaminated waters.* Environ. Eng. Sci. 17, 257-265.

Kim, K. and Logan, B.E. 2001. *Microbial reduction of perchlorate in pure and mixed culture packed-bed bioreactors.* Wat. Res. 35, 3071-3076.

Kirk, A.B., 2006. *Perchlorate, why it matters.* Analytica. Chimica. Acta. 567(1), 4–12.

Korenkov, V.N., Romanenko, V.I., Kuznetsov, S.I., and Voronnov, J.V. 1976. *Process for purification of industrial waste waters from perchlorates and chlorates.* U.S. patent 3,943,055.

Lee K.C., and Rittmann, B.E., 2002. *Applying a novel autohydrogenotrophic hollow-fiber membrane biofilm reactor for denitrification of drinking water.* Wat. Res. 36,2040–2052.

Lin, X., Roberts, D.J., Hiremath, T., Clifford, D.A., Gillogly, T., Lehman, S.G. 2007. *Divalent cation addition (Ca2+ or Mg2+) stabilizes biological treatment of ion-exchange spent brine*. Environ. Eng. Sci. 24, 725–735.

Lehman, S.G., Badruzzamana, M., Adhama, S., Roberts, D., Clifford, D. 2008 *Perchlorate and nitrate treatment by ion exchange integrated with biological brine treatment.* Wat. Res. 42, 969-976.

Logan, B.E. 1998. *A review of chlorate and perchlorate respiring microorganisms.* Bioremed. J. 2, 69-79.

Logan, B.E. 2001a. *Assessing the outlook for perchlorate remediation.* Environ. Sci. Technol. 35, 482A-487A.

Logan, B.E., Kim, K., and Price, S. 2001b. *Perchlorate degradation in bench- and pilot-scale ex-situ bioreactors*. In A. Leeson, B.M. Peyton, J.L. Means, and V.S. Magar, eds., Bioremediation of Inorganic Compounds. Battelle Press, Columbus, OH. pp. 303-308.

Logan, B.E., Wu, J., Unz, R.F. 2001c. *Biological perchlorate reduction in high-salinity solutions.* Wat. Res. 35, 3034-3038.

Logan, B.E., and LaPoint, D. 2002. *Treatment of perchlorate- and nitrate-contaminated groundwater in an autotrophic, gas phase, packed-bed bioreactor.* Wat. Res. 36, 3647-3653.

Losi, M.E., Giblin, T., Hosangadi, V., Frankenberger, JR., W.T. 2002. *Bioremediation of perchlorate-contaminated groundwater using a packed bed biological reactor.* Bioremed. J. 6, 97-103.

Massachusetts Department of Environmental Protection (MADEP), 2006. *http://www.mass.gov/dep/water/drinking/percinfo.htm#info.*

Malmqvist, A., Welander, T., Gunnarsson, L. 1991a. *Anaerobic growth of microorganisms with chlorate as an electron-acceptor.* Appl. Environ. Microbiol. 57, 2229-2232.

Malmqvist, A., Welander, T., Moore, E., Ternstrom, A., Molin, G., Stenstrom, I.M. 1991b. *Ideonella dechloratans gen. nov., sp. nov., a new bacterium capable of growing anaerobically with chlorate as an electron acceptor.* Syst. Appl. Microbiol. 17,58–64.

Matos, C.T., Velizarov, S., Crespo, J.G., Reis, M.A.M. 2006. *Simultaneous removal of perchlorate and nitrate from drinking water using the ion exchange membrane bioreactor concept.* Wat. Res. 40, 231-240.

McAdam, E.J. and Judd, S.J. 2008 *Biological treatment of ion-exchange brine regenerant for re-use: A review.* Sep. Purif. Technol. 62, 264-272.

McCarty, P.L. 1975. Stoichiometry of biological reactions. Prog. Wat. Technol. 7,157–172.

Michaelidou, U., Achenbach, L.A., and Coates, J.D. 2000. *Isolation and characterisation of two novel (per)chlorate-reducing bacteria from swine waste lagoons*. In E.T. Urbansky, ed., Perchlorate in the Environment, New York, Kluwer Academic/Plenum Publishers. pp. 271-283.

Miller, J.P. and Logan, B.E. 2000. *Sustained perchlorate degradation in an autotrophic, gas-phase, packed-bed bioreactor.* Environ. Sci. Technol. 34, 3018-3022.

Min, B., Evans, P. J., Chu, A.K., Logan, B.E. 2004. *Perchlorate removal in sand and plastic media bioreactors*. Wat. Res. 38(1): 47-60.

Nagata, S., Sasaki, H., Oshima, A., Takeda, S., Hashimoto, Y., Ishida, A. 2005 *Effect of proline and K+ on the stimulation of cellular activities in Escherichia coli K-12 under high salinity.* Biosci. Biotechnol. Biochem. 69(4), 740-746.

National Aeronautics and Space Administration. Jet Propulsion Laboratory. (2006) Per*chlorate (ClO4-) treatment technologies literature review. Operable Unit 1 expanded treatability study*. EPA ID# CA9800013030. Pasadena, CA.

Nerenberg, R. and Rittmann, B.E. 2002. *Perchlorate as a secondary substrate in a denitrifying hollow-fiber membrane biofilm reactor.* Wat. Sci Technol. WS 2,259-265.

Nerenberg, R. and Rittmann, B.E. 2004. *Hydrogen-based, hollow-fber membrane biofilm reactor for reduction of perchlorate and other oxidized contaminants.* Wat. Sci. Technol. 49(11), 233-230.

Nerenberg, R., Kawagoshi, Y., Rittmann, B.E. 2008. *Microbial ecology of a hydrogen-based membrane biofilm reactor reducing perchlorate in the presence of nitrate or oxygen.* Wat. Res. 42,1151-1159.

Okeke, B.C., Giblin, T., and Frankenberger, W.T. (2002). *Reduction of perchlorate and nitrate by salt tolerant bacteria*. Environ. Pollut. 118, 357-363.

Oren, A. 1999a. *Microbiology and biochemistry of hypersaline environments.* CRC Press. N.Y., N.Y.

Oren, A. 1999b. *Bioenergetics aspects of halophilism.* Microbiol. Mol. Bio. Rev. 63(2),334-348.

Patel, A., Zuo, G., Lehman, S.G., Badruzzaman, M., Clifford, D.A., Roberts, D.J. (2008) *Fluidized bed reactor for the biological treatment of ion-exhange brine containing perchlorate and nitrate.* Wat. Res. 42, 4291-4298.

Park, C. and Marchand, E.A. 2006. *Modelling salinity inhibition effects during biodegradation of perchlorate.* J. Appl. Microbiol. 101, 222-233.

Perlmutter, M.W., Britto R., Cowan J.D., Patel M., Jacobs A., Logan B., and Craig M. (2000). *Bioremediation of perchlorate-contaminated groundwater at naval weapons reserve plant McGregor,* Texas. Proc. National Defense Industrial Association Proceedings, 26th Environmental Symposium and Exhibition, Long Beach, California, March 27 - 30, 2000.

Purolite 2010. Accessed 8-16-10. *http://www.purolite.com/RelId/606893/ISvars/default /Perchlorate_Removal.htm*

Quastel, J.H., Stephenson, M., and Whetham, M.D. 1925. *Some reactions of resting bacteria in relation to anaerobic growth.* Biochem. J. 14, 304-317.

Rice, C.W. and Hempfling, W.P. 1978. *Oxygen limited continuous culture and respiratory energy conservation in Escherichia coli.* J. Bacteriol. 134, 115-124.

Rikken, G.B., Kroon, A.G.M., and Van Ginkel, C.G. 1996. *Transformation of (per)chlorate into chloride by a newly isolated bacterium: Reduction and dismutation.* Appl. Microbiol. Biotechnol. 45, 420-426.

Robert, H., Marrec, C.L., Blanco, C., Jebbar, M. 2000. *Glycine betaine, carnitine, choline enhance salinity tolerance and prevent the accumulation of sodium to a level inhibiting growth of Tetragenococcus halophila.* Appl. Environ. Microbiol. 66,509–517.

Sahu, A.K., Conneely, T., Nusslein, K., Ergas, S.J. 2009. *Hydrogenotrophic denitrification and perchlorate reduction in ion exchange brines using membrane biofilm reactors.* Biotechnol. Bioeng. 104(3), 483-491.

Shrout, J. D.and Parkin, G. F. 2002. *Isolation of autotrophic, hydrogen-utilizing, perchlorate-reducing bacteria. Appl. Microbiol.* Biotechnol. 67, 261-268.

Siemens, 2010. Accessed 8-26-10. *http://www.water.siemens.com/en/applications/drinking_water_treatment/perchlorate-*removal/Pages/default.aspx

Song, Y. 2004. *Respiratory pathways used by perchlorate-respiring microorganisms.* The Pennsylvania State University Ph.D. dissertation.

Song, Y., and Logan, B.E. 2004a. *Inhibition of aerobic respiration and dissimilatory perchlorate reduction using cyanide.* FEMS Microbiol. Lett. 239, 229-234.

Song, Y., and Logan, B.E. 2004b. *Effect of O2 exposure on perchlorate reduction by Dechlorosoma sp. KJ.* Wat. Res. 38, 1626-1632.

Srinivasan, R. and Sorial, G.A. (2009) T*reatment of perchlorate in drinking water: A critical review.* Sep. Purif. Technol. 69, 7-21.

Steinberg, L.M., Trimble, J.J., Logan, B.E. 2005. E*nzymes responsible for chlorate reduction by Pseudomonas sp. are different from those used for perchlorate reduction by Azospira sp.* FEMS Microbiol. Lett. 247, 153-159.

Stenklo, K., Thorell, H.D., Bergius, H., Aasa, R., and Nilsson, T. 2001. *Chlorite dismutase from Ideonella dechloratans.* J. Biol. Inorg. Chem. 6, 601-607.

Stepanyuk, V.V., Smirnova, G.F., Klyushnikova, T.M., Kanyuk, N.I., Panchenko, L.P., Nogina, T.M., and Prima, V.I. 1992. *New species of the Acinetobacter Genus-Acinetobacter hermotoleranticus sp.* Nov. Mikrobiologiya 61, 490-500.

Stouthamer, A.H. 1967. *Nitrate reductase in Aerobacter aerogens 1. Isolation and properties of mutants strains blocked in nitrate assimilation and resistant against chlorate.* Arch. Microbiol. 56, 68-75.

Stouthamer, A.H. 1977. *Energetic aspects of the growth of microorganisms, p. 285-315.* In B.A. Haddock and W.A. Hamilton (ed.), Microbial energetics. Cambridge University Press, Cambridge, United Kingdom.

Sutton, P.M. 2006. *Bioreactor configurations for ex-situ treatment of perchlorate: a review.* Wat. Environ. Res. 78(13), 2417-2427.

Ting, D., Howd, R.A., Fan, A.M., Alexeeff, G.V. 2006. *Development of a health-protective drinking water level for perchlorate.* Environ. Health Perspect. 114 (6), 881–886.

Togna A.P., Guarini W.J., Frisch S., Vecchio M.D., CliffMurray J.P., Tolbert D.E. 2001. *Case study of ex-situ biological treatment of perchlorate-contaminated groundwater. The sixth international symposium on* In Situ and On-site Bioremediation, June 4-7, San Diego, CA, USA.

United States Environmental Protection Agency, 2008. *Federal Facilities Restoration and Reuse, Perchlorate Available from: http://www.epa.gov/swerffrr/documents/perchlorate.htm*

Van Ginkel, C.G., Kroon, A.G.M., and Van Vijk, R.J. 1995. *Process for the degradation of chlorite.* U.S. Patent 5,891,339.

Van Ginkel, C.G., Plugge, C.M., and Stroo, C.A. 1995. *Reduction of chlorate with various energy substrates and inocula under anaerobic conditions.* Chemosphere 31, 4057-4066.

Van Ginkel, C.G., Rikken, G.B., Kroon, A.G.M., Kengen, S.W.M. 1996. *Purification and characterization of chlorite dismutase: A novel oxygen-generating enzyme.* Arch. Microbiol. 166, 321-326.

Van Ginkel, S.W., Ahn, C.H., Badruzzaman, M., Roberts, D.J., Lehman S.G., Adham, S.S., Rittmann, B.E. 2008. *Kinetics of nitrate and perchlorate reduction in ion-exchange brine using the membrane biofilm reactor (MBfR).* Wat. Res. 42,4197-4205.

Van Ginkel, S.W., Lamendella, R., Kovacik, W.P., Santo Domingo, J.W., Rittmann, B.E. 2010. *Microbial community structure during nitrate and perchlorate reduction in ion exchange brine using the hydrogen-based membrane biofilm reactor (MBfR).* Biores. Technol. 101(10), 3747-3750.

Venkatesan, A.K., Sharbatmaleki, M., Batista, J.R. 2010. *Bioregeneration of perchlorate-laden gel-type anion-exchange resin in a fluidized bed reactor.* J. Haz. Mat. 177, 730–737.

Ventosa, A., Nieto., Oren, A. 1998. *Biology of moderately halophilic aerobic bacteria.* Microbiol. Mol. Biol. Rev. 62(2), 504-544.

Wallace, W., Ward, T., Breen, A., and Attaway, H. 1996. *Identification of an anaerobic bacterium which reduces perchlorate and chlorate as Wolinella succinogenes.* J. Ind. Microbiol. 16, 68-72.

Wallace, W., Beshear, S., Williams, D., Hospadar, S., and Owens, M. (1998). *Perchlorate reduction by a mixed culture in an up-flow anaerobic fixed bed reactor.* J. Ind. Microbiol. Biotechnol. 20, 126-131.

Wang, C., Lippincott, L., Meng, X. 2008. *Feasibility and kinetics study on the direct bio-regeneration of perchlorate laden anion-exchange resin.* Wat. Res. 42, 4619-4628.

Wang, C., Lippincott, L., Yoon, I.H., Meng, X. 2009. *Modeling, rate-limiting step investigation, and enhancement of the direct bio-regeneration of perchlorate laden anion-exchange resin.* Wat. Res. 43, 127-136.

Weisenberger, S., Schumpe, A. 1996. *Estimation of gas solubilities in salt solutions at temperatures from 273 K to 363 K.* J. Amer. Inst. Chem. Eng. 42(1), 298-300.

Wolterrink, A.F., Jonker, A.B., Kengen, S.W., and Stams, A.J. 2002. *Pseudomonas chloritidismutans sp. nov., a non-denitrifying chlorate-reducing bacterium.* Int. J. Syst. Evol. Microbiol. 52,2183-2190.

Wu, J., Unz, R.F., Zhang, H.S., and Logan, B.E. (2001). *Persistence of perchlorate and the relative numbers of perchlorate- and chlorate-respiring microorganisms in natural waters, soils, and wastewater.* Bioremed. J. 5, 119-130.

Xiao, Y., Roberts, D.J., Zuo, G., Badruzzaman, M., Lehman, S.G. 2010. *Characterization of microbial populations in pilot-scale fluidized-bed reactors treating perchlorate- and nitrate-laden brine.* Wat. Res. 44, 4029-4036.

Xu, J., Trimble, J.J., Logan, B.E. 2002. *Perchlorate reduction and denitrification pathways are separate in perchlorate reducing bacteria.* In The 102nd general meeting of American Society for Microbiology, Salt Lake City, Utah, USA.

Xu, J., Song, Y., Min, B., Steinberg, L., Logan, B.E. 2003. *Microbial degradation of perchlorate: principles and application.* Environ. Eng. Sci. 20(5), 405-422.

Yoshie, S., Ogawa, T., Makino, H., Hirosawa, H., Tsuneda, S., Hirata, A. (2006) *Characteristics of bacteria showing high denitrification activity in saline wastewater.* Lett. Appl. Microbiol. 42, 277–283.

Zhang, H.S., Bruns, M.A., and Logan, B.E. 2002. *Perchlorate reduction by a novel, chemolithoautotrophic, hydrogen-oxidizing bacterium.* Environ. Microbiol. 4, 570-576.

Ziv-El M.C. and Rittmann B.E. 2009. *Systematic evaluation of nitrate and perchlorate bioreduction kinetics in groundwater using a hydrogen-based membrane biofilm reactor.* Wat. Res. 43,173-181.

In: Bioremediation
Editor: Alexander C. Mason
ISBN: 978-1-61122-730-7

Chapter 3

Sustainable Bioremediation, Industrial Ecology, and Public Policies: New Challenges for Chile

A. Godoy-Faúndez[1], L. Reyez-Bozo[2], W. Montecinos-Bustamante[3] and S. Quiroz-Valenzuela[4]

[1]Universidad Andrés Bello, Facultad de Ecología y Recursos Naturales, Santiago, Chile
[2]Universidad Andrés Bello, Escuela de Industrias, Santiago, Chile
[3]Universidad Andrés Bello, Escuela de Obras Civiles, Santiago, Chile
[4]Universidad Santo Tomás, Escuela de Biotecnología, Santiago, Chile

Abstract

In an international context, developed and developing countries must face global challenges such as economic crises and global warming. Latin-America and the Caribbean are highly vulnerable to the effects of global warming and must understand that facing these environmental issues require changes in public policies, green technologies, green jobs, environmental education, and international public affairs. The main challenge of developing countries is to reach a sustainable level of development without compromising said development. Solving conflicts between environment, industry, and society requires multidisciplinary approaches, with a strong focus on the best available technology, science and public policies.

Different approaches based on science and technology could give enough support to decision-makers to solve national and international concerns related to global warming, especially about historic relations among the industry, the environment, and communities. These interactions have been complicated due to environmental impacts, where topics range from exploitation of resources to cultural disturbances. For industries, a tipping point is to reach sustainable production while avoiding increasing the level of pollutants, an area where Bioremediation processes can be exploited as an environmental tool.

The use of innovative bioremediation, based on sustainability concepts, is necessary due to recent changes in waste management regulations around the world, especially in a

Post-Kyoto scenario, which aims to build up a low-carbon economy with sustainable development. These approaches must take biotechnologies and sustainable engineering concepts forward incorporating a vision of industrial ecology, which focuses on a more integrated interaction among industries by practical, sustainable, and economic schemes, instead of treatment at the end of the pipe.

The bioremediation of polluted environments is a major environmentally friendly technology which can be employed by several kinds of industries around the world. The advantages of this technology are its feasibility to be applied to a wide range of environments and contaminants. Its disadvantage is the need of preliminaries and feasibility studies at a laboratory scale, using biotechnological tools, and its escalation to industrial processes. Currently, not much literature on sustainable bioremediation technologies to be applied together with industrial ecology exists. This chapter will relate bioremediation and industrial ecology, provide a review of the state of the art related research, provide successful cases of bioremediation applications in several industries, and show how industrial ecology could complement these approaches.

In particular, two cases of successful application will be described: one about bioremediation by means of composting fuel-contaminated desert-mining soils in the Atacama Desert (Chile) and the other in the use of biosolids as a flotation reagent inside the Chilean mining industry. Future needs for sustainable bioremediation and industrial ecology will be identified, and a discussion about technological management, public policies, and regulatory considerations to support environmental tools will be included. The advantages and disadvantages of industrial ecological tools for environmental management will also be discussed.

I. Introduction

Currently, the concepts of environment, sustainability, or sustainable edvelopment are not part of the Latin-American and Caribbean cultures, partly due to the brief span of their environmental history. An example of this is Chile. Since the 50s, the concept of environment was introduced due to influences of conservationism associated with academic-intellectual national elites. These elites formed an eco-spiritual movement close to a socialist ideology as a natural outcome of end of the 60s and early 70s culture. The military governments established in the 70s and 80s persecuted the thinkers associated with socialist ideologies, resulting in a dispersal of members and ideas leading to the hibernation for more than 10 years of these environmental concepts. Nevertheless, economic reforms in 1980s based in the adoption of the neoliberal conception of market was based on the exploitation of natural resources. These reforms helped boost economic development and led to the awakening of environmental thinkers by creating a new institutional network. New institutions, allowed the establishment of various institutes, research centers, and research as a response to the neoliberal economic model imposed for economic revival. Introducing ecological concepts, their vision relied on indiscriminate exploitation of resources, mainly due to public discussion on profit and environmental issues related to agriculture and mining resources (Vallejos 1994). In the 1990s, new intellectuals emerged associated with NGOs in the promotion of democracy and social justice, rescuing their social influence from original socialist thinkers. This influence meant a turning point in the collective conscience about the environment by introducing a social component instrumentally used according to personal purposes.

The "institutionalization", created by new intellectuals disappointed with the form of socialism, adopted equality issues and developmental concerns, putting an end to the ideological point of view. However, it is recognized that these trends have set the overall development of the new legal institutions which, since the return to democracy and the incorporation of skilled professionals to high public positions and private management, led to the development of environmental laws, like the Law on Environmental Bases No. 19,300, in Chile. This law has set a turning point in the development of national public policy as a model to be followed in Latin-America and the Caribbean, defining environment as a "system consisting of natural and artificial nature of physical, chemical or biological, socio-cultural and their interactions, in permanent modification by human or natural, and conditions governing the existence and development of life in its many manifestations," (Government of Chile, 1994); a definition that includes socio-cultural systems in interaction with the physical, chemical, and biological systems, recognizing this global system as part of an underlying critical natural capital on issues of sustainability. Nevertheless, these legal frameworks and institutions have been not enough to solve environmental conflicts among governments, industry, and civil society.

Today, a Framework Convention on Climate Change, and the importance to resolve environmental concerns, has encouraged the region to think about their role and commitment as developing countries on fighting Global Warming and adaptating to change under new approaches capable of designing protocols for implementing environmental policies. Technical properties of this new approach are determined by the growing relationship between industry and universities to implement improvements in manufacturing processes, minimizing the impacts on the environment, as well as environmental investments promoting energy efficiency and renewable energy use. Moreover, more incentives are necessary to promote and support private initiatives for the solution of environmental conflicts. These initiatives might incorporate approaches that rely on sustainability concepts to solve problems among government and industry. This tool could be an incentive for the promotion of accountability with a focus on environmental issues, paving the scientific basis on which the knowledge of this generation may in turn have an effect (Pohl, 2008) on new public policies based in interdisciplinarity of environmental science and not on ideology (Kwa, 2008). As aforementioned, visions that have prevailed in environmental policy and its historical development have four main streams of thought: Conservationist, Ecological, Environmentalist, and finally the economical vision due to neoliberal socio-economic system (Aldunate, 2001). Independently or merged, these visions cannot by themselves propose a model of sustainable development. A holistic approach based on Sustainability is necessary as a first step, allowing the promotion of new green technologies, starting as remediation treatments applied to major industrial activities to conserve environmental resources.

Road From Intensive Treatments of Waste to Green Remediation and Sustainable Remediation

In response to publicized discoveries of toxic chemicals, considered hazardous waste disposed in landfills and drinking water sources, the industry of remediation born in the late 1970s under governmental pressure by laws that were passed, at a state and national level, supported by environmental regulatory agencies to oversee and sometimes fund the cleanups. In the beginning, the 1960s, the behaviour of industry was to build mono-landfills for the use of new wastes characterized as hazardous with high costs associated them. During the thirty years of its use, this approach was complemented with actions relying on clean production combating pollution by end-of-pipe technology, ie. treatment of wastes and polluting streams involve treating water, air, noise, and solid wastes. These actions involved a wide range of strategies and technological tools like biological, physical, and chemical systems, used for treating water and soils, as well as various composting or disposal methods. Nevertheless, since the 1990s these treatment technologies focused on intensive treatment of waste without excluding environmental impacts to air, water, land, and ecosystems, as well as expenditure associated to energy (Figure 1).

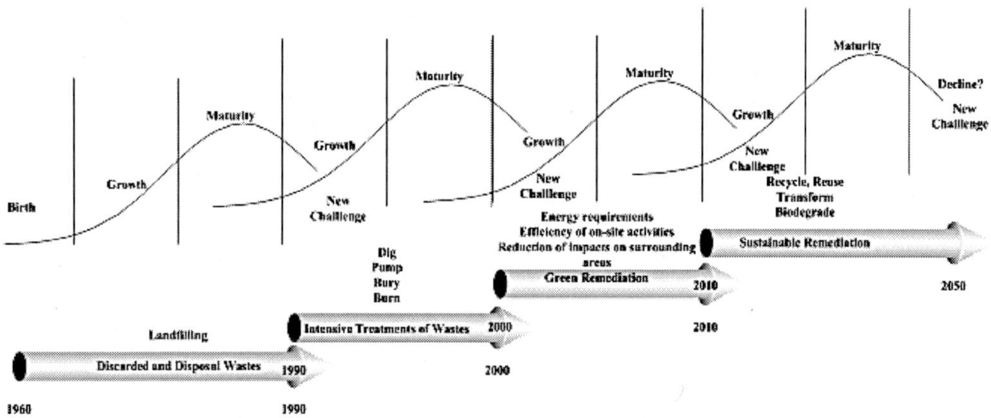

Figure 1. Evolution of the thinking about wastes and cleanups: Transforming our thought process (adapted from SURF 2009).

The intensive treatments of waste have been applied to contaminated soil, sediment, bedrock, and sludge, which may be treated using a wide range of available techniques (Table 1) and through different approaches categorised as in-situ or ex-situ, depending on the location of the treatment process in reference to the location of the contamination (CLU-IN 2007). Ex situ treatments have been considered highly successful due to shorter time requirements for decontamination and a facilitated optimization of operational parameters obtained mainly at lab scale and transferred to pilot scale. Additionally, these treatments allow adding external agents used as enhancers to remediation of the target contaminants i.e bioremediation that under aerobic and anaerobic conditions with the possibilities are used i.e. oxygen, nutrients, and inocula to stimulate autochthonous microbial communities or surfactants to facilitate the bioavailability of the pollutant target. Although there is a greater control on application, these treatments require the containment of contaminants, its transport and final disposal on hazardous waste landfills (HWL) increasing the costs.

On other hand, in situ treatments do not require additional costs for containment and confinement but present a high variability and thus difficulties in achieving a decontamination end point as well as a difficult control process and operational verification. These disadvantages involve large costs in monitoring post applications.

Table 1. Treatment technologies applied to Soil, Sediment, Bedrock, and Sludge

	Relative Overall Cost & Performance							
	Development Status	**Treatment Train**	**Operation and Maintenance Intensive**	**Capital**	**System Reliability & Maintainability**	**Relative Costs**	**Time**	**Availability**
	In Situ Biological Treatment							
Bioventing	AA	AA	AA	AA	AA	AA	A	AA
Enhanced Bioremediation	AA	AA	BA	A	A	AA	A	AA
Phytoremediation	AA	AA	AA	AA	BA	AA	BA	A
	In Situ Physical/Chemical Treatment							
Chemical Oxidation	AA	AA	BA	A	A	A	AA	AA
Electrokinetic Separation	AA	BA	BA	A	A	BA	A	A
Fracturing	AA	A	A	BA	A	A	A	AA
Soil Flushing	AA	AA	BA	A	A	A	A	AA
Soil Vapor Extraction	AA	BA	BA	A	AA	AA	A	AA
Solidification/Stabilization	AA	AA	A	BA	AA	AA	AA	AA
	In Situ Thermal Treatment							
Thermal Treatment	AA	BA	BA	BA	AA	A	AA	AA
	Ex Situ Biological Treatment (assuming excavation)							
Biopiles	AA	AA	AA	AA	AA	AA	A	AA
Composting	AA	AA	AA	AA	AA	AA	A	AA
Landfarming	AA	AA	AA	AA	AA	AA	A	AA
Slurry Phase Biological Treatment	AA	BA	BA	BA	A	A	A	AA
	Ex Situ Physical/Chemical Treatment (assuming excavation)							
Chemical Extraction	AA	BA	BA	BA	A	A	A	AA
Chemical Reduction /Oxidation	AA	A	A	BA	AA	A	AA	AA
Dehalogenation	AA	A	BA	BA	BA	BA	A	A
Separation	AA	A	BA	A	AA	A	AA	AA
Soil Washing	AA	BA	BA	BA	AA	A	AA	AA
Solidification/Stabilization	AA	AA	A	BA	AA	AA	AA	AA
	Ex Situ Thermal Treatment (assuming excavation)							
Hot Gas Decontamination	BA	AA	BA	BA	AA	AA	AA	A
Incineration	AA	AA	BA	BA	A	BA	AA	AA
Open Burn/Open Detonation	AA	AA	BA	BA	AA	AA	AA	AA
Pyrolysis	AA	AA	BA	BA	BA	BA	AA	AA
Thermal Desorption	AA	AA	BA	BA	A	A	AA	AA
	Containment							
Landfill Cap	AA	AA	A	BA	AA	AA	BA	AA
Enhancements/Alternatives	AA	AA	A	BA	AA	AA	BA	AA
	Other Treatment							
Excavation, Retrieval, Off-Site Disposal	AA	AA	AA	AA	AA	?	AA	AA

[1] AA= Above Average; A=Average; BA=Below Average; N/A= Not Applicable; I/D=Insufficient Data, ?=Level of Effectiveness Highly dependent of specific contaminant and its application.

Table 2. Treatment technologies applied to Ground Water, Surface Water, and Leachate

	Relative Overall Cost & Performance							
	Development Status	**Treatment Train**	**Operation and Maintenance Intensive**	**Capital**	**System Reliability & Maintainability**	**Relative Costs**	**Time**	**Availability**
	In Situ Biological Treatment							
Enhanced Bioremediation	AA	AA	BA	A	A	AA	?	AA
Monitored Natural Attenuation	AA	AA	BA	A	A	AA	?	AA
Phytoremediation	AA	AA	AA	AA	BA	AA	BA	A
	In Situ Physical/Chemical Treatment							
Air Sparging	AA	AA	AA	AA	AA	AA	AA	AA
Bioslurping	AA	A	AA	AA	A	AA	A	AA
Chemical Oxidation	AA	AA	BA	A	A	A	AA	AA
Directional Wells (enhancement)	AA	AA	A	BA	A	A	A	AA
Dual Phase Extraction	AA	BA	BA	BA	A	A	A	AA
4.37 Thermal Treatment	AA	BA	BA	BA	A	A	AA	AA
Hydrofracturing Enhancements	AA	A	AA	AA	AA	A	A	AA
In-Well Air Stripping	AA	A	A	BA	A	A	BA	AA
Passive/Reactive Treatment Walls	AA	AA	A	BA	AA	A	BA	AA
	Ex Situ Biological Treatment							
Bioreactors	AA	AA	A	BA	AA	AA	BA	AA
Constructed Wetlands	AA	AA	AA	AA	A	AA	BA	AA
	Ex Situ Physical/Chemical Treatment (assuming pumping)							
Adsorption/ Absorption	AA	AA	BA	A	A	BA	BA	AA
Advanced Oxidation Processes	AA	A	BA	BA	A	A	BA	AA
Air Stripping	AA	A	BA	A	AA	AA	BA	AA
Granulated Activated Carbon/Liquid Phase Carbon Adsorption	AA	A	BA	A	AA	A	BA	AA
Groundwater Pumping/Pump & Treat	AA	A	BA	BA	AA	BA	BA	AA
Ion Exchange	AA	A	BA	BA	AA	A	BA	AA
Precipitation/Coagulation/Flocculation	AA	A	A	BA	AA	A	BA	AA
Separation	AA	A	BA	BA	AA	BA	AA	AA
Sprinkler Irrigation	AA	AA	AA	AA	AA	BA	BA	AA
	Containment							
Physical Barriers	AA	AA	A	BA	AA	AA	BA	AA
Deep Well Injection	AA	AA	AA	AA	A	AA	BA	AA
	Air Emissions/Off-Gas Treatment							
Biofiltration	AA	N/A	AA	AA	?	AA	AA	BA
High Energy Destruction	BA	N/A	I/D	I/D	BA	A	I/D	A
Membrane Separation	BA	N/A	I/D	I/D	BA	A	I/D	A
Oxidation	AA	N/A	AA	AA	AA	AA	I/D	AA
Scrubbers	AA	N/A	A	BA	AA	AA	I/D	AA
Vapor Phase Carbon Adsorption	AA	N/A	AA	AA	AA	AA	I/D	AA

[1] AA= Above Averga; A=Average; BA=Below Average; N/A= Not Applicable; I/D=Insufficient Data, ?=Level of Effectiveness Highly dependent of specific contaminant and its application.

The same wide ranges of techniques are available for contaminated ground water, surface water, and leachate (Table 2), categorised as in-situ or ex-situ, depending on the location of the treatment (CLU-IN 2007). With respect to the treatments, the structure of costs will depend on specific sites conditions. The biological treatments have shown a lower structure of costs associated with remediation treatments compared with physical and chemical treatments that will depend on the use of energy, water, land and ecosystem's resources.

The decision making analysis between several types of technologies includes not only operational cost and time necessary to achieve a friendly clean up of contaminated sites, but must incorporate the initial investment as well as the resources of industry that can be used to apply approaches such as industrial ecology and amortized costs. Furthermore, these structures of cost will be dependent on the type of contaminant or its mixtures with other contaminants present in an environmental accident (i.e. oil spill, extension of contamination, type of source contaminated soil, water, underwater, environmental risk, human exposure, and others). In general, the tendency has been to apply in-situ treatments because they are cheaper, but they present a high uncertainty about the results effectiveness. On the other hand, ex-situ treatments are more expensive, but they offer a better control of the process monitoring (USEPA 2002).

Regarding bioremediation treatments, bio-venting has been widely used as an in-situ bioremediation approach, where aerobic biodegradation is enhanced not only by adding microorganism as inocula but complemented with supplier nutrients water, air or oxygen, and surfactants if required to achieve the removal of the target contaminant or mixture of contaminants (Dejonghe et al. 2001, USEPA 2002). Today, Phytoremediation has been used as a complementing approach to the bioremediation process when contaminated soil, sediment, bedrock, and sludge are present. Phytoremediation , which is defined as the process by which plants remove environmental pullutants by accumulating, degrading or rendering them harmless (Cunningham et al. 1996), is an attractive alternative to avoid expensive disposal in HWL (Macek et al. 2000, Kechavarzi et al. 2007). Biopiles, landfarming, and composting (in open an area or in vessels) have been widely used as ex-situ bioremediation approaches to treat oil spills. Landfarming was the first approach used as an ex-situ treatment. The approach involves soil excavation and relocation in a remote site in thin layers with the support of agriculture machinery that allows plough and irrigates contaminated soils to increase the contact surface between microbial communities, interface soil-water, and pollutants with appropriate moisture content and periodic aeration (Guerin 1999). Biopiles, used particularly in the oil industry (Guerin 2000), involve soil excavation and relocation in remote sites where contaminated soils are aerated by mixing pretreated soils and inocula to enhance contaminant biodegradation.

Finally, bioremediation by composting is currently under investigation due to systematic and efficient applications to contaminated soils especially in-vessel. Composting involves excavated soils mixed with green waste or sludge (green amendment), which provides an appropriate ratio carbon (nitrogen or phosphorous) to balance and stimulate, as well as residential microbial communities in green amendment and contaminated soils (Antizar-Ladislao et al. 2004). In oil spills with the presence of commercial products, the rate of biodegradation would depend on biological and physicochemical factors, present in site (Antizar-Ladislao et al. 2005, 2006a) and on several physicochemical properties related to contaminant (i.e. hydrophobicity, volatility, polarity), solid matrix (i.e. organic matter, porosity) (Mader et. al 1997), and bioavailability (Antizar-Ladislao et al. 2006b, Semple et al. 2003). Indeed, fuel oils spills such as real consecutive accidental spills will depend on the nature and amount of contaminants (aliphatic and aromatic hydrocarbon mixtures) present (Mrayyan and Battikhi 2005).

Until only recently, environmental issues related to impacts and the use of resources have moved the remediation technologies including bioremediation to an environmental-friendly application known as green remediation. Green Remediation is improving and protecting human health, and the environment is looking to maximize the net environmental benefit, reducing or eliminating the risks and deleterious impacts of contaminated sites. These cleanups may be through the consideration of remedy energy requirements, efficiency of on-site activities, and reduction of impacts on surrounding areas.

The U.S. Environmental Protection Agency, through its website Clean-Up Information (CLU-IN 2010), have enounced the five core elements of Best Management Practice (BMP).

1. Energy requirements of the treatment system: "BMPs focus on opportunities to improve energy efficiency of onsite equipment and buildings, meet electricity demands of treatment systems through renewable energy instead of fossil-fuel based energy, and reduce consumption of petroleum-based fuels for routine vehicles and heavy machinery".
2. Air emissions: "BMPs focus on opportunities to further reduce emission of greenhouse gas and criteria pollutants (ozone, particulate matter, carbon monoxide, nitrogen dioxide, sulfur dioxide, and lead)".
3. Water requirements and impacts on water resources: "BMPs focus on opportunities to reduce quantities of water consumed (intensity) during cleanup, re-use treated water, store or divert stormwater for beneficial use, and preserve natural hydraulic conditions".
4. Land and ecosystem impacts: "BMPs focus on opportunities to preserve natural land features, maintain open space, sequester carbon, enhance biodiversity, increase wildlife habitat, and minimize surface and subsurface disturbance".
5. Material consumption and waste generation: "BMPs focus on opportunities to reduce waste generation, recycle spent products, reuse materials, salvage items for donation or resale, beneficially use industrial byproducts, and purchase environmentally preferred products".

In relation to Bioremediation treatments, the BMPs are focused on achieving an efficiency in energy and natural resource consumption through optimizing the initial design of a bioremediation system at lab scale prior to being implemented to pilot scale. An early bench-scale treatability and feasibility studies testing soil, sediment, bedrock, and sludge samples collected from the target treatment area will allow us to determine:

- The characteristics of contaminants onsite as well as its by- products of degradation, other metabolic products, and autochthonous microbial communities
- Potential delivery methods and dispersion characteristics of pollutants, including the use of biodegradable or biological surfactants
- Any need for supplemental technologies to destroy contaminants in hot spots or areas anticipated to involve lengthy periods of microbial acclimation.
- Possibilities of biodegradation mechanisms as well as potential microbial cultures or enhanced autochthonous microbial communities, chemical substrates associated to aerobic or anaerobic conditions and amendments (most suitable reagents and optimal

concentrations or proportions). Indeed, natural resource efficiencies also are gained by conducting an onsite pilot test that evaluates methods for delivering the selected substrate or amendment.

In recent years, independently of any stream that will probably be a series of equally acceptable treatment options, terms such as Green Remediation and Sustainable Remediation are often used interchangeably. Nevertheless, Sustainable Remediation encompasses a broader approach. These visions about new treatments say, "*environmental option is to minimise treatment needs to minimise end-of-pipe interventions and to maximise the Cleaner Production thinking that goes on upstream to save resources*". Looking at the traditional approaches to managing wastes, the movement has turned from Green Remediation to Sustainable Remediation.

In conclusion, societies are looking for new ways to minimize environmental impacts, even on treatment technologies, so that human activity can become more sustainable and rethinking how behaviour, reliance on technology, and consumption of energy impact the environment, society, and economy. Therefore, sustainable remediation must be defined and maximized by approaches such as:

1. Minimize or eliminate energy consumption or the consumption of other natural resources.
2. Minimize or eliminate natural resource consumption or replacement.
3. Reduce or eliminate dispersion of pollutants as well as releases to the environment such as emissions to the air.
4. Harness or mimic a natural process and promote use of biodegradable reagents.
5. Reuse or recycling of land or otherwise undesirable materials.
6. Encourage the use of remedial technologies that permanently destroy contaminants.

To the next years in a framework of global warming and climate change, sustainable remediation is a new approach and challenge due to current and future practices of an industry related to cleaning up those parts of the environment impacted by poor industrial practices of the past. In addition, sustainability concepts have the potential to minimize the deleterious environmental and human health side effects of remediation to make better decisions and not just end-of-pipe strategies.

ROLE OF INDUSTRIAL ECOLOGY AND SUSTAINABLE CONSTRUCTION IN SUSTAINABLE BIOREMEDIATION

Despite the alternatives to conduit green remediation processes by physical, chemical, and biological methods, the bioremediation could be a real chance to achieve sustainable remediation. Sustainable Bioremediation can use innovative reagents from non-traditional sources i.e. biodegradable or bio surfactants that can significantly reduce consumption of chemical reagents sometimes more recalcitrant than pollutants. Indeed, bioremediation mediated by microbial communities can replace the use of virgin natural resources used by waste products in such approaches as bioremediation by composting, landfarming, or biopiles.

These bioremediation systems can particularly benefit from the use of commercial waste such as municipal biosolids, sawdust, and paper sludge that can be included in recipes for soil amendments or used to enhance activity of indigenous microbial populations. This approach can also be integrated with phytoremediation to encourage contaminant degradation and volatilization while enriching soil for revegetation in significantly disturbed areas such as mining sites. These fluxes of wastes can be explained by the relationships among several industrial activities. These flow networks are known as industrial ecology. Industrial ecology was conceived as a global industrial activity that can be thought of the same way as a biological ecosystem and that in its ideal form would strive toward integration of activities and cyclization of resources (Graedel 1996). Indeed, an emerging field of industrial ecology is Industrial Symbiosis which engages traditionally separate industries in a collective approach to competitive advantage involving physical exchange of materials, energy, water, and/or by-products. A successful application of this concept will depend on collaboration and the synergistic possibilities offered by geographic proximity among industrial activities (Chertow 2000). To achieve a real application of Sustainable Bioremediation, or any sustainable remediation process of wastes, it is necessary to not forget the role of the Sustainable Construction of the facilities that finally support the removal of pollutants or the final site of disposal.

Brownfields are closely related to construction due to old industrial sites that are usually composed by buildings-structures which made possible a productive activity within them while in operation. Considering the latter, it is straightforward to consider that the sustainable remediation of a brownfield should consider aspects and approaches of sustainable construction to be performed in a proper way. Sustainable remediation, according to SURF, is broadly defined as a remedy or combination of remedies whose net benefit on human health and the environment is maximized through the judicious use of limited resources. For MassDEP 2010, the terms sustainable remediation and green remediation are often used interchangeably but sustainable remediation covers a wider approach. Green remediation tries to maximize the net environmental benefit of a cleanup through consideration of remedy energy requirements, efficiency of on-site activities, and the reduction of impacts on surrounding areas. Sustainable Remediation includes these benefits but also brings social and economic benefits into consideration.

Sustainable construction, according to UNEP 2010, is a way for the building industry to move towards achieving sustainable development, taking into account environmental, socio-economic, and cultural issues. Kibert, 2008 indicates that the terms high performance, green, and sustainable construction are often used interchangeably, however, the term sustainable construction most comprehensively addresses the ecological, social, and economical issues of a building in the context of its community.

From the definitions given in the latter paragraphs it is possible to see that both terms, sustainable remediation and sustainable construction, deal with environmental/ecological (usually known as green), economic, and social factors. The environmental factors are usually well established, mainly because they can be evaluated from a technical viewpoint. In the case of economic and social factors, they are more difficult to be evaluated in quantitative terms. The main ecological components considered for both concepts are shown in table 3, where it is possible to see that most of them are common in response to the main global environmental concerns.

Table 3. Ecological elements of green remediation and green construction

Green remediation *	Green construction **
Energy requirements of the treatment system	Siting
Air emissions	Energy efficiency
Water requirements and impacts on water resources	Water Efficiency
Land and ecosystem impacts	Materials Efficiency
Material consumption and waste generation	Occupant Health and Safety
Long-term stewardship actions	Building Operations and Maintenance
	Waste Reduction

Source: * Taken from USEPA 2008; ** Adapted from Kibert 2008, USEPA 2010, LEED 2009.

When going further into the components of each element taken from Kibert 2008, USEPA 2008, 2010, and Leed 2009, and mentioned in table 3, it is possible to establish that many of them are common for both sustainable remediation and construction. The latter due to the fact that buildings/structures are always present in industrial sites and therefore, many strategies applying to sustainable buildings apply also to sustainable remediation.

Thus, the relation between industrial site and buildings comes in two main possible ways:

- Existing building and /or structures as part of an industrial site which should be remediated
- New temporary buildings/structures needed as part of the remediation process of an industrial site

Table 4. Ecological elements of green construction and its application to existing buildings requiring remediation works

Element	Requirement
Siting	Use engineered barriers to reduce air emissions from on-site construction demolition equipment and from trucking waste materials.
	Impose idling restrictions on construction equipment.
	Sequence work to minimize double-handling of materials.
Energy Efficiency	Incorporate renewable energy sources, such as wind or solar, into treatment systems.
	Use energy efficient systems and office equipment in job trailer.
Materials Efficiency	Identify salvage options for materials from existing structures.
	Consider reuse options for existing structures.
	Consider structural reuse of slabs or foundations.
	Crush existing structures to optimize scrap recovery and produce fill materials.
	Grind waste wood and other organics for on-site use.
Waste Reduction	Identify recycling options for waste and debris, such as metal and C&D.

Source: Taken from Illinois 2008.

In both cases, green building strategies can be applied in order to obtain sustainable remediation goals. In the first case, when a brownfield with existing buildings must be remediated, the green building strategies mentioned in Table 4 can be applied, based in the elements of green construction indicated in Table 3.

In the second case, when new buildings/structures are needed for the remediation process, those structures should comply with the elements associated with green construction, mentioned in Table 3, and expanded in Table 5. When existing buildings are present, the strategies mentioned in Table 4 should also be applied.

Table 5. Ecological elements of green construction and its application to new temporary buildings required to perform remediation works

Element	Requirement
Siting	Minimize the footprint of new buildings/structures within the site, optimizing the plan size of the building, and controlling the movement of machines within and outside the site.
	Maintain the natural stormwater flow of the site, controlling the quality and quantity of the stormwater, reducing imperviousness, and managing stormwater runoff.
	Minimize the heat island effect produced by the new buildings/structures.
	Minimize the light pollution produced by the new buildings/structures.
Energy Efficiency	Design building envelope to maximize energy performance.
	Use on-site renewable energy supply.
	When connected to the grid, get energy from providers based on renewable energies.
	Periodical check of building's systems.
	No use of ozone depleting substances for heating, ventilating, air conditioning and refrigeration (HVAC&R) systems.
Water Efficiency	Reduce the potable water consumption and wastewater production using high efficiency fixtures, on-site wastewater treatment, collecting rainwater and, in case is necessary, local vegetation, which need less or no water.
Materials Efficiency	Provide collection and storage place of recyclables for the new buildings' occupants.
	Design for Deconstruction, i.e. when the building finish its function in the place, their parts can be reused somewhere else.
	In the design and construction of the new temporary buildings, use recycled, reused materials, regional, rapidly renewable, and environmentally certified materials.
Occupant Health and Safety	Use of low-emitting materials.
	User's controllability of thermal and lighting systems.
Building Operations and Maintenance	To keep track on the green features of the building.
Waste Reduction	Reduce the amount of waste produce during the construction.
	Provide users with alternatives to minimize their waste production.

Source: Adapted from Kibert 2008, USEPA 2010, LEED 2009.

SUSTAINABLE BIOREMEDIATION AND INDUSTRIAL ECOLOGY INSIDE CHILEAN MINING INDUSTRY

Chile is a South American regional economic leader mainly considered a mining country due to the continuous investments in mining industry. Currently, Chile has become the main producer of copper in the world, representing 40% of our commercial balance and 9% of our Gross National Product (Ministerio de Relaciones Exteriores, 2008) thanks to CODELCO (Copper Corporation, a governmental industry) and continuous private investments inside the mining industry (i.e. BHP-Billiton; Minera Escondida, 2007). Since the early 1900s, diverse environmental impacts due to continuous excavations, dust production, noise emission by heavy machinery, production of acid mine drainage, and accumulation of hazardous wastes (organic and inorganic wastes) have disturbed land and ecosystems through habitat modifications as a consequences of these activities and its type of exploitation between open pit and underground mines . In Chilean history, relationships among the mining industry, society, and the environment have been complex due to continuous environmental impacts arising from abandoned mine sites and to confrontations and cultural disturbances (Vallejos, 1994). Nevertheless, the general perception is that the Chilean population has eventually accepted the mining industry as a very important development engine (Ministerio de Relaciones Exteriores, 2008).

During the last decade, and despite the economic benefits that the mining industry offers to Chile, environmental concerns and their regulations are relevant to the government, which has the mission of designing and implementing environmental policies based on the "environmental impact assessment system," (Government of Chile, 1994) linked to specific regulation on closed and abandoned mine sites, and production of hazardous wastes that already exist. Finally this year, the Chilean Ministry of Environment has been created with the mission of designing protocols to implement environmental policies based in our "environmental law," linked to specific regulations.

Today, a big challenge is design protocols that include the main and common environmental impacts as well as measures to treat contaminated wastes including different phases and matrix as contaminated soils, facilitating an appropriate application of the environmental policies among administrations, industry, and citizens. Therefore, new protocols incorporating new approaches about green technologies must be relying on basic research. Nevertheless, in the Chilean mining industry, the use of basic scientific research is not considered as part of its economic process but within environmental management can support finding a solution promoting the sustainability through new technologies for help in the cleaning by Universities-industry alliances.

In a global context, the mining industry is looking for new strategies and products to reduce environmental impacts and to achieve greener and sustainable processes. Chile, a developing country, must evolve to an industrial society but rely on holistic models that can create fluxes of available energy and material resources required among industries to operate effectively with lower environmental impacts. Approaches such as industrial ecology (IE) and ecological or sustainable engineering (SE) will allow a balance between human activities and the use of ecosystem resources (MEA, 2005) to favor a sustainable development addressed to resolve more complex environmental issues (Tilley, 2003). Therefore, an important challenge as a transition economy in a global framework is to align the main industrial activities such as mining, agriculture, and municipal wastewater treatment with incentives at technical levels related with public affairs.

The next cases show that a bioremediation process by composting can be linked to industrial ecology strategies inside the mining industry to achieve a sustainable bioremediation process.

A. Bioremediation of Desert Soils Contaminated with Fuel Oils

The main wastes of the mining industry are the huge solid deposits such as tailings storage facilities, sterile piles, and lixiviation piles, among others. However, a current environmental concern constitutes the unseen pollution of small spills of oil that occur during repair and maintenance of machinery, as well as accidents. The continuous fuel spills have been absorbed mainly by desert native soils at Atacama Region (Chile) and sawdust as a low-cost locally available sorbent material to control environmental pollution. This results in large amounts of fuel-contaminated materials. These materials have accumulated on HWL over time and need to be cleaned-up. Chilean legislation treats this fuel-contaminated material as hazardous waste and therefore should be contained or treated (Ministerio de Salud, Chile, 2004), including alternatives such as bioremediation, an environmental friendly technique for decontamination.

Bioremediation could be a good approach to be applied as a sustainable remediation process due to it being conducted through the direct action of microbial communities that use such organic compounds as an exogenous source of carbon and energy, converting them to more stable and innocuous forms for the environment at lab scale under aerobic and anaerobic conditions, physical and chemical parameters, concentration, and quality of the contaminant (Dua et al. 2002, Rieger et al. 2002, Iwamoto and Nasu 2001). There are few reports in related literature referring to treatment (biodegradability) of complex mixtures once spills have occurred over pristine soils (Dua et al. 2002). The difficulty of studies on oil spills is due to high rates of uncertainty to determine kinetics of degradation and speeds of remediation as tools of prediction for feasibility of treatment (Mulder et al. 2001) and the final residual concentration post-treatment (Nocentini et al. 2000).

An innovative approach to achieve a sustainable bioremediation could be to use reagents from non-traditional sources that can significantly reduce consumption of natural resources while beneficially using various waste products. Land-based systems and in or ex situ bioreactors can particularly benefit from the use of commercial waste. Different wastes by other industries contains common by-products such as municipal biosolids, wood ash, sawdust, and paper sludge that can be included in recipes for soil amendments to enhance activity of indigenous microbial populations. Bioremediation by composting could be a good approach to be applied to Chilean contaminated desert soils that can also be integrated with phytoremediation to encourage contaminant degradation and volatilization while enriching soil for revegetation in significantly disturbed areas such as mining sites. At present, few studies are found on the applicability of composting as a bioremediation process applied to contaminated desert mining soils at Atacama Desert. These soils typical have low organic matter content, a high mineral content, and a high salinity (Mader et al. 1997). Following the Iraqi invasion of Kuwait in the early 1990s (Balba et al. 1998), bioremediation treatments were implemented to treat more than 20 Mm^3 of oil accumulated in the soils using a bulking agent (i.e. wood chips), NPK fertilizers, and controlling the moisture content for enhancing the biodegradation rate of heavily contaminated desert soils. As current Chilean legislation contemplates bioremediation technologies, such as composting as a valid alternative practice to environmental decontamination (Ministry of Health, 2004), our approach was to determine the feasibility of replicating the Kuwait experience to desert mining soils in Chile implemented at laboratory scale. The composting proves to be effective in biodegrading PAHs (Antizar-Ladislao, 2006; Canet, 2001; Antizar-Ladislao, 2004; Cai, 2007, chlorophenols (Laine, 1997), polychlorinated biphenyls, explosives, and petroleum hydrocarbons at laboratory and/or field-scales (Namkoong et al. 2002, Jorgensen et al. 2000).

In Atacama Desert, desert soils are classified as extremely saline (greater than 10 mmhos cm-1,) with a high content of metals. For the purpose of the present investigation, Minera

Escondida Ltd. provided soil (main contaminant present, diesel fuel, and hydrocarbons in the C10-C19 range) and contaminated sawdust (main contaminant present, fuel oil hydrocarbons in the C4-C13 range) previously kept for two years in an HWL exposed with daily oscillations of temperature between 0°C during the night and above 40°C during the day. Autochthonous microbial communities were assumed to be acclimated to cold and warm temperatures and to the presence of pollutants. 30 laboratory-scale cylindrical aerated composting reactors at five ratios of contaminated soil to contaminated sawdust from sawmills, S:SD 1:0, 3:1, 1:1, 1:3, and 0:1 were operated continuously for 56 day's temperatures, and moisture content was close to 50%. Atmospheric air warmed to 60°C was introduced to each reactor and circulated through an internal perforated piping to ensure sufficient oxygen concentration in the reactors (Figure 1). Sampling was conducted in triplicate after 0, 14, 28, 42, and 56 days and analyzed following standard methods (USEPA, 2007).

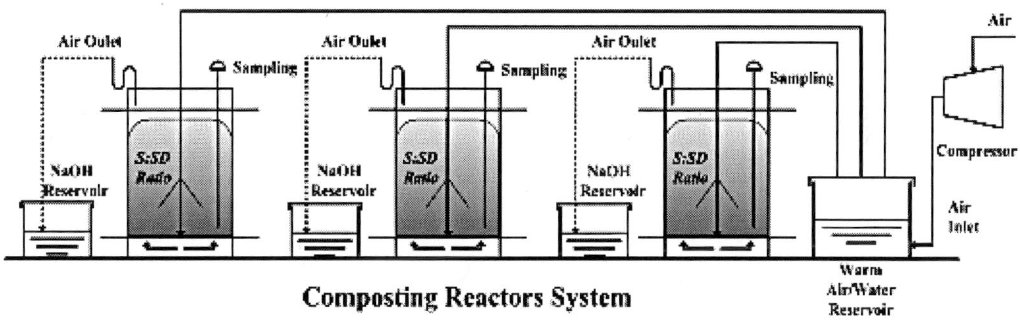

Figure 2. Experiment at laboratory scale (Godoy et al. 2009).

Following 56 days of treatment, different biodegradation curves were obtained in all reactors, with most of the biodegradation occurring during the first 42 days of treatment. Biodegradation curves indicated a similar trend with smooth decreases reaching steady state and contaminant removal between 35% and 60% after 42 days of continuous treatment. Reactors with higher levels of sawdust presented higher removal rates (Figure 3). Higher removal rates, within the range of 70-80% have been previously obtained treating an aged coal-tar contaminated soil amended with green waste under optimal conditions of water moisture and nutrients (Antizar-Ladislao et al., 2006). The results of this research indicate that bioremediation of an aged and weathered contaminated mixture of desert mining soil and sawdust with fuel oil is feasible. This study recommended an S: SD ratio 1:3 and a correct nutrient balance in order to achieve a maximum overall hydrocarbon removal of fuel oil in the weathered and aged contaminated wastes (Antizar-Ladislao et al., 2006; Atlas and Bartha, 1997; Godoy-Faúndez et al., 2008).

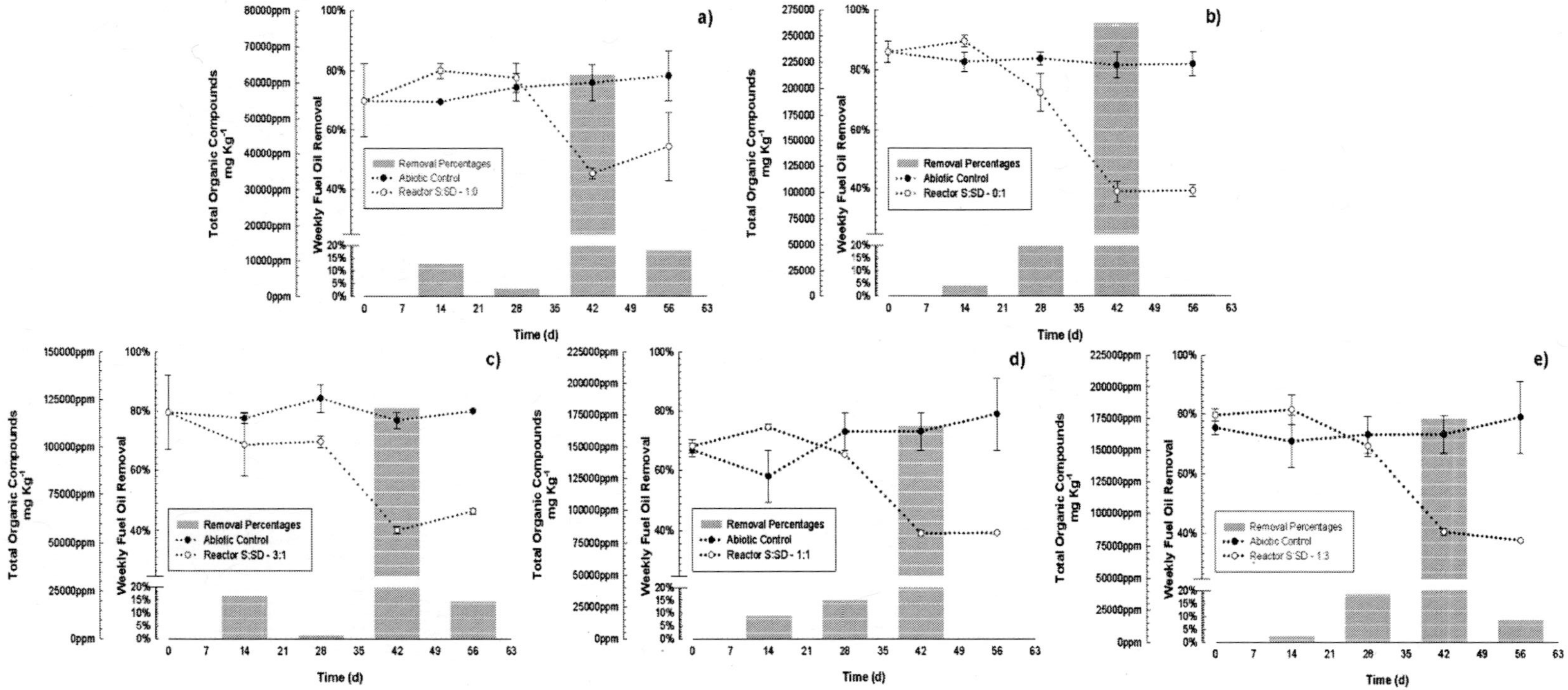

Figure 3. Removal curves and weekly fuel oil removal percentages calculated as the remotion percentage among two consecutive measurements for every ratio Soil: Sawdust (S:SD) by week: a) Reactor with 100% soil; b) Reactor with 100% Sawdust; c) Reactor with Ratio S:SD-3:1; c) Reactor with Ratio S:SD-1:1; c) Reactor with Ratio S:SD-1:3.

Direct extraction and purification of DNA from soils was conducted. Terminal-restriction fragment length polymorphism and Biolog Ecoplate^{tm} tests were conducted to find Operational Taxonomic Units (OUT) through the detection of restriction fragments, richness (S), and diversity expressed as the Shannon-Wiener index (H′). Direct counts and the percentage of viable bacteria indicated that count numbers were high at the beginning of the treatment and that no significant changes were observed during the course of the treatment (over 10^6 cells gr^{-1}) with over 60% of the total number of cells. Cutting-edge biomonitoring tools (molecular and metabolic profiles) were applied to the Atacama desert soil samples, yielding contradictory results , indicating that the changes in the number of phylotypes were not in agreement with the changes in the metabolic activity (Godoy-Faúndez et al., 2009). Metabolic profiles presented correlations with the removal curves, indicating that the microbial activity in composting reactors increased with treatment duration in agreement with the biodegradation curves, and thus higher microbial activities resulted in higher fuel removal rates. It is suggested that the metabolic profiles and not the molecular results can be an important monitoring tool since they provide information on the metabolic activity based on differential consumption of organic sources (Godoy-Faúndez et al., 2009). In conclusion, our results indicate that the bioremediation of an aged contaminated mixture of desert mining soil and sawdust with fuel oil is feasible and that the autochthonous microbial communities adapted to fuel contaminated soils are capable of removing pollutants at a low rate coupling wastes from different industrial activities such as forestry industry.

B. Biosolids as Foaming Reactive for Green Froth Flotation Processes

The most advanced technology used to obtain copper concentrates from sulphide minerals at large-scale operations is froth flotation (FF). Froth flotation uses various chemical reagents to increase performance and selectivity of the flotation process commonly classified as collectors, depressants, activators, modifiers, and froth reagents. During the process, mineralogical species of interest with particle sizes in micrometers are retrieved from a mixture of materials and from an aqueous suspension with the aid of aeration and controlled mixing. After the collection step mainly associated with surface phenomena induced by collector reagents, froth flotation takes place mediated by tensoactive substances that are adsorbed on the air/water interface. Without these reagents, which control the wettability of solid surfaces, the electrochemistry of the solution, the dispersion and aggregation of solid particles, and also the generation and foam stability, it would be impossible to recover the particles containing the mineral species of interest (Herrera-Urbina, 2003). Tensoactive substances, such as methyl isobutyl carbinol (MIBC), are used as froth flotation reagents to increase selectivity of extraction at large-scale operations. However, MIBC is an expensive reagent that poses environmental risks due to potential health effects as well as hazardous environmental pollution.

One of the most important challenges facing the large-scale mineral processing copper industry is keeping operating costs as low as possible and with an acceptable balance between environmental and social impacts associated with their operations (Cox, 2007). Therefore, the mining industry needs new environmental-friendly reagents for FF. A feasible and novel approach to solve this need is the incorporation of concepts from the Industrial Ecology to mining processes such as integrated management of available materials either inside mine plants or derived from other industrial and depurative activities. For example, biosolids (BS), also known as sewage sludge, a byproduct of the treatment of municipal wastewater, are reusable organic materials that may be used as cheap environmentaly-friendly reagents for FF processes as a possible foaming agent (Guellil et al., 2001; Park et al., 2008).

A. Biosolids, A Waste with Properties as Foam Reagent: A Preliminary Assessment

By the year 2013, Chile will reach the treatment of at least 98% of household wastewater according to statistics and projections of the Superintendencia de Servicios Sanitarios (Superintendency of Sanitary Services). The main technology used and implemented at national level, is the activated sludge, representing 51% (SISS, 2007). This is characterized by the generation of large volumes of organic waste called sludge or biosolids (BS). In the metropolitan area of Santiago alone, with six million inhabitants, about 70,000 tons of biosolids are generated annually (SISS, 2007).

BS from wastewater treatment plants from central zone of Chile were obtained after performed activated sludge technology. Samples were physically and chemically analyzed in certified laboratories containing 31.7% C, 4.4% N, and 55.0% OM (dry basis). The pH and electrical conductivity in suspension were 7.5 and 7.8 mS cm^{-1}, respectively with presence of metals and metalloids (Zn, K, Mg, Mn, Na, P, Ni, Se As, Cd, Cr, Hg, Mo, and Pb) at several concentrations. The content of humic substances were 3.0% fulvic acid (FA) and 13.8% humic acids (HA). In our experiments, a representative reagent of humic substances was used as a commercial salt of humic acid (CHA, Sigma-Aldrich^{TM}) that according to Pandey et al. (1999), with to characteristic composition is: 44.67% organic carbon, 5.87% hydrogen, 4.88% total nitrogen, 43.9% oxygen, and 0.58% ash, with a total acidity of 12.3 mol/kg with traces of metals (Al, Ba, Ca, Cu, Fe, Mg, Na, S, Sr).

To determinate the properties of BS to be applied as foam reagent, BS and CHA were analyzed by Fourier Transform Infrared spectrometer (FT-IR) according to literature (CRC Handbook of Chemistry and Physics, 2007-2008; Grube et al., 2003; Senesi et al., 2003; Smidt and Schwanninger, 2005). IR absorbance spectra of BS was assigned to CO stretching of polysaccharides, Si-O silicates, and mineral clay impurities forming complexes with similar IR spectrum similar found in HS (Senesi et al., 2003). About surface tension measurements performed to BS, HS and MIBC solutions at concentrations of 0, 1, 10, 25, 50 and 100 g L^{-1}; 0, 0.1, 1, 5, 10, and 25 g L^{-1} and 0, 0.1, 0.5, 1, 2.5, 5.0, 7.5, and 15.0 g L^{-1} respectively. All reagents tested showed tensoactive activity in the entire range of concentrations measured (Table 6). Corrected concentration by sedimented fractions, BS had a similar behaviour to MIBC; therefore, from an operational standpoint, the use of BS as frother is feasible considering equivalent concentrations to those of MIBC. However, the use of BS should consider a preconditioning stage that permits separating soluble and colloidal fractions that show a tensoactive effect from those sedimentable fractions.

Table 6. Surface tension measurements for pH 7 and 10 by different concentrations of biosolids (BS) and Methyl Isobutyl Carbinol (MIBC). Table shows average values (n>=4) and standard error (S.E)

Concentration* (g/L)	pH 7		pH 10	
	Biosolids	**MIBC**	**Biosolids**	**MIBC**
	γ (mN/m) ± S.E.	γ (mN/m) ± S.E	γ (mN/m) ± S.E	γ (mN/m) ± S.E
0	**72.1** ± 0.10	**72.1** ± 0.10	**72.1** ± 0.10	**72.1** ± 0.10
0.1	**65.8** ± 0.08	**70.8** ± 0.05	**64.1** ± 0.05	**70.5** ± 0.21
0.5	**62.0** ± 0.09	**65.0** ± 0.06	**56.5** ± 0.06	**63.7** ± 0.09
1.0	**54.4** ± 0.11	**59.9** ± 0.08	**52.0** ± 0.08	**59.0** ± 0.18
2.5	**47.4** ± 0.05	**51.0** ± 0.18	**45.9** ± 0.05	**48.8** ± 0.15
5.0	**44.2** ± 0.05	**44.5** ± 0.08	**43.8** ± 0.05	**43.2** ± 0.05
7.5	**42.9** ± 0.08	**40.7** ± 0.05	**42.5** ± 0.07	**39.5** ± 0.05
15.0	**40.5** ± 0.08	**40.7** ± 0.05	**40.2** ± 0.07	**39.5** ± 0.05

* The biosolids concentration is corrected according to the sedimented fraction of these substances.

About frother capability and foam stability by Bikerman (Bikerman, 1973; Dey and Bhattacharya, 1998; Blute et al., 2009) at concentrations of 0.1, 1, 5, and 10 g L^{-1} at pH 10 and 7, was determined by dynamic method as forming foam continuously produced by injection of differential air (Gupta et al., 2007). The static stability was quantified as the time required for that foam produced to come down completely once the gas flow is turned off (Figure 4). Significant amounts of foam were found at different concentrations of MIBC, BS, and CHA (Figure 4). BS and CHA show a higher foam volume at pH 7. All cases show that the volumes of foam have a linear dependence on gas flow as described by Blute et al. (2009). Furthermore, CHA and BS showed a tensoactive activity at different stabilities. According to Spyridopoulos and Simons (2004), the effect of natural organic matter (mainly humic substances) on the stability of a liquid film between two adjacent bubbles, at low concentrations (2 mg L^{-1}) is the coalescence of bubbles reducing or inhibiting with increasing the coalescence's time, with a lower effect for concentrations above 25 mg L^{-1}. Therefore, BS can be used as a feasible frother reagent inside the mining industry through industrial ecology.

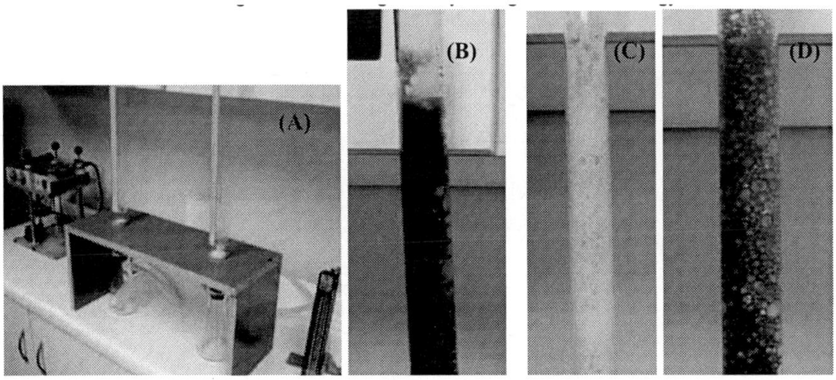

Figure 4. Measurement of Frother Capability by Bikerman method for pH 10 by concentrations of 5 g/L of Commercial Humic Acid (CHA), Biosolids (BS) and Methyl isobutyl carbinol (MIBC) and 3 L/min air flow. (A) Experimental System. (B) Results for CHA. (C) Results for MIBC. (D) Results for BS.

B. Biosolids, A Waste with Properties as Collector for Copper Sulphide Ores: A Preliminary Assessment

FT-IR analysis shows that biosolids have different functional groups that support differential degrees of reactivity to the final structure and allowing its interaction as well as organic and inorganic components. Due to these properties, BS have been nationally and internationally attempted to get economic benefits from wastewater treatment plants considering its physical and chemical properties. Its potential application in agriculture and forestry as improving and recovering degraded soils to rehabilitation of massive mine waste has also been evaluated (U.S. EPA, 1995; Santibañez et al., 2008). Previously to our studies, BS has not yet been considered in the recovery of certain mineralogical species through froth flotation in the minerals industry. The flotation reagents used currently represent a high cost for large-scale mineral processing operations. Therefore, more efficient and economical flotation reagents as collectors and frothers may aid in solving environmental and social issues arising from other sources that would be considerably relevant to the minerals industry, and also would create a great social impact.

In testing Biosolids as a collector of Copper sulphide ore (CSO), the CSO was obtained from División El Soldado, Anglo American Chile. This ore was crushed and freed from chemical collector reagents and industrial frothers. Further, samples of chalcopyrite (Cpy) obtained from División El Teniente, Codelco Chile and pyrite (Py) from a deposit in north-central Chile were ground in a ball mill, homogenized, and sieved in a ROTAP. The elemental chemical analysis of El Soldado copper sulphide ore reports a total content of 7,415 mg kg^{-1} copper and 38,140 mg kg^{-1} iron, while its mineralogical analysis shows the presence of copper sulphide in the form of Cpy (0.7%), chalcocite (0.26%), bornite (0.21%), enargite (0.01%), and covellite (0.01%). The Py content was 0.07%. Other observed sulphides were molybdenite (0.03%) and sphalerite (0.01%). The nonmetallic gangue of the sample was 98.35%.

The elemental chemical analysis of Cpy reports 33.2% copper and 27.6% iron, whereas Py reports 0.2 % copper and 45.1% iron. The contents of copper and iron for Py and Cpy are in according to values reported by Bulatovic (2007). Mineralogical analysis indicates that Cpy contains copper sulphides like chalcopyrite (89.80%) and bornite (8.72%) as well as minor amounts of pyrite (0.64%) and such iron oxides as magnetite (0.22%), limonite (0.16%), and hematite (0.2). The nonmetallic gangue content of the chalcopyrite sample was 0.23%. On the other hand, the purity of the pyrite sample was found to be 98.73%, with minor contents of magnetite (0.21%) and nonmetallic gangue (1.07%).

A first assay to determine the collector properties, the zeta potential (Zeta Meter, Inc. USA) of CSO, Cpy, and Py was determined as a function of pH without adding chemical reagents as well as standard dosage of industrial collector chemical reagents (ICCR) (mercaptan (P3) 11 g ton^{-1}; potassium dialkyl dithiophosphates (Lib-k) 16 g ton^{-1}; sodium isobutyl xanthate 5 g ton^{-1}). In addition, the zeta potential of CSO conditioned with BS was measured at dosages of 5%, 10%, and 25% (w/w, dry basis) and CHA at dosages of 0.7%, 1.5%, and 4.3% (w/w, dry basis). Both dosages are equivalent when they are expressed in grams of humic substances per ton of mineral.

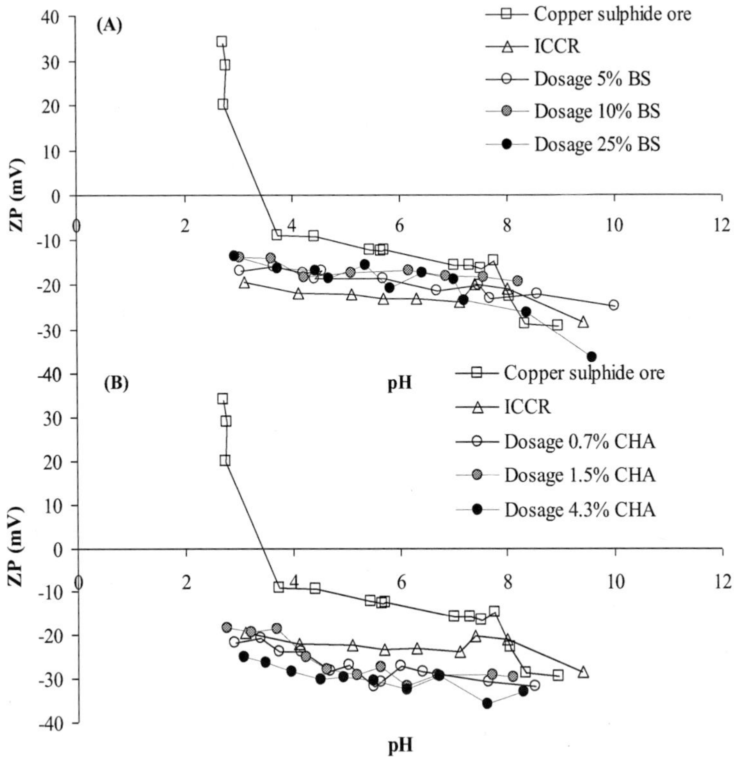

Figure 5. The zeta potential of copper sulphide ore as a function of pH. (A) Without adding chemical reagents, with a standard dosage of industrial chemical collector reagents (ICCR), and with biosolids (BS) dosages of 5%, 10% and 25% (w/w, dry basis)). (B) Without adding chemical reagents, with a standard dosage of ICCR and with commercial humic acid (CHA) at dosages of 0.7%, 1.5% and 4.3% (w/w, dry basis).

Our results shown that the zeta potential of BS was negative over the entire pH range investigated whereas the mixture of ICCR made the zeta potential of CSO more negative in the same range of pH. Therefore, the mixtures of ICCR are more effective in making the zeta potential of copper sulphide ore more negative. Despite of the aforementioned, all BS dosages show an interaction with CSO making its zeta potential more negative (Figure 2A). About CHA, this reagent makes the zeta potential more negative with respect to the mixture of ICCR (Figure 2B). According to Laskowski and Yu (1994), this change could be due to adsorption of soluble species derived from BS and CHA that can change the zeta potential of a mineral making this more negative.

Samples of Cpy and Py were conditioned, at pH 10, with 1.5% (w/w, dry basis), CHA, and potassium dihydrogen orthophosphate (KH_2PO_4; 0.2% as total P, w/w, dry basis) for FT-IR analysis. These dosages are equivalent to 10% BS. The zeta-potential of BS showed an electronegative behaviour inside pH range from 2 to 10. The physical and chemical analyses of BS indicate that humic substances (approximately 17%) and phosphorus compounds (approximately 21,210 mg kg^{-1} of total P) would be the main negative species present. Studies from La Farfana wastewater treatment plant (Santiago, Chile) indicate that over 90% of the phosphorus present in these substances is inorganic orthophosphate (Escudey et al., 2004).

FT-IR results on samples of Cpy and Py conditioned with CHA and orthophosphate show that the main Cpy bands are in the region 1,000 to 1,100 cm^{-1} corresponding to vibrations of copper sulphide (Mustafa et al., 2004). The response to orthophosphate could indicate that orthophosphate is capable of interacting chemically with the mineral surface. Therefore, these substances would only act physically with the surface of the mineral. The results show that CHA can act physically through hydrogen bonds or Van der Waals forces (Alvarez-Puebla and Garrido, 2005). By extending these results to BS, which contains humic substances and also phosphate functional groups, it would be possible to predict physical and chemical interactions of BS with a copper sulphide ore.

Finally, to determine the effect of different dosages of BS, CHA, and ICCR on the flotability of CSO, Cpy, and Py, the film flotation technique described by Fuerstenau et al. (1991) and Diao and Fuerstenau (1991) was used. This technique allows the determination of the hydrophilic and hydrophobic fractions of a mineral species added to different mixtures of water:alcohol at ratios 1:0, 3:1, 1:1, 1:3, and 0:1.

The dosages of ICCR used were: P3 11 g ton^{-1}, Lib-k 16 g ton^{-1}, and xanthate 5 g ton^{-1}. Dosages of biosolids were 5%, 10%, and 25% and CHA at dosages of 0.7%, 1.5%, and 4.3% equivalent when they are expressed in grams of humic substances per ton of mineral at pH 10. The floated mineral was determined by differential weight between total mass of material tested and the mass of sedimented mineral. Surface tension results for water were:ethanol mixtures of 100%, 75%, 50%, 25%, and 0% water, correspond to 71.9, 32.4, 25.2, 22.5, and 16.8 mN m^{-1}, respectively.

Film flotation results show that for the different conditions, the percentage of floated material (CSO, Cpy, and Py) decreased rapidly decreasing surface tension. The statistical analysis shows that the percentage of hydrophobic fraction varies significantly with the type of material (CSO, Cpy, and Py), with the flotation reagent used (without addition of reagents or ICCR) and also with the percentage of water.

Different dosages of CHA (0.7%, 1.5%, and 4.3%) and biosolids (5%, 10%, and 25%) could change the floatability of CSO, Cpy, and Py. The addition of CHA increases the floatability of CSO to 13%, showing a behavior similar to that obtained with ICCR. BS increases the hydrophobicity of CSO particles at different dosages. A dosage of 5% BS yields 26.6% of floated mineral, while the dosages of 10% and 25% of BS yield 36.2% and 35.6%, respectively.

In the case of Cpy film flotation, results indicate that 1.5% CHA does not change its hydrophobicity, whereas 10% biosolids increases the hydrophobicity of Cpy, floating 21% of this sulphide mineral. In contrast, ICCR floats 43% of this mineral being adsorbed on the particles of Cpy increasing their hydrophobicity. To Py, the dosage of 1.5% CHA increases its hydrophobicity to 17%, and the dosage of ICCR and that of 10% BS, yields 40% of Py floated. For Cpy and Py, ICCR can float a similar percentage (approximately 40%) of these minerals, indicating that ICCR do not act selectively. However, BS is more selective towards Py. The multifactorial ANOVA for different dosages of CHA and BS indicates that the percentage of floated minerals varies significantly with the type of mineral (CSO, Cpy, and Py), and the percentage of minerals at different dosages of CHA and BS ($P < 0.05$) varies with the flotation reagent used (CHA or BS) ($P < 0.05$), and also with the percentage of water ($P < 0.05$).

In conclusion, the complexity of the functional groups on the BS surfaces and the various mineral species present in the CSO used in this study (sulphides of copper, iron, molybdenum, and nonmetallic species, among others), make it difficult to predict the behavior of the different interactions between BS and the surface of copper sulphide ores.

However, the results show that the functional groups present on BS and CHA can interact with the surface of CSO, Py, and Cpy, making these mineral surfaces more hydrophobic and an alternative to complementing traditional reagents used inside mining industry.

MODEL OF SUSTAINABLE BIOREMEDIATION FOR CHILEAN MINING INDUSTRY

The use of fluxes of wastes such as sawdust from sawmills of the forestry industry as well as BS from the wastewater industry allows the use of them inside the mining industry to yield greening mining industry and Sustainable Bioremediation. BS are nutrient-rich organic materials resulting from the treatment of domestic sewage in a treatment facility and it has been applied as amended to soils previously. When treated and processed, these residues can be recycled and applied as fertilizer to improve and maintain productive soils and stimulate plant growth. In this study, they can be applied as complementing bioremediation treatments with sawdust as well as used as collector and frother reagents.

Preliminary assessment of these substances in industrial flotation processes would have multiple advantages over the current process (Table 7). Use of biosolids on froth flotation process shows a low cost, innocuous and safe for human health, greater selectivity in the recovery of valuable metals, and tailings can be operated in a safe way.

At laboratory scale, preliminary assay on Denver cell (Figure 3) shows that biosolids can recover a concentrate rich in iron, while humic substances can recover a concentrate rich in copper. Then, biosolids can be used on mineral processing as collector and froth reagents.

Table 7. Comparison between use of biosolids or industrial chemical reagents on froth flotation processes

	Research	Current Process
Production Costs	Low	High
Source of reagents	Natural - organic	Chemistry
Toxicity of reagents (collectors and frothers)	Innocuous for mineral processing and safe for human health and the environment: organic biodegradable	Toxic, flammable and irritating to eyes and respiratory system: persistent organic
Use of reagents in industrial operation	Easy and safe, with low environmental stress for staff	More carefully because of its toxicity, generating job stress (odors)
Operational advantages of reagents	Greater selectivity in the recovery of valuable metals, greater range of applications (sulphide and not-sulphide minerals), pH varying operational.	Low selectivity in the recovery of valuable metals, applicable to sulphide mineral in alkaline pH operational.
Disposition of collectors and foaming reagents downstream flotation plant	Tailings deposited in safe ways. Facilitate environmental rehabilitation processes deposits, an environmentally and socially relevant feature.	In tailings deposit, this would be harmful to living organism (microorganisms and vegetation).

Figure 6. Froth flotation test in Denver cell. (A) Copper sulphide ore + commercial humic acid. (B) Copper sulphide ore + biosolids. (C) Copper sulphide ore + industrial reagents (collectors and frothers).

In Sydney, Australia, different options for biosolid handling were studied by lifecycle assessment to compare its environmental impacts by local dewatering, use as lime amendment, centralized drying system, and thermal drying (Peters and Lundie, 2001). The authors have shown the influence of technology choices on system configurations as the more significant influence on the environmental profile of biosolid processing to be used finally as amendment in agricultural application. This study did not show feasibility for using biosolids in association to other industrial activities, unless minimization of trucking distances, content of solids, and avoidance of coal sourced electrical energy as a sustainable approach are achieved.

Studies done in Stockholm about management of biodegradable waste driven to minimize the environmental impact were based mainly on final disposal and alternative treatments such as incineration and landfilling compared to systems with significant nutrient recycling such as large-scale composting, anaerobic digestion, and separate collection and utilization of urine (Björklund et al., 1999). A main conclusion was that the final use of biosolids will depend on metal contents, a typical problem of wastewater treatment. According to our results, taking advantage of the aforementioned results, biosolids could be used directly from wastewater treatment plants in mining processes as froth flotation.

Indeed, the non-reactive fraction of these materials can be used directly on sustainable management of tail to incorporate processes such as phytoremediation and support bioremediation treatments of desert mining soils contaminated with fuel oils (Figure 7).

Like the chemical reagents used as frothers in the mining industry, HS and BS are capable of changing the surface tension of a solution and can also temporarily generate and stabilize foams. The pH strongly affects the change in the surface tension obtained with the species studied (HS, BS, and MIBC). Considering equivalent concentrations and solution conditions, only BS can obtain a change of surface tension similar to what is obtained with MIBC. At doses below 4 g/L, BS showed to be more surface-active than MIBC, irrespective of pH tested. Furthermore, BS showed Bikerman parameters of the same order of magnitude to those obtained for MIBC in all concentrations and air flows tested.

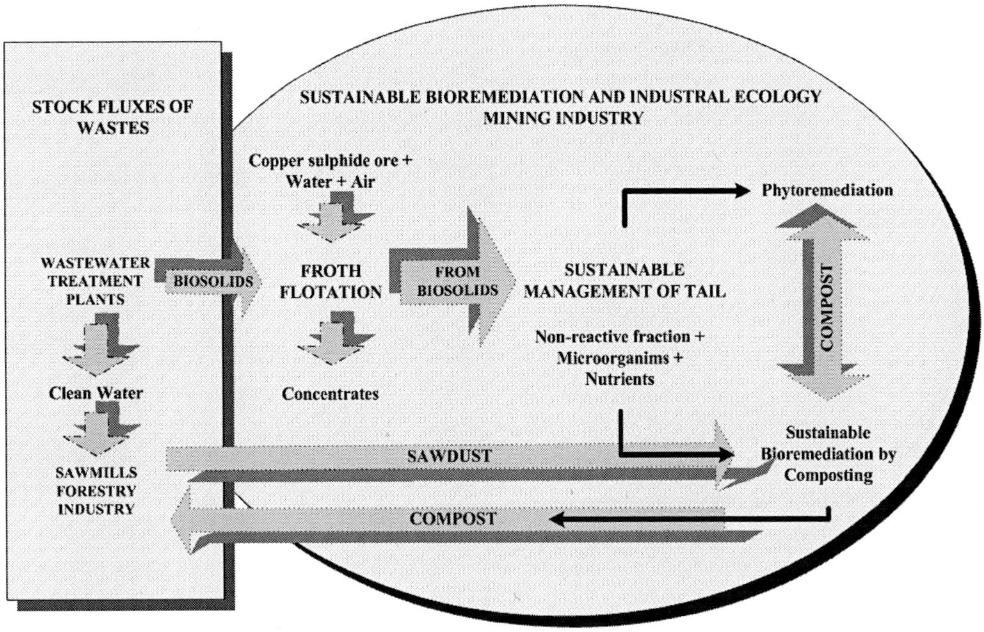

Figure 7. Sustainable model for the mining industry.

Biosolids of different origins have diverse chemical compositions (macronutrients, metals, and metalloids), physical properties and functional groups. FT-IR analysis show that BS present negatively charged functional groups (OH- y –COO-), giving them gelling characteristics and make them difficult to dry. These substances are capable of producing foam and have hydrophobic regions that might be responsible for the tensoactive effect observed.

From an operational standpoint, BS could be used as a foaming agent for FF at equivalent MIBC concentrations. However, for an effective use of these substances as frothers, it is recommended to consider a preconditioning stage, which permits separating soluble and colloidal fractions that show a tensoactive effect from those sedimentable fractions. Considerations such as sustainable engineering based on lifecycle assessment applied biosolid´s processing, energetic efficiency concepts, green buildings concepts applied to design of (bio)processes, and industrial ecology to create connections among industrial activities, i.e., the mining industry, will allow a building of a closed circuit from the cradle to the cradle and not just from the cradle to the grave. Therefore, this study shows the feasibility for using new environmentally-friendly organic reagents (i.e., biosolids) that are considered wastes from the municipal wastewater treatment systems inside the mining industry. Further research is, however, needed in order to scale current laboratory assays to operational mining scales.

The use of BS and CHA as potential collectors for CSO was evaluated through zeta potential measurements, FT-IR analysis, and film flotation experiments. Zeta potential measurements show that BS and CHA can interact (physically and chemically) with the surfaces of a CSO. Such interactions make CSO more electronegative in the entire pH range investigated. The zeta potential of CSO in the presence of CHA and BS is similar to that obtained with a mixture of collector reagents currently used in the mineral processing industry. FT-IR analysis revealed that CHA does not interact chemically with Cpy and Pyrite

Py. For a CSO, it can be concluded that BS may act chemically and physically on the surfaces of sulphide minerals.

Environmental Regulations Considerations: A Brief Overview of Their Applications to Bioremediation

To coupling the Industrial Ecology with Bioremediation as a valid cleaning technology that can be used to treat some waste produced during the normal processes and operations of some industries, it is necessary to consider certain environmental regulations . Obviously, the cleaning process has an associated expense, which is why companies will try to avoid it, unless there is some benefit that outweighs the expenses. The US society decided in 1976 that environmental protection and restoration was an important benefit and thus created a series of regulations to classify contaminated waste, giving the EPA the authority to clean it with charge to whoever produced it (Timian and Connolly 1996). Those rules created a framework to effectively clean contaminated materials, either by the company that produced it or by the EPA. Latin-America can learn many lessons from the US experience, which we discuss here.

An effective regulation must (1) define what is contaminated and what it is not, (2) decide how to handle the contaminated material, and (3) establish responsibilities, incentives, and punishments.

1. Defining Contamination

Usually environmental regulations will define whether something is hazardous and if it needs special handling. Once something has been declared hazardous there are two options: to store it or clean it. The first option is clearly cheaper and simpler, but not always the best. It is unlikely that any company would apply the cleaning treatment without being forced or having been given an incentive, since it has a cost. There is also another option: to change the production system to make sure they will not end up with contaminated material afterward, which can also be expensive. The definition of contaminated material is not trivial, because it marks the point where companies need to assess all the alternatives mentioned before, clearly affecting their procedures.

Bioremediation and other cleaning procedures would not be used unless some material is declared hazardous first, requiring special treatment.

2. Handling Contaminated Material

Once a material has been declared contaminated, a new question arises: what to do with it? Safe storage, dilution, and different procedures are available. Bioremediation can be a suitable alternative if the cost is competitive, results are comparative, and if regulation considers it. The US regulation required "the best technology available" to be used, but in many cases there was no such proof, preventing the use of bioremediation (Timian and Connolly 1996).

The utilization of genetically modified organisms (GMOs) for bioremediation must also be considered. These types of organisms can be safe and allow for the optimizing of the bioremediation process in many cases, because they can adapt microorganisms to particular conditions required by some materials. Nevertheless, GMOs are not yet considered safe, which may result in decontaminated waste considered hazardous by the presence of GMOs utilized in the cleaning process.

3. Accountability

Incentives and punishments are necessary, not only to ensure regulations will be followed, but to allow new and better technologies, such as bioremediation, to arise. Thus, incentives must be focused on development, testing, and standardization of technologies to help companies comply with regulations.

Environmental concerns related to hazardous waste are considered on current Chilean legislation (Ministry of Health 2004). Our existing legislation contemplates the integral management, manipulation, and disposal of hazardous waste, but does not include bioremediation technologies or a coherent or clear guide to implement the existing normative. In the present year, the Ministry of Environment has been created in Chile with the mission of designing protocols to implement environmental policies. Nevertheless, there is also a lack of technical expertise within the field of the application of bioremediation technologies supporting the new challenges to sustainable remediation. The feasibility study presented in this chapter provides evidence that the application of in-vessel composting as a bioremediation technology to treat hazardous wastes is a real treatment alternative to decontaminate wastes as well as create an industrial ecology among different industries. The application of sustainable bioremediation technology is encouraged by economic pressures since the disposal cost of hazardous wastes to relieve the costs related to new environmental friendly reagents as well as by regulatory pressures over the mining industry, as they are enforced to present plans of closing and abandonment of mines.

Final Remarks: Challenges in Environmental Public Policies for Latin-America and Caribbean and Bioremediation Regulation

Citizens and especially the segment with higher education and therefore higher income do not have a proper grasp, or misunderstand the concepts related to sustainability. Today, concepts such as accountability and sustainability are not considered as a competence acquired in their higher education training programs. Academic accountability, defined as incorporating social and academic concepts that allow the promotion of a sustainable science through environmental education, must include values such as respect and care from the community, environmental and ecological integrity, as well as social justice and economy.

Briefly, the approaches toward understanding the environment have been implemented throughout higher education, governmental institutions, and environmental history mainly influenced by ideological tendencies over technical approaches in public affairs (Fields,

1999). The evolution and understanding of the environment has been done by a constant exchanging of ideas through the influence of various expressions of social policies embodied today in a legal framework that defines the concept of environment.

The discussion in Latin-America and the Caribbean has been focused on how to improve the interaction between government, industry, and civil society, through new public-private partnerships. However, international, environmental, and social scenarios related to the Framework Convention on Climate Change, Global Warming, democracy, and social equity present a new challenge about our compromise to reach stringent goals regarding abatement of greenhouse gases without compromising our development as well as achieving sustainable development. The Economic and Social Commission for Latin America and the Caribbean (ECLAC) reported a decrease by 7% in the region's forest cover between 1990 and 2005. Nearly 20% of global greenhouse gas emissions resulting from deforestation and degradation as results of loss of forest cover (7%) related to biodiversity loss. Furthermore, our region has also experienced an increase of nearly 41% of carbon dioxide emissions from the burning of fossil fuels. Therefore, our region needs a new call for adopting a "proactive" approach to stepping up efforts to advance environmental sustainability, especially in light of the new challenges posed by climate change.

The bicentennial for Latin America and the Caribbean is an opportunity to take and implement environmental-public policies as a legacy of quality of life for future generations according to the concepts of sustainable development. These public policies encourage us to create an equilibrium among recovered public spaces as central elements of the construction of cities and neighborhoods, think in a smart grid based on renewable resources, incorporate green technologies, as well as maintaining a sustainable use of our ecosystem services. This set of public policies, independent of a political system, must translate the will to build new habitats with strategic vision that reflect our spirit in the twenty-first century, relying on policies of social equity and economic growth based primarily on the identification of our "natural capital"(Ekins, 2003).

Latin-America and the Caribbean, a region with a variety of environments and landscapes, need to identify primarily its "natural capital" and "ecosystem resources," seen as an environmental function for sustainable development while not compromising its development for the coming years (Jaffe et al. 2005). Such capital and ecosystem resources are not restricted to natural resource exploitation, but rather define itself as a function of value creation on the basis of goods and services that are based on their understanding of natural resources, social, and cultural factors that make Latin-America and the Caribbean a region with an identity distinct to developed countries. As a result of history, relations between industry, environment, and civil society have not been smooth, leaving an apparent lack of environmental regulations, and socio-cultural disruption. The consideration of new models of coexistence, placing human beings in this environment, will come only from the historical analysis of relations towards the future in line with idiosyncrasies incorporating these new challenges towards new public policies.

Finally, the vision of maintaining an environment-society balance based only on governmental regulations or rules of the market will lead to the transfer of market imperfections to public policies (Bretschger, 2005). This is due to the fact that the environment, despite being a valuable public good, open and lacking market price, exploitation of resources, growth, public health, adaptation to change, and abatement of green house gases will produce new internal social conflicts. These conflicts may be resolved through instruments of government, where the trend towards the resolution of these conflicts has been influenced increasingly by technological and scientific evidence on the economic compensation of negative externalities. At this level, sustainability models that promote greening the industrial process and sustainable remediation supported by tools that promote technological innovation will bring responses to the rational use of natural resources to sustain viable economic growth, avoiding their indiscriminate exploitation (Rehfelda et al. 2007).

Therefore, advanced scientific and technological innovations related to environmental matters could support the development of new environmental policies for the resolution of these conflicts. In conclusion, new methods of environmental management and environmental policies based on sustainability concepts are needed. This approach would allow for the meeting of the most stringent environmental and regulatory pressures in a framework towards reaching the goal of sustainable development and a low carbon economy.

In conclusion, it needs a goal that will allow the understanding of the main barriers in technology policy and management, including technological innovation and R&D policy. Barriers must allow green technologies, as sustainable-engineering technologies create the capacity to resolve current and future environmental problems in developing countries such as Latin-America and the Caribbean, but especially focus in Chile, a new OCDE country as model. Each country needs to develop its own model on the policy implications of feasibility to implement and develop new environmental policies based on science and technology. These models could allow us to reach the goal of sustainable development and a low carbon economy by promoting green technologies in our economies characterized by natural resources, climate, infrastructure, engineering needs, social, and political aspects in technical decision making. Our challenge is to develop frameworks of analysis for understanding the causes and effects of current events to contribute to sustainable development and a low carbon economy through better understanding and integration of local systems, characterized by its natural resources, climate, infrastructure needs, engineering, social and political, for the service of decision making techniques to the public policy level, linking science, technology, society, and education. These frameworks of analysis must be flexible to face global climate changes.

References

Aldunate Balestra, Carlos, *El factor ecológico: Las mil caras del pensamiento verde*, Chile: LOM, 2001

Alvarez-Puebla, R.A., Garrido, J.J., Effect of pH on the aggregation of a gray humic acid in colloidal and solid states, *Chemosphere* 59 (2005): 659–667.

Antizar-Ladislao B., Lopez-Real J.M., Beck A.J. (2004) Bioremediation of polycyclic aromatic hydrocarbon (PAH)-contaminated waste using composting approaches, *Crit. Rev. Env. Sci. Tecnol.* 34: 249-289

Antizar-Ladislao B., Lopez-Real J., Beck A.J. (2006a) Degradation of polycyclic aromatic hydrocarbons (PAHs) in an aged coal-tar contaminated soil under in-vessel composting conditions, *Environ. Pollut.* 141: 459-468.

Atlas R. and Bartha R. (1998) *"Ecología microbiana y microbiología ambiental"*. 4ª Edición. Cap.13: Interacciones microbianas con contaminantes xenobióticos e inorgánicos. Cap 14: Ensayos de biodegradabilidad y seguimiento de la bioremediación de contaminantes xenobióticos.

Balba, M.T., R. Al-Daher, N. Al-Awadhi, H. Chino, H. Tsuji. 1998 Bioremediation of oil-contaminated desert soil: The Kuwaiti experience, *Environ. Int.* 24: 163-173.

Bhatt M., Cajthaml T., Sasek V. (2002) *Mycoremediation of PAH-contaminated soils, Folia Microbiol* (Praha) 47: 255-258.

Bikerman, J.J. 1973. *"Foams"*. Springer-Verlag, New York.

Blute, I., Pugh, R.J., Van de Pas, J. and Callaghan, I. 2009. "Industrial manufactured silica nanoparticle sols. 2. Surface tension, particle concentration, foam generation and stability". *Colloids and Surfaces A: Physicochem. Eng. Aspects*. 337: 127-135.

Björklund, A., Bjuggren, C., Dalemo, M., and Sonesson, U. 1999. "Planning biodegradable waste management in Stockholm". *J. Industrial Ecology*, 3(4), 43-58.

Bulatovic, M., *Handbook of flotation reagents: chemistry, theory and practice: flotation of sulphides ores,* Elsevier, (2007), 446 p.

Bretschger, Lucas."Economics of technological change and the natural environment: How effective are innovations as a remedy for resource scarcity?". *Ecological Economics* 2005, 54:148–163

Cai Q.Y., C.H. Mo, Q.T. Wu, and Q.Y.Zeng, A. Katsoyiannis, J.F. Ferard. 2007. Bioremediation of polycyclic aromatic hydrocarbons (PAHs)-contaminated sewage sludge by different composting processes, *J. Hazard. Mat.* 142: 535-542.

Canet R., J.G. Birnstingl, D.G. Malcolm, J.M. Lopez-Real and A.J. Beck. 2001. Biodegradation of polycyclic aromatic hydrocarbons (PAHs) by native microflora and combinations of white-rot fungi in a coal-tar contaminated soil, *Bioresour. Technol.* 76: 113-117

Chertow, Marian 2000. Industrial Symbiosis: Literature and Taxonomy. *Annu. Rev. Energy Environ.* 2000. 25:313–37

CLU-IN. 2007 Technology Innovation Program, United States Environmental Protection Agency. *http://www.clu-in.org/techfocus/*

CLU-IN. 2010. Best Management Practice (BMP) Toolkit - Core Elements: Energy [online]. Available from: *http://www.clu-in.org/greenremediation/index.cfm [accessed 7 July 2010]*

Cox, J. 2007. Mapa tecnológico de la Gran Minería. Cluster Minero de Innova Chile CORFO *"Hacia un programa de Innovación para el Cluster Minero"*.

CRC Handbook of Chemistry and Physics. 2007-2008. Infrared correlation charts, Section 9: *Molecular Structure and Spectroscopy.*

Cunningham S.D., Andersson T.A., Schwab A.P., Hsu F.C. (1996) Phytoremediation of soils contaminated with organic pollutants, *Adv. Agronomy* 56: 55-113.

Dejonghe W., Boon N., Seghers D., Top E., Verstraete W. (2001). Bioaugmentation of soils by increasing microbial richness: missing links, *Environ. Microbiol.* 3: 649-657.

Dey, S. and Bhattacharya, S. 1998. "Comparative Performance Analysis of Frothers: A Review of Available Methodologies". *Miner. Process. Extr. M.* 18(2): 201-213.

Diao, J., Fuerstenau, D.W., Characterization of the wettability of solid particles by film flotation: 2. *Theoretical analysis, Colloid Surface*, 60 (1991): 145-160.

Dua M., Singh A., Sethunathan N. and Johri A. (2002). "Biotechnology and bioremediation: successes and limitations". *Appl Microbiol Biotechnol* 59:143–152.

Ekins, Paul "Identifying critical natural capital Conclusions about critical natural capital". *Ecological Economics*, 2003, 44:277-292

Escudey, M., Galindo, G., Avendaño, K., Borchardt, D., Chang, A., Briceño, M., Distribution of phosphorus forms in Chilean soils and sewage sludge by chemical fractionation and 31P-NMR, *J. Chil. Chem. Soc.* 49(2) (2004): 107-110.

Fields, S. "The price of gold in Chile", *Environmental Health Perspectives*. 1999. 114:A536-539.

Fuerstenau, D.W., Diao, J., Williams, M.C., Characterization of the wettability of solid particles by film flotation: 1. *Experimental investigation, Colloid Surface* 60 (1991): 127-144.

Godoy-Faúndez, A., Antizar-Ladislao, B., Franke-Whittle, I., Turrión-Gómez, J.L. and Perez-Roepke, L. Chapter N°3: Bioremediation of diesel fuel-contaminated soils in the Atacama Desert, Chile. *Book"Leading-Edge Environmental* Biodegradation Research".www.novapublishers.com, binding: Hardcover, Pub. Date: 2008 - 1st Quarter, ISBN: 1-60021-903-9, Status: PP.

Godoy-Faundez, A., Antizar-Ladislao, B., Reyes-Bozo, L., Camaño, A. & Sáez-Navarrete, C. 2008 Bioremediation of contaminated mixtures of desert mining soil and sawdust with fuel oil by aerated in-vessel composting in the Atacama Region (Chile). *J Hazard Mater,* 151, 649-57.

Godoy-Faúndez A., B.Antizar-Ladislao, L. Reyes-Bozo and C.Sez-Navarrete. 2010. *Unseen metabolic plasticity of microbial communities in Desert Soils at Atacama Desert allows successful bioremediation of soils contaminated with high organic load* (Submmitted).

Godoy-Faúndez A., B.Antizar-Ladislao, L. Reyes-Bozo and C.Sez-Navarrete. 2009. Environmental Leadership Model and Public Affairs: Relation Mining Industry-University in Bioremediation of Desert Soils at Atacama Desert. Securing the Future. *Mining and Metal & Environment in a Sustainable Society* and 8th International Conference on Acid Rock Drainage. Skelleftea, Sweden, June 22-29, 2009.

Government of Chile, 1994. *Ministerio Secretaria General de la Presidencia, Ley de Bases del Medioambiente, Chile*: 1994

Graedel, T.L. 1996. On the concepto of industrial ecology. *Annu. Rev. Energy Environ.* 21:69–98

Grube, M., Lin, J.G., Lee, P.H. and Kokorevicha, S. 2006. "Evaluation of sewage sludge-based compost by FT-IR spectroscopy". *Geoderma* 130: 324-333.

Guellil, A., Boualam, M., Quiquampoix, H., Ginestet, P., Audio, J.M. and Block, J.C. 2001. "Hydrolysis of wastewater colloidal organic matter by extra-cellular enzymes extracted from activated sludge flocs". *Water Sci. Technol.* 43(6): 33-40.

Guerin T.F. (1999) Bioremediation of phenols and polycyclic aromatic hydrocarbons in creosote contaminated soil using ex-situ landtreatment, *J. Hazard. Materials*, 1999, 65: 305-315.

Guerin T.F. (2000) The differential removal of aged polycyclic aromatic hydrocarbons from soil during bioremediation, *Environ. Sci. Pollut. Res.,* 2000, 7: 19-26.

Gupta, A.K., Banerjee, P.K., Mishra, A. and Pradip, P.S. 2007. "Effect of alcohol and polyglycol ether frothers on foam stability, bubble size and coal flotation". *Int. J. Miner. Process.* 82: 126–137.

Herrera-Urbina, R. 2003. "Recent developments and advances in formulations and applications of chemical reagents used in froth flotation". *Miner. Process. Extr.* M. 24: 139-182.

Illinois EPA. 2008. Illinois' Greener Cleanups Matrix: How to Maximize the Environmental Benefits of Site Remediation. Illinois Environmental Protection Agency. [online]. Available from: *http://www.epa.state.il.us/land/greener-cleanups/matrix.pdf [accessed 7 July 2010]*

Iwamoto, T. & Nasu, M. (2001) Current bioremediation practice and perspective. *Journal of Bioscience and Bioengineering*, 92, 1-8.

Jaffe, Adama; Newell, Richard; Stavins, Robert, A tale of two market failures: Technology and environmental policy. *Ecological Economics* 2005, 54: 164– 174

Jorgensen, K.S. J. Puustinen, A.M. Suortti. 2000. Bioremediation of petroleum hydrocarbon-contaminated soil by composting in biopiles, *Environ. Pollut.* 107: 245-254.

Kechavarzi C., Pettersson K., Leeds-Harrison P., Ritchie L., Ledin S. (2007) Root establishment of perennial ryegrass (L-perenne) in diesel contaminated subsurface soil layers, *Environ. Pollut.* 145: 68-74.

Kibert, C. 2008. *Sustainable Construction: Green Building Design and Delivery*, 2nd Edition. John Wiley & Sons, Hoboken, NJ, USA.

Kwa, Chunglin "Interdisciplinarity and Postmodernity in the Environmental Sciences". *Environmental Sciences, History and Technology*, 2008, 21,4: 331-344

Laskowski, J.S., Yu, Z., *The effect of humic acids on the emulsion flotation of inherently hydrophobic minerals*. III LatinAmerican Congress on Froth Flotation, Universidad de Concepción, Concepción-Chile, 1994.

LEED. 2009. LEED 2009 for New Construction and Major Renovations. United States Green Building Council. [online]. Available from: *http://www.usgbc.org/ShowFile.aspx?DocumentID=7244 [accessed 7 July 2010]*

Macek T., Mackova M., Kas J. (2000) Exploitation of plants for the removal of organics in environmental remediation, *Biotechnol. Adv.* 18: 23-34.

Mader B.T., UweGoss K., Eisenreich S.J. (1997) Sorption of nonionic, hydrophobic organic chemicals to mineral surfaces, *Environ. Sci. Technol.* 31: 1079-1086.

MassDEP. 2010. Cleanup of Sites & Spills. Massachusetts Department of Environmental Protection. [online]. Available from: *http://www.mass.gov/dep/cleanup/priorities/susrem.htm [accessed 7 July 2010]*

MEA, 2005. Ecosystems and Human Well-Being: Our Human Planet. Summary for Decision Makers. Millennium Ecosystem Assessment. Island Press. Washington D.C.

Ministerio de Relaciones Exteriores (Ministry for Foreign Affairs). 2008. Dirección General de Relaciones Economicas Internacionales y Departamento de Estudios e Informaciones (2008*) Evolución de Saldo de la Balanza Comercial 1999-2007.*, Santiago, República de Chile .

Minera Escondida. 2007. Minera Escondida, certified with ISO 14001. Atacama, Chile. *www.escondida.cl/Escondida/ingles/index.html.*

Ministerio de Salud (Ministry of Health). 2004. *Decreto Supremo No. 148: Reglamento Sanitario Sobre Manejo de Residuos Peligrosos* (Sanitary normative on handling hazardous wastes), Santiago, República de Chile. Santiago, República de Chile, 2004.

Mrayyan B., Battikhi M.N. (2005) Biodegradation of total organic carbons (TOC) in Jordanian petroleum sludge, *J. Hazard. Mater.* 120: 127-134.

Mulder H., Breure A., and Rulkens W. (2001). “Prediction of complete bioremediation periods for PAH soil pollutants in different physical states by mechanistic models”. *Chemosfere* 43:1085-1094.

Mustafa, S., Hamid, A., Naeem, A., Xanthate adsorption studies on chalcopyrite ore, Int. J. Miner. Process. 74 (2004): 317– 325.

Nagaraj, D.R., Brinen, J.S., SIMS study of adsorption of collectors on pyrite. *Int. J. Miner. Process.* 63 (2001): 45–57.

Namkoong, W., E.Y. Hwang, J.S. Park, J.Y. Choi, Bioremediation of diesel-contaminated soil with composting, *Environ. Pollut.* 119 (2002) 23-31.,

Nocentini M., Pinelli D. and Fava F. (2000). Bioremediation of soil contaminated by hydrocarbon mixtures: the residual concentration problem. *Chemosphere*. 41:1115-1123.

Park, C., Novak, J.T., Helm, R.F., Ahn, Y., Esen, A. 2008. “Evaluation of the extracellular proteins in full-scale activated sludges”. *Water Res.* 42: 3879-3889.

Pandey, A.K., Pandey, S.D., Misra, V., Viswanathan, P.N. 1999. “Formation of soluble complexes of metals with humic acid and its environmental significance”. *Chem. Ecol.* 16(4): 269-282.

Peters, G.M. and Lundie, S. 2001. “Life-Cycle Assessment of Biosolids Processing Options”. *J. Industrial Ecology* 5 (2): 103 – 121

Pohl, Christian , “From science to policy through transdisciplinary research”. *Environmental Science & Policy*. 2008, 11: 46–53.

Rehfelda, Katharina-Maria ; Rennings, Klaus ; Ziegler, Andreas, “Integrated product policy and environmental product innovations: An empirical analysis” *Ecological economics,* 2007, 61,: 91–100

Rieger P., Meier H., Gerle M., Vogt U., Groth T. and Knackmuss H. (2002). “Xenobiotics in the environment: present and future strategies to obviate the problem of biological persistence”. *J. Biotechnol.* 94: 101–123.

Santibáñez, C., Verdugo, C., Ginocchio, R., Phytostabilization of copper mine tailings with biosolids: Implications for metal uptake and productivity of Lolium perenne. *Sci. Total Environ.,* 395 (2008):1-10.

Semple K.T., Morriss A.W.J., Paton G.I. (2003) Bioavailability of hydrophobic organic contaminants in soils: fundamental concepts and techniques for analysis, *Eur. J. Soil Sci.* 564: 1-10

Senesi, N., D’Orazio, V. and Ricca, G. 2003. “Humic acids in the first generation of EUROSOILS”. *Geoderma* 116: 325-344.

Smidt, E. and Schwanninger, M. 2005. “Characterization of Waste Materials Using FTIR Spectroscopy: Process Monitoring and Quality Assessment”. *Spectrosc. Lett.* 38(3): 247-270.

Spyridopoulos, M.T. and Simons, S.J.R. 2004. “Particulate interactions in water flotation froths: a micro-mechanistic approach”. *Int. J. Miner. Process.* 74: 115-120.

Sustainable Remediation Forum (SURF). 2009.Sustainable remediation white paper—Integrating sustainable principles, practices, and metrics into remediation projects. *Remediation Journal*. VOl 19, Issue 3, Pp. 1520-6831

Sustainable Remediation Forum (SURF), "Integrating sustainable principles, practices, and metrics into remediation projects", *Remediation Journal*, 19(3), pp 5 - 114, editors P. Hadley and D. Ellis, Summer 2009.

Superintendencia de Servicios Sanitarios (SISS), *Informe anual de coberturas urbanas de servicios sanitarios 2007,* SISS, Santiago, (2007), 19 pp.

Tilley, D.R., 2003. "Industrial ecology and ecological engineering: opportunities for symbiosis". *J. Industrial Ecology* 7(2): 13-32.

Timian, S. J., and D. M. Connolly. (1996). "The regulation and development of bioremediation." *Risk - Health Safety & Environment,*: 279-290 (sole1)

Vallejos, J.P. "Historia y minería en Chile: estudios y fuentes". *América Latina en la Historia Económica,* 1994. 1: 65-88.

USEPA 1995, *Sewage Sludge Incineration* AP-42.

USEPA (2002) U.S. *Environmental Protection Agency, Application, Performance, and costs of biotreatment technologies for contaminated soils* EPA/600/R-03/037, USEPA, Washington, DC.

USEPA 2008. Green Remediation: Incorporating Sustainable Environmental Practices into Remediation of Contaminated Sites. United States Environmental Protection Agency. [online]. Available from: *http://www.clu-in.org/download/remed/green-remediation-primer.pdf* [accessed 7 July 2010]

USEPA. 2010. Principles for Greener Cleanups. United States Environmental Protection Agency. [online]. Available from: *http://www.epa.gov/oswer/greencleanups/principles.html [accessed 7 July 2010]*

USEPA. 2010. Components of Green Building. United States Environmental Protection Agency. [online]. Available from: *http://www.epa.gov/greenbuilding/pubs/components.htm#sustain [accessed 7 July 2010]*

USEPA. 2010b. Green Remediation Best Management Practices: Bioremediation. [online]. Available from: *http://www.clu-in.org/greenremediation/docs/GR_factsheet_biorem_32410.pdf [accessed 7 July 2010]*

UNEP. 2010. Sustainable Building and Construction. United Nations Environmental Program. Division of Technology, Industry and Economy. International Environmental Technology Centre (IETC). [online]. Available from: *http://www.unep.or.jp/ietc/Activities/Urban/sustainable_bldg_const.asp [accessed 7 July 2010]*

Wang, X., Interfacial electrochemistry of pyrite oxidation and flotation II. FTIR studies of xanthate adsorption on pyrite surfaces in neutral pH solutions. *J. Colloid Interf. Sci.*, 171 (1995): 413-428.

In: Bioremediation
Editor: Alexander C. Mason ISBN: 978-1-61122-730-7

Chapter 4

BACTERIA AND THEIR GENES CODING FOR ENZYMES INVOLVED IN BIOREMEDIATION OF PETROLEUM HYDROCARBONS

Diogo Jurelevicius and Lucy Seldin
Departamento de Microbiologia Geral, Instituto de Microbiologia Prof. Paulo de Góes (IMPPG), Centro de Ciências da Saúde (CCS), Universidade Federal do Rio de Janeiro (UFRJ), RJ, Brasil

Petroleum is part of our biosphere for much longer than a thousand years. Petroleum is a complex mixture formed by oxygen-, nitrogen-, and sulfur-containing compounds and hydrocarbons (Planckaert, 2005). Quantitatively and economically, hydrocarbons are the most important compounds of the petroleum mixture (up to 85%). These compounds lack functional groups, since they are consisted exclusively of carbon and hydrogen which are characterized by apolar C-C and C-H bonds. They are distinguished according to the presence of π-bonds and cyclic structures and are classified as n-alkanes, alkenes or alkynes, and alicyclic or aromatic hydrocarbons, respectively (Prince & Atlas, 2005). Hydrocarbon compounds can either have a biological origin, such as waste or metabolic overflow compounds, structural elements, and defense mechanisms, derived from plants, algae, animals, and microbes, or a chemical origin, through the incomplete combustion of organic matter (van Beilen *et al.*, 2003b; Norse, 2005). Actually, hydrocarbons are widespread in low concentrations in the environment; but oil spillages increase their concentration, making them toxic for most living forms (Philp *et al.*, 2005a).

Many microorganisms have evolved or acquired the ability to utilize these highly reduced compounds for biomass and energy production, since they have been coexisting for billions of years in the environment in strict contact with hydrocarbon compounds (Dua *et al.*, 2002; Yakimov *et al.*, 2007). Despite the low chemical reactivity of hydrocarbons and the special biochemical reactions needed for their degradation, hydrocarbon-utilizing microorganisms are found in almost all ecosystems ever studied, including extreme environments, such as Arctic and Antarctic sites (van Hamme *et al.*, 2003; Kuhn *et al.*, 2009).

Numerous hydrocarbon compounds can be used as carbon and energy sources by microorganisms (Lovley & Lloyd, 2000; Leung *et al.*, 2007). Straight and branched chain alkanes, alkenes, alicyclic, heterocyclic, and aromatic hydrocarbons are examples of them (Ratledge, 1984). It is noteworthy that the complete microbial degradation of petroleum hydrocarbons gives rise to non-toxic compounds, such as water and carbon dioxide (CO_2), and cell biomass (mainly proteins) that can securely enter into the food chain (Atlas, 1995a). However, recalcitrant hydrocarbons are difficult to be degraded by microorganisms, probably because of the lack of a solubilizing enzyme prior to the oxidative degradation (Johnsen *et al.*, 2005).

Nearly 200 bacterial, cyanobacterial, algal, and fungal genera, representing more than 500 species and strains, have already been described as presenting the ability to use at least one kind of hydrocarbon (Prince, 2005). Understanding the catabolic pathways and the enzymes responsible for hydrocarbon degradation is an effective means to define important factors for the efficient clean-up of pollutants. It is widely known that hydrocarbon-utilizing microorganisms share at least three distinguishing inherent characteristics: (i) the presence of an efficient system of hydrocarbon uptake, (ii) the presence of a group-specific oxygenase, so as to activate the almost inert hydrocarbons, and (iii) an inducer which promotes a positive response of the organism to hydrocarbon [Rosenberg & Gutnick, 1977 (apud Philp *et al.*, 2005b)].

Hydrocarbon Degradation in the Domain Bacteria

Almost seventy aerobic bacterial genera have already been described harboring species which possess the ability to grow in expense of at least one kind of hydrocarbon. These include Gram-negative bacteria belonging to Alpha, Beta, and Gammaproteobacteria, Bacteroidetes, Gram-positive spore-forming bacteria, and high G+C content Gram-positive strains (Prince, 2005). These bacteria are in general metabolically versatile and utilize a large range of organic substrates for growth (Nazina *et al.*, 2000; Larkin *et al.*, 2005; Moore *et al.*, 2006; Yakimov *et al.*, 2007). They seem to be more adapted to oligotrophic environments, having developed uptake systems to quickly overcome the environmental changes, rather than specialized in hydrocarbon degradation (Johnsen *et al.*, 2005; Peng *et al.*, 2008). However, some microorganisms are known as obligate hydrocarbonoclastic bacteria (OHCB). They are mainly an ecophysiologically unusual group of marine bacteria that are obligate hydrocarbon utilizers (Head *et al.*, 2006; Yakimov *et al.*, 2007).

Despite being distributed geographically worldwide, these prokaryotes are often found to have similar specialized enzymatic systems and metabolic pathways for the utilization of hydrocarbon compounds. In general, hydrocarbons are first activated through the action of oxygenases, and then are metabolized until they become precursors of the TCA cycle. Aliphatic hydrocarbons are degraded via constitutive β-oxidation, and aromatic compounds often require the induction of suitable enzymes for catabolism and are degraded by either ortho or meta cleavage of the ring structure (van der Meer *et al.*, 1992; Stroud *et al.*, 2007).

The identification of the bacteria and genes responsible for catalyzing biogeochemical reactions is one of the major goals of microbial ecology. It is important to collect information about the prevalence and distribution of the different pathways or enzymes responsible for degradation of hydrocarbon compounds in different bacteria. Among these bacteria, examples of the most prominent and generally found bacterial genera growing in contaminated sites are summarized in Table 1. The genera possessing hydrocarbonoclastic species are also shown in this table. The state of knowledge about genes and enzymes responsible for the hydrolysis of petroleum hydrocarbons will be presented in the following section.

Table 1. Bacterial genera associated with petroleum hydrocarbon degradation [a]

Class	Genera	Reference
Gram-negative bacteria		
Gammaproteobacteria	*Pseudomonas*	Lovely, 2003
	Stenotrophomonas	Juhasz *et al.*, 2000
	Acinetobacter	Ratajczak *et al.*, 1998
	Alcanivorax [b]	Yakimov *et al.*, 1998
	Alkanindiges [c]	Borgan *et al.*, 2003
	Oleiphilus [b]	Golyshin *et al.*, 2002
	Oleispira [b]	Yakimov *et al.*, 2003
	Cycloclasticus [b]	Niepseron *et al.*, 2009
	Marinobacter	Deppe *et al.*, 2005
	Neptunomonas	Hedlund *et al.*, 1999
	Thalassolituus [b]	Yakimov *et al.*, 2004
Betaproteobacteria	*Burkholderia*	Marín *et al.*, 2001
	Comamonas	Goyal & Zylstra, 1997
Alphaproteobacteria	*Sphingomonas*	Pinyakong *et al.*, 2003
	Oceanicaulis	Yakimov *et al.*, 2007
Bacterioidetes	*Yeosuana* [d]	Kwon *et al.*, 2006
Gram-positive bacteria		
Firmicutes	*Bacillus*	Cunha *et al.*, 2006
	Paenibacillus	Daane *et al.*, 2002
	Geobacillus	Tourova *et al.*, 2008
	Planomicrobium [d]	Engelhardt *et al.*, 2001
Actinobacteria	*Rhodococcus*	Whyte *et al.*, 2002
	Mycobacterium	Cerniglia, 2003
	Microbacterium	Alvares *et al.*, 2008
	Nocardia	Alvares *et al.*, 2008
	Gordonia	Alvares *et al.*, 2008
	Dietzia	von der Weid *et al.*, 2007
	Thermoleophilum [c]	Yakimov *et al.*, 2003

[a] Only the most common genera found in contaminated aquatically or terrestrial sites were cited.
[b] Marine Obligate Hydrocarbonoclastic Bacteria.
[c] Terrestrial Obligate Hydrocarbonoclastic Bacteria.
[d] These genera possess at least one species described as OHCB.

The Oxygenases: The First Step for Aerobic Bacterial Hydrocarbon Degradation

The catabolic versatility and the genetic plasticity observed within the bacterial group play essential roles in the carbon cycle. However, for the degradation of organic matter, microorganisms depend on oxygenases which catalyze the selective incorporation of oxygen atoms into many organic substrates, trigging the beginning of their degradation pathways.

Oxygenases are enzymes responsible for the incorporation of one or both oxygen atoms from the molecular oxygen (O_2) to their substrates (Harayama *et al.*, 1992; Hayaishi, 2004).

The oxygenases are a class of enzymes that are essential to the metabolism of endogenous or exogenous compounds of all aerobic living forms (Hayaishi, 2008). They are involved in the synthesis and degradation of hormones, vitamins, amino acids, sugars, and in dispense of adventives metabolic compounds, such as drugs, toxins, xenobiotic compounds, among others (Hayaishi, 2005). In petroleum biodegradation, these enzymes are responsible for the beginning of the aromatic hydrocarbon degradation through aromatic ring hydroxylation. The latter process reduces the molecule stability preparing it to a subsequent hydrolysis of the aromatic ring, and by catalyzing the ring-fission reaction (Harayama *et al.*, 1992). In microbial degradation of aliphatic hydrocarbons, the oxygenases are responsible for catalyzing the activation of these almost chemically inert compounds, which is the first step towards degradation (Gibson & Parales, 2000; Rojo, 2005; van Beilen *et al.*, 2006; Rojo, 2009; Haddock, 2010).

In prokaryotes, different oxygenases related to oxidation of hydrocarbons are known. Monooxygenases, such as alkane-monooxygenase (AlkB) and enzymes of the cytochrome P-450 family, are generally found in bacteria able to hydroxylate aliphatic hydrocarbons (Smits *et al.*, 1999; van Beilen *et al.*, 2003b; Liu *et al.*, 2006; van Beilen *et al.*, 2006). On the other hand, dioxygenases, such as those encoded by the *nah*, *bph*, *tod*, *nod*, *nid*, and *nir* genes, are related to monocyclic and polycyclic aromatic hydrocarbon oxidation (Larkin *et al.*, 1999; Saito *et al.*, 1999; Khan *et al.*, 2001; Habe & Omori, 2003; Nyyssönen *et al.*, 2006; Parales *et al.*, 2008). Another group of oxygenases called bacterial multicomponent monooxygenases (BMMs) are related to the degradation of short alkanes, alkenes, and aromatic compounds (Sazinsky & Lippard, 2006; Parales *et al.*, 2008; Notomista *et al.*, 2009). It is noteworthy that in studying the catabolic genes coding for degradative enzymes, or the enzymes themselves, it is possible to directly investigate microbes with a specific biodegradation potential in the environment (Nyyssönen *et al.*, 2006). However, for most of these compounds, more than one pathway or gene coding for degradation enzymes have been identified, giving rise to the question about their ecological significance.

The Bacterial Species and the Genes Coding for Aliphatic Hydrocarbons Degradation

Aliphatic hydrocarbons are important carbon and energy sources for many bacteria. These carbon and hydrogen-containing compounds are open-chain methane derivatives, which are both non-aromatic and non-cyclic organic compounds (Stroud *et al.*, 2007). They may be saturated (alkanes – saturated hydrocarbons with single C-C bonds) or unsaturated (alkenes – unsaturated hydrocarbons containing double C=C bonding; and alkynes – unsaturated hydrocarbons containing a triple C≡C bond), straight-chain, simple branched or highly branched compounds. However, all aliphatic hydrocarbons are insoluble and hydrophobic molecules ranging from gaseous, such as methane and ethane, through liquid to 40 or more carbon atoms that are solid at physiological temperature (Watkinson & Morgan, 1990; Stroud *et al.*, 2007).

Straight chain alkanes (from C_{10} to C_{18}) are the most readily assimilated aliphatic hydrocarbons. Utilization of long-chain alkanes (higher than C_{19}) for growth is less common, since these compounds are progressively less soluble with the increase of the chain length. On the other hand, shorter chain alkanes (ranging from C_5 to C_{10}) are more soluble and consequently more toxic (van Beilen *et al.*, 1994a; Rojo, 2009). Compounds such as isoalkanes with a single methyl side chain can be used for growth and, like the straight chain alkanes, be incorporated into cell components (for example, the lipids). However, isoalkanes with branched chains at both ends of the molecule tend not to be utilized immediately, and those with multiple branching are even less promptly utilized (Ratledge, 1984; Stroud *et al.*, 2007).

Bacteria have developed efficient strategies through specialized systems of enzymes and metabolic pathways in order to utilize alkane as carbon and energy sources (van Beilen *et al.*, 2003a; Wentzel *et al.*, 2007). In most cases the oxidation of alkanes is catalyzed by complex hydroxylation systems formed by an oxygenase and electron carriers (Hamamura *et al.*, 2001; van Beilen & Funhoff, 2005; van Beilen & Witholt, 2005). The most common pathway for alkane oxidation is through an attack of monooxygenase to one of the methyl groups of the molecule, generating an alkanol. This alcohol is further metabolized to its corresponding aldehyde and fatty acids by alcohol and aldehyde dehydrogenases, respectively (Rojo, 2005). However, sub-terminal group oxidation of alkanes could equally occur. In this case, ketones and even a corresponding secondary alcohol might be produced. These ketones are degraded through the introduction of an oxygen molecule into the chain, giving a corresponding ester which is hydrolyzed to produce an alkanol shorter in two carbon atoms than the original alkane (Ratledge, 1984, van Beilen *et al.*, 2003b).

Alkene compounds can be degraded through the action of the same enzymes responsible for the beginning of the alkane oxidation. However, these compounds can be attacked in two ways: (i) at one of the terminal carbon atoms, or (ii) at their double bond, although the double bond position affects alkene degradability since terminal alkenes are more easily degradable than alkenes with an internal double bond (Stroud *et al.*, 2007). Consequently, based on these two alkene oxidation possibilities, a variety of products may be formed (Ratledge, 1984). For example, with the oxidation of methyl groups, either alkenol, fatty acids, primary or secondary alcohols, or methyl ketones can be produced. Moreover, during the oxidation of the double bonds, an epoxide or a diol can be generated (Stroud *et al.*, 2007).

The best studied complex hydroxylation system for the degradation of middle-chain-length alkanes (C_5 - C_{12}) or long-chain-length alkanes (greater than C_{12}) is the 'alkane hydroxylase system' (Marín *et al.*, 2003; van Beilen & Funhoff, 2007; Rojo, 2009). This system consists of three components: a monooxygenase transmembrane named alkane hydroxylase (AlkB), and two soluble proteins which act as electron carries between NADH and the hydroxylase, the rubredoxin (AlkG), and rubredoxin reductase (AlkT) (Kok *el at.*, 1989; van Beilen *et al.*, 1994a, Vomberg & Klinner, 2000; van Hamme *et al.*, 2003). The AlkB transfers one oxygen atom from O_2 to one terminal methyl group of the alkane molecule, while the other oxygen is reduced to H_2O by the electron transferred by the rubredoxin complex (Rojo, 2005; van Beilen & Witholt, 2005). This system was first discovered and studied in *Pseudomonas putida* GPo1 growing with middle-chain-length alkane expenses (van Beilen *et al.*, 1994a; van Beilen *et al.*, 1994b). Homologues systems were found later in *Rhodococcus* strains and in *Acinetobacter* sp. ADP1 able to grow in the presence of middle-chain-length up to long-chain-length alkanes (greater than C_{12}) (Ratajczak *et al.*, 1998; Whyte *et al.*, 2002). The gene coding for alkane hydroxylase in *Acinetobacter* sp. ADP1 showed only 41% similarity when compared to *Pseudomonas putida* GPo1 *alkB*, and was previously called *alkM* (Ratajczak *et al.*, 1998). Subsequent experiments based on molecular biology showed that the genes coding for alkane hydroxylase (AlkB) were present in many alkane-degrading bacteria (Kohno *et al.*, 2002; Kloos *et al.*, 2006). For example, some alkane-utilizing bacteria belonging to Alpha, Beta, and Gammaproteobacteria, high G+C content Gram-positive strains, and Gram-positive spore-forming bacteria have *alkB* homologues (Smits *et al*, 1999; Vomberg & Klinner, 2000; Marín *et al*, 2001; Smits *et al.*, 2002; Cunha *et al.*, 2006; Marchant *et al.*, 2006; Tourova *et al.*, 2008). Some strains, such as *Rhodococcus erythropolis* Q15 and *Alcanivorax borkumensis* SK2, may possess up to five different copies of *alkB* in genome. Each of these copies is expressed in specific conditions and the resulting protein can catalyze the reaction of oxidation of different chain length alkanes (Marín *et al.*, 2003; Whyte *et al.*, 2002; van Beilen *et al.*, 2003b; Sabirova *et al.*, 2006).

However the actual status of knowledge about the *alkB* gene diversity is unknown. Kuhn and co-workers (2009) found new *alkB* genes when contaminated and non-contaminated marine sediments in Admiralty Bay, King George Island, Antarctica, were analyzed. In addition, they observed that the sequences generated from Antarctic samples were more similar between each other than to those sequences obtained from reference gene banks, and that a greater diversity of *alkB* genes are found in pristine samples when compared to contaminated ones (Kuhn *et al.*, 2009). Similar results were obtained by Wasmund *et al.* (2009). They concluded that the Timor Sea, in Australia, is a rich source of novel alkane hydroxylase (AlkB) enzymes, since they found AlkB protein sequences only poorly related to previously identified AlkB sequences.

Some alkane-degrading bacteria do not possess AlkB homologues or at least do not possess AlkB homologues capable of oxidizing all length range of alkanes utilized by these bacteria. For instance, *Rhodococcus erythropolis*, which can grow at the expense of alkanes ranging from C_{10} to C_{30} (Milekhina *et al.*, 1998), has no integral membrane alkane hydroxylases able to act on middle-length-chain alkanes (Whyte *et al.*, 2002). This species must have another system to activate these alkane compounds.

van Beilen and co-workers (2006) found that an alkane hydroxylase related to Class I cytochrome P450 of the CYP153 family is common among bacteria, such as mycobacteria, rhodococci, and *Proteobacteria*, able to degrade alkanes but without integral membrane alkane hydroxylases. These soluble cytoplasmatic monooxygenases, linked to ferredoxin and ferredoxin reductase electron carriers, might be important to bacterial hydroxylation of middle-length-chain (C_5-C_{10}) alkanes. In addition, an alkane-degrading *Sphingomonas* sp. strain HXN-200 possesses up to five copies of a gene related to CYP153 (van Beilen *et al.*, 2006). However, bacterial P-450 systems involved in alkane oxidation are, as of yet, poorly studied. Since 2001 when Maier and co-workers isolated the first enzyme CYP153 related to alkane hydroxylation in *Acinetobacter calcoaceticus* EB104, another P-450-like enzyme has been found in bacteria able to utilize middle-length-chain alkanes. This new class of P-450 that acts on alkanes was discovered in *Rodococcus rubber* DSM 44319. It was named CYP116B3 and is a new member of Class IV P-450 (Roberts *et al.*, 2002; Liu *et al.*, 2006). Class II P-450 from the CYP52 family is frequently found in many types of yeast able to degrade alkanes, as *Candida maltosa, Yarrowia lipolytica*, and *Debaryomyces hansenii* (van Beilen *et al.*, 2003b).

While the study on alkane-degradation advances, more is discovered about the routes of degradation. In *Acinetobacter* sp. M1, that is able to grow on middle to long-chain length alkanes, it is proposed that a soluble dioxygenase acts on C_{10} to C_{30} alkane and C_{12}-C_{20} alkene degradation (Maeng *et al.*, 1996; Tani *et al.*, 2001). Another *Acinetobacter* sp. strain DSM 17874 presents two genes coding for alkane hydroxylase AlkB that are involved on the hydrolysis of n-alkanes with a chain length from C_{10} to C_{20} carbons atoms (Throne-Holst *et al.*, 2006). However, this strain is able to grow in expense of alkanes up to tetracontane ($C_{40}H_{82}$), which are solid at 30ºC, the optimum temperature for strain growth. Through the screening of a transposon mutant library, Throne-Holst *et al.* (2007) identified a gene coding for a putative flavin-binding monooxygenase which is involved in the degradation of *n*-alkanes with carbon chain lengths longer than C_{20}. This gene was named *almA*, and homologues to it is found in other long-chain-length n-alkane degrader *Acinetobacter* strains.

Table 2. Enzymes involved in aerobic bacterial oxidation of aliphatic hydrocarbons

Enzymes type	Compound length oxide[a]	Described in:	References
BMM sMMO[b]	From short to middle-chain-length alkane	Alpha and Gammaproteobacteria	McDonald *et al.*, 2006
BMM pMMO[b]	From short to middle-chain-length alkane	Alpha and Gammaproteobacteria	McDonald *et al.*, 2006
BMMs BMO[b]	From short to middle-chain-length alkane	*Pseudomonas butanovora*	Doughty *et al.*, 2006
BMMs PMO[b]	From short to middle-chain-length alkane	Actinobacteria	Kotani *et al.*, 2003
CYP P-450[c]	Middle-chain-length alkanes	Proteobacteria and Actinobacteria	Maier *et al.*, 2001; van Beilen *et al.*, 2006; Liu *et al.*, 2006
AlkB	From middle to long-chain-length alkane	Proteobacteria, Actinobacteria and Firmicutes	van Beilen *et al.*, 2003b; Rojo, 2005; van Beilen & Funhoff, 2007
Diox[d]	Long-chain-length alkanes up to C_{30}	*Acinetobacter* sp. M1	Maeng *et al.*, 1996
AlmA	Long-chain-length alkanes up to C_{32}	*Acinetobacter* strains	Throne-Holst *et al.*, 2007
LadA	Long-chain-length alkanes up to C_{36}	*Geobacillus termodenitrificans* NG80-2	Feng *et al.*, 2007

[a] The alkane length oxidized by each enzyme is based on data available.
[b] BMMs (Bacterial multicomponent monooxygenases) are presented below.
[c] CYP P-450 is a member of the cytochrome P-450 family.
[d] Dioxygenase of *Acinetobacter* sp. M1.

In general, the enzymes able to oxidize alkanes larger than C_{20} seem to be very different from the previously described AlkB and cytrochrome P-450 (Rojo, 2009). In 2007, other type of oxygenase related to bacterial long-chain-n-alkane utilization was isolated in a thermophilic strain of *Geobacillus termodenitrificans* NG80-2. This strain is able to grow in expense of long-length-chain alkanes (C_{15}-C_{36}) but does not present any AlkB homologues or P-450-like enzymes acting on alkanes. Oppositely, this strain has an extracellular monooxygenase named long-chain alkane monooxygenase (LadA), which is coded by genes located in a plasmid (Feng *et al.*, 2007). This enzyme acts as a binary complex with coenzyme FMN (mononucleotide flavine), which transforms the alkane in a corresponding alcohol (Li *et al.*, 2008). Table 2 summarizes the enzymes involved in the hydrolysis of alkane compounds found in aerobic bacteria.

The Bacterial Species and the Genes Coding for Aromatic Hydrocarbons Degradation

Aromatic hydrocarbons are a family of hydrocarbons characterized by the presence of at least one benzene ring (Planckaert, 2005). They can be monocyclic aromatic hydrocarbons, (MAH) such as benzene, xylene, and toluene, or polycyclic aromatic hydrocarbons (PAH) containing two or more fused benzene rings in linear, angular, and cluster arrangements, such as naphthalene, anthracene, and benzo[α]pyrene (Cerniglia, 2003). These compounds are usual contaminants of soils and groundwater, and their environmental presence is worrying because of their mutagenic, ecotoxic, and carcinogenic potential (Habe & Omori, 2003; Jacques *et al.*, 2007; Chauhan *et al.*, 2008).

The removal of aromatic hydrocarbons from polluted sites by microorganisms represents a prospective solution to environmental problems, and many bacteria species are able to oxidize aromatic hydrocarbons ranging in size from toluene to benzo[a]pyrene (Gibson & Parales, 2000). However, this family of compounds shows different biodegradability rates. As observed in aliphatic hydrocarbons, while the solubility and volatility decrease, the hydrophobicity increases proportional to the increase of fused benzene rings (Juhasz & Naidu, 2000). Low-molecular-mass monocyclic aromatic hydrocarbons (benzene, toluene, and xylene), considered the most toxic compounds in petroleum, are readily biodegraded by many bacteria (Atlas, 1995b). The more complex the hydrocarbon structure, the more resistant is its degradation (Leahy & Colwell, 1990). Low-molecular-weight polycyclic aromatic hydrocarbons such as naphthalene (two fused rings), phenanthrene (three fused rings, angular arrangements), and anthracene (three fused rings, linear arrangements) are also usually readily degraded while high-molar-mass PAHs as pyrene, benzo[α]pyrene, and benz[α]anthracene (four or more fused rings) resist to extensive microbial degradation (Goyal & Zylstra, 1997; Cerniglia, 2003). Only a low number of bacterial isolates can grow using high-molecular-weight PAH, probably because of the very low biodisponibility of these compounds (Johnsen *et al.*, 2005). These compounds are considered to be oxidized by a consortium of microbes or by isolated bacteria growing with an alternative carbon source (Juhasz & Naidu, 2000).

In aerobic biodegradation of aromatic hydrocarbons, bacteria oxidize the aromatic rings by the enzymatic incorporation of atmospheric oxygen into the substrate to form diols. Following reactions result in the formation of precursors of tricarboxylic acid cycle (TCA cycle) intermediates (Juhasz & Naidu, 2000; Johnsen *et al.*, 2005). General comparison among the main routes of aromatic hydrocarbon degradation showed that initial pathways are carried out by different enzymes, but that compounds are transformed into a limited number of intermediates (Chaudry & Chapalamadugu, 1991). Naphthalene degradation systems are generally used as a model for aerobic aromatic degradation. In these pathways, the first step involves the incorporation of two oxygen atoms derived from molecular oxygen into one aromatic nucleus by dioxygenases, decreasing the compound's stability. The reaction product is a cis-1,2-naphthalene dihydrodiol. A dehydrogenase acts on this compound generating a 1,2-dihydroxynaphthalene. Subsequent oxidations allow the opening of one aromatic ring, resulting in salicylaldehyde and pyruvic acid. Then, the salicylaldehyde is transformed in salicylate by salicylaldehyde dehydrogenase. The latter is metabolized generating precursors of the tricarboxylic acid (TCA) cycle through catechol or gentisate metabolic pathways (Goyal & Zylstra, 1996; Habe & Omori, 2003; Jacques *et al.*, 2007; Chauhan *et al.*, 2008).

The first step of the aromatic degradation or the initial ring oxidation is usually the rate-limiting step in the biodegradation reaction (Cerniglia, 1992). In aromatic degradation, this step is mainly catalyzed by a multicomponent dioxygenase, belonging to a family of aromatic-ring-hydroxylating dioxygenases (RHDs), which generally consist of a reductase, a ferredoxin, and an iron-sulfur protein (Harayama, 1997; Habe & Omori, 2003). This multicomponent enzyme system transfers electrons from NADH to the terminal dioxygenases, which are composed of a large (α) and a small (β) subunit (Zylstra & Gibson, 1989; Kim *et al.*, 2006).

RHDs systems have been extensively studied in many microorganisms, as the initial reaction mostly determines the range of aromatic substrates that can be degraded (Kim *et al.*, 2006). This range is determined mainly by the α-subunit of terminal dioxygenase (Furukawa *et al.*, 2004). In addition, the origin of many dioxygenases is determined through the analysis of the nucleotide or amino acids sequences from the α-subunit (Pieper *et al.*, 2004). Comparison of the amino acid sequences of α-subunits (ISPα) to terminal oxygenase revealed that they form a family of diverse but evolutionarily related sequences (Witzig *et al.*, 2006). However, dioxygenases involved in aromatic degradation in Gram-negative bacteria are only distantly related to aromatic dioxygenases from Gram-positive bacteria (Pieper *et al.*, 2004).

Several operons coding for aromatic catabolic enzymes have been characterized. Many of them are present in large, self-transmissible plasmids, such as TOL and NAH plasmids, coding for toluene and naphthalene degradation pathways in *Pseudomonas* species, respectively (Jussila *et al.*, 2007). This fact might have contributed to the dissemination of this property among prokaryotes (Ma *et al.*, 2006; Akhmetov *et al.*, 2008). Homologous genes coding for naphthalene degradation pathway are found either in the NAH plasmid or in the chromosome of *Pseudomonas stutzeri* strain AN10 (Bosch *et al.*, 1999).

A great amount of genetic information about aromatic degradation was first obtained studying Gram-negative bacteria, especially *Pseudomonas* species. For example, almost all data concerning toluene and xylene oxidation (genes found in large catabolic plasmids named TOL) was described in bacteria belonging to this genus. This plasmid contains xylene monooxygenase (XMO) genes (*xylM* and *xylA*) that code for the two subunits responsible for monohydroxylating the methyl side chain of toluene and its derivatives. This is the first step for toluene degradation (Susuki *et al.*, 1991). The subunit XylM is a monooxygenase transmenbrane related to alkane hydroxilase AlkB (25% of homology) while XylA acts as an electron carrier (Susuki *et al.*, 1991; Bühler *et al.*, 2000). However, other pathways for toluene degradation by *Pseudomonas* species are known, such as the metabolic pathway coded by *tod* operon. This operon is located in a catabolic island in the chromosome of *Pseudomonas putida* F1 (Eaton, 1997). In this case, toluene is oxidized by a three component enzyme system, called toluene dioxygenase (TDO), where a two subunit terminal dioxygenase (*todC1C2*) adds two atoms of molecular oxygen into the aromatic nucleus of toluene (Zylstra & Gibson, 1989). Other aromatic compounds, such as styrene, biphenyl, cumene, and n-alkyl benzene, are also metabolized by *tod* pathway of *Pseudomonas putida* F1 (Cho *et al.*, 2000).

Another type of dioxygenase that acts in aromatic hydrocarbons is the biphenyl dioxygenase (BPDO). BPDO is responsible for the beginning of the biphenyl degradation by a dihydroxylation in carbon positions 2 and 3 of the aromatic compound (or at the 2,3 position of the molecule) (Erikson & Mondello, 1992). Different from other aromatic dioxygenases, BPDO (encoded by *bphA1A2A3A4*) is a three-component enzyme constituted by four subunits: a large subunit (BphA1) and a small subunit (BphA2) of the terminal dioxygenase, an electron transfer chain constituted by a ferredoxin (BphA3), and a ferredoxin reductase (BphA4) (Furukawa *et al.*, 2004; Takeda *et al.*, 2004). The *bph* genes are found in both Gram-negative and Gram-positive bacteria. Although they posses only low similarity between each other, it is suggested that they are all derived from the same ancestral origin (Kosono *et al.*, 1997; Springael *et al.*, 2001). They are related to *tod* genes (Astúrias *et al.*, 1994). The *bph* genes are found on bacterial chromosomes of many Gram-negative bacteria (Furukawa *et al.*, 1989; Hayase *et al.*, 1990), but are also found in conjugative transposons (Furukawa *et al.*, 2004). While in Gram-negative bacteria the *bph* genes are clustered in a single locus and in a single set, in Gram-positive bacteria, such as *Rhodococcus*, diverse biphenyl degradation genes that encode multiple isoenzymes for each metabolic step are distributed among multiple clusters, sometimes found in plasmids. For example, in *Rhodococcus erythropolis* TA421, part of the *bph* genes are located on a linear plasmid (Kosono *et al.*, 1997). In addition, in *Rhodococcus* sp. strain RHA1, the *bph* genes are found in at least two different plasmids (pRHL1, pRHL2) and also in the chromosome (Kitagawa *et al.*, 2001; Sakai *et al.*, 2002). Amino acid sequences of *bph* genes from *Rhodococcus* strains share homology to aromatic ring-hydroxylating dioxygenases from Gram-negative bacteria. However, higher identity was observed with the toluene dioxygenase (*tod*) of *Pseudomonas putida* F1 than with other *bph* genes from Gram-negative bacteria (Astúrias *et al.*, 1994; Masai *et al.*, 1995; Yang *et al.*, 2007).

Sphingobium yanoikuyae B1 is a biphenyl degrader that can also grow in expense of toluene, naphthalene, phenanthrene, anthracene, and other high-molecular-weight HPAs (Gibson, 1999). The genome of this strain possesses at least six sets of putative oxygenase genes (Ferraro *et al.*, 2007). However, it is possible that BPDO from *S. yanoikuyae* B1 catalyzes the dioxygenation not only of biphenyl, but also of polycyclic aromatic hydrocarbons containing up to five fused rings, such as benzo[*a*]pyrene (Ferraro *et al.*, 2007; Yu *et al.*, 2007; Haddock, 2010).

Like the metabolic pathways for monocyclic aromatic hydrocarbon degradation, the routes of polycyclic aromatic degradation were previously described in *Pseudomonas*. Naphthalene dioxygenases (NDO) are enzymatic systems that oxidize bi- and tri-cyclic PAH substrates, such as naphthalene, phenanthrene, and anthracene (Peng *et al.*, 2008). NDO was first described in NAH7 plasmid from *Pseudomonas putida* G7. NAH7 carries three operons related to strain G7 naphthalene degradation: *nah* operon coding for upper pathway (naphthalene to salicylate), *sal* operon coding for lower pathway (salicylate to cathecol), and the third codes the regulatory systems of naphthalene pathways (Habe & Omori, 2003). The multicomponent dioxygenase is encoded by *nahAaAbAcAd* in *nah* operon. In addition, other genes coding for NDO were described in *Pseudomonas* species. Gene *dox* in *Pseudomonas* sp. C18 (Denome *et al.*, 1993), *pah* in *P. putida* OUS82 (Takizawa *et al.*, 1994), and *nod* in *P. putida* NCIB 9816 (Kurkela *et al.*, 1988) are usually called "classical *nah*-like genes" as they all possess about 90% of sequence similarity to *nah* genes from NAH7 plasmid of *P. putida* G7. These genes were named in accordance to the substrate utilized by the strain (*phn*= phenanthrene; *pah*= PAH), although all strains also grow with naphthalene as a substrate (Habe & Omori, 2003). In addition, the genes *nag* identified in *Ralstonia* sp. U2 and found in *Comamonas testosteroni* GZ42 are correlated to those classical *nah* genes (Zhou *et al.*, 2001).

NDO different from those coded by the classic "*nah*-like genes" were found in other Gram-negative bacteria. The genes *phd* of *C. testosteroni* GZ39 and *phn* of *Burkholderia* sp. RP007 are quite different from those of classical *nah*-like genes (Goyal & Zylstra, 1996; Laurie & Lloyd-Jones, 1999). Another NDO coded by *phn* genes of the PAH-degrader *Cycloclasticus* sp. A5 was shown to be different from the *nah-nod-dox-pah* group (Kasai *et al.*, 2003). Also, genes coding for PAH degradation in *Sphingomonas* are different from genes found in other genera (Pinyakong *et al.*, 2003). These genes seem to be located in a large complex chromosomal gene cluster that encodes enzymes for various aromatic hydrocarbon pathways, such as naphthalene, biphenyl, and xylene degradation (Ferraro *et al.*, 2007). This gene complex is found in aromatic hydrocarbon degrading sphingomonads, such as *Sphingobium yanoikuyae* B1, *Novosphingomonas aromaticivorans* F199, and *Sphingobium* sp. strain P2 (Romine *et al.*, 1999; Pinyakong *et al.*, 2003; Ferraro *et al.*, 2007).

Metabolic pathways for PAH-degradation in Gram-positive bacteria are less studied, although it is postulated that Gram-positive bacteria may be potentially more important for environmental biodegradation than Gram-negative isolates mainly related to high-molecular-weight PAH (Brezna *et al.*, 2003; Cerniglia, 2003; Quatrini *et al.*, 2007). The ability to grow using PAH as carbon and energy sources have been shown in many Gram-positive bacteria, including *Rhodococcus*, *Mycobacterium*, *Nocardioides*, *Terrabacter,* and *Paenibacillus* (Saito *et al.*, 1999; Kasuga *et al.*, 2001; Daane *et al.*, 2002; Brezna *et al.*, 2003; Larkin *et al.*, 2005). In *Paenibacillus naphthalenovorans,* a naphthalene degrading strain isolated from petroleum hydrocarbon contaminated sediment and salt march, no homologous genes such as *nah*, *phd*, and *nid* (NOD from *Mycobacterium vanbaalenii* PYR-1) were found (Daane *et al.*, 2001). The authors believe that novel genes not described so far might act on PAH degradation in this bacterium.

Gram-positive phenanthrene-utilizing *Nocardioides* sp. KP7 which degrades phenanthrene via phthalate pathway possesses NDO coded by a set of genes called *phd* with only 60% of similarity among sequences of other dioxygenases (Saito *et al.*, 1999; 2000). In *Rhodococcus* sp. NCIMB12038, the genes *nar* were described as coding for PAH metabolic pathways. It has been demonstrated that NarAa and NarAb (encoding for α and β subunit of NDO respectively) possess only 30% of amino acid homology to the corresponding *P. putida* NDO (Larkin *et al.*, 2005). Gene *nid* has already been charactcrized in pyrene-utilizing *Mycobacterium vanbaalenii* PYR-1, where a dioxygenase named NidAB2 initiates the degradation pathway (Khan *et al.*, 2001). NidA and NidB proteins are only 40 and 56% similar to the corresponding isofunctional enzymes in *Nocardioides* sp. strain KP7 and *Pseudomonas* sp. strain NCIMB12038, respectively (Kim *et al.*, 2006). Unexpectedly, another PAH-hydroxylase found in *Mycobacterium vanbaalenii* PYR-1 called NidA3B3 showed a higher nucleotide identity to *phd* dioxygenase from *Nocardioides* sp. strain KP7 than to *nidA* previously described in strain PYR-1 (Kim *et al.*, 2007).

Only a low number of high-molecular-weight PAH-utilizing bacteria have been isolated so far (Johnsen *et al.*, 2005). HMW PAHs-utilizing bacteria such as *Mycobacterium vanbaalenii* PYR1 (Kim *et al.*, 2006), *Sphingomonas* sp. strain CHY-1 (Jakoncic *et al.*, 2007), and *Sphingobium yanoikuyae* B1(Yu *et al.*, 2007) are able to use low-molecular-weight HPA and also oxidize large aromatic hydrocarbons such as benzo[*a*]pyrene, benzo[*a*]anthracene, and chrysene. The pathways of HMW PAH degradation have already been described in the above mentioned bacteria and the same genes that code for LMW PAH degradation in *M. vanbaalenii* PYR1 and in *S. yanoikuyae* B1 (*nid* and *bph* genes, respectively) act also in HMW-PAH (Kim *et al*, 2006; Ferraro *et al.*, 2007; Yu *et al.*, 2007). The same is observed in *Sphingomonas* sp. strain CHY-1 where a dioxygenase coded by the *phn* genes acts on at least 8 PAH compounds including four-ring chrysene and benz[a]anthracene, and on the five-ring benzo[a]pyrene (Jakoncic *et al.*, 2007).

Other HMW PAHs-utilizing bacteria, such as *Stenotrophomonas maltophilia* strain VUN 10003, *Pseudomonas aeruginosa, Burkholderia cepacia, Flavobacterium* sp., *Cycloclasticus* sp., and actinomycetes from the genera *Mycobacterium*, *Gardonia*, *Rhodococcus*, among others, have been isolated. Although the number of high-molecular-weight PAH-utilizing isolates has been increasing in the past 10 years, information concerning the metabolism of HMW PAHs is still to be obtained (Chauhan *et al.*, 2008).

Bacterial Multicomponent Monooxygenases (BMMs), a Soluble Complex that Acts in Microbial Degradation of Aromatic and Aliphatic Hydrocarbons

Short alkanes (C1-C8), alkenes, and aromatic hydrocarbons are converted to alcohols, phenol, and expoxides, respectively, by enzymes called bacterial multicomponent monooxygenases (BMMs) (Sazinsky & Lippard, 2006; Notomista *et al.*, 2009). BMMs are soluble multicomponent enzymes from a family of non-heme, diiron proteins. They use the carboxylate-bridged diiron centre of their hydroxylases to convert a variety of hydrocarbons to hydroxylate in a reaction that requires NAD(P)H as an electron donor (Leahy *et al.*, 2003; Notomista *et al.*, 2003; Lippard, 2005). Each BMM contains several proteins, but at least a hydroxylase, a reductase, and a small regulatory component are required for the efficient coupling of electron consumption with hydrocarbon oxidation (Lippard, 2005, Murray & Lippard, 2007). However, usually BMMs contain three or four general components: a dimeric hydroxylase protein formed by two or three subunits ($\alpha_2\beta_2$ or $\alpha_2\beta_2\gamma_2$), a FAD-containing redutase that shuttles electron from NADH to the diiron center of hydroxylase (N component), a small effector or coupling protein (R subunit), and in some cases a Rieske-type ferredoxin protein mediating the electron transfer to diiron center of the hydroxylase complex (Leahy *et al.*, 2003; Sazinsky & Lippard, 2006). It is probable that BMMs use very similar catalytic mechanisms, although their substrates specificity can vary a lot (Notomista *et al.*, 2003).

All known BMMs so far are transcribed in a single operon that codes for four to six polypeptides (Notomista *et al.*, 2003). The evolutionary relationship of BMMs was established comparing the operon subunits structure based on α, β, N, and R conserved components and operon organization. Five phylogenetic BMM groups have been defined, and they are presented hereafter. However, it is important to state that the BMMs classification has been correlated to the bacterial phenotype (ability to grow using specifics hydrocarbon) and not to their phylogeny, which suggests that these pathways might have been evolved through horizontal gene transfer (Notomista *et al.*, 2003). Group 1 (PH phenol hydroxylases) and group 2 (toluene-benzene monooxygenases) are involved in aromatic hydrocarbons hydroxylation (Haddock, 2010), while group 3 (MMO methane monooxygenase), group 4 (AMO alkene monooxygenase), and group 5 (THMO tetrahydrofluran/propane monooxygenase) are involved in converting mainly short aliphatic hydrocarbons (Coleman *et al.*, 2006; Sazinski & Lippard, 2006). It is likely that enzymes characterized from the strains isolated so far represent only a small fraction of the total biochemical diversity and biotechnological potential of the BMMs family (Galvão *et al.*, 2005).

- BMMs from group 1: the three-component phenol hydroxylase (PHs) group

The BMMs from group 1 are composed of enzymatic systems related to three-component phenol hydroxylases (PHs). These include the phenol hydroxylases as Dmp phenol hydroxylase from *Pseudomonas* sp. strain CF600, and enzymes involved in toluene hydroxylation, as toluene-2-monooxygenase (T2MO) from *Burkholderia vietnamiensis* G4 (Notomista *et al.*, 2003; Parales *et al.*, 2008). These enzymes are able to oxidize phenol and methyl-substituted phenols. In addition, T2MO (*tomA1A2A3A4A5*), which is encoded on self-transmissible plasmid TOM of *B. vietnamiensis* G4 (Shields *et al.*, 1995), is able to incorporate one atom of oxygen from dioxygen in more than a single position of the aromatic ring in two consecutive monooxygenation reactions (Cafaro *et al.*, 2004), and is able to hydroxylate hydrocarbons, such as benzene, toluene, naphthalene, and other unactivated hydrocarbons including short-chain alkenes (Yeager *et al.*, 1999; McClay *et al.*, 2000). However, some PHs exhibit a substrate range similar to that of T2MO (Johnson & Olsen, 1995; Teramoto *et al.*, 1999; Leahy *et al.*, 2003). This group of BMMs has already been described in Beta and Gammaproteobacteria, such as the PHs from *Pseudomonas*, *Comamonas,* and *Ralstonia*, and T2MO from *Pseudomonas* and *Burkholderia* (Johnson & Olsen, 1995; Teramoto *et al.*, 1999; Leahy *et al.*, 2003; Cafaro *et al.*, 2004, Parales *et al.*, 2008).

- BMMs from group 2: the four-component alkene monooxygenase and aromatic ring monooxygenases

The BMMs from group 2 are characterized by four component alkene monooxygenases (AMO) from *Xanthobacter* strain Pys (Zhou *et al.*, 1999) and by closely related aromatic ring monooxygenases that exhibit overlapping substrate specificity. These latter enzymes catalyze the monohydroxylation of benzene, toluene, and phenol compounds (Leahy *et al.*, 2003).

The first studied aromatic monooxygenase in this group was the toluene 4-monooxygenase (T4MO), encoded by the genes *tmoABCDE* in *Pseudomonas mendocina* KR1 (Pikus *et al.*, 1996). Studies have shown that T4MO possesses a wide substrate specificity, which includes ethylbenzene and C_3 to C_8 alkenes (McClay *et al.*, 2000). Other aromatic monooxygenases related to T4MO were found in Alpha, Beta, and Gammaproteobacteria. BMM toluene 3-monooxygenase (T3MO) is coded by the genes *tbuA1UBVA2C* located in the chromosome of *Ralstonia pickettii* PkO1 (Byrne *et al.*, 1995). It is also found in *Burkholderia cepacia* AA1 and is another toluene monooxygenase belonging to group 2 that is able to hydroxylate toluene, benzene, ethylbenzene, and other hydrocarbons (Notomista *et al.*, 2003; Parales *et al.*, 2008). While T4MO from *P. mendoncina* KR1 catalyzes the hydroxylation of C4 position of toluene, T3MO from *R. pickettii* PkO1 catalyzes the hydroxylation of C3 (Parales *et al.*, 2008; Haddock, 2010). Oppositely to that observed with T3MO and T4MO, a group 2 toluene/*o*-xylene monooxygenase (ToMO) from *Pseudomonas stutzeri* OX1 (genes *touABCDEF*) exhibits relaxed region specificity for the positions of substrate hydroxylation (Bertoni *et al.*, 1996; Bertoni *et al.*, 1998; Barbieri *et al.*, 2001). However, this BMM can also hydroxylate benzene and other aromatic hydrocarbons (Haddock, 2010). It is known that all toluene monooxygenase producing strains can catalyze the expoxidation of short-chain-alkenes (McClay *et al.*, 2000).

Alkene monooxygenase coded by chromosomal genes *aamABCDEF* from *Xanthobacter* strain Pys (XAMO) is closely related to the toluene/benzene monooxygenase family (Leahy *et al.*, 2003; Notomista *et al.*, 2003). This enzyme can also catalyze the aromatic monohydroxylation of toluene, benzene, and phenol (Zhou *et al.*, 1999). It has been already shown that XAMO presents more sequence similarity to monooxygenases like toluene (TMO), benzene (BMO), and methane (MMO) monooxygenases, than to alkene monooxygenase from *Rhodococcus coralinus* B-276 (Zhou *et al.*, 1998).

- BMMs from group 3: methane monooxygenase (MMO)

BMMs from group 3 are composed by a soluble methane monooxygenase, which catalyze one of the most difficult reactions observed in nature: the hydroxylation of highly stable methane compounds to methanol (Kopp & Lippard, 2002; Leahy *et al.*, 2003). MMOs are found in all bacteria that use methane as carbon and energy sources for growth, generally Alpha and Gammaproteobacteria (Murrell *et al.*, 2000; Kopp & Lippard, 2002). In methanotrophic bacteria, two types of MMO are observed. One of them, a copper-containing membrane bound enzyme is called particulate methane monooxygenase (pMMO), which is coded by the genes *pmoCAB*. Two copies of this operon are generally present in the genome of the majority of methanotrophs (Murrell *et al.*, 2000). The other MMO is a soluble iron-containing enzyme named soluble methane monooxygenase (sMMO) that is coded by *mmoXYBZC* genes (McDonald *et al.*, 1997; van Beilen & Funhoff, 2007). Almost all methanotrophs possess pMMO while only a few of them also have sMMO. In strains possessing both MMOs, the sMMO expression occurs in low copper availability conditions (Rojo, 2009). Although less widespread in methanotrophs, the sMMOs are more studied and understood than pMMO. It is known that sMMOs generally exhibit broad specificities, catalyzing the insertion of oxygen into a number of different hydrocarbon compounds, including substituted methanes, alkanes, cycloalkanes, alkenes, haloalkenes, and aromatic and heterocyclic hydrocarbons (Sazinski & Lippard, 2006; Rojo, 2009).

Pseudomonas butanovora, a bacterium that can grow with C_2-C_8 alkanes, possesses a BMM related to soluble butane monooxygenase (sBMO) coded by the genes *bmoXYBZDC* that is highly related to sMMO from methanotrophs (Sluis *et al.*, 2002; Rojo, 2009). Another strain, a propane-utilizing *Gordonia* sp. TY-5 possesses a propane monooxygenase also similar to sMMO (Kotani *et al.*, 2007). However, the propane monooxygenase of *Gordonia* sp. TY-5 belongs to group 5 of BMMs, as described below.

- BMMs from group 4: Amo Alkene Monoxygenases (AMOs)

The forth group of BMMs is characterized by Amo alkene monooxygenase (AMO) described in *Rhodococcus coralinus* B-276. This soluble aliphatic alkene hydroxylase is coded by the genes *amoABCD* (Ensign, 2001; Perry & Smits, 2006). Another BMM, propene monooxygenase (PMO) coded by the genes *pmoABCD* in *Mycobacterium* sp. strain M156, is closed related to Amo alkene monooxygenase from *R. coralinus* B-276. It is proposed that PMO is a second member of BMM from group 4 (Chion *et al.*, 2005; Perry & Smits, 2006). AMOs related to PMO have been described in the propene-assimilating strain *Gordonia* B-276 (Saeki & Furuhashi, 1994) and in ethene-assimilating strain *Mycobacterium* JS60 (EtnABCD) (Coleman & Spain, 2003) and in strain *Nocardioides* sp. JS614 (EtnABCD) (Mattes *et al.*, 2005).

- BMMs from group 5: tetrahydrofuran/propane monooxygenases (THMOs)

BMMs from group five are composed mainly by a tetrahydrofuran monooxygenase (Thm) described in *Pseudonocardia* sp. SK1 (Thiemer *et al.*, 2003), and by propane monooxygenase (Pmr) related to sMMO that were found in *Gordonia* sp. TY-5 (Kotani *et al.*, 2003). Propane monooxygenase related to Pmr has already been found in propane-utilizing *Mycobacterium* sp. TY-6 and *Pseudonocardia* sp. TY-7 (Kotani *et al.*, 2006). Besides the Thm and Pmr, monooxygenases found in the genome of *Bradyrhizobium japonicum* USDA 110, *Rhodobacter sphaeroides* 2.4.1, and *Rubrivivax gelatinosus* PM1 are related to group 5 BMMs (Coleman *et al.*, 2006).

Conclusion

Aerobic utilizing-hydrocarbon bacteria possess a group of enzymes, called oxygenases, which initiate hydrocarbon degradation through the introduction of one or both oxygen atoms from dioxygen into their substrates. Although different enzymes have already been described, it is possible that oxygenases characterized from bacterial strains so far represent only a small fraction of their biochemical diversity and biotechnological potential in the natural environment.

References

Akhmetov, L.I., Filonov, A.E., Puntus, I.F., Kosheleva, I.A., Nechaeva, I.A., Yonge, D.R., Petersen, J.N. & Boronin, A.M. (2008). *Horizontal transfer of catabolic plasmids in the process of naphthalene biodegradation in model soil systems*. Microbiology. 77(1), 23-32.

Alvarez, V.M., dos Santos, S.C.C., Casella, R.C., Vital, R.L., Sebastián, G.V. & Seldin, L. (2008). *Bioremediation potential of a tropical soil contaminated with a mixture of crude oil and production water*. J Microbiol Biotechnol. 18(12), 1966-1974.

Asturias, J.A., Diaz, E. & Timmis, K.N. (1994). *The evolutionary relationship of biphenyl dioxygenase from Gram-positive Rhodococcus globerulus P6 to multicomponent dioxygenases from Gram-negative bacteria.* Gene. 156(1), 11-18.

Atlas, R.M. (1995a). *Bioremediation of petroleum pollutants.* Int Biodeterior Biodegr. 35, 317-327.

Atlas, R.M. (1995b). *Petroleum biodegradation and oil spill bioremediation.* Marine Pollut Bull. 31, 4-12.

Barbieri, P. Arenghi, F.L.G., Bertoni, G. Bolognese, F. & Galli, E. (2001). *Evolution of catabolic pathways and metabolic versatility in Pseudomonas stutzeri OX1*. Antonie van Leeuwenhoek. 79, 135-140.

Bertoni, G., Bolognese, F., Galli, E. & Barbieri, P. (1996). *Cloning of the genes for and characterization of the early stages of toluene and o-xylene catabolism in Pseudomonas stutzeri OX1.* Appl Environ Microbiol. 62, 3704-3711.

Bertoni, G., Martino, M., Galli, E. & Barbieri, P. (1998). *Analysis of the gene cluster encoding toluene/o-xylene monooxygenase from Pseudomonas stutzeri OX1.* Appl Environ Microbiol. 64, 3626-3632.

Bogan, B.W., Sullivan, W.R., Kayser, K.J., Derr, K.D., Aldrich, H.C. & Paterek, J.R. (2003). *Alkanindiges illinoisensis gen. nov., sp. nov., an obligately hydrocarbonoclastic, aerobic squalane-degrading bacterium isolated from oilfield soils.* Int J Syst Evol Microbiol. 53, 1389-1395.

Bosch, R., García-Valdés, E. & Moore, E.R.B. (1999). *Genetic characterization and evolutionary implications of a chromosomally encoded naphthalene-degradation upper pathway from Pseudomonas stutzeri AN10.* Gene. 226(1), 149-157.

Brezna, B., Khan, A.A. & Cerniglia, C.E. (2003). *Molecular characterization of dioxygenases from polycyclic aromatic hydrocarbon-degrading Mycobacterium spp.* FEMS Microbiol Lett. 223(2), 177-183.

Bühler, B., Schmid, A., Hauer, B. & Witholt, B. (2000). *Xylene monooxygenase catalyzes the multistep oxygenation of toluene and pseudocumene to corresponding alcohols, aldehydes, and acids in Escherichia coli JM101.* J Biol Chem. 275(4), 10085-10092.

Byrne, A.M., Kukor, J.J., & Olsen, R.H. (1995). *Sequence analysis of the gene cluster encoding toluene-3-monooxygenase from Pseudomonas picketti PKO1.* Gene. 154, 65-70.

Cafaro, V., Izzo, V., Scognamiglio, R., Notomista, E., Capasso, P., Casbarra, A., Pucci, P. & Di Donato, A. (2004). *Phenol hydroxylase and toluene/o-xylene monooxygenase from Pseudomonas stutzeri OX1: Interplay between two enzymes.* Appl Environ Microbiol. 70(4), 2211-2219.

Cerniglia, C. (1992). *Biodegradation of polycyclic aromatic hydrocarbons.* Biodegradation. 3(2-3), 351-368.

Cerniglia, C.E. (2003). *Recent advances in the biodegradation of polycyclic aromatic hydrocarbons by Mycobacterium species. In V. Sasek (ed.). The utilization of bioremediation to reduce soil contamination: problems and solutions.* Kluwer Academic Publishers. 51-73.

Chaudry, G.R. & Chapalamadugu, S. (1991). *Biodegradation of halogenated organic compounds.* Microbiol Rev. 55, 59-79.

Chauhan, A., Fazlurrahman, Oakeshott, J.G. & Jain, R.K. (2008). *Bacterial metabolism of polycyclic aromatic strategies for bioremediation.* Indian J Microbiol. 48, 95-113.

Chion, C.K.C.K., Askew, S.E. & Leak, D.J. (2005). *Cloning, expression, and site-directed mutagenesis of the propene monooxygenase genes from Mycobacterium sp.* strain M156. Appl Environ Microbiol. 71(4), 1909-1914.

Cho, M.C., Kang, D-O., Yoon, B.D. & Lee, K. (2000). *Toluene degradation pathway from Pseudomonas putida F1: substrate specificity and gene induction by 1-substituted benzenes.* J Ind Microbiol Biotechnol. 25(3), 163-170.

Coleman, N.V. & Spain, J.C. (2003). *Epoxyalkane: coenzyme M transferase in the ethene and vinyl chloride biodegradation pathways of Mycobacterium strain JS60.* J Bacteriol. 185(18), 5536-5545.

Coleman, N.V., Bui, N.B. & Holmes, A.J. http://www3.interscience.wiley.com/ journal/118567447/abstract - c1#c1. (2006). *Soluble di-iron monooxygenase gene diversity in soils, sediments and ethene enrichments.* Environ Microbiol. 8(7), 1228-1239.

Cunha, C.D., Rosado, A.S., Sebastián, G.V., Seldin, L. & von der Weid, I. (2006). *Oil biodegradation by Bacillus strains isolated from the rock of an oil reservoir located in a deep-water production basin in Brazil.* Appl Microbiol Biotechnol. 73(4), 949-959.

Daane, L.L., Harjono, I., Zylstra, G.J. & Häggblom, M.M. (2001). *Isolation and characterization of polycyclic aromatic hydrocarbon-degrading bacteria associated with the rhizosphere of salt marsh plants.* Appl Environ Microbiol. 67(6), 2683-2691.

Daane, L.L., Harjono, I., Barns, S.M., Launen, L.A., Palleroni, N.J. & Häggblon, M.M. (2002). *PAH-degradation by Paenibacillus spp. and description of Paenibacillus naphthalenovorans sp. nov., a naphthalene-degrading bacterium from the rhizosphere of salt marsh plants.* Int J Syst Evol Microbiol. 52, 131-139.

Denome, S.A., Stanley, D.C., Olson, E.S. & Young, K.D. (1993). *Metabolism of dibenzothiophene and naphthalene in Pseudomonas strains: complete DNA sequence of an upper naphthalene catabolic pathway.* J Bacteriol. 175(21), 6890-6901.

Deppe, U., Richnow, H.H., Michaelis, W. & Antranikian, G. (2005). *Degradation of crude oil by an arctic microbial consortium.* Extremophiles. 9, 461-470.

Doughty, D.M., Sayavedra-Soto, L.A., Arp, D.J. & Bottomley, P.J. (2006). *Product repression of alkane monooxygenase expression in Pseudomonas butanovora.* J Bacteriol. 188(7), 2586-2592.

Dua, M., Singh, A., Sethunathan, N. & Johri, A.K. (2002). *Biotechnology and bioremediation: successes and limitations.* Appl Microbiol Biotechnol. 59, 143-152.

Eaton, R.W. (1997). *p-Cymene catabolic pathway in Pseudomonas putida F1: Cloning and characterization of DNA encoding conversion of p-cymene to p-cumate.* J. Bacteriol. 179, 3171-3180.

Engelhardt, M.A., Daly, K., Swannell, R.P. & Head, I.M. (2001). *Isolation and characterization of a novel hydrocarbon-degrading, Gram-positive bacterium, isolated from intertidal beach sediment, and description of Planococcus alkanoclasticus sp. nov.* J Appl Microbiol. 90(2), 237-247.

Ensign, S.A. (2001). *Microbial metabolism of aliphatic alkenes.* Biochemistry. 40(20), 5845-5853.

Erickson, B.D. & Mondello, F.J. (1992). *Nucleotide sequencing and transcriptional mapping of the genes encoding biphenyl dioxygenase, a multicomponent polychlorinated-biphenyl-degrading enzyme in Pseudomonas strain LB400.* J Bacteriol. 174(9), 2903-2912.

Feng, L., Wang, W., Cheng, J., Ren, Y., Zhao, G., Gao, C., Tang, Y., Liu, X., Han, W., Peng, X., Liu, R. & Wang, L. (2007). *Genome and proteome of long-chain alkane degrading*

Geobacillus thermodenitrificans NG80-2 isolated from a deep-subsurface oil reservoir. Proc Natl Acad Sci USA. 104(13), 5602-5607.

Ferraro, D.J., Brown, E.N., Yu, C-L., Parales, R.E., Gibson, D.T. & Ramaswamy, S. (2007). *Structural investigations of the ferredoxin and terminal oxygenase components of the biphenyl 2,3-dioxygenase from Sphingobium yanoikuyae B1.* BMC Struct Biol. 7, 10.

Furukawa, K., Hayase, K., Taira, K. & Tomizuka, N. (1989). *Molecular relationship of chromosomal genes encoding biphenyl/polychlorinated biphenyl catabolism: some soil bacteria possess a highly conserved bph operon.* J Bacteriol. 171(10), 5467-5472.

Furukawa, K., Suenaga, H. & Goto, M. (2004). *Biphenyl dioxygenases: Functional versatilities and directed evolution.* J Bacteriol. 186(16), 5189-5196.

Galvão, T.C., Mohn, W.W. & de Lorenzo, V. (2005). *Exploring the microbial biodegradation and biotransformation gene pool.* Trends Biotechnol. 23(10), 497-506.

Gibson, D.T. & Parales, R.E. (2000). *Aromatic hydrocarbon dioxygenases in environmental biotechnology.* Curr Opin Biotechnol. 11, 236-243.

Gibson, D.T. (1999). *Beijerinckia sp strain B1: a strain by any other name.* J Ind Microbiol Biotechnol. 23(4-5), 284-293.

Golyshin, P.N., Chernikova, T.N., Abraham, W.R., Lunsdorf, H., Timmis, K.N. & Yakimov, M.M. (2002). *Oleiphilaceae fam. nov., to include Oleiphilus messinensis gen. nov., sp. nov., a novel marine bacterium that obligately utilizes hydrocarbons.* Int J Syst Evol Microbiol. 52, 901-911.

Goyal, A.K. & Zylstra, G.J. (1996). *Molecular cloning of novel genes for polycyclic aromatic hydrocarbon degradation from Comamonas testosteroni GZ39.* Appl Environ Microbiol. 62, 230-236.

Goyal, A.K. & Zylstra, G.J. (1997). *Genetics of naphthalene and phenanthrene degradation by Comamonas testosteroni GZ39.* J Ind Microbiol Biotechnol. 19(5-6), 401-407.

Habe, H. & Omori, T. (2003). *Genetics of polycyclic aromatic hydrocarbon metabolism in diverse aerobic bacteria.* Biosci Biotechnol Biochem. 67(2), 225-243.

Haddock, J.D. (2010). *Aerobic Degradation of Aromatic Hydrocarbons: Enzyme Structures and Catalytic Mechanisms. K. N. Timmis (ed.).* Handbook of Hydrocarbon and Lipid Microbiology. Springer-Verlag Berlin Heidelberg. 4-74.

Hamamura, N., Yeager, C.M. & Arp, D.J. (2001). *Two distinct monooxygenases for alkane Oxidation in Nocardioides sp. strain CF8.* Appl Environ Microbiol. 67, 4992-4998.

Harayama, S., Kok, M. & Neidle, E.L. (1992). *Functional and evolutionary relationships among diverse oxygenases.* Annu Rev Microbiol. 46, 565-601.

Harayama, S. (1997). *Polycyclic aromatic hydrocarbon bioremediation design.* Curr Opin Biotechnol. 8(3), 268-273.

Hayaishi, O. (2008). *From oxygenases to sleep.* J Biol Chem. 283(28), 19165-19175.

Hayaishi, O. *Fifty years of oxygen activation.* (2005). J Biol Inorg Chem. 10, 1-2.

Hayaishi, O. *Oxygenases.* (2004) Encyclopedia of Biological Chemistry. 3, 178-182.

Hayase, N., Taira, K. & Furukawa, K. (1990). *Pseudomonas putida KF715 bphABCD operon encoding biphenyl and polychlorinated biphenyl degradation: cloning, analysis, and expression in soil bacteria.* J Bacteriol. 172(2), 1160-1164.

Head, I.M., Jones, M.D. & Röling, W.F.M. (2006). *Marine microorganisms make a meal of oil.* Nat Rev Microbiol. 4(3), 173-182.

Hedlund, B.P., Geiselbrecht, A.D., Bair, T.J. & Staley, J.T. (1999). *Polycyclic aromatic hydrocarbon degradation by a new marine bacterium, Neptunomonas naphthovorans gen. nov., sp. nov.* Appl Environ Microbiol. 65(1), 251-259.

Jacques, R.J.S., Bento, F.M. & Camargo, F.A.O. (2007). *Biodegradação de hidrocarbonetos aromáticos policíclicos.* Ciência e Natura, UFSM. 29(1), 7-24.

Jakoncic, J., Jouanneau, Y., Meyer, C. & Stoja, V. (2007). *The crystal structure of the ring-hydroxylating dioxygenase from Sphingomonas CHY-1.* FEBS J. 274, 2470-2481.

Johnsen, A.R., Wick, L.Y. & Harms, H. (2005). *Principles of microbial PAH-degradation in soil.* Environ Pollut. 133, 71-84.

Johnson, G.R. & Olsen, R.H. (1995). *Nucleotide sequence analysis of genes encoding a toluene/benzene-2-monooxygenase from Pseudomonas sp. strain JS150.* Appl Environ Microbiol. 61, 3336-3346.

Juhasz, A.L. & Naidu, R. (2000). *Bioremediation of high molecular weight polycyclic aromatic hydrocarbons: a review of the microbial degradation of benzo[a]pyrene.* Int Biodeterior Biodegradation. 45(1-2), 57-88.

Juhasz, A.L., Stanley, G.A. & Britz, M.L. (2000). *Microbial degradation and detoxification of high molecular weight polycyclic aromatic hydrocarbons by Stenotrophomonas maltophilia strain VUN 10,003.* Lett Appl Microbiol. 30, 396-401.

Jussila, M.M., Zhao, J., Suominen, L. & Lindström, K. (2007). *TOL plasmid transfer during bacterial conjugation in vitro and rhizoremediation of oil compounds in vivo.* Eviron Pollut. 146(2), 510-524.

Kasai, Y., Shindo, K., Harayama, S. & Misawa, N. (2003). *Molecular characterization and substrate preference of a polycyclic aromatic hydrocarbon dioxygenase from Cycloclasticus sp. strain A5.* Appl Environ Microbiol. 69(11), 6688-6697.

Kasuga, K., Habe, H., Chung, J.S., Yoshida, T., Nojiri, H., Yamane, H. & Omori, T. (2001). *Isolation and characterization of the genes encoding a novel oxygenase component of angular dioxygenase from the Gram-positive dibenzofuran-degrader Terrabacter sp. strain DBF63.* Biochem Biophys Res Commun. 238, 195-204.

Khan, A.A., Wang R.-F., Cao, W.-W., Doerge, D.R., Wennerstrom, D. & Cerniglia, C.E. (2001). *Molecular cloning, nucleotide sequence, and expression of genes encoding a polycyclic aromatic ring dioxygenase from Mycobacterium sp. strain PYR-1.* Appl Environ Microbiol. 67, 3577-3585.

Kim, S.J., Kweon, O., Freeman, J.P., Jones, R.C., Adjei, M.D., Jhoo, J.W., Edmondson, R.D. & Cerniglia C.E. (2006). *Molecular cloning and expression of genes encoding a novel dioxygenase involved in low- and high-molecular-weight polycyclic aromatic hydrocarbon degradation in Mycobacterium vanbaalenii PYR-1.* Appl Environ Microbiol. 72(2), 1045-1054.

Kim, S.J., Kweon, O., Jones, R.C., Freeman, J.P., Edmondson, R.D. & Cerniglia, C.E. (2007). *Complete and integrated pyrene degradation pathway in Mycobacterium vanbaalenii PYR-1 based on systems biology.* J Bacteriol. 189, 464-472.

Kitagawa, W., Miyauchi, K., Masai, E. & Fukuda, M. (2001). *Cloning and characterization of benzoate catabolic genes in the Gram-positive polychlorinated biphenyl degrader Rhodococcus sp. strain RHA1.* J Bacteriol. 183(22), 6598-6606.

Kloos, K., Munch, J.C. & Schloter, M. (2006). *A new method for the detection of alkane-monooxygenase homologous genes (alkB) in soils based on PCR-hybridization.* J Microbiol Methods. 66(3), 486-496.

Kohno, T., Sugimoto, Y., Sei, K. & Mori, K. (2002). *Design of PCR primers and gene probes for general detection of alkane-degrading bacteria.* Microbes Environ. 17(3), 114-121.

Kok, M., Oldenhuis, R., van der Linden, M.P.G., Raatjes, P., Kingma, J., van Lelyveld, P.H. & Withold, B. (1989). *The Pseudomonas oleovorans alkane hydroxylase gene.* J Biol Chem. 284(10), 5435-5441.

Kopp, D.A. & Lippard, S.J. (2002). *Soluble methane monooxygenase: activation of dioxygen and methane.* Curr Opin Chem Biol. 6, 568-576.

Kosono, S., Maeda, M., Fuji, F., Arai, H. & Kudo, T. (1997). *Three of the seven bphC genes of Rhodococcus erythropolis TA421, isolated from a termite ecosystem, are located on an indigenous plasmid associated with biphenyl degradation.* Appl Environ Microbial. 63(8), 3282-3285.

Kotani, T., Yamamoto, T., Yurimoto, H., Sakai, Y. & Kato, N. (2003). *Propane monooxygenase and NAD(+)-dependent secondary alcohol dehydrogenase in propane metabolism by Gordonia sp. strain TY-5.* J Bacteriol. 185, 7120-7128.

Kotani, T., Kawashima, Y., Yurimoto, H., Kato, N. & Sakai, Y. (2006). *Gene structure and regulation of alkane monooxygenases in propane-utilizing Mycobacterium sp. TY-6 and Pseudonocardia sp.* TY-7. J Biosci Bioeng. 102(3), 184-192.

Kuhn, E., Bellicanta, G.S. & Pellizari, V.H. (2009). *New alk genes detected in Antarctic marine sediments.* Environ Microbiol. 11(3), 669-73.

Kurkela, S., Lehväslaiho, H., Palva, E.T., & Teeri, T.H. (1988). *Cloning, nucleotide sequence and characterization of genes encoding naphthalene dioxygenase of Pseudomonas putida strain NCIB9816.* Gene. 73, 355-362.

Kwon, K.K., Lee, H.S., Jung, H.B., Kang, J.H. & Kim, S.J. (2006). *Yeosuana aromativorans gen. nov., sp. nov., a mesophilic marine bacterium belonging to the family Flavobacteriaceae, isolated from estuarine sediment of the South Sea, Korea.* Int J Syst Evol Microbiol. 56(4), 727-732.

Larkin, M.J., Allen, C.C.R., Kulakov, L.A. & Lipscomb, D.A. (1999). *Purification and characterization of a novel naphthalene dioxygenase from Rhodococcus sp. strain NCIMB12038.* J Bacteriol. 181, 6200-6204.

Larkin, M.J., Kulakov, L.A. & Allen, C.C.R. (2005). *Biodegradation and Rhodococcus.* Curr Opin Biotechnol. 16, 282–290.

Laurie, A.D. & Lloyd-Jones, G. (1999). *The phn genes of Burkholderia sp. strain RP constitute a divergent gene cluster for polycyclic aromatic hydrocarbon catabolism.* J Bacteriol. 181, 531-540.

Leahy, J.G & Colwell, R.R. (1990). *Microbial degradation of hydrocarbons in the environment.* Microbiol Mol Biol Rev. 54(3), 305-315.

Leahy, J.G., Batchelor P.J. & Morcomb, S.M. (2003). *Evolution of the soluble diiron monooxygenases.* FEMS Microbiol Rev. 27(4), 449-479.

Leung, K.T., Nandakumar, K., Sreekumari, K., Lee, H. & Trevors, J.T. (2007). *Biodegradation and bioremediation of organic pollutants in soil.* In: Modern Soil Microbiology. J.D. van Elsas, J. Jansson & J. T. Trevors (ed.) 2nd ed., Taylor & Francis Group, CRC Press. 521-552.

Li, L., Liu, X., Yang, W., Xu, F., Wag, W., Feng, L., Bartlam, M., Wang, L. & Rao, Z. (2008). *Crystal structure of long-chain alkane monooxygenase (LadA) in complex with coenzyme FMN: unveiling the long-chain alkane hydroxylase.* J Mol Biol. 376, 453-465.

Lippard, J.S. (2005). *Hydroxylation of C–H bonds at carboxylate-bridged diiron centres.* Phil Trans R Soc A. 363(1829), 861-877.

Liu, L., Schmid, R.D. & Urlacher, V.B. (2006). *Cloning, expression, and characterization of a self-sufficient cytochrome P450 monooxygenase from Rhodococcus ruber DSM44319.* Appl Microbiol Biotechnol. 72(5), 876-882.

Lovley, D.R. & Lloyd, J.R. (2000). *Microbes with a mettle for bioremediation.* Nature Biotechnol. 18, 600-601.

Lovley, D.R. (2003). *Cleaning up with genomics: Applying molecular biology to bioremediation.* Nature. 1, 35-44.

Ma, Y., Wang, L. & Shao, Z. (2006). *Pseudomonas, the dominant polycyclic aromatic hydrocarbon-degrading bacteria isolated from Antarctic soils and the role of large plasmids in horizontal gene transfer.* Environ Microbiol. 3, 455–465.

Maeng, J.H., Sakai, Y., Ishige, T., Tani, Y. & Kato, N. (1996). *Diversity of dioxygenases that catalyze the first step of oxidation of long-chain n-alkanes in Acinetobacter sp. M-1.* FEMS Microbiol Lett. 144(2-3), 177-182.

Maier, T., Förster, H.H., Asperger, O. & Hahn, U. (2001). *Molecular characterization of the 56-kDa CYP153 from Acinetobacter sp. EB104.* Biochem Biophys Res Commun. 286(3), 652-658.

Marchant, R., Sharkey, F.H., Banat, I.M., Rahman, T.J. & Perfumo, A. (2006). *The degradation of n-hexadecane in soil by thermophilic geobacilli.* FEMS Microbiol Ecol. 56(1), 44-54.

Marín, M.M., Smits, T.H.M., van Beilen, J.B. & Rojo, F. (2001). *The alkane hydroxylase gene of Burkholderia cepacia RR10 is under catabolite repression control.* J. Bacteriol. 183(14), 4202-4209.

Marín, M.M., Yuste, L. & Rojo, F. (2003). *Differential expression of the components of the two alkane hydroxylases from Pseudomonas aeruginosa.* J. Bacteriol. 185(10), 3232-3237.

Masai, E., Yamada, A., Healy, J.M., Hatta, T., Kimbara, K., Fukuda, M. & Yano, K. (1995). *Characterization of biphenyl catabolic genes of gram-positive polychlorinated biphenyl degrader Rhodococcus sp. strain RHA1.* Appl Environ Microbiol. 61(6), 2079-2085.

Mattes, T.E., Coleman, N.V., Spain, J.C. & Gossett, J.M. (2005). *Physiological and molecular genetic analyses of vinyl chloride and ethene biodegradation in Nocardioides sp. strain JS614.* Arch Microbiol. 183(2), 95-106.

McClay, K., Fox, B.G. & Steffan, R.J. (2000). *Toluene monooxygenase-catalyzed epoxidation of alkenes.* Appl Envioron Microbiol. 66(5), 1877-1882.

McDonald, I.R., Uchiyama, H., Kambe, S., Yagi, O. & Murrell, J.C. (1997). *The soluble methane monooxygenase gene cluster of the trichloroethylene-degrading methanotroph Methylocystis sp. strain M.* Appl Environ Microbiol. 63:1898-1904.

McDonald, I.R., Miguez, C.B., Rogge,G., Bourque,D., Wendlandt, K.D., Groleau, D. & Murrell, J.C. (2006). *Diversity of soluble methane monooxygenase-containing methanotrophs isolated from polluted environments.* FEMS Microbiol Lett. 255, 225–232.

Milekhina, E.I., Borzenkov, I.A., Zvyagintseva, I.S., Kostrikina, N.A., & Belyaev, S.S. (1998). *Characterization of a hydrocarbon-oxidizing Rhodococcus erythropolis strain isolated from an oil field [J].* Microbiology. 67(3), 271-274.

Moore, E.R.B, Tindall, B.J., Martins dos Santos, V.A.P., Pieper, D.H., Ramos, J-L. & Palleroni, N.J. (2006). *Nonmedical: Pseudomonas.* Prokaryotes. 6, 646-703.

Murray, L.J. & Lippard, J.S. (2007). *Substrate trafficking and dioxygen activation in bacterial multicomponent monooxygenases.* Acc Chem Res. 40, 466-474.

Murrell, J.C., Gilbert, B. & McDonald, I.R. (2000). *Molecular biology and regulation of methane monooxygenase.* Arch Microbiol. 173, 325-332.

Nazina, T.N., Tourova, T.P. Poltaraus, A.B., Navikiva, E.V., Ivanova, A.E., Grigorryan, A.A., Lysenko, A.M. & Belyav, S.S. (2000). *Physiological and phylogenetic diversity of thermophilic spore-forming hydrocarbon-oxidizing bacteria from oil fields.* Microbiology. 69(1), 96-102.

Niepceron, N., Niepceron, M., Portet-Koltalo, F., Merlin, C., Motelay-Massei, A., Barray, S. & Bodilis, J. (2009). *Both Cycloclasticus spp. and Pseudomonas spp. as PAH-degrading bacteria in the Seine estuary (France).* FEMS Microbiol Ecol. 71(1), 137-147.

Norse, D. (2005). *Non-point pollution from crop production: Global, regional and national issues.* Pedosphere. 15(4), 499-508.

Notomista, E., Cafaro, V., Bozza, G., & Di Donato, A. (2009). *Molecular determinants of the regioselectivity of toluene/o-xylene monooxygenase from Pseudomonas sp. strain OX1.* Appl Environ Microbiol. 75(3), 823–836.

Notomista, E., Lahm, A., Di Donato, A. & Tramontano, A. (2003). *Evolution of bacterial and archaeal multicomponent monooxygenases.* J. Mol. Evol. 56, 435–445.

Nyyssönen, M., Piskonen, R. & Itävaara, M. (2006). *A targeted real-time PCR assay for studying naphthalene degradation in the environment.* Microbiol Ecol. 16, 127-134.

Parales, R.E. Parales, J.V. Pelletier, D.A. & Ditty, J.L. (2008). *Diversity of microbial toluene degradation pathways.* Adv Appl Microbiol. 64, 1-73.

Peng, R-H., Xiong, A-S., Xue, Y., Fu, X-Y., Gao, F., Zhao, W., Tian, Y-S. & Yao, Q-H. (2008). *Microbial biodegradation of polyaromatic hydrocarbons.* FEMS Microbiol Rev. 32(6), 927-955.

Perry A. & Smith, T.J. (2006). *Protocol for mutagenesis of alkene monooxygenase and screening for modified enantiocomposition of the epoxypropane product.* J Biomol Screen. 11(5), 553-556.

Philp. J.C., Bamforth, S.M., Singleton, I. & Atlas, R.M. (2005a). *Environmental pollution and restoration: a role for bioremediation. In: Atlas, R.M. & Philp, J.C. (eds). Bioremediation: Applied microbial solutions for real-world environmental cleanup.* ASM Press, Washington, D.C. 1-49.

Philp, J.C., Whiteley, A.S., Ciric, L. & Bailey, M.J. (2005b). *Monitoring bioremediation. In: Atlas, R.M. & Philp, J.C. (eds). Bioremediation: Applied microbial solutions for real-world environmental cleanup.* ASM Press, Washington, D.C. 237-268.

Pieper, D.H., dos Santos, V.A.P.M. & Golyshin, P.N. (2004). *Genomic and mechanistic insights into the biodegradation of organic pollutants.* Curr Opin Biotechnol. 15(3), 215-224.

Pikus, J.D., Studts, J.M., Achim, C., Kauffmann, K.E., Munck, E., Steffan, R.J., McClay, K. & Fox, B.G. (1996). *Recombinant toluene-4-monooxygenase: catalytic and Mossbauer studies of the purified diiron and Rieske components of a four-protein complex.* Biochemistry. 35, 9106-9119.

Pinyakong, O., Habe, H. & Omori, T. (2003). *The unique aromatic catabolic genes in sphingomonads degrading polycyclic aromatic hydrocarbons (PAHs).* J Gen Appl Microbiol. 49, 1-19.

Planckaert, M. (2005). *Oil Reservoirs and Oil Production.* In: Petroleum Microbiology. B. Ollivier & M. Magot (ed.) 1st ed., Washington, ASM Press, pp. 1-20.

Prince, R.C. & Atlas, R.M. (2005). *Bioremediation of marine oil spills. In: Atlas, R.M. & Philp, J.C. (eds). Bioremediation: Applied microbial solutions for real-world environmental cleanup.* ASM Press, Washington, D.C., pp. 269-292.

Prince, R.C. (2005). *The microbiology of marine oil spill bioremediation.* In: Petroleum Microbiology. B. Ollivier & M. Magot (ed.) 1st ed., Washington, ASM Press, pp. 215-238.

Quatrini, P., Scaglione, G., De Pasquale, C., Riela, S. & Puglia, A.M. (2008). *Isolation of Gram-positive n-alkane degraders from a hydrocarbon-contaminated Mediterranean shoreline.* J Appl Microbiol. 104(1), 251-259.

Ratajczak, A., Geissdörfer, W. & Hillen, W. (1998). *Alkane hydroxylase from Acinetobacter sp. strain ADP1 is encoded by alkM and belongs to a new family of bacterial integral-membrane hydrocarbon hydroxylases.* Appl Environ Microbiol. 64(4), 1175-1179.

Ratledge, C. (1984). *Microbial conversions of alkanes and fatty acids.* JAOCS. 61(2), 447-453.

Roberts, G. A., Grogan, G., Greter, A., Flitsch, S.L. & Turner N.J. (2002). *Identification of a new class of cytochrome P450 from a Rhodococcus sp.* J Bacteriol. 184(14), 3898-908.

Rojo, F. (2005). *Specificity at the end of the tunnel: Understanding substrate length discrimination by the AlkB alkane hydroxylase.* J Bacteriol. 187(1), 19-22.

Rojo, F. (2009). *Degradation of alkane by bacteria.* Environ Microbiol. 11(10), 2477-2490.

Romine, M.F., Stillwell, L.C., Wong, K-K., Thurston, S.J., Sisk, E.C., Sensen, C., Gaasterland, T., Fredrickson, J.K. & SaVer, J.D. (1999). *Complete sequence of a 184-kilobase catabolic plasmid from Sphinomonas aromaticivorans F199.* J Bacteriol. 181, 1585-1602.

Rosenberg, E. & Gutnick, D.L. (1977). *Oil tankers and pollution: A microbiological approach.* Ann Rev Microbio. 31, 379-396.

Sabirova, J.S., Ferrer, M., Regenhardt, D., Timmis, K.N. & Golyshin, P.N. (2006). *Proteomic insights into metabolic adaptations in Alcanivorax borkumensis induced by alkane utilization.* J Bacteriol. 188, 3763-3773.

Saeki, H. & Furuhashi, K. (1994*). Cloning and characterization of a Nocardia corallina B-276 gene cluster encoding alkene monooxygenase.* J Ferment Bioeng. 76(6), 399-406.

Saito, A., Iwabuchi, T. & Harayama, S. (1999). *Characterization of genes for enzymes involved in the phenanthrene degradation in Nocardioides sp. KP7.* Chemosphere. 38, 1331-1337.

Saito, A., Iwabuchi, T. & Harayama, S. (2000). *A novel phenanthrene dioxygenase from Nocardioides sp. strain KP7: Expression in Escherichia coli.* J Bactreriol. 182(8), 2134-2141.

Sakai, M., Masai, E., Asami, H., Sugiyama, K., Kimbara, K. & Fukuda, M. (2002). *Diversity of 2,3-dihydroxybiphenyl dioxygenase genes in a strong PCB degrader, Rhodococcus sp. strain RHA1.* J Biosci Bioeng. 93, 421-427.

Sazinsky, M.H. & Lippard, S.J. (2006). *Correlating structure with function in bacterial multicomponent monooxygenases and related diiron proteins*. Acc Chem Res. 39, 558-566.

Shields, M.S., Reagin, M.J., Gerger, R.R., Campbell, R. & Somerville, C. (1995). *TOM, a new aromatic degradative plasmid from Burkholderia (Pseudomonas) cepacia G4*. Appl Environ Microbiol. 61, 1352-1356.

Sluis, M.K., Sayavedra-Soto, L.A. & Arp, D.J. (2002). *Molecular analysis of the soluble butane monooxygenase from 'Pseudomonas butanovora'*. Microbiology 148, 3617-3629.

Smits, T.H., Balada, S.B., Witholt, B. & van Beilen, J.B. (2002). *Functional analyses of alkane hydroxylases from gram-negative and gram-positive bacteria*. J Bacteriol. 184(6), 1733-1742.

Smits, T.H., Röthlisberger, M., Witholt, B. & van Beilen, J.B. (1999). *Molecular screening for alkane hydroxylase genes in Gram-negative and Gram-positive strains*. Environ Microbiol. 1(4), 307-317.

Springael, D., Ryngaert, A., Merlin, C., Toussaint, A. & Mergeay, M. (2001). *Occurrence of Tn4371-related mobile elements and sequences in (chloro)biphenyl-degrading bacteria*. Appl Environ Microbiol. 67(1), 42-50.

Stroud, J.L., Paton, G.I. & Semple, K.T. (2007). *Microbe-aliphatic hydrocarbon interactions in soil: implications for biodegradation and bioremediation*. J Appl Microbiol. 102(5), 1239-1253.

Suzuki, M., Hayakawa, T., Shaw, J.P., Rekik, M. & Harayama, S. (1991). *Primary structure of xylene monooxygenase: similarities to and differences from the alkane hydroxylation system*. J Bacteriol. 173(5), 1690-1695.

Takeda, H., Yamada, A., Miyauchi, K., Masai, E. & Fukuda, M. (2004). *Characterization of transcriptional regulatory genes for biphenyl degradation in Rhodococcus sp. strain RHA1*. J Bacteriol. 186(7), 2134-2146.

Takizawa, N., Kaida, N., Torigoe, S., Moritani, T., Sawada, T., Satoh, S., & Kiyohara, H. (1994). *Identification and characterization of genes encoding polycyclic aromatic hydrocarbon dioxygenase and polycyclic aromatic hydrocarbon dihydrodiol dehydrogenase in Pseudomonas putida OUS82*. J Bacteriol. 176, 2444-2449.

Tani, A., Ishige, T., Sakai, Y. & Kato, N. (2001). *Gene structures and regulation of the alkane hydroxylase complex in Acinetobacter sp. strain M-1*. J Bacteriol. 183(5), 1819-1823.

Teramoto, M., Futamata, H., Harayama, S. & Watanabe, K. (1999). *Characterization of a high-affinity phenol hydroxylase from Comamonas testosteroni R5 by gene cloning and expression in Pseudomonas aeruginosa PAO1c*. Mol Gen Genet. 262, 552-558.

Thiemer, B., Andreesen, J.R. & Schräder, T. (2003). *Cloning and characterization of a gene cluster involved in tetrahydrofuran degradation in Pseudonocardia sp. strain K1*. Arch Microbiol. 179, 266-277.

Throne-Holst, M., Markussen, S., Winnberg, A., Ellingsen, T.E., Kotlar, H-K., & Zotchev, S.B. (2006). *Utilization of n-alkanes by a newly isolated strain of Acinetobacter venetianus: the role of two AlkB-type alkane hydroxylases*. Appl Microbiol Biotechnol. 72(2), 353-360.

Throne-Holst, M., Wentzel, A., Ellingsen, T.E., Kotlar, H-K. & Zotchev, S.B. (2007). *Identification of novel genes involved in long-chain n-alkane degradation by Acinetobacter sp. Strain DSM 17874*. Appl Environ Microbiol. 73(10), 3327-3332.

Tourova, T.P., Nazina, T.N., Mikhailova, E.M., Radionova, T.A., Ekimov, A.N., Mashukova, A.V. & Poltaraus, A.B. (2008). *alkB homologs in thermophilic bacteria of the genus Geobacillus*. Mol Biol, 42(2), 217-226.

van Beilen, J.B., Wubbolts, M.G. & Witholt, B. (1994a). *Genetic of alkane oxidation by Pseudomonas oleovorans.* Biodegradation. 5, 161-174.

van Beilen, J.B., Kingma, J. & Witholt, B. (1994b). *Substrate specificity of the alkane hydroxylase of Pseudomonas oleovorans GPo1.* Enzyme Microb Technol. 16, 904-911.

van Beilen, J.B., Duetz, W.A., Schmid, A. & Witholt, B. (2003a). *Practical issues in the application of oxygenases.* Trends Biotechnol. 21(4), 170-177.

van Beilen, J.B., Li, Z., Duetz, W.A., Smits, T.H.M. & Witholt, B. (2003b). *Diversity of alkane hydroxylase systems in the environment.* Oil Gas Sci Technol. 58, 427-440.

van Beilen, J.B. & Funhoff, E.G. (2005). *Expanding the alkane oxygenase toolbox: new enzymes and applications.* Curr Opin Biotechnol. 16, 308-314.

van Beilen, J.B. & Witholt, B. (2005). *Diversity, function, and biocatalytic applications of alkane oxygenases.* In: B. Ollivier & M. Magot (ed.) Petroleum Microbiology. 1st ed., Washington, ASM Press. 259-275.

van Beilen, J.B., Funhoff, E.G., van Loon, A., Just, A., Kaysser L., Bouza, M., Holtackers, R., Röthlisberger, M., Li, Z. & Witholt, B. (2006). *Cytochrome P450 alkane hydroxylases of the CYP153 family are common in alkane-degrading Eubacteria lacking integral membrane alkane hydroxylases.* Appl Environ Microbiol. 72(1), 59-65.

van Beilen, J.B. & Funhoff, E.G. (2007). *Alkane hydroxylases involved in microbial alkane degradation.* Appl Microbiol Biotechnol. 74(1), 13-21.

van der Meer, J.R., de Vos, W.M., Harayama, S, & Zehnder, A.J. (1992). *Molecular mechanisms of genetic adaptation to xenobiotic compounds.* Microbiol Mol Biol Rev. 56(4), 677-694.

van Hamme, J.D., Singh, A. & Ward, O.P. (2003). *Recent advances in petroleum microbiology.* Microbiol Mol Biol Rev. 67(4), 503-549.

Vomberg, A. & Klinner, U. (2000). *Distribution of alkB genes within n-alkane-degrading bacteria.* J Appl Microbiol. 89, 339-348.

von der weid, I., Marques, J.M., Cunha, C.D., Lippi, R.K., dos Santos, S.C.C., Rosado, A.S., Lins, U. & Seldin, L. (2007). *Identification and biodegradation potential of a novel strain of Dietzia cinnamea isolated from a petroleum-contaminated tropical soil.* Syst Appl Microbiol. 30(4), 331-339.

Wasmund, K., Burns, K.A., Kurtböke, D.I. & Bourne, D.G. (2009). *Novel alkane hydroxylase gene (alkB) diversity in sediments associated with hydrocarbon seeps in the Timor Sea, australia.* Appl Environ Microbiol. 75(23), 7391-7398.

Watkinson, R.J. & Morgan, P. (1990). *Physiology of aliphatic hydrocarbon-degrading* Micro-organisms. Biodegradation. 1, 79-92.

Wentzel, A., Ellingsen, T.E., Kotlar, H.K., Zotchev, S.B. & Throne-Holst, M. (2007). *Bacterial metabolism of long-chain n-alkanes.* Appl Microbiol Biotechnol. 76(6), 1209-1221.

Whyte, L.G., Smits, T.H., Labbé, D., Witholt, B., Greer, C.W. & van Beilen, J.B. (2002). *Gene cloning and characterization of multiple alkane hydroxylase systems in Rhodococcus strains Q15 and NRRL B-16531.* Appl Environ Microbiol. 68(12), 5933-5942.

Witzig, R., Junca, H., Hecht, H.J. & Pieper, D.H. (2006). *Assessment of toluene/biphenyl dioxygenase gene diversity in benzene-polluted soils: Links between benzene biodegradation and genes similar to those encoding isopropylbenzene dioxygenases.* Appl Environ Microbiol. 72(5), 3504-3514.

Yakimov, M.M., Golyshin, P.N., Lang, S., Moore, E.R., Abraham, W.R., Lunsdorf, H. & Timmis, K.N. (1998). *Alcanivorax borkumensis gen nov., sp. nov., a new, hydrocarbon-degrading and surfactant-producing marine bacterium.* Int J Syst Bacteriol. 48, 339-348.

Yakimov, M.M., Giuliano, L., Gentile, G., Crisafi, E., Chernikova, T.N., Abraham, W.R., Lunsdorf, H., Timmis, K.N. & Golyshin, P.N. (2003). *Oleispira antarctica gen. nov., sp. nov., a novel hydrocarbonoclastic marine bacterium isolated from Antarctic coastal sea water.* Int J Syst Evol Microbiol. 53(3), 779-785.

Yakimov, M.M., Giuliano, L., Denaro, R., Crisafil, E., Chernikova, T.N., Abraham, W-R., Luensdorf, H., Timmis, K.N. & Golyshin, P.N. (2004). *Thalassolituus oleivorans gen. nov., sp. nov., a novel marine bacterium that obligately utilizes hydrocarbons.* Int J Syst Evol Microbiol. 54, 141-148.

Yakimov, M.M., Timmis, K.N. & Golyshin, P.N. (2007). *Obligate oil-degrading marine bacteria.* Curr Opin Biotechnol. 18(3), 257-66.

Yang, X., Liu, X., Song, L., Xie, F., Zhang, G. & Qian, S. (2007). *Characterization and functional analysis of a novel gene cluster involved in biphenyl degradation in Rhodococcus sp. strain R04.* J Appl Microbiol. 103(6), 2214-2224.

Yeager, C.M., Bottomley, P.J., Arp, D.J. & Hyman, M.R. (1999). *Inactivation of toluene 2-monooxygenase in Burkholderia cepacia G4 by alkynes.* Appl Environ Microbiol. 65, 632-639.

Yu, C.L., Liu, W., Ferraro, D.J., Brown, E.N., Parales, J.V., Ramaswamy, S., Zylstra, G.J., Gibson, D.T. & Parales, R.E. (2007). *Purification, characterization, and crystallization of the components of a biphenyl dioxygenase system from Sphingobium yanoikuyae B1.* J Ind Microbiol Biotechnol. 34, 311-324.

Zhou, N.Y., Fuenmayor, S.L. & Williams, P. (2001). *nag genes of Ralstonia (formerly Pseudomonas) sp. Strain U2 encoding enzymes for gentisate catabolism.* J Bacteriol. 183, 700-708.

Zhou, N-Y., Jenkins, A., Chion, C.K.N.C.K. & Leak, D.J. (1998). *The alkene monooxygenase from Xanthobacter Py2 is a binuclear non-haem iron protein closely related to toluene 4-monooxygenase.* FEBS Lett. 430(3), 181-185.

Zhou, N-Y., Jenkins, A., Chion, C.K.N.C.K. & Leak, D.J. (1999). *The alkene monooxygenase from Xanthobacter strain Py2 is closely related to aromatic monooxygenases and catalyzes aromatic monohydroxylation of benzene, toluene, and phenol.* Appl Environ Microbiol. 65(4), 1589-1595.

Zylstra, G.J. & Gibson, D.T. (1989). *Toluene degradation by Pseudomonas putida F1: Nucleotide sequence of the todC1C2BADE genes and their expression in E. coli.* J Biol Chem. 264, 14940-14946.

In: Bioremediation
Editor: Alexander C. Mason

ISBN: 978-1-61122-730-7

Chapter 5

Bioremediation of Polycyclic Aromatic Hydrocarbons

***Belinda Hewitt*[1], *Naresh Singhal*[*1] *and Simon Swift*[2]**
[1]Department of Civil & Environmental Engineering, University of Auckland, Auckland, New Zealand
[2]Department of Molecular Medicine & Pathology, University of Auckland, Auckland, New Zealand

Bioremediation has emerged as a promising treatment technology for restoring contaminated environments, where organisms degrade contaminants to non harmful end products. A growing field of research is investigating the way in which bioremediation can be optimised and applied to widely occurring pollutants in the soil column. As a common by-product of the petroleum industry, Polycyclic Aromatic Hydrocarbons (PAHs) have become a ubiquitous soil pollutant. Bioremediation of PAHs has a number of potential advantages over traditional contaminated site treatment approaches, which are often intrusive, labour intensive, and may themselves produce harmful by-products. A number of commonly occurring bacteria have been identified which can use PAHs as a sole carbon source for metabolic activity. A wide range of methods have been investigated to enhance the bacterial degradation of pollutants in both laboratory trials and broad field scale application. Research indicates that the principal limiting factors to bioremediation involve pollutant/microbe interaction and the ability of an organism to catabolise a pollutant. PAHs are highly hydrophobic and may often be sorbed deep within the soil column; they must be drawn out of soil pores before they may be accessed by microbes. Research has demonstrated that the use of surface acting agents including 'biosurfactant' is effective in increasing hydrocarbon solubility in soil and consequently, bioremediation activity. Recent research is increasingly focusing on the genetics of metabolic pathways by which organisms catabolise pollutants. Certain strains of commonly occurring bacteria contain genes which encode enzymes that degrade PAHs into non harmful end products while attaining carbon for metabolic processes. Enhancing bacterial growth, distribution and the expression of PAH degradation gene pathways may significantly enhance bioremediation activity. In order to characterise and optimise all aspects of bioremediation, novel research technologies are being developed in order to monitor and characterise the bioremediation process in soil. There is rapidly growing

* Corresponding author: Tel: +64-9-3737599 ext 84512; Fax: +64-9-3737462; Email: n.singhal@auckland.ac.nz

potential to monitor bacterial activity and gene expression in-situ with the use of novel genetic techniques. In recent research, 'biosensor' organisms have been created which transcribe genes for fluorescence or bioluminescence such as *gfp* (green fluorescent protein) and *lux* (luciferase) as reporters for activities such as growth and bioremediation. The development of this field is an exciting approach to developing an improved understanding of in-situ bacterial activity and advancing the application bioremediation in the field.

1. Introduction

Hydrocarbon contamination poses an increasing threat to the global environment. In particular, Polycyclic Aromatic Hydrocarbons (PAHs), also referred to as Polynuclear Aromatic Hydrocarbons, are a widespread soil pollutant, having accumulated primarily as incompletely combusted by-products of intensified fossil fuel use (Van Metre, Mahler et al. 2009). The development of specific remediation approaches to remove or detoxify hydrocarbons in soil has become an important issue for the consideration of modern science and engineering. Conventional treatment approaches often tend to be labour intensive, costly, invasive, and may themselves produce undesirable end products. A growing field of research is investigating the way in which living systems of biological degradation can be optimised for effective field application to PAH contaminated sites.

As PAH contamination has become subject to growing international concern and regulation, the development of appropriate remediation technologies has become increasingly important. Traditionally, remediation approaches involved the removal and storage of contaminated soil, or an ex-situ chemical or biological treatment. Increasingly, modern science and technology is investigating the processes of microbial activity as a system for broad in-situ biodegradation of PAHs, or bioremediation. A large number of ubiquitously occurring microorganisms have been identified which are able to utilise PAHs as a carbon source for cellular activity (Habe and Omori 2003), where through a series of oxidative reactions PAHs are metabolised into non-harmful end products. This process presents a number of attractive advantages over traditional treatment approaches. Bioremediation of PAH contaminated soil is potentially less intrusive, less expensive, and less labour-intensive than other treatment approaches (Boopathy 2000). Microbial degradation can facilitate the catabolism of PAHs without requiring or producing potentially harmful reactants and end-products.

In soil environments, the rate at which microorganisms can proceed with biodegradation reactions is limited by a number of biotic and abiotic factors (Hart 1996; Zhang, Cheng et al. 2006). These limitations must be overcome in order to engineer these systems for broad scale field application. This chapter will discuss the way in which bioremediation research can characterise and overcome these limits to develop effective, broad-scale field applications for biodegradation in the soil column.

1.1. Polycyclic Aromatic Hydrocarbon Pollution

PAHs are a large class of common pollutants which are commonly present in derivatives of oil, petroleum, coal and wood preservatives. These compounds are also a product of industrial processes operating at elevated temperatures, such as petroleum refining and wood preservation processes (Wilcke 2000). PAHs occur ubiquitously in the natural environment as a result of both biotic and abiotic reactions; however the major sources of PAH accumulation are anthropogenic and result largely from ineffective transport and storage of petroleum products. Although developments in international legislation and technology have meant that the transport and combustion of PAH compounds is regulated with increasing effectiveness, the point as well as non-point source release of PAHs into air, water and soil environments continues to emerge as a particularly pertinent issue of global concern. The US EPA has classified 16 PAHs as priority pollutants, illustrated in Figure 1.

PAHs are defined as a class of compounds composed of three or more fused aromatic rings with the general formula: $C_{4n+2}H_{2n+4}$ (n = number of aromatic rings). For the purpose of many remediation studies and applications, naphthalene (two fused aromatic rings only) is also considered to behave as a PAH. PAHs are characteristically recalcitrant and hydrophobic in nature, exhibiting high rates of sorption within soil pores (Johnsen, Wick et al. 2005). Sorption and partitioning of PAHs within the soil column means that they react at slower rates than would be expected from their chemical structure. These characteristics imply that PAHs present particularly difficult targets for the development of effective soil remediation technology. The exact chemical nature of each PAH differs with its molecular structure, where generally PAH hydrophobicity and toxicity increase with structural complexity and molecular weight (refer to Table 1).

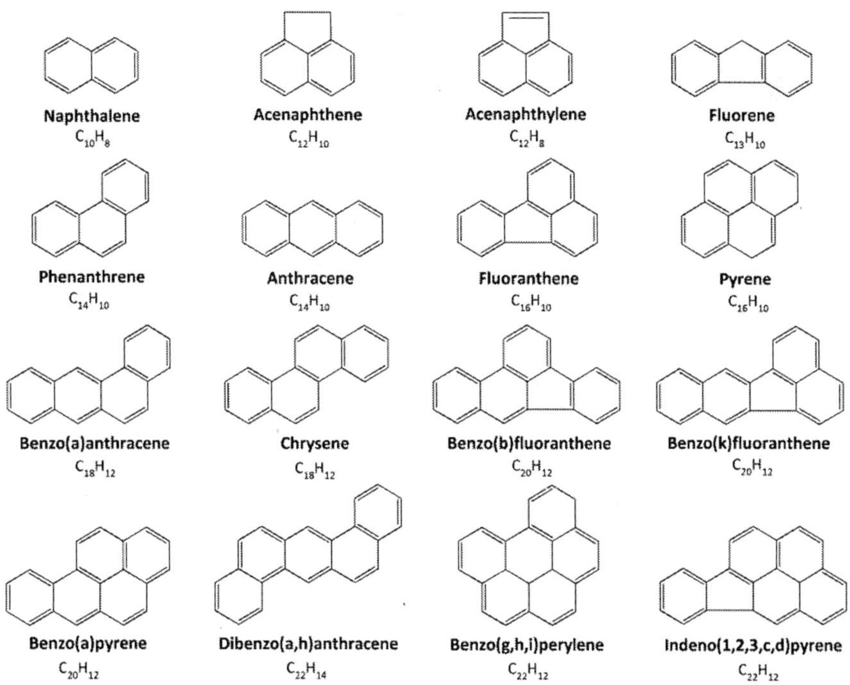

Figure 1. 16 Priority PAHs regulated by the US EPA (modified from Rogers, Ong et al. 2002).

PAHs are a priority pollutant in the soil environment due to their wide range of toxic effects. The potential for PAHs present in soil to leech into groundwater drinking supplies is a major concern. Due to their hydrophobicity and low reactivity, PAHs are highly persistent in natural environments with a tendency to biomagnify and accumulate in organisms and food chains (Samanta, Singh et al. 2002; Jonker, Van der Heijden et al. 2007). While the particular effects and toxicity of each PAH is structurally dependent, a growing body of scientific research has revealed a wide range of mutagenic, carcinogenic and deleterious developmental effects associated with the exposure of humans and mammals to a wide range of PAHs (Boström, Gerde et al. 2002; Samanta, Singh et al. 2002; Haritash and Kaushik 2009).

Table 1. Solubility and toxic equivalent factors for US EPA Priority PAHs (developed from Nisbet and LaGoy 1992). The TEF expresses toxicity in relation to the most toxic form of dioxin 2,3,7,8-tetrachlorodibezodioxin

Compound	No. of Carbon Rings	Pure Solid Aqueous Solubility (μg/L)	Toxic Equivalent Factor (TEF)
Naphthalene	2	31690	0.001
Acenaphthylene	3	3930	0.001
Fluorene	3	1690	0.001
Phenanthrene	3	1300	0.001
Anthracene	3	45	0.01
Fluoranthene	4	260	0.001
Pyrene	4	140	0.001
Chrysene	4	1.8	0.01
Benzo[a]anthracene	4	2	0.1
Benzo[b]fluoranthene	5	14	0.1
Benzo[k]fluoranthene	5	4.3	0.1
Benzo[a]pyrene	5	3.8	1
Dibenz[a,h]anthracene	5	.5	5
Indeno[1,2,3-cd]pyrene	6	.53	0.1
Benzo[g,h,i]perylene	6	.26	0.01

2. Biodegradation of Polycyclic Aromatic Hydrocarbons

An increasing number of microorganisms in the soil column have been identified with the genetic capacity to degrade PAHs. Species of bacteria, fungi, algae and cyanobacteria have been isolated which are capable of degrading PAHs with up to four fused aromatic rings as a sole carbon source, as well as use several more complex PAHs in cellular co-metabolism (Haritash and Kaushik 2009). Bioremediation technologies often focus on utilising populations of PAH degrading microorganisms already present in the soil column, which tend to be much more prevalent in areas where PAHs are highly concentrated (Cerniglia 1993).

2.1. Biology of PAH Biodegradation

The three key mechanisms for microbial biodegradation of PAHs are metabolism, co-metabolism, and mineralization, all of which degrade PAHs one aromatic ring at a time. These mechanisms involve two major reaction pathways: (1) the breakdown of the polycylic aromatic ring and (2) the degradation of the compound into PAH metabolites and/or carbon dioxide. The anaerobic degradation of PAHs is less well observed, but research suggests that this reaction proceeds through the hydrogenation, rather than oxygenation, of the aromatic ring (Bamforth and Singleton 2005). In order to perform PAH degrading activities, a microorganism must have genes to produce enzymes for specific biochemical degradation pathways. These genes are regulated through systems of cellular feedback which regulate the production of degradation enzymes in the presence of a PAH (Habe and Omori 2003).

Microbial species may degrade PAHs in cellular metabolism as a part of basic or secondary metabolic processes. Species which degrade PAHs for basic (primary) metabolism are of particular interest in bioremediation research as they utilise a specific PAH as a sole carbon source for cellular metabolism. However, no microbial strains have been identified which are able to metabolise PAHs with greater than four aromatic rings for basic cellular metabolism. Conversely, microbial species which utilise co-metabolite (secondary) pathways for pollutant degradation may be less reliant on a target contaminant and as such less effective biodegraders; however some bacterial species have demonstrated the ability to degrade more complex PAH compounds (Ye, Siddiqi et al. 1995).

2.2. Degradation of PAHs by Bacteria

A diverse range of ubiquitously occurring bacterial species have been identified which are capable of PAH biodegradation. These bacteria contain the necessary genetic elements to produce specific enzymes for metabolism of a particular PAH through basic or secondary metabolic pathways. These bacteria are widely distributed in the soil environment, and as such native populations of bacteria can be useful in in-situ bioremediation applications. Predominantly, bacterial biodegradation of PAHs is an aerobic process. The basic aerobic pathway for bacterial metabolism has been well characterised (Grishchenkov, Townsend et al. 2000). Firstly, the rate-limiting step of this pathway involves the oxidation of the PAH benzene ring by dioxygenase enzymes to form *cis*-dihydrodiols. Secondly, dihydrodiols are dehydrogenated into dihydroxylated intermediates. These intermediates can be further metabolised into catechols, and finally carbon dioxide and water. Specific metabolic pathways for simple PAHs, such as naphthalene and phenanthrene, have been defined in some bacterial species (e.g. *Pseudomonas* and *Rhodococcus*). The vast majority of bacteria which metabolise PAHs utilise dioxygenase enzymes; however less common pathways have been identified in some species (e.g. *Mycobacterium*), utilising alternative enzymes such as cytochrome P monoxygenase (Bamforth and Singleton 2005).

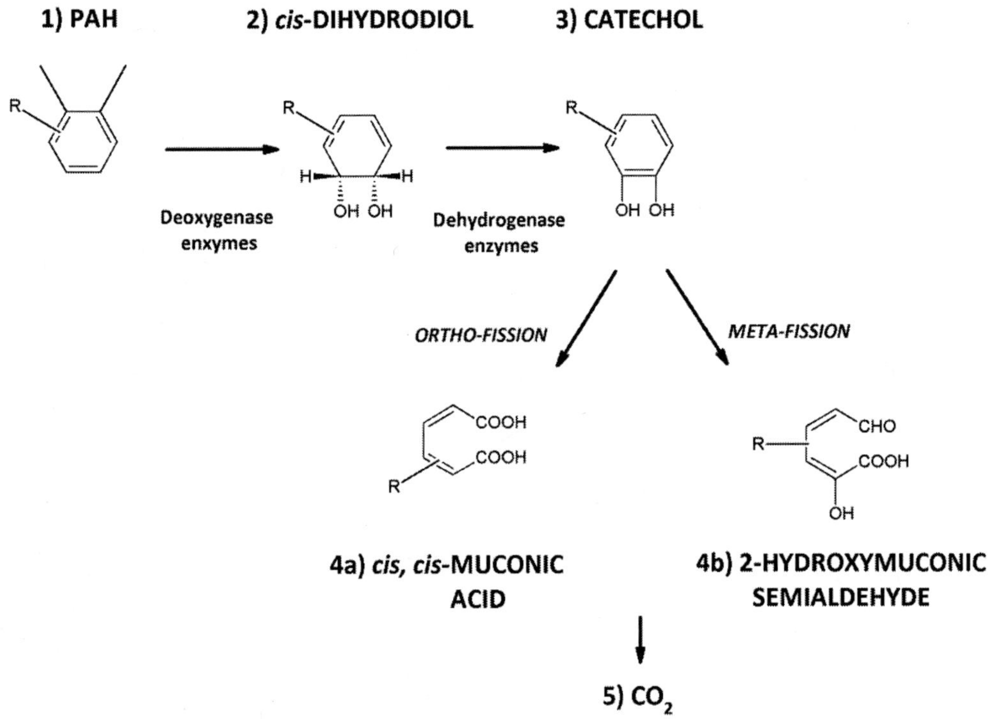

Figure 2. Aerobic bacterial oxidation of PAHs (developed from Rochkind, Blackburn et al. 1986).

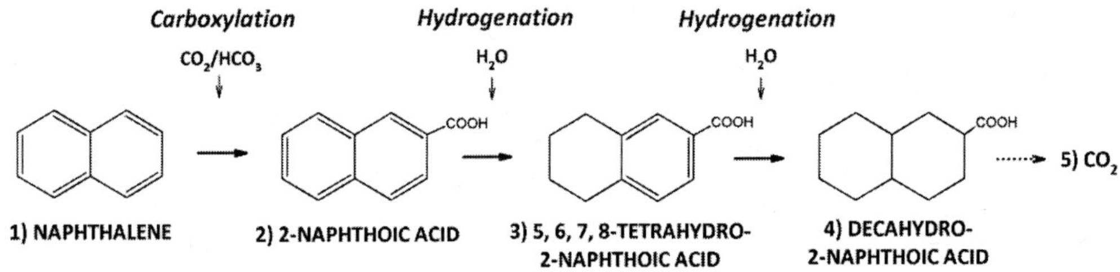

Figure 3. Anaerobic degradation of naphthalene by bacteria (modified from Rochkind, Blackburn et al. 1986).

Although the potential for anaerobic metabolism of PAHs has been identified in some bacteria, these processes have not been well characterised, and their extent and efficiency is largely unknown. Subsurface areas of the soil column may develop anaerobic zones due to the depletion of oxygen during aerobic microbial activity. Where molecular oxygen is absent, microbes may utilise other electron acceptors including ferrous iron, sulphate and nitrate (Chakraborty and Coates 2004). The proposed mechanism by which anaerobic metabolism of PAHs proceeds involves carboxylation of the aromatic ring into 2-naphthoic acid, which is reduced in a stepwise manner through hydrogenation reactions into (initially) decaclin-2-carboxylic acid and (subsequently) decahydro-2-naphthoic acid (refer to Figure 3). Other mechanisms for anaerobic digestion under varying conditions have been proposed but not well characterised, such as the degradation of naphthalene under sulphate-reducing conditions to naphthol intermediates through hydroxylation. Anaerobic reactions proceed to an order of magnitude which is significantly less than that of aerobic biodegradation, and may also be sensitive to the presence of co-pollutants such as heavy metals (Grishchenkov, Townsend et al. 2000; Bamforth and Singleton 2005).

2.3. Degradation of PAHs by Fungi

The two major groups of fungi which degrade PAHs for cellular metabolism can be classified as (i) ligninolytic and (ii) non-ligninolytic (refer to Figure 4). Non-ligninolytic fungi make up the majority of these species, do not grown on lignin-based compounds, and do not produce enzymes for lignin metabolism simultaneously to producing enzymes for PAH metabolism. Initially, these fungi oxidise the PAH aromatic ring into an arene oxide through a cytochrome P monoxygenase catalysed reaction. Secondly, the arene oxide is hydrated into a *trans*dihydrodiol through an epoxide-hydrolase catalysed reaction. Phenol derivatives may also be formed from arene oxides through non-enzymatic rearrangement, which may then be involved in processes of sulfation, methylation, or conjugation. Although non-ligninolytic fungi are rarely able to completely mineralise PAHs, these end-products are less harmful and more soluble than in their original state.

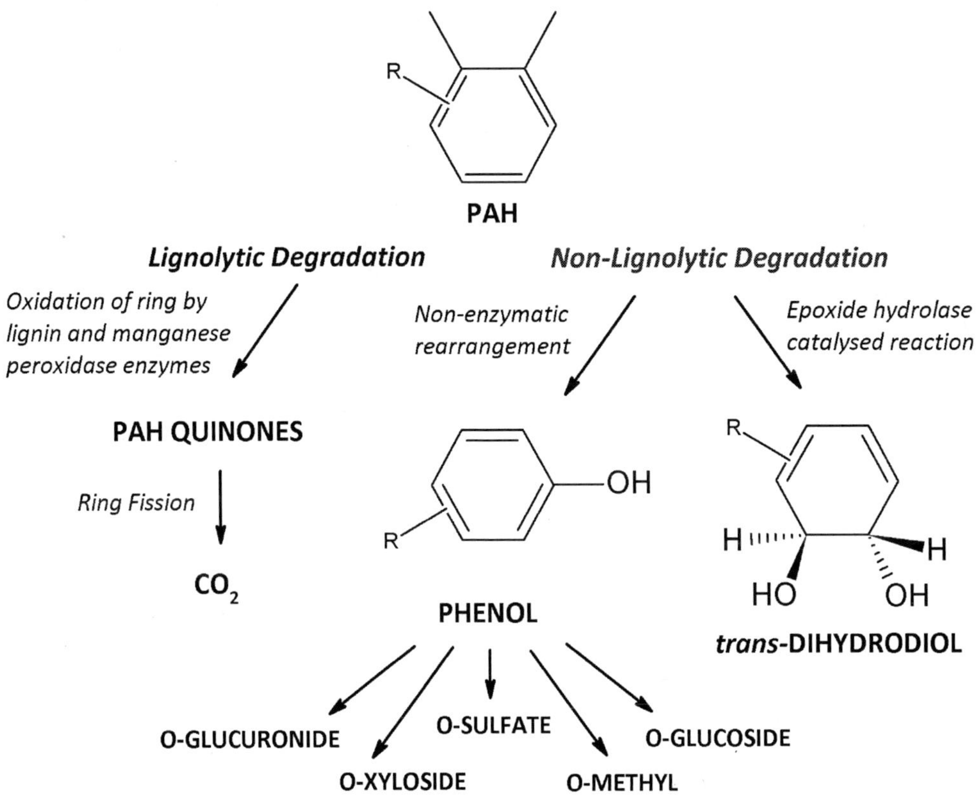

Figure 4. Fungal degradation of PAHs (modified from Bamforth and Singleton 2005).

Ligninolytic fungi, or 'white-rot fungi,' produce lignin peroxidise or laccase enzymes which allow them to oxidise lignin based compounds for cellular metabolism. Under ligninolytic conditions, peroxidase and laccase enzymes are capable of oxidising PAHs simultaneously to lignin. These enzymes generate free radicals by donating one electron to oxidise the PAH ring, producing PAH-quinones and acids. This degradation pathway is of particular interest as it is not highly substrate-specific and is able to degrade more highly recalcitrant PAHs than other metabolic systems, offering significant advantages to bacterial systems of degradation. Enzymes are produced extra cellularly and hence may diffuse further into the soil column. White rot fungi have been identified which are capable of metabolising PAHs with high molecular weights, such as benzo[*a*]pyrene (Cerniglia 1997; Eggen and Majcherczyk 1998).

3. Limiting Factors of PAH Biodegradation in Soil

PAH biodegradation processes are governed by a range of physical, chemical and biological factors which place crucial limits on microbial degradation activity. Although a wide range of ubiquitous microbial species have been identified which are capable of degrading PAHs for cellular metabolism, if native populations of these microorganisms are present at a source of contaminant, the rate at which degradation reactions can proceed may be inhibited by limits to pollutant/microbe interaction and a range of environmental conditions.

3.1. Contaminant Bioavailability

Pollutant/microbe interaction is a vital rate limiting factor to the biodegradation process. With few exceptions, microorganisms must be in direct contact with PAH molecules in order for cellular metabolism to occur. PAHs can be a problematic target pollutant for biodegradation-based treatment approaches, due to their high hydrophobicity and tendency for deep adsorption within soil pores and organic sediments. Even at sites where a PAH contaminant is present at very high concentrations, soil organisms may be strongly limited in their ability to access these molecules for cellular digestion (Zhang, Cheng et al. 2006). A growing body of literature suggests that soil microorganisms can only access and metabolise PAHs when they are present in the aqueous phase, whereas PAHs sorbed onto or partitioned in solid phase cannot be accessed (Masten and Davies 1997; Johnsen, Wick et al. 2005). Some microorganisms are able to produce surfactants to help facilitate solubilization of hydrophobic contaminants such as PAHs (Deziel, Paquette et al. 1996). Transfer of a PAH to the aqueous phase is influenced by a range of factors including solubility, soil moisture content, temperature, pH, and capacity for adsorption in a particular soil environment (Johnsen, Wick et al. 2005).

PAH sorption to soil pores or particle surfaces may be a key limiting factor in contaminant bioavailability in restricting mass transfer to the aqueous phase. There is no wide agreement in prevailing research as to what extent microbes are able to access sorbed PAHs. Generally, PAH adsorption in soil increases with surface area and cation exchange capacity, and is also affected by environmental conditions (Weissenfels, Klewer et al. 1992). Soil moisture content is a major determining factor in contaminant sorption and further aspects of bioavailability. Inadequate hydration of soils may inhibit bioavailability by restricting the dissolution of PAHs and limiting their transport through the soil column. Low water content will also have effects on the transport of nutrients and bacteria through the soil column (Providenti, Lee et al. 1993).

3.2. Physical and Environmental Factors

A wide range of physical and environmental factors can have a complex and often highly interactive range of physical and biological effects on systems of biological degradation. While some bacterial species may perform biodegradation activity under a range of conditions, others may have very specific requirements for cellular growth. All microbial cells require a range of nutrients for cellular growth and function. While heterotrophic PAH biodegrader species may obtain carbon for cellular respiration partially or entirely from PAH compounds, they also require sufficient levels of phosphorous and nitrogen, and various species may also require a wide range of other nutrients including sulphur, magnesium, potassium, calcium and other trace elements (Ward, Loehr et al. 1995). Sufficient water content is necessary for all cellular activities, although excess water can inhibit oxygen transport and limit aerobic growth. The oxygen requirement of microorganisms for aerobic metabolism of PAHs is widely variable. Molecular oxygen is utilised as both an electron acceptor in cellular respiration, and as a substrate for oxygenase-catalysed PAH biodegradation reactions. Therefore, many applied biodegradation treatment technologies seek to enhance soil aeration through venting (air injection), tilling, or composting with bulking agents. The redox potential of soils as a measure of available electron acceptors is an important consideration in evaluating the potential for biodegradation processes.

Physio-chemical characteristics of the soil environment, in particular temperature and pH, are important determining factors in contaminant solubility/bioavailability, sorption, volatilisation, microbial species community structure, and the rate of biodegradation activity. Most bacteria exhibit optimal growth in pH neutral conditions (pH 7), while many fungi can tolerate a wider pH range, particularly more acidic conditions at around 4-6 pH units (Ruiz-Aguilar, Fernández-Sánchez et al. 2002). Temperature is a major determining factor for microbial metabolic rate, where lowered temperatures may inhibit growth and raised temperatures may inactivate cellular enzymes. Generally, most aerobic metabolism of PAHs occurs under mesophilic conditions (10°-40°C). Within this range, rates of growth and biodegradation generally increase with greater temperature. Thermophilic bacteria have also been identified as effective biological degraders in temperatures of 40°-60°C (Eweis, Ergas et al. 1998). Real-world temporal and spatial fluctuations in soil column physio-chemical characteristics may cause significant variation in bacterial community structure and rates of biodegradation.

Interference to systems of biodegradation may arise as PAHs often occur as point-source pollution in the presence of other more readily accessible co-pollutants such as aliphatic hydrocarbon compounds, benzene, toluene, ethylbenzene and xylene (BTEX) compounds. Heavy metal co-pollutants may prevent the effect the efficiency of biodegradation activity by slowing rates of anaerobic oxidation and microbial growth (Bamforth and Singleton 2005). A wide range of environmental factors and seasonal effects may cause temporal and spatial fluctuation in rates of biodegradation activity at a particular site. Avoiding these fluctuations and optimising conditions for cellular growth is an essential aspect of establishing effective biodegradation systems for field application in treating PAHs.

4. Overcoming Limits to PAH Biodegradation in Soil

4.1. PAH Bioavailability

A growing body of scientific research is investigating the way by which limits to PAH bioavailability can be addressed by soil remediation technologies. Surface acting agents, or 'surfactants,' may increase bioavailability of PAHs by facilitating their transfer in aqueous phase (Khalladi, Benhabiles et al. 2009). A wide range of surface acting compounds exist, and can collectively be described as amphiphilic molecules which have both hydrophobic and hydrophilic domains. Amphiphilic molecules have a strong affinity to partition at the interface of zones which have different polarity and hydrogen bonding, such as boundaries between water and air or oil. At this partition, surfactants tend to lower surface tension and may disperse these phases and/or solubilise interacting compounds such as PAHs (Bai, Brusseau et al. 1997). PAHs which are sorbed into the soil column may be mobilised into aqueous phase and become available to organisms. Figure 5 illustrates surfactant behaviour. At adequate concentrations, surfactants form micelle structures (spheroid or lamellar), and solutes are partitioned into the hydrophobic micelle interior (Guha and Jaffé 1996). At lower concentrations than are necessary to form micelles, surfactants may still increase PAH solubility by lowering surface tension.

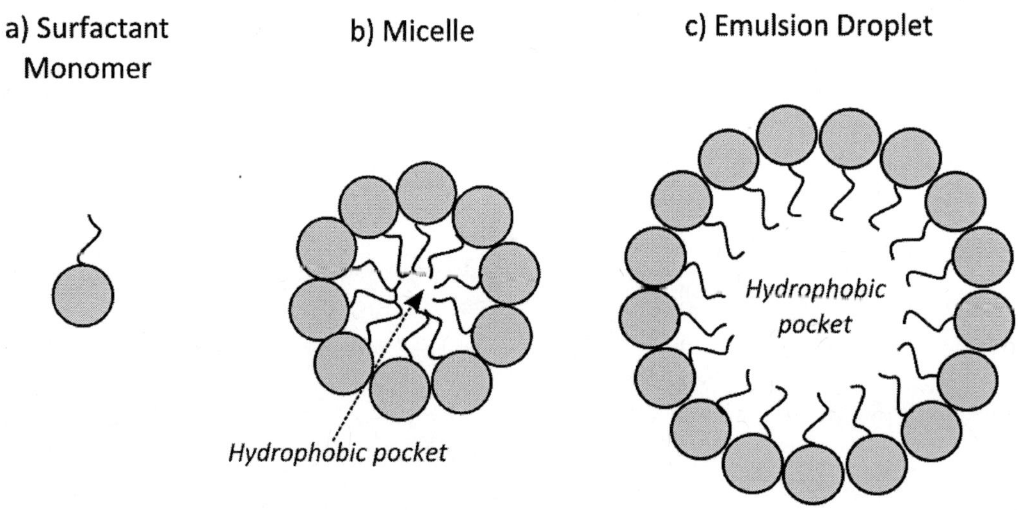

Figure 5. Surfactant formations in solution.

Certain types of surface acting compounds which are sourced from biological origins can be referred to as 'biosurfactants.' These surfactants are of particular interest to the development of soil remediation technologies as they tend to exhibit less negative environmental effects including lower levels of toxicity, higher biodegradability, and better tolerance of a wide range of temperature, pH and salinity (Bai, Brusseau et al. 1997).

Biosurfactants have been identified which demonstrate distinct biochemical advantages over their synthetic counterparts, including low interfacial tensions and lower concentrations required for micelle formation. Certain biosurfactants can be biologically synthesised using bacterial cultures and renewable sources such as glucose and glycerol (Wu, Yeh et al. 2008).

It is important to consider the full range of effects which surfactants may have in the aqueous environment. The effect of a surfactant on bacterial cell surface behaviour is highly specific to the particular combination of surfactant and bacterial species. Depending on the specific nature of a surfactant and the microbial cell surface, surface acting agents may alter cell surface hydrophobicity, increase cell-surface repulsion, or improve micellar sequestration of PAHs, thereby reducing contaminant bioavailability and limiting biodegradation activity. Overall, the use of surfactants and biosurfactants has strong potential for field applications in optimising biodegradation treatment technologies.

4.2. Enhancing Microbial Growth and PAH Degradation Activity

The application of in-situ biodegradation processes has been accompanied by a wide range of engineered technologies which focus on ways to enhance the growth rates and intrinsic metabolic degradation activity of microbial species at contaminated sites. Where some PAH contaminated sites are capable of 'intrinsic bioremediation' by endemic microbial populations, this alone is rarely a reliable or economically feasible option. Not all sites will have endemic populations of microorganisms which have the necessary genetic elements to degrade pollutants; pre-existing populations of such organisms may only establish concentrated populations over many years. 'Bioaugmentation' technologies address the limits to microbial populations by adding cultured microbes to a contaminated site. There is a strong potential for this approach to be combined with techniques that enhance bioavailability by delivering bacteria in liquid or foam biosurfactant. Bioaugmentation may also be applied to ex-situ biodegradation treatment processes, such as bioreactors. Even if capable biodegrading microorganisms are present at a site, their cellular metabolism is limited by the availability of electron acceptors (see section 3.2 above). 'Biostimulation' is the process by which compounds are added to a site in order to facilitate enhanced rates of microbial metabolism, usually nutrients or co-substrates (Hamdi, Benzarti et al. 2007).

Once a population of biodegrading bacteria has become established in a contaminated soil column, their growth and biomass production (metabolic activity) is limited by the expression of the genetic elements which regulate the production of pollutant degrading cellular compounds (usually oxygenase enzymes). Increasingly, modern molecular biology is able to identify and characterise these specific genes and their regulation pathways, such as the pathways involved in naphthalene degradation through bacterial oxygenation (Figure 2). Salicylate, the intermediary product in this two-stage biodegradation pathway, is known to stimulate the expression of genes which produce bacterial deoxygenase enzymes, forming a positive feedback system for degradation gene expression once PAH byproducts are present. A simplified overview of a cellular feedback system for the bacterial production of PAH degradation enzymes is illustrated in Figure 1Figure 6. There is potential to stimulate biodegradation activity in the soil column through the engineered application of liquid salicylate to contaminated sites which have established populations of suitable microbial degraders (Tian, Ma et al. 2003).

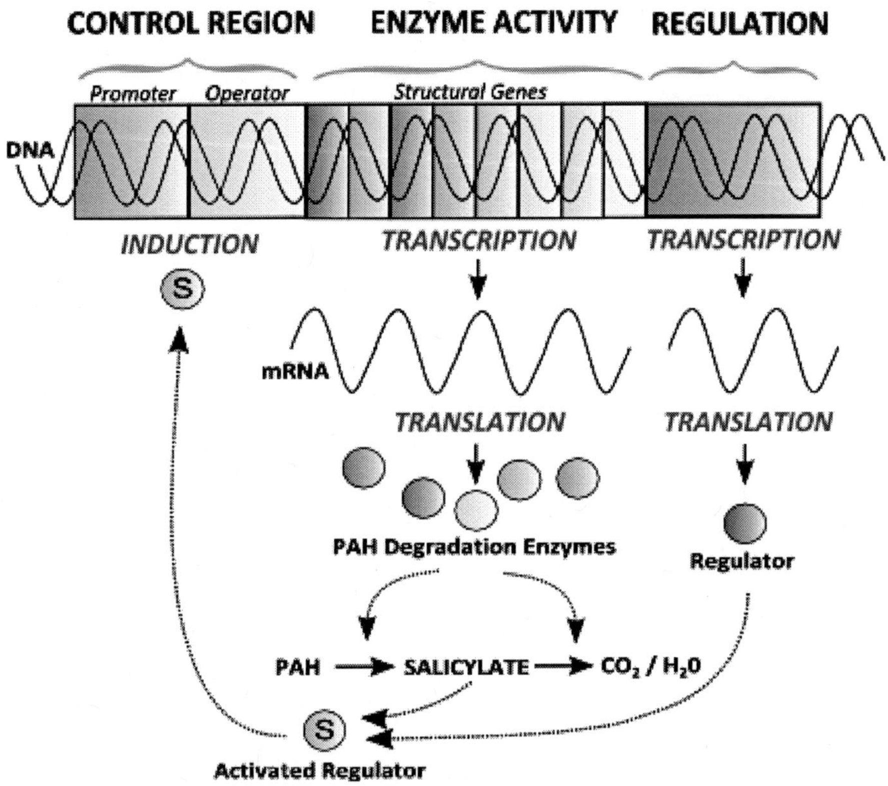

Figure 6: Simplified overview of genetic feedback system involved in aerobic bacterial PAH degradation. A control region on the bacterial genome directs the expression of structural genes which which code for PAH degradation enzymes (synthesised via transcription and translation of mRNA). A regulation region of the genome codes for regulator proteins, which interact with an intermediary PAH degradation product (salicylate). This complex binds to a site on the control region; signalling it to enhance expression of PAH degradation enzymes in a positive feedback system.

5. Bioremediation Approaches for PAH Contaminated Soil

Soil bioremediation is the process by which pre-existing systems of biological degradation activity are enhanced in engineered treatment processes to remove a target contaminant. The key objectives for bioremediation processes involve enhancing rates of microbial growth and biomass, and increasing cellular metabolism for higher rates of biodegradation. Many of these approaches seek to overcome these limits. A range of soil bioremediation treatment approaches have been identified as effective means for degrading high concentrations of PAHs in the soil column.

5.1. Land Treatment

Also referred to as *solid-phase treatment* or *landfarming,* land treatment describes the application of agricultural practices in the decomposition of organic waste. This simple process involves spreading waste over a non-contaminated soil surface, then tilling the site to mix the waste into the upper soil column while enhancing aeration and microbial activity. Land farming became a popular approach to treating hazardous and non-hazardous organic waste such as PAHs over the first half of the twentieth century. Removal rates of 90-99% have been documented for total petroleum hydrocarbons (Ward, Loehr et al. 1995). Traditionally, land treatment relied upon processes of volitization and leaching as key mechanisms for pollutant removal from a contaminated soil site. However, these processes have become discouraged as modern environmental regulations have recognised the wide range of environmental effects caused by contaminant transport systems. Biological removal is the key process involved in modern land treatment applications. More detailed information regarding land treatment processes can be found in Eweis, Ergas et al.(1998) and Ward, Loehr et al.(1995).

5.2. Composting

Soil composting involves the oxidation of wet organic solids to more biologically stable forms through aerobic biological processes. Waste is piled in mounds and highly biodegradable organic material such as manure and sawdust may be added to increase metabolic potential. In some cases, additional microbial populations may be stimulated through the addition of sewerage sludge which is rich in electron acceptors. The release of heat in conditions with high organic content and low moisture causes a 'heat-trap' effect in the waste mounds, where elevated temperature conditions often exceed 55°C (Eweis, Ergas et al. 1998). These temperatures are useful in enhancing the biological removal of hazardous compounds such as PAHs. Modern treatment approaches have allowed composting treatment treatments to be optimised for enhanced PAH biodegradation (refer to Antizar-Ladislao, Lopez-Real et al. 2006 for further information).

5.3. Bio-Venting

The bio-venting process introduces air movement through surface soils to stimulate the aerobic microbial degradation of organic contaminants such as PAHs into less hazardous end products. Air movement is induced in the soil column through the placement of air wells, and in some cases air recovery wells. Additionally, organic nutrients and moisture may also be applied to soil in order to further reduce the limits to biological activity (refer to Eweis, Ergas et al. 1998; Frutos, Escolano et al. 2010 for further information regarding bio-venting treatment processes and applications).

5.4. Soil Vapour Extraction

The soil vapour extraction process removes volatile organic compounds for the soil column through the installation of shallow ventilation wells and the application of negative pressure. Contaminants are carried out of the soil column in vapour form, and separated into liquid and vapours for removal and further treatment. This process is a remediation process which may be effective for volatile PAHs such as gasoline constituents, but does not facilitate the degradation of these contaminants (refer to Ward, Loehr et al. 1995 for further information).

5.5. Ex-Situ Treatment

Ex-situ treatment processes are applied to a pollutant after contaminated soil has been physically removed from a site and transported to a treatment facility. Treatment technologies may induce degradation of contaminants through chemical or biological processes. A number of ex-situ biological treatment technologies for contaminants such as PAHs take place in either open or closed bioreactors. Reactors may include tanks, lined lagoons, and soil piles, which are designed to maximise biological degradation processes. These treatment technologies are subject to similar rate limiting factors imposed in in-situ biological remediation systems (Rochkind, Blackburn et al. 1986; Ward, Loehr et al. 1995).

6. Systems for Monitoring PAH Biodegradation Activity

As the potential for applied bioremediation in addressing soil pollution has become more apparent over recent years, there has been a growing emphasis on the need to characterise the complex array of factors which influence biodegradation processes in order to engineer feasible treatment strategies. Increasingly, the importance of monitoring biodegradation processes at a cellular level has been recognised. Technologies for monitoring cellular processes of biodegradation can be applied in laboratory trials in order to optimise and validate engineered treatment approaches, or in the field to monitor real-time in-situ treatment applications. Rapid advances in the fields of cellular technology which have allowed for the development of novel methods to monitor microbial growth, distribution, activity and gene expression, may be applied to biodegradation in both these scenarios.

6.1. Fluorescent Proteins as Microbial Labels

In recent years advances in recombinant gene technology have allowed molecular biologists to develop 'biosensor' microorganisms, the activity and distribution of which can be monitored through their production of fluorescent or bioluminescent proteins. Such organisms are engineered with genes which encode the production of reporter proteins as a part of cellular function. Fluorescence and bioluminescence may be expressed constantly as a part of basic cellular growth (constitutive expression), or linked directly to the expression of genes associated to a particular activity (promoter linked expression). Biosensor organisms have been engineered to respond to a wide range of conditions (Yagi 2007).

Fluorescent and luminescent proteins are now available as genetic tools in the development of 'biosensor' microorganisms. A wide range of these proteins have been developed which are compatible with many bacterial species, along with an equally diverse range of advantages and disadvantages. Bioluminesence is associated with the bacterial luciferase or '*lux*' gene, which produces light independently through luciferase-catalyzed reactions. Bioluminesence is attractive in that it does not require an excitation source for signal response, and a real-time indication of biosensor activity (Kohlmeier, Mancuso et al. 2007). Systems of fluorescence produce proteins which require external excitation to produce a signal response. These systems may be favored in biosensor development as they involve much simpler genetic pathways than those involved in luminescence, and a wide variety of emission wavelengths are available (Leveau and Lindow 2002; Jansson 2003). Common fluorescent proteins include green fluorescent protein (Gfp) and red fluorescent protein (DsRed), with genes originally sourced from jellyfish and coral respectively. Unlike luminescence, fluorescence systems can be easily quantified at a cellular level in terms of single cell response using microscopy or flow cytometry techniques.

In recent years, a growing range of biosensor organisms have been developed and applied to a range of microbiological applications. For example, biosensors can be used to identify and measure the presence of toxic compounds, or indicate the expression of genes associated with environmental stresses and cellular responses such as sporulation and biofilm formation (Gu and Chang 2001; Lee, Villaume et al. 2003; Yagi 2007). Advances in this technology mean it is entirely feasible to apply biosensors in measuring and characterizing PAH biodegradation activity; for example by associating the bacterial production of Gfp with the production with naphthalene deoxygenase enzyme. Multiple systems of fluorescence and/or luminescence may potentially be used in combination to indicate interacting factors, such as correlating cell growth with biodegradation activity.

6.2. Monitoring Living Systems

Traditionally, systems for monitoring biodegradation activity have been limited to the measurement of soil redox potential, 'before and after' analyses of PAH contaminant concentrations and toxicology assays. These measures do not provide any detail on the range of complex processes that take place in systems of biodegradation at environmental and cellular levels. The development of reliable bioremediation technologies applicable to field environments has been stalled by these limits to understanding microbial processes and how they can be optimised. The development of biosensor technology offers an attractive alternative to traditional methods of monitoring and characterising biodegradation activity. While this technology is still in its infancy, rapid advances in molecular technology indicate its strong potential as a means to shed light on processes of PAH metabolism. Fluorescence and bioluminescence may be measured as a population response or in some scenarios at a cellular level using a range of technologies including epifluoresence microscopy, digital imaging, and flow cytometry (Maraha, Backman et al. 2004). There is strong potential for the development of technology which is capable of collecting measurements of in-situ fluorescence or bioluminescence, for application in soil column bioremediation trials without the requirement of destructive sampling which may interfere with microbial activity. However, the use of biosensor organisms is largely limited to controlled laboratory modelling scenarios. Genetically modified organisms are unlikely to be authorised for release into the environment and therefore their application in actual field trials is limited.

7. Conclusions

Engineered systems of microbial biodegradation present a promising strategy for addressing the increasingly widespread problems associated with PAH soil pollution. Biodegradation processes are becoming increasingly well defined by modern research as performed by variety of ubiquitous microorganisms in a broad range of environmental conditions. A growing number of engineered treatment approaches have been developed in order to overcome the array of complex and problematic limits to microbial biodegradation activity caused by environmental conditions, pollutant bioavailability, microbial distribution, activity and gene expression.

Although many of these processes involved in PAH biodegradation are not yet well understood, advancements in recombinant gene technology present the very real possibility of monitoring and defining biodegradation systems down to a cellular level. Application of biosensor technology to bioremediation trials and the development of efficient fluorescence imaging systems is important aspect in understanding the activity and distribution of biosensor organisms. Ultimately, cellular technology is a vital aspect of developing an understanding of the biodegradation process that will allow for the development of reliable engineered systems for field application.

References

Antizar-Ladislao, B., J. Lopez-Real, et al. (2006). "Bioremediation of polycyclic aromatic hydrocarbons (PAH) in an aged coal-tar-contaminated soil using different in-vessel composting approaches." *Journal of Hazardous Materials* 137(3): 1583-1588.

Bai, G., M. Brusseau, et al. (1997). "Biosurfactant-enhanced removal of residual hydrocarbon from soil." *Journal of contaminant hydrology* 25(1-2): 157-170.

Bamforth, S. and I. Singleton (2005). "Bioremediation of polycyclic aromatic hydrocarbons: current knowledge and future directions." *Journal of Chemical Technology & Biotechnology* 80(7): 723-736.

Boopathy, R. (2000). "Factors limiting bioremediation technologies." *Bioresource Technology* 74(1): 63-67.

Boström, C., P. Gerde, et al. (2002). "Cancer risk assessment, indicators, and guidelines for polycyclic aromatic hydrocarbons in the ambient air." *Environmental health perspectives* 110(Suppl 3): 451.

Cerniglia, C. (1993). "Biodegradation of polycyclic aromatic hydrocarbons." *Current opinion in Biotechnology* 4(3): 331-338.

Cerniglia, C. (1997). "Fungal metabolism of polycyclic aromatic hydrocarbons: past, present and future applications in bioremediation." *Journal of Industrial Microbiology and Biotechnology* 19(5): 324-333.

Chakraborty, R. and J. D. Coates (2004). "Anaerobic degradation of monoaromatic hydrocarbons." *Applied microbiology and biotechnology* 64(4): 437-446.

Deziel, E., G. Paquette, et al. (1996). "Biosurfactant production by a soil Pseudomonas strain growing on polycyclic aromatic hydrocarbons." *Applied and environmental microbiology* 62(6): 1908.

Eggen, T. and A. Majcherczyk (1998). "Removal of polycyclic aromatic hydrocarbons (PAH) in contaminated soil by white rot fungus Pleurotus ostreatus." *International Biodeterioration & Biodegradation* 41(2): 111-117.

Eweis, J. B., S. J. Ergas, et al. (1998). *Bioremediation Principles,* WCB/McGraw Hill.

Frutos, F. J. G., O. Escolano, et al. (2010). "Bioventing remediation and ecotoxicity evaluation of phenanthrene-contaminated soil." *Journal of Hazardous Materials* In Press, Corrected Proof.

Grishchenkov, V., R. Townsend, et al. (2000). "Degradation of petroleum hydrocarbons by facultative anaerobic bacteria under aerobic and anaerobic conditions." *Process Biochemistry* 35(9): 889-896.

Gu, M. and S. Chang (2001). "Soil biosensor for the detection of PAH toxicity using an immobilized recombinant bacterium and a biosurfactant* 1." *Biosensors and Bioelectronics* 16(9-12): 667-674.

Guha, S. and P. Jaffé (1996). "Biodegradation kinetics of phenanthrene partitioned into the micellar phase of nonionic surfactants." *Environ. Sci. Technol* 30(2): 605-611.

Habe, H. and T. Omori (2003). "Genetics of polycyclic aromatic hydrocarbon metabolism in diverse aerobic bacteria." *Bioscience, biotechnology, and biochemistry* 67(2): 225-243.

Hamdi, H., S. Benzarti, et al. (2007). "Bioaugmentation and biostimulation effects on PAH dissipation and soil ecotoxicity under controlled conditions." *Soil Biology and Biochemistry* 39(8): 1926-1935.

Haritash, A. and C. Kaushik (2009). "Biodegradation aspects of polycyclic aromatic hydrocarbons (PAHs): a review." *Journal of Hazardous Materials* 169(1-3): 1-15.

Hart, S. (1996). "In situ bioremediation: Defining the limits." *Environmental Science & Technology* 30(9): 398.

Jansson, J. (2003). "Marker and reporter genes: illuminating tools for environmental microbiologists." *Current Opinion in Microbiology* 6(3): 310-316.

Johnsen, A., L. Wick, et al. (2005). "Principles of microbial PAH-degradation in soil." *Environmental pollution* 133(1): 71-84.

Jonker, M., S. Van der Heijden, et al. (2007). "Predicting PAH bioaccumulation and toxicity in earthworms exposed to manufactured gas plant soils with solid-phase microextraction." *Environ. Sci. Technol* 41(21): 7472-7478.

Khalladi, R., O. Benhabiles, et al. (2009). "Surfactant remediation of diesel fuel polluted soil." *Journal of Hazardous Materials* 164(2-3): 1179-1184.

Kohlmeier, S., M. Mancuso, et al. (2007). "Bioreporters: gfp versus lux revisited and single-cell response." *Biosensors and Bioelectronics* 22(8): 1578-1585.

Lee, H., J. Villaume, et al. (2003). "Monitoring and classification of PAH toxicity using an immobilized bioluminescent bacteria." *Biosensors and Bioelectronics* 18(5-6): 571-577.

Leveau, J. H. J. and S. E. Lindow (2002). "Bioreporters in microbial ecology." *Current Opinion in Microbiology* 5(3): 259-265.

Maraha, N., A. Backman, et al. (2004). "Monitoring physiological status of GFP-tagged Pseudomonas fluorescens SBW25 under different nutrient conditions and in soil by flow cytometry." *FEMS Microbiology Ecology* 51(1): 123-132.

Masten, S. and S. Davies (1997). "Efficacy of in-situ for the remediation of PAH contaminated soils." *Journal of contaminant hydrology* 28(4): 327-335.

Nisbet, I. C. T. and P. K. LaGoy (1992). "Toxic equivalency factors (TEFs) for polycyclic aromatic hydrocarbons (PAHs)." *Regulatory Toxicology and Pharmacology* 16(3): 290-300.

Providenti, M., H. Lee, et al. (1993). "Selected factors limiting the microbial degradation of recalcitrant compounds." *Journal of Industrial Microbiology and Biotechnology* 12(6): 379-395.

Rochkind, M. L., J. W. Blackburn, et al. (1986). *Microbial Decomposition of Chlorinated Aromatic Compounds*. Cincinnati, Ohio, United States Environmental Protection Agency.

Rogers, S., S. Ong, et al. (2002). "Natural attenuation of polycyclic aromatic hydrocarbon-contaminated sites: review." *Practice Periodical of Hazardous, Toxic, and Radioactive Waste Management* 6: 141.

Ruiz-Aguilar, G. M. L., J. M. Fernández-Sánchez, et al. (2002). "Degradation by white-rot fungi of high concentrations of PCB extracted from a contaminated soil." *Advances in Environmental Research* 6(4): 559-568.

Samanta, S. K., O. V. Singh, et al. (2002). "Polycyclic aromatic hydrocarbons: environmental pollution and bioremediation." *Trends in Biotechnology* 20(6): 243-248.

Tian, L., P. Ma, et al. (2003). "Impact of the presence of salicylate or glucose on enzyme activity and phenanthrene degradation by Pseudomonas mendocina." *Process Biochemistry* 38(8): 1125-1132.

Van Metre, P., B. Mahler, et al. (2009). "PAHs underfoot: Contaminated dust from coal-tar sealcoated pavement is widespread in the United States." *Environ. Sci. Technol* 43(1): 20-25.

Ward, C. H., R. C. Loehr, et al. (1995). *Bioremediation*, American Academy of Environmental Engineers.

Weissenfels, W. D., H.-J. Klewer, et al. (1992). "Adsorption of polycyclic aromatic hydrocarbons (PAHs) by soil particles: influence on biodegradability and biotoxicity." *Applied microbiology and biotechnology* 36(5): 689-696.

Wilcke, W. (2000). "Polycyclic Aromatic Hydrocarbons (PAHs) in Soil-a Review." *Journal of plant nutrition and soil science* 163(3): 229-248.

Wu, J.-Y., K.-L. Yeh, et al. (2008). "Rhamnolipid production with indigenous Pseudomonas aeruginosa EM1 isolated from oil-contaminated site." *Bioresource Technology* 99(5): 1157-1164.

Yagi, K. (2007). "Applications of whole-cell bacterial sensors in biotechnology and environmental science." *Applied microbiology and biotechnology* 73(6): 1251-1258.

Ye, D., M. Siddiqi, et al. (1995). "Degradation of polynuclear aromatic hydrocarbons by Sphingomonas paucimobilis." *Environ. Sci. Technol* 30(1): 136-142.

Zhang, X., S. Cheng, et al. (2006). "Microbial PAH-Degradation in Soil: Degradation Pathways and Contributing Factors." *Pedosphere* 16(5): 555-565.

In: Bioremediation
Editor: Alexander C. Mason

ISBN: 978-1-61122-730-7

Chapter 6

INTERACTION OF COPPER OR CHROMIUM WITH YEASTS: POTENTIAL APPLICATION ON POLLUTED ENVIRONMENTAL CLEAN UP

Liliana B. Villegas'[1, 2], María J. Amoroso[1, 2, 3] and Lucía I.C. de Figueroa [1, 4]

[1]PROIMI - CONICET, Av. Belgrano y Pje. Caseros, 4000, Tucumán, Argentina
[2]Facultad de Ciencias de la Salud. Universidad del Norte Santo Tomás de Aquino
[3]Microbiología General, Facultad de Bioquímica, Química y Farmacia, Universidad Nacional de Tucumán, Argentina
[4]Microbiología Superior, Facultad de Bioquímica, Química y Farmacia, Universidad Nacional de Tucumán, Argentina

ABSTRACT

Heavy metal pollution is one the most serious environmental problems. Mining and industrial activity are the main sources of heavy metal contamination. Heavy metals, such as copper (Cu) and chromium (Cr), play an important role as trace elements in biochemical reactions but these heavy metal ions are toxic at higher concentrations, producing serious public health problems.

Traditional technologies for the removal of these heavy metals are very expensive and may have several disadvantages. New techniques are required for reducing their concentrations to environmentally acceptable levels at low costs.

Microorganisms are potent bioremediators. To date, studies on heavy metal resistant have mainly concerted in bacteria. Nevertheless, one of the most ubiquitous biomass types utilized on a large scale by man for centuries are yeasts, moreover this microorganism is an inexpensive, readily available source of biomass. Furthermore, yeast cells are able to accumulate a broad range of heavy metals to varying degrees under a wide range of external condition.

Recent researches have made emphasis to the yeasts isolated from contaminated area. This allows isolating indigenous yeasts able to live in these unfavorable environments.

Candida fukuyamaensis and *Rhodotorula mucilaginosa* were isolated from a copper filter at a mine plant. These yeasts show multiple heavy metal tolerance. They are able to remove Cu(II) of the culture medium by a process of bioaccumulation in the cytoplasm. The resistance mechanism in the yeasts is through sequestration in cytoplasmic of copper ions rather than avoidance or compartmentalization. However, the copper accumulation profiles of both strains are different: whereas the metal is later released by *C. fukuyamaensis*. *R. mucilaginosa* has the capacity to keep copper inside the cell. The toxicity of copper is evidenced for the morphological changes and the increased oxidative stress response by both strains. However the copper bioaccumulation by *R. mucilaginosa* is increased by adaptation of the strain under controlled aeration

On the other hand, *Lecythophora* sp. NGV-1, *Candida sp.* NGV-9 and *Aureobasidium pullulans* VR-8 isolated from tannery effluents and from a nickel-copper mine, show Cr(VI) tolerance . In these cases, the yeast Cr(VI) tolerant is due to the reduction of Cr(VI) to Cr(III) and no bioaccumulation. In vivo Cr(VI) toxicity is 1,000-fold more cytotoxic than Cr(III). Apart from its toxicity, Cr(VI) is highly soluble and thus mobile and biologically available in the ecosystems. In contrast, Cr(III) has a high affinity for organic compounds resulting in the formation of complexes that precipitate as amorphous hydroxides. The reduction of Cr(VI) to Cr(III) is an interesting approach for environmental clean up.

The ability of metal processing by yeasts can be used to concentrate, remove and recover metals from streams and could enhance the efficiency of wastewater treatment processes. This capacity depends of the specific interaction of strain and heavy metal. Different mechanisms of heavy metal resistance may affect different bioremediation strategies.

INTRODUCTION

Daily the societies cope with the need to protect to the environment. The industrial activities, which are necessary for the modern life, generate a series of environmental dangers by its effluent and its emanations. The majority of the industrial wastes contain harmful substances of different nature, being the heavy metals one of the most abundant and toxic pollutants present in them. The heavy metals are not chemical not biological degraded, as the majority of the organic compounds, therefore they tend to circulate, thus, they can remain for a long time in the nature.

When they have been released through the atmosphere or into waterways, they end up as diffuse pollutants in soils and sediments. The accumulation of these compounds, especially in urban areas, can to generate the filtration towards the underground waters: the aquiferous that constitute the reservation of drinking water.

The accumulation of high concentration of heavy metals affects the biological diversity causing serious ecological problems and negative influence on soil fertility. When the accumulation is in aquiferous and superficial waters can cause diseases and up to the death of animals with the consistent economic loss. Also they can to produce the increase in the food chain exposing to the population to the daily consumption of unknown quantities of pollutants, which puts in risk the human health (Soares et al. 2003; Malik 2004; Machado et al. 2010).

HEAVY METALS DEFINITION

The term "heavy metals" is often used as a group name for metals and semimetals that have been associated with contamination and potential toxicity or ecotoxicity. The use of this term is inconsistently reflected for discrepancy in the scientific literatura and lead to general confusion because tendency to assume that all so-colled "heavy metals" have highly toxic or ecotoxic properties. The term has never been defined by any authoritative body such as IUPAC (International Union of Pure and Applied Chemistry). The most ancient definition of heavy metals is based upon the density of the elemental form of the metal, and he classifies heavy metals as those metals with elemental densities above 7 g ml^{-1}. Over the years, this definition has been modified by various authors, and there is no consistency because no relationship can be found between density and any of the various physicochemical concepts that have been used to define "heavy metals" and the toxicity or ecotoxicity attributed to them (Duffus 2002).

Some heavy metal can be toxic to various life forms; others are considered essential, but they can become toxic at higher doses. For that reason, some researchers propose to classify to the heavy metals in two categories, according to physiological point of view (Valls and Lorenzo 2002): essential, but harmful at high concentration (typically, Fe, Mn, Zn, Cu, Co, Ni, Cr) and toxic (e.g. Hg, Cd). Further, interactions with heavy metals depend not only on the particular element but also on its chemical speciation. This means that the toxicity depend strongly on the particular form in which the element is present in the system named species: electronic or oxidation state, and/or complex or molecular structure. For example Cr(III) may be an essential element for normal glucose metabolism, but Cr(VI) is genotoxic and carcinogenic. Diverse chemical species of an element possess different physics and chemistry properties that influence in bioavailability and solubility, affecting the absorption and the elimination and therefore in the toxicity characteristics of them (Templeton et al, 2000; Hursthouse, 2001).

COPPER

In humans, copper is third in abundance amongst the essential heavy metals, after Fe and Zn (Barceloux 1999b). Copper is a metal able to cycle between two redox states, oxidized Cu(II) and reduced Cu(I). All organisms require copper as a catalytic cofactor for biological process such as respiration, iron transport, oxidative stress protection, normal cell growth and development, etc. (Puig and Thiele 2002).

Although copper is an essential heavy metal for metabolism at lower concentrations, it is toxic to organisms, either eukaryotes or prokaryotes, above specific threshold concentrations and it is a severe public health problem (Arnesano et al. 2001). Likewise, its toxicity is exploited in agriculture as commercial fungicides and algicides.

The chronic consumption of excessive amounts of copper was associated with the development of acute gastrointestinal symptoms (nausea, vomiting, abdominal pain). The medicinal use of copper as an emetic was discontinued because of its toxicity. The health-based guidelines for drinking water in many countries ranges from 1–2 mg Cu L^{-1}.

Currently, there is considerable debate over the scientific basis for these guidelines because of the lack of both human and animal data on copper toxicity (Barceloux 1999b).

Cu (II) is known to be one of the most widespread heavy metal contaminant of the environment. Copper enters aquatic systems primarily through mining and several industrial activities. These systems are often the source of drinking water (Salomons, 1995; Audry et al., 2004).

CHROMIUM

In nature the most common and stable oxidation states of this element are Cr(VI) and Cr(III) (Barceloux 1999b; Iyer et al. 2004). Cr(III) is an essential trace element well known for its particular role in the maintenance of normal carbohydrate metabolism in mammals and yeasts (Debski et al. 2004). It has also been suggested that this ion is involved in the tertiary structure of proteins and the conformation of cell RNA and DNA (Gulan Zetic *et al.* 2001; Zayed and Terry 2003). However, at high concentrations, Cr(III) showed negative effects on cellular structures. In contrast, Cr(VI) is always toxic to living organisms causing allergies, eczema, irritations and respiratory track disorders, and it is a strong mutagenic and carcinogenic agent. This toxic action is due to the fact that Cr(VI) complexes can easily penetrate cellular membranes and undergo immediate reduction reactions leading to the formation of various reactive intermediates. These intermediates are harmful to cell organelles, proteins and nucleic acids (Raspor et al. 2000; Plaper et al. 2002). In vivo Cr(VI) toxicity is 1,000-fold more cytotoxic than Cr(III) (Biedermann and Landolph 1990; Barceloux 1999a). Apart from its toxicity, Cr(VI) is highly soluble and thus mobile and biologically available in the ecosystems. On the contrary, Cr(III) has a high affinity for organic compounds resulting in the formation of complexes that precipitate as amorphous hydroxides (Palmer and Wittbrodt 1991).

Contamination by chromium compounds is caused by a variety of industrial applications eventually leading to heavy pollution of soil, surface water and atmosphere. Major sources of chromium contamination include the metal finishing industry, petroleum refineries, iron and steel industries, textile dyeing and leather tanning.

Kalbitz y Wennrich (1998) studied the mobility of accumulated heavy metals, as Copper and Chromium, in sediments contaminated of the Mulde river in Germany. The acid-soluble concentrations were up to 0.8 for Cr and 0.36 mg g-1 for Cu. The mobility or/ and translocation of heavy metals to deeper caps of the soil and to the underground waters, concern elements with a low and with a high affinity to organic matter. The mobility process of the heavy metals depends: a) of the total concentration of the metal in the sediments, b) of the intrinsic mobility of the heavy metal, c) of the pH of soil and d) of other characteristics of the soil like the concentration of organic matter dissolved in it. It is necessary to emphasize that both heavy metals, Cu and Cr have high affinity to the organic matter of the soil and the mobility determined for Cu was from 33 to 173 g ha^{-1} $year^{-1}$ and for Cr was from 31 to 78 g g ha^{-1} $year^{-1}$ (Kalbitz and Wennrich, 1998).

BIOREMEDIATION

In the developed countries the heavy metals removal by chemical traditional methods, such as oxidation and reduction, chemical precipitation, filtration, electrochemical treatment, evaporation, ion-exchange and reverse osmosis. Theses processes are very expensive and may have several disadvantages, such as unpredictable metal ion removal, high reagent requirements and the generation of toxic sludge, which are often difficult to dewater and require extreme caution when disposing of them and secondary environmental pollution (Silóniz et al. 2002). These disadvantages accent even more when the metal concentrations to removing are too low. The situation became more critical in developing countries where legislation and water treatment facilities are poor. As a result, it is important to found new techniques to reduce heavy metal concentrations up to acceptable environmental levels at low costs and to protect the human health.

The remediation of environments contaminated with biological methods, bioremediation, offers high specificity in the removal of the heavy metal of interest with operational flexibility. A great variety of microorganisms like bacteria, yeast, algae and fungi have been found into contaminated sites with heavy metals and they are now understudy as bioremediators. Bioremediation processes with microbial biomass are of interest due to their eco-friendly nature and to be a cost-effective alternatives in comparison to physicochemical technique above mentioned (Silóniz et al. 2002; Malik, 2004).

The types of microbial biomass of interest can be pragmatically those that can be easily obtained in large quantities. One of the most ubiquitous biomass types utilized on a large scale by man for centuries are yeasts. Furthermore, yeast cells retain their ability to accumulate a broad range of heavy metals to varying degrees under a wide range of external condition. Earlier observations noting the capacity of yeast strains to sequester heavy metals have been largely scattered and inconsistent, focusing mainly on the nutritional requirements or toxicology of the yeast cell (Volesky and May-Phillips 1995). Recent researches have made emphasis to the yeasts isolated from contaminated area. This allows isolating indigenous yeasts able to live in these unfavorable environments. Yeasts can contribute to the understanding of metal uptake and toxicity, and how metal resistance may be achieved in eukaryotic cells. In a biotechnological context, yeasts may be useful in the metal-containing effluents treatment.

CU(II) AND CR(VI)-RESISTANT YEASTS

Sampling and Isolation

In the last decades great importance has had the study of the diversity of the biological systems and especially, the different forms of life adapted to specific ecosystems as those wherefrom the samples were extracted. The high concentrations of heavy metals influenced negatively on microbial soil, water and sediment communities (Romero *et al.*, 1999; Turpeinen *et al.*, 2004; Branco *et al.*, 2005; Pal et al., 2005).

There is not many information about this yeast populations in Argentina therefore, Villegas et al (2005; 2008) worked in it. These authors worked with samples collected from 3 contaminated areas in Argentina: from wastewater sediment of a copper filter mine plant located in Tucumán province; from tannery ponds, situated in Nonogasta, La Rioja province, and sediment from a nickel-copper mine in Virorco, San Luis province.

Total chromium and copper concentration were performed by atomic absorption spectrophotometry, the results are summarized in the table 1. This table also shows the number of isolations with colonies and cellular morphology typical to yeasts obtained in every zone. These isolations were obtained following the method described for (Villegas *et al.* 2004). These polluted area allowed to isolate indigenous yeasts able to live in these unfavourable environments.

The isolates from copper filter mine plant were labelled RCL, from tannery samples were named NGV (Nonogasta) and from mine samples VR (Virorco) followed by a serial number.

Table 1. Sampling zone, total copper and chromium concentrations, Number of isolations (Villegas 2006)

Sediments	Province	Copper mg g^{-1}	Chromium mg g^{-1}	number of Isolations
Copper filter mine plant	Tucumán	0.52 – 75.30	ND*	11
Tannery ponds	La Rioja	0.02 – 0.08	0.75 – 1.50	13
Nickel-copper mine	San Luís	3 - 28	0.008 – 0.010	8

*ND:Not detected

Qualitative Screening for Heavy Metal Tolerance on Solid Medium

Heavy metals tolerance of the isolates was tested using the method of agar diffusion previously described for actinomycetes (Amoroso *et al.* 1998) and modified to the screening yeasts by Villegas *et al* (2004). Experiments were performed with the following heavy metals separately: Cu(II) (as $CuSO_4$), Ni(II) (as $NiSO_4$), Cd(II) (as $CdCl_2$) and Cr(VI) (as $K_2Cr_2O_7$). Solutions tested with these salts were prepared in the range from 1 to 10 mM (final concentration). *Saccharomyces cerevisiae* ATCC 32051 was used as reference strain. This assay was carried out to find out if the isolates were capable of tolerating one or more of the heavy metals studied. Yeast growth was assumed as a qualitative parameter of heavy metal tolerance. Growth up to the ditch was considered tolerant to the concentration supplied, while no growth was sensitive. The size of growth inhibition gave an approximate idea about the level of tolerance in cases where no complete tolerance was observed. The results were expressed with crosses using 4 categories with cruces: from non growth (-:non-heavy metal-tolerant yeast) to abundant growth (+++: highly-heavy metal-tolerant yeast).

As can be observed in table 2, the order of tolerance for isolates from copper filter mine plant was Cu(II) > Ni(II) > Cd(II) ≥ Cr(VI). Similar results were obtained with yeasts isolated from nickel-copper mine. Theses findings were agreement with yeasts isolated from sewage sludge (Balsalobre *et al.* 2003).

Table 2. Qualitative analysis for heavy metal tolerance yeast (Villegas 2006)

A- Copper filter mine plant

Isolations	Heavy metal concentration (mM)											
	Cu(II)			Ni(II)			Cr(VI)			Cd(II)		
	1	5	10	1	5	10	1	5	10	1	5	10
RCL-1	+++	+++	++	+++	++	+	++	++	+	++	++	+
RCL-2	+++	++	++	+++	++	+	++	+	-	++	++	+
RCL-3	+++	+++	+++	+++	+++	++	+++	+++	++	+++	++	++
RCL-4	+++	++	++	+++	++	+	++	++	+	+++	++	+
RCL-5	+++	++	++	+++	++	++	++	++	++	++	+	-
RCL-6	+++	++	++	+++	++	+	++	+	-	+++	++	++
RCL-7	+++	+++	++	+++	++	+	+++	++	+	+++	++	++
RCL-8	+++	+++	++	+++	++	++	++	+	-	++	+	-
RCL-9	+++	++	++	+++	++	+	++	+	+	++	++	+
RCL-10	+++	++	++	++	+	-	++	+	-	+++	+++	++
RCL-11	+++	+++	+++	+++	+++	+++	+++	++	++	+++	+++	+++

B- Tannery ponds

Isolations	Cu(II)			Ni(II)			Cr(VI)			Cd(II)		
	1	5	10	1	5	10	1	5	10	1	5	10
NGV-1	+++	+++	++	+++	++	++	+++	+++	+++	+++	++	++
NGV-2	+++	++	++	++	++	+	+++	++	+	++	+	-
NGV-3	+++	+++	++	+++	+++	++	+++	++	+	++	++	+
NGV-4	+++	++	+	++	++	+	+++	++	++	+++	++	+
NGV-5	+++	++	++	+++	++	++	+++	++	+	++	+	-
NGV-6	+++	++	+	+++	++	+	++	++	+	++	++	+
NGV-7	+++	+++	++	+++	++	+	+++	+++	++	+++	++	+
NGV-8	+++	++	++	+++	++	++	+++	++	++	++	++	-
NGV-9	+++	++	++	+++	+++	++	+++	+++	+++	+++	+++	++
NGV-10	+++	++	+	++	+	-	+++	+++	++	++	++	+
NGV-11	+++	++	++	+++	+++	++	+++	++	++	+++	++	+
NGV-12	+++	++	+	+++	++	+	+++	++	++	+++	++	+
NGV-13	+++	+++	++	+++	++	+	++	+	+	++	+	-

C- Nickel-copper mine.

Isolations	Cu(II)			Ni(II)			Cr(VI)			Cd(II)		
	1	5	10	1	5	10	1	5	10	1	5	10
VR-1	+++	++	++	++	++	+	+++	++	++	++	++	+
VR-2	+++	+++	++	+++	++	++	+++	+	+	+++	++	++
VR-3	+++	++	++	+++	++	++	+++	++	+	++	+	-
VR-4	+++	++	++	+++	++	+	+++	++	++	++	++	+
VR-5	+++	++	++	+++	+++	++	++	+	+	++	+	-
VR-6	+++	++	++	+++	++	+	++	+	-	+++	++	+
VR-7	+++	+++	++	+++	+++	++	++	+	-	++	-	-
VR-8	+++	+++	++	+++	+++	++	+++	+++	+++	+++	+++	++

(-) non-heavy metal tolerant yeast; (+) scanty heavy metal tolerant; (++) moderate heavy metal tolerant; (+++) highly-heavy metal tolerant yeast.

In contrast, the isolates obtained from the tannery ponds showed highest tolerance to Cr(VI), followed by Cu(II), Ni(II) and Cd(II).The reference strain, *S. cerevisiae* ATCC 32051, was less tolerant to heavy metal tested than the isolates obtained from sediment samples. Based on these results RCL-3, RCL-11, NGV-1, NGV-9 and VR-8 were selected for further Cr(VI) or Cu(II) removal assaying (Villegas 2006; Villegas *et al.* 2005; 2008).

Copper is the principal heavy metal contaminant in mine plant and from nickel-copper mine, the isolations obtained there presented major tolerance to the Cu(II). Those isolates from tannery ponds had major tolerance to Cr(VI), the main heavy metal in this area. These results are in agreement with other authors who reported that yeast strains isolated of environments contaminated with heavy metals were more tolerant to them than yeast collection with similar characteristics in laboratory test (Ksheminska *et al.* 2003; Ramírez-Ramírez *et al.* 2004; López-Archilla *et al.* 2004; Srivatava and Thakur 2006). These finding suggest that these natural strains microorganisms, probably, developed mechanisms that permit them to adjust to hostile conditions like that wherefrom isolated.

Characterization of Selected Isolates

Selected isolates were characterized by conventional yeast identification assays described by Yarrow (1998) and molecular test, DNA sequence analysis of the D1/D2 region of the 26S rRNA gene, using NL1 and NL4 primers (Kurtzman and Robnett 1998). According to the results only three isolations were characterized up to species level RCL-3, RCL-11 and VR8 which were identified like of whereas *Candida fukuyamaensis*, *Rhodotorula mucilaginosa* and *Aureobasidim pullulans* in that order. However, these tests were insufficient for identification at species level to NGV-1 and NGV-9 therefore was assumed to correspond to the *Lecythophora* and *Candida* genus respectively (Villegas *et al.* 2005, 2008, 2009).

These yeast and yeasts-like genera are normally found in terrestrial habitats, in the earth or on plant surfaces. Many authors have isolated these types of yeast from environments contaminated with heavy metals, e.g. *R. mucilaginosa* strains isolated from a leachate of a uranium mineral heap as well as from municipal wastes (Silóniz et al. 2002b; Balsalobre et al. 2003); Candida strains were isolated from tannery and industrial waste (Baldi *et al.*, 1990; Pepi and Baldi 1992; Dönmez and Aksu 2001; Ramírez-Ramírez *et al.* 2004), Lecythophora were isolated from an environment with low pH and high heavy metal concentrations, demonstrating their ability to tolerate and grow in such extreme habitats (López-Archilla *et al.* 2004), whereas A. pullulans was isolated from a site contaminated with Pb, Cd and Zn (Bewley 1980) and from stainless steel industry effluents (Santos and Linardi 2001).

Quatitative Cu(II) or Cr(VI) Toxicity

Before evaluating the toxicity of the Cu (II) or Cr(VI) on yeast strain, some parameters must be analyzed, as the composition of the culture medium where the yeasts will grow, the pH influence on metal uptake by yeasts and the temperature effects because the heavy metal accumulation can be metabolism dependent.

In the Figure 1 shows profile of the growth by *R. mucilaginous* RCL-11 in YEPD liquid medium containing (in g l^{-1}): glucose, 20; yeast extract, 10; peptone, with different concentrations of Cu(II). In it, can see the growth did not show differences in relation with the control (absence of Cu(II)). On the other hand when YNB-glucose medium was used, medium containing (in g l^{-1}): yeast nitrogen base (YNB), 6.7; glucose, 20, the growth decreased when heavy metal concentration increased in the culture medium (Villegas 2006). Consequently in heavy metal removal studies, minimal medium as YNB-glucose would be used instead of the complex media, as YEPD, because are inadequate due to the high concentration of organic components that absorb the metal ions. As it was mentioned previously copper and chromium have high affinity to organic matter diminishing the bioavailability of them. Other authors obtained the same conclusions in Cr(VI) toxicity studies in actinomyces (Laxman and More 2002; Polti *et al.* 2007)

It is necessary to bear that in mind that when liquid media cultures are employed, the availability of the nutrients and of the toxic substance, as heavy metals is better than in solid media culture. Therefore the heavy metal concentrations are lower in the first conditions.

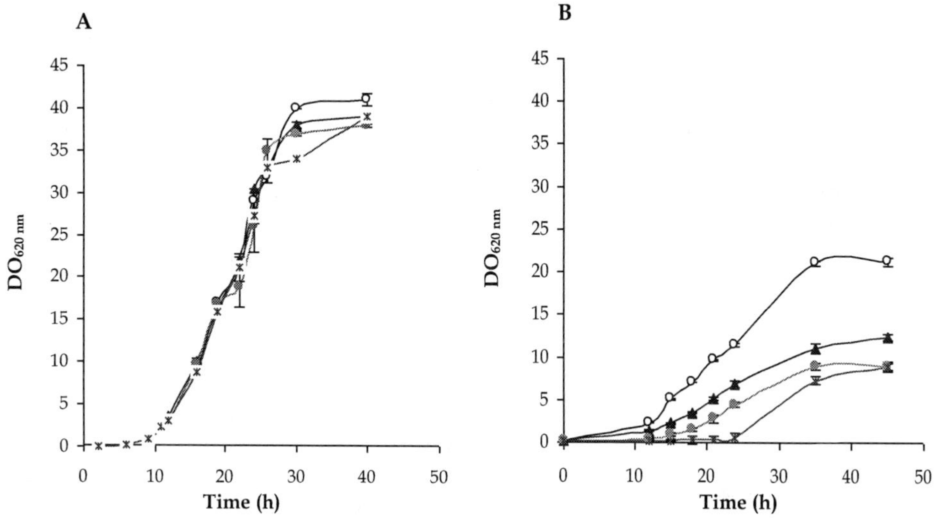

Figure 1. Effect of Cu(II) concentration on the cell growth of: R. mucilaginosa RCL-11. (A) YEPD and (B) YNB- glucose medium: (o) 0.0 (control); (▲) 0.1; (•) 0.5 and (*) 1 mM Cu(II) (Villegas 2006).

In relation to pH, earlier studies on heavy metals biosorption have shown that pH is the single most important parameter affecting the biosorption process (Brady and Duncan 1994, Aksu and Dönmez 2001). It is known that extreme pH generally decrease the rate of metal uptake: while at low pH the ions H + battle with the metallic ions for the combination sites of the cells, at high pH the metallic ions can rainfall for the formation of hydroxides or insoluble oxides, as well as to a wide range of hidroxi complexes. These compounds reduce the concentration of free ions in solution and therefore the bioavailability of them. The range of pH between 4,0 - 8,0 was most accepted as the ideal one for the metal removal by the majority of the microorganisms (Fourest and Roux, 1992). Later studies with yeast realized by Brady and Duncan (1994) found that for *S. cerevisiae*, the range of pH between 5,0 - 9,0 was the most favourable for the accumulation of copper, diminishing roughly to acid pH . Marques *et. al* (1999) worked with flocculent and non-flocculent yeast, isolated from wastes of Portuguese breweries. They showed that, in the absence of pH adjustment, the optimum initial pH for the removal of three cations: (Cu(II), Cd(II) and Pb(II)), was in the 4.5-5.5 range for both types of biomass. However, a gradual pH increase was observed during the removal process, up to a final equilibrium value of 7.0-8.0. On the other hand, the optimum pH value for Cu(II) or Ni(II) accumulation by *Candida* spp was determinate as 4 (Dönmez and Azku, 2001). Other work with collection strain of *Aspergillus niger* showed ability for heavy metals accumulation, but the optimum pH changed with the metal in study (Dursun *et al.,* 2003)

To determine pH ideally for the removal of copper or chromium culture of *C. fukuyamaensis* RCL-3 and *R. mucilaginosa* RCL 11 in presence of cu(II) and *Candida* sp. NGV-9 in presence of Cr(VI) medium with or without buffer were used. The pH was buffered at 5.0 and 7.0 with 50 mM Tris-succinate. This buffer does not possess the property of chelate metals (Ross and Parkin, 1989). In the absence of pH adjustment, initial pH was of 6.0 and diminished with the growth of the yeast, reaching values of 3.0 ± 0.5 after 12 h approximately. Under this condition, the values of copper uptake for both yeasts used were fluctuating and very near to the initial values. In the pH adjustment at 5.0 and 7.0 while, *C. fukuyamaensis* RCL-3 showed 44 ± 4 y 38 ± 3 % of copper uptake in that order, *R. mucilaginosa* RCL-11presented 30 ± 5 % 15 ± 2 % respectively. Results similar for Cr(VI) removal by *Candida* sp. NGV-9 were obtained with pH adjustment but in the absence of pH adjustment also presented Cr(VI) removal, major than pH 7 and minors than to pH 5. In spite of the differences, in all the cases at pH 5 better removals were found (Villegas 2006).

Temperature effects on heavy metal removal are restricted to metabolism-dependent metal accumulation. At low temperature (0-5°C) little or no metal is sequestered through metabolic processes by viable biomass. Most laboratory experiments are carried out in the temperature range 20-30°C, which has been reported optimal for metal accumulation and for the growth of the majority of the yeasts (Brady and Duncan 1994).

On the basis of these studies the conditions selected to evaluate the Cu(II) and Cr(VI) removal on yeasts were: YNB-glucose liquid medium, incubated in an orbital shaker (250 rpm) at 30 °C in buffered medium at pH 5.0.

Cu(II) Toxicity on Yeasts

Effect of Cu(II) concentration on *C. fukuyamaensis* RCL-3 and *R. mucilaginosa* RCL-11 was analyzed by means of the following parameters: biomass production (by cell count and dry weight) specific growth rate and glucose consumption. The growth of both yeasts decreased according to the increment of Cu(II) (Fig 2 and 3). Under these conditions of work *R. mucilaginosa* RCL-11 was the most affected. The growth of it diminished 23, 41 and 50 % in the presence of 0.1; 0.2 and 0,5 mM of Cu(II) respectively when dry weight was determined. Nevertheless, it is possible to observe clearly that the results of cellular viability demonstrated a major impact of Cu(II) concentration on this strain. For example, the number of viable cells at the end of the exponential phase of growth, in the presence of 0.5 mM of Cu(II), produced a reduction of viable cells of *R. mucilaginosa* RCL-11 about 90 % in relation to the control (absence of Cu(II) in the culture medium). This yeast reached the maximum value of CFU ml^{-1} at 30 h of incubation and later it was kept constant (Fig 2).

The growth rate of both yeasts decreased according to the increment of Cu(II) concentration. The specific growth rate (μ decreased in relation to the control about 40 and 50% for both yeasts with the Cu(II) presence in culture medium, but significant variations were not observed at the different Cu(II) concentrations (Fig. 4 A). These results indicate that the cells divide to similar speed in the presence of 0.1 or 0.5 mM of Cu(II).

Although the increase of this heavy metal concentration reduces significantly the cellular viability, suggesting that as the Cu(II) concentrations is increased in the culture medium the cells would have a more short life.

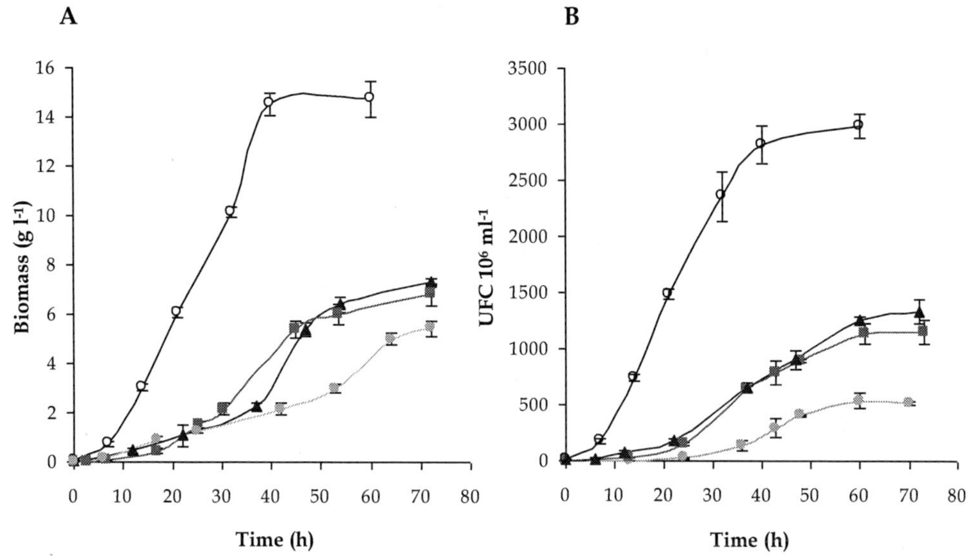

Figure 2. Effect of different Cu(II) concentration on the C. fukuyamaensis RCL-3 growth: (A) dry weight and (B) cell count (o) 0.0 (control); (▲) 0.1; (■) 0.2 and (•) 0.5 mM Cu(II) (Villegas et al. 2005; Villegas 2006).

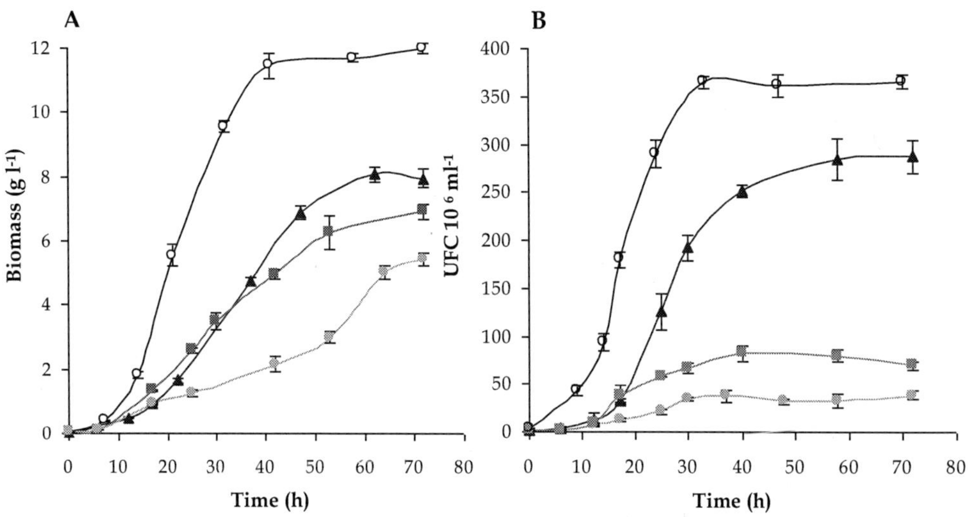

Figure 3. Effect of different Cu(II) concentration on the R. mucilaginosa RCL-11 growth: (A) dry weight and (B) cell count (o) 0.0 (control); (▲) 0.1; (■) 0.2 and (•) 0.5 mM Cu(II) (Villegas et al. 2005; Villegas 2006).

In relation of the specific glucose consumption maximums values of specific glucose consumption by both yeasts were reached in the exponential phase and then remained constant. *C. fukuyamaensis* RCL-3 showed similar values of consumption of glucose in the presence of 0.1 and 0.2 mM of Cu(II) and did not present significant variations in relation to the control. However, in presence of 0.5 mM of Cu(II) decreased the glucose consumption about 20 %, possibly because under this cation concentration the toxicity is bigger. On the other hand, *R. mucilaginosa* RCL-11 showed values of specific glucose consumption practically linear with the increase of the concentration of the Cu(II) in the culture medium. Both parameters presented decreases of 15, 35 and 55 % approximately, in presence of 0.1; 0.2 and 0.5 mM of Cu (II), respectively (See Fig. 4 B) (Villegas 2006).

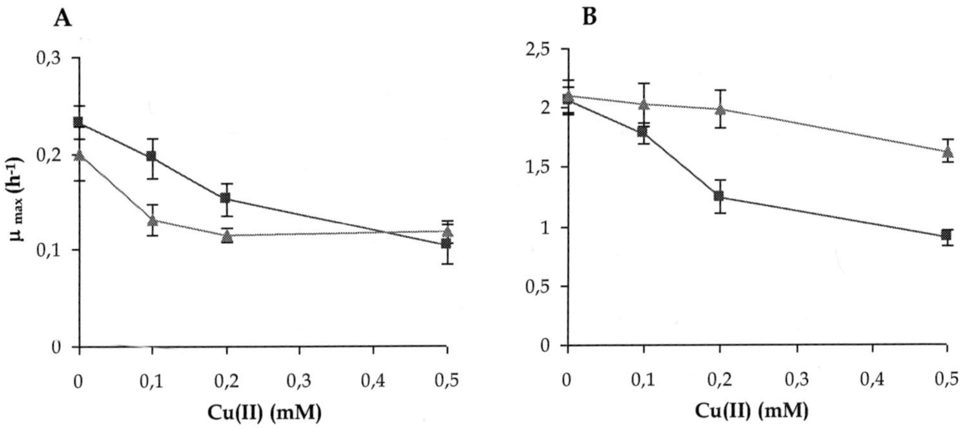

Figure 4. Effect of different Cu(II) concentration on (A) specific growth rate and (B) specific glucose consumption; (▲) C. fukuyamaensis RCL-3 (■) R. mucilaginosa RCL-11 (Villegas 2006).

The decrease of the viability in presence of Cu(II) did not have correlation with the decrease of specific growth rate and consumption of glucose maybe because a most important metabolic cellular activity to offset of Cu(II) toxic effects was necessary.

The obtained results coincide with the values of μmax determined for collection strain as *Schizosaccharomyces pombe*, *Saccharomyces cerevisiae*, *Kluyveromyces marxianus*, *Candida* sp and for yeast isolated from municipal wastes as *Cryptococcus laurentii, Debaryomyces hansenii*, *P. guilliermondii*, *R. aurantiaca*, *R. mucilaginosa* and *Torulaspora delbrueckii.* (Dönmez and Aksu, 1999; Balsalobre *et. al.* 2003).

Cu(II) Uptake by Yeasts

Figure 5 shows the intra and extracellular copper concentration during yeast growth at 0.5 mM of Cu(II). So much *C. fukuyamaensis* RCL-3 as *R. mucilaginosa* RCL-11 showed the skill to copper uptake from culture medium. However the bioaccumulation kinetics profile of Cu(II) by both yeasts showed significant differences: whereas the maximum values of copper uptake were achieved by *C.fukuyamaensis* RCL-3, this value was maintained inside the cells and librated later to the supernatant of the culture, *R. mucilaginosa* RCL-11 stored the accumulated metal up to the end of the experiment. The behaviour of both yeasts was similar at all of copper concentration tested (Villegas *et al.* 2005).

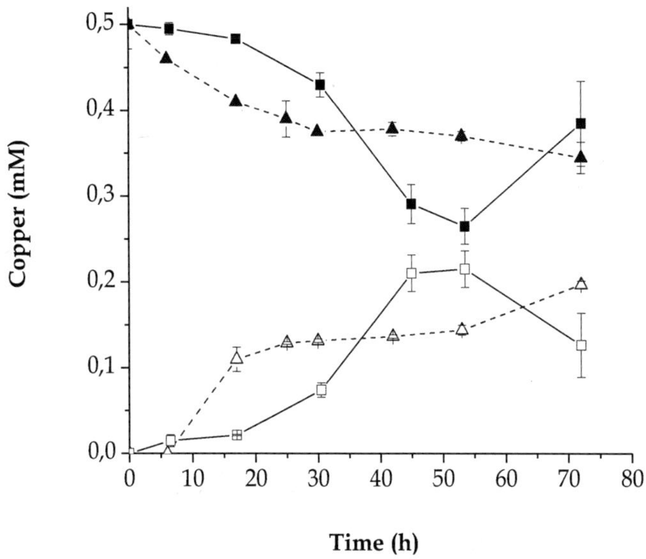

Figure 5. Intra and extracellular copper concentration of yeasts cultures at 0.5 mM Cu(II) initial concentrations. (□) C. fukuyamaensis. RCL-3 intracellular, (■) C. fukuyamaensis. RCL-3 extracellular, (Δ) R. mucilaginosa RCL-11 intracellular and (▲) R. mucilaginosa RCL-11, extracellular copper concentration (Villegas et al. 2005).

For methods of molecular biology, copper extrusion pump that would drive copper out of the cell in *C. albicans* was found. This pump was identified in Golgi compartment membrane of *S cerevisiae*; whereas in *C. albicans* it was located also in the plasmatic membrane causing the unusual copper tolerance of this strain in the relation to *S. cerevisiae* (Weissman *et al.* 2000). However, the mechanism of exclusion of copper, observed in *C. fukuyamaensis* RCL-3, was one of the first informed in yeasts, only have been reported in vegetable and mammal cells before (Clemens 1999; Linder 2001).

It is interesting to notice that *C. fukuyamaensis*. RCL-3 belongs to an ascomycetous genus while *R. mucilaginosa* RCL-11 to a basidiomycetous one. The difference in cell wall structure, cytology, physiology, protein and enzyme contents, etc., between ascomycetous and basidiomycetous yeasts are well established (Moore 1998). For this reason it can be assumed that the feature of *C. fukuyamaensis* RCL-3 to liberate copper to the extracellular space and of *R. mucilaginosa* RCL-11 to keep the metal inside the cell are related to their phylogenetic differences. After this work, some authors suggested that although a high copper concentration induces an efflux mechanism in *Yarrowia lipolytica* (also belongs to an ascomycetes genus), the released copper becomes entrapped in the periplasm and in other parts of the cell wall (Ito *et al.* 2007).

The specific bioaccumulation analysis of copper (copper uptake per gram of biomass) by *C. fukuyamaensis* RCL-3 and *R. mucilaginosa* RCL-11 showed that the bioaccumulation increased with the initial copper fed in the medium. After 17 h of incubation in medium containing 0.5 mM Cu(II), values of 11.5 and 8.0 mg Cu g biomass^{-1} for *C. fukuyamaensis* RCL-3 and *R. mucilaginosa* RCL-11, respectively, were obtained. However, the specific bioaccumulation decreased with time indicating that both yeasts possess a high capacity to take up copper from the medium initially, which was eventually limited by Cu(II) concentration and culture age. Soares *et al.* (2002) reported that the age of the culture effects metal uptake. (Fig. 5A and B).

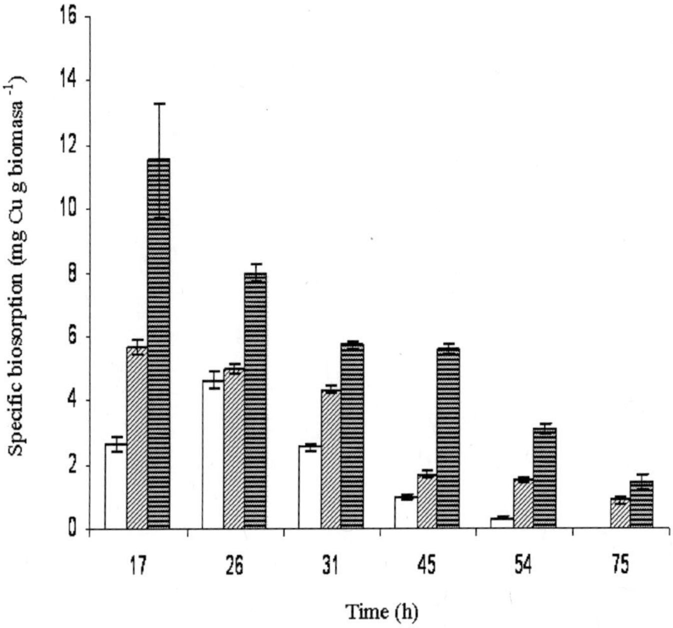

Figure 6. Specific bioaccumulation (mg Cu g biomass -1) at different Cu(II) concentrations on Candida fukuyamaensis RCL-3: (□) 0.1; ((□) 0.2 and (■) 0.5 mM Cu(II) (Villegas et al. 2005).

The trait that enables cells to use copper is, at the same time, a threat to the organism when the metal is present at high concentrations, since it catalyzes the production of reactive oxygen species (ROS) via the Fenton-reaction (Halliwell and Gutteridge 1990; Puig and Thiele 2002). These forms of oxygen are highly damaging towards cellular constituents, including DNA, lipids and proteins (Imlay 2003).

Cells prokaryotes and eukaryotes possess copper homeostasis to maintain an adequate non toxic of copper. This is achieved diminishing the concentrations of the free copper ion and therefore avoiding its toxic effects. In the baker's yeast *S. cerevisiae*, extracellular Cu(II) is reduced to Cu(I) via the action of the cell surface metalloreductase Fre1 , and transported across the plasma membrane through the high and low affinity copper transporters (Hasset and Kosman 1995; Koch *et al.* 1997). After crossing the plasmatic membrane, cells induce the biosynthesis of small soluble proteins known as "copper-chaperones", which bind copper atoms efficiently and have the ability to distribute it to specific intracellular targets. The copper chaperone Atx1 delivers copper to Golgi compartment for newly synthesized proteins; the copper chaperona Sodl carry copper to Cu,Zn superoxide dismutase (Sodl) and the copper chaperone Cox17 transport to mitochondria for its incorporation to cytochrome c oxidase (Cco) (Puig and Thiele 2002; Rees and Thiele 2004)

Copper uptake observed by *C. fukuyamaensis* RCL-3 y *R. mucilaginosa* RCL-11, increased with the initial concentration of the ion present in the medium, possibly because increased progressively the interactions with the bond sites of low affinity of copper (Fig 5). Reduction and transport of copper are not gratuitous process, but rather are associated with a cost of energy, this justify that the decrease of the consumption of glucose in both yeasts is minor to the decrease of the viability (Fig 2, 3).

On the other hand, there was demonstrated that these microorganisms present positive and negative regulation copper-sensitive and specific of the expression of genes involved in copper homeostasis, in response to its fluctuating concentrations in the environment. For example, these genes are repressed when cells are grown in the presence of copper and highly activated during copper starvation (Labbé et al. 1997). This phenomenon would explain the decrease of the specific bioaccumulation of copper with the time in *C. fukuyamaensis* RCL-3 and *R. mucilaginosa* RCL-11 (Fig. 6).

COPPER BIOACCUMULATION FOR INDUCTION

Dönmez and Aksu (1999; 2001) reported the adaptability of several yeasts to high copper concentrations for successive incubation with increasing Cu(II) concentrations in the growth medium. These adapted yeast accumulated Cu(II) more efficient than non-adapted cells. The level of Cu(II) accumulation depended on the metal concentration and yeast strain and best results were obtained with *Candida* sp.: 14.8 mg of copper per g biomass at 5 mM Cu(II) in complex medium.

Silóniz et al. (2002) found that increasing concentrations of copper affected physiological parameters and bioaccumulation capacity of viable *Pichia guilliermondii*, isolated from sewage sludge in Spain. The growth rate of non-adapted cells of this yeast decreased with increasing copper concentrations, whereas Cu-induced cells of *P. guilliermondii* maintained constant biomass yield and growth rate, independently of the external copper concentration.

Studies with *R. mucilaginosa* RCL-11 in sequential batch cultures carried out by increasing Cu(II) concentrations in five sequential steps: 0; 0.1; 0.2; 0.5 and 1 mM with and without aeration controlled. This strain increased Cu(II) bioaccumulation capacity by adaptation to high copper concentrations, being able to grow to 1 M, concentration to which not growth was obtained by not adapted cells. With pre-adapted cells in successive cultures with 0.2, 0.5 and 1 mM of Cu(II) in without aeration, bioaccumulation was 50, 25 and 30%, respectively. When this study was realized in fermentor, under controlled aeration, copper removal was 50% compared to controls regardless of the initial Cu(II) concentration (Villegas *et al.* 2009b). As was mentioned above, non adapted cells of *R. mucilaginosa* RCL-11 diminished the specific Cu(II) bioaccumulation with the time, while with pre-adapted cells of the same strain, specific bioaccumulation increased with increasing initial Cu(II) concentration in the culture medium through time until reaching maximum values at exponential growth phase, and remained constant later. At an initial Cu(II) concentration of 1 mM maximum Cu(II) accumulation was 13 mg per g of biomass after 24 h under without aeration and 17 mg after 45 h under controlled aeration. Cells grown under controlled oxygen conditions showed 30% more copper bioaccumulation and 10% glucose consumption when compared with cells grown without aeration.

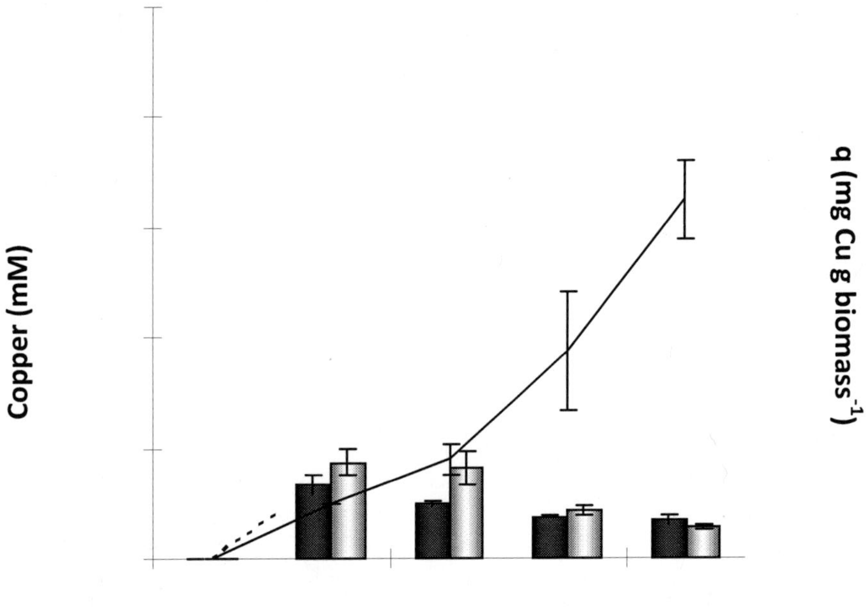

A

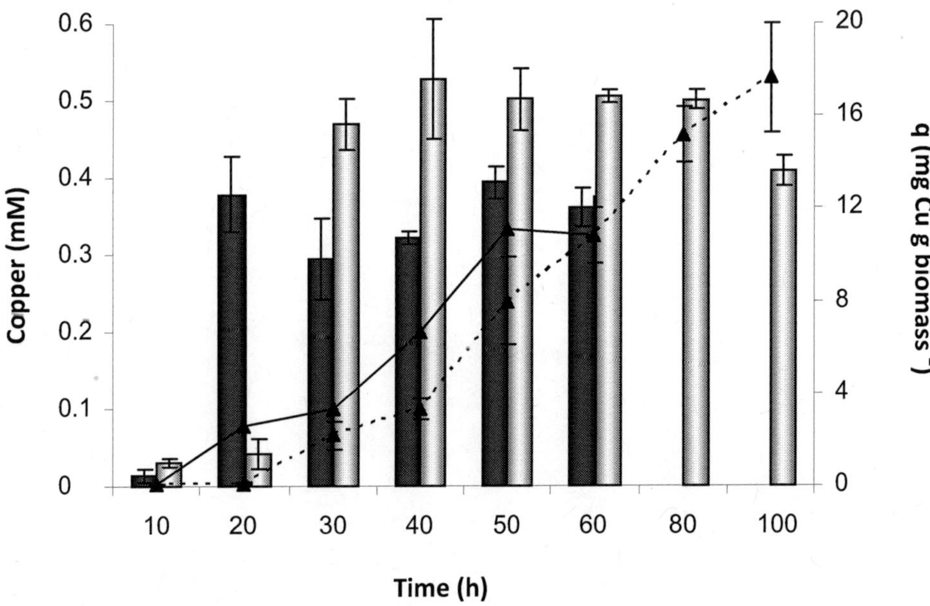

B

Figure 7. Intracellular copper concentration of R. mucilaginosa RCL-11 cultures with controlled (----) and without (____) aeration. Specific bioaccumulation, q (mg Cu g biomass–1), at different initial Cu(II) concentrations: (A) 0.1; (B) 0.2; (C) 0.5 mM and (D) 1 mM Cu(II). (▒) controlled and (■) without aeration.

Copper Intracellular Localization

Not many studies about morphology changes produced by heavy metals have been made in yeasts. For instance, in presence of uranium, *S. cerevisiae* presented deposits of this metal as fine needle-like crystals on the cell wall and membrane (Volesky and May-Phillips 1995). Other studies carried out in vanadate-tolerant yeast, Hansenula polymorpha, grown in the presence of vanadate induced changes in cellular ultrastructure (Mannazzu *et al.* 1998). Electron microscopic images of *P. guilliermondii* ATCC 201911 incubated in presence of Cr(VI) revealed significant changes with the observation of dark entities located preferentially in the vacuole (Ksheminska *et al.* 2005) Scanning electron microscopy indicated that accumulation of cadmium in the Y11 did not cause any visible effects on cell morphology and transmission electron microscopy analysis showed that plasmolysis and thickened cell wall were not observed in all of the cells (Li and Yuan, 2008).

In relation of copper, only was reported the work of Ramsay and Gadd (1997) who determined the absence of compartmentalization mechanism in copper detoxification by yeasts and Silóniz *et al.* (2002) determined in cells of *P. guilliermondii* grown in presence of Cu(II), that this metal ion was located into the cell uniformly. Villegas *et al.* (2009a) realized a deeper study showing the sequential Cu(II) 0.5 mM effects on the cellular structures and their possible location for *C. fukuyamaensis* RCL-3 and *R. mucilaginosa* RCL-11 obtained at 24, 48 and 72 h incubation and they were compared with control cells. After 24 h incubation both yeasts showed cytoplasmatic lysis, moreover *C. fukuyamaensis* RCL-3 evidenced also a cell wall degradation and irregular shape. After 48 h incubation the photographs of both yeasts reveal the effect of copper on cellular structures. Significant changes in cellular interior contrast and density could be observed after interaction with copper, indicating the intracellular deposition of the metal in cytoplasm as big dark entities (grains) (Fig. 8 A and 9 A).

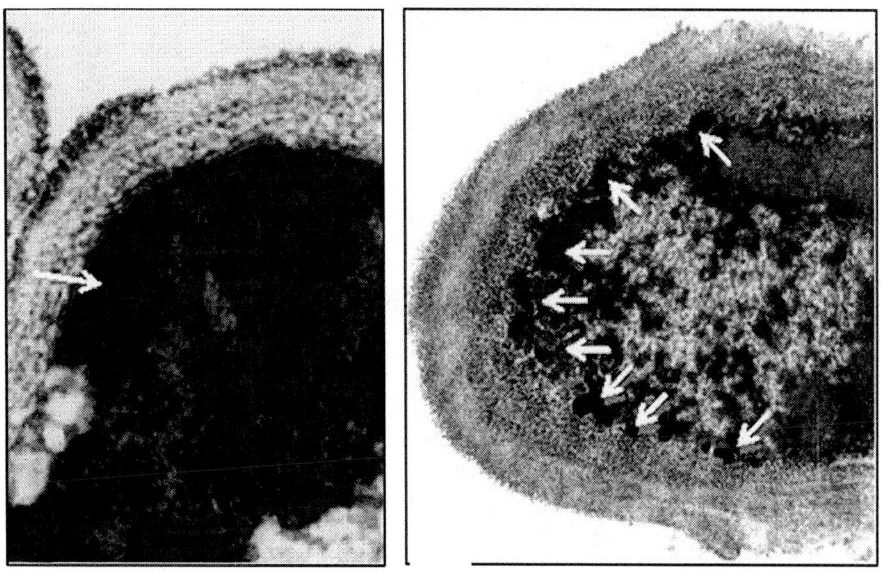

Figure 8 . Transmission electron images (TEM) of C. fukuyamaensis RCL-3; grown in the presence of Cu(II) 0.5 mM A: 48 h; B: 72 h. Arrows indicate bodies electron-dense. 33.300X (Villegas et al 2009 a).

At 72 h incubation *R. mucilaginosa* presented grains much more located into the cytoplasm compared with other one keeping the characteristical morphology as the control. Few *C. fukuyamaensis* RCL-3 cells presented the dark bodies, which were mainly located next to the cell wall and presented abundant cell broken (Fig. 8B and 9B). The electron-dense bodies found in *C. fukuyamaensis* RCL-3 and *R. mucilaginosa* RCL-11, would correspond to intracellular copper deposits.

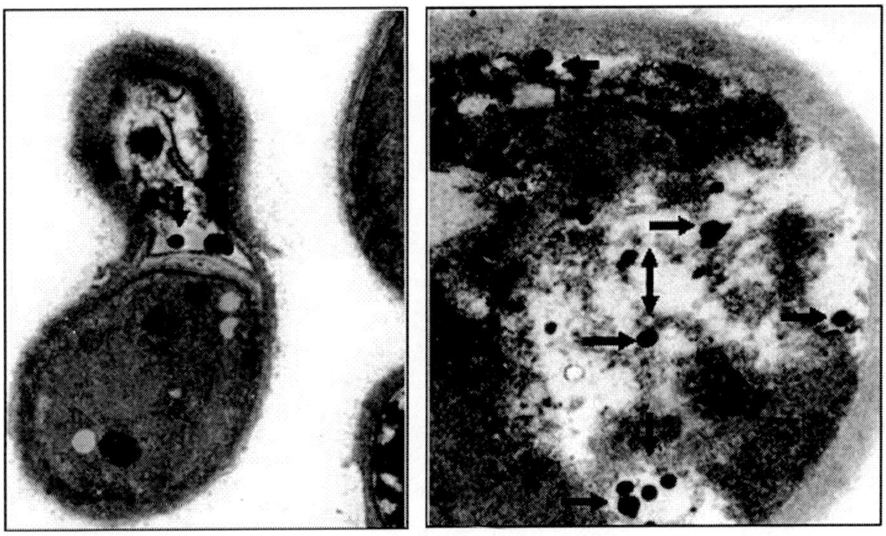

Figure 9. Transmission electron images (TEM) of R. mucilaginosa RCL-11; grown in the presence of Cu(II) 0.5 mM A: 48 h; B: 72 h. Arrows indicate bodies electron-dense. 33.300X (Villegas et al 2009 a).

Size and shape properties of yeast cells incubated with and without Cu(II) 0.5 mM were compared by Scanning Electron Microscopy (SEM). Surprising while *C. fukuyamaensisi* RCL-3 did not reveal any differences compared with the control images (Figure 6 A and B), *R. mucilaginosa* RCL-11 cells grown in presence of Cu(II) 0.5 mM (Figure 6 D) were shown bigger (2.6 x 2 μm approximately) than the control images, which presented a size of 2.3 x 1.3 μm (Villegas *et al.* 2009 a)

COPPER EFFECT ON OXIDATIVE STRESS

Yeasts, like other aerobic organisms, generate ROS endogenously as by-products of metabolic processes. These forms of oxygen are highly damaging towards cellular constituents, including DNA, lipids and proteins (Imlay 2003). ROS can also be formed by exposure of cells to either ionizing radiation, redox-cycling chemicals present in the environment or by exposure to heavy metals. Through these mechanisms, all aerobically growing organisms are continuously exposed to reactive oxidants, under physiological conditions, endogenous antioxidant cellular defence systems are adequate to maintain ROS at basal, unharmful level and repair damages. These oxidative molecules are detoxified via superoxide dismutases (SODs), which in two steps dismutate the superoxide to O_2 and H_2O_2.

Subsequently, the H_2O_2 is detoxified by catalases (CATs). These enzymes create the first and the most important line of antioxidant defence (Lushchak and Gospodaryov 2005). When ROS production is accelerated or there is a defect in the antioxidant system ROS, an oxidative stress results (Costa and Moradas-Ferreira 2001).

Abe *et al.* (2001) studied the SOD and CAT activities in deep-sea and copper tolerant yeast, *Cryptococcus* sp. N6. Total SOD activity was 25 and 100 times higher when this strain was grown in Cu(II) 1 and 10 mM, respectively, in relation to growth in absence of Cu(II). In native polyacrylamide gel, *Cryptococcus* sp. N6 showed a single band with SOD activity. In relation to CAT activity, it was only 1.4 and 1.1 times superior. Endogenous SOD and CAT activity from *C. fukuyamaensis* RCL-3 and *R. mucilaginosa* RCL-11 increased in free cell extracts, which was mainly related to the high Cu(II) concentration in the culture medium, as the yeast had to cope with high ROS levels. This increase was more than twice when the initial Cu(II) concentration was 0.5 mM compared to the control (Fig. 10 A). While, total SOD activity demonstrated an unimportant increase in relation to the incubation time, CAT activity, in both yeasts demonstrated increases with time. The values of activity of this enzyme were major in *C. fukuyamaensis* RCL-3 but *R. mucilaginosa* RCL-11 was the most affected by the presence of the metal. Furthermore, when *C. fukuyamaensis* RCL-3 was incubated with Cu(II) 0.5 mM, the CAT activity was only 1.5 times more than the control at 72 h of incubation, while from *R. mucilaginosa* RCL-11 was 15 times more than its control at the same incubation time (Fig 10 B) (Villegas *et. al* 2009a).

In opposition, Fujs *et al.* (2005) found different results with *C. Intermedia,* the CAT activity decreased 30% when the cells were treated with Cu(II) but the total SOD activity was not significantly altered.

It is well known that the yeasts contain two intracellular SODs; MnSOD and CuZnSOD; CuZnSOD is located predominantly in the cytoplasm and it can act as both antioxidant and as a copper chaperone (Jamienson 1998). Miura *et al.* (2002) suggested that the expression of superoxide dismutase is involved in the high tolerance of yeast to copper. Some authors as Ito *et al.* (2007) and Teh *et al.* (2008) reported the induction of this enzyme in *Yarrowia lipolytica* and *Cryptococcus* sp. N6 grown in the presence of Cu(II) in complex medium respectively. On the other hand, the MnSOD has generally been found in the mitochondria matrix. Miuran *et al.* 2002 analyzed the SOD expression in a copper-tolerant yeast, *Cryptococcus* sp. N6. Using cell extracts and native polyacrylamide gel electrophoresis (PAGE), two distinct bands exhibiting SOD activity appeared on native PAGE: one band, with higher mobility, appeared when the cells were grown without $CuSO_4$, and the other band appeared when the cells were grown with 10 mM CuSO4. Cells grown with 3 mM $CuSO_4$ produced both SOD isoforms.

Moreover, *C. fukuyamaensis* RCL-3 presented two bands with SOD activity but *R. mucilaginosa* RCL-11 showed a single band. All the bands presented different electrophoretic patterns in relation to the MnSOD from *Escherichia coli* and to the CuZnSOD from bovine serum, included as reference enzyme (Fig 10 C). After the treatment with H_2O_2, CuZnSOD from bovine was the only one that presented inhibition (Villegas *et al.* 2009 a).

Moore *et al.* 1989, informed the absence of CuZnSOD in pigmented yeasts and they attributed this observation to the carotenoid pigments presence that act as antioxidant. However, Hernández-Saavedra (2003) demonstrated the existence of CuZnSOD in *Udeniomybes puniceus*, a pigmented yeast, which activity increased with the Cu(II) presence. They could also confirm the lack of CuZnSOD in *R. mucilaginosa* and *R. graminis*, which demonstrated a single band with SOD activity. These results would explain the absence of CuZnSOD activity from yeast extract of *R. mucilaginosa* RCL-11. This enzyme was not detected in *C. fukuyamaensis* RCL-3 either. It is also possible that the main defence mechanism of *C. fukuyamaensis* RCL-3 against copper could be to extrude the cytoplasmatic copper, as it was referred before, therefore the expression of the CuZnSOD would not be necessary.

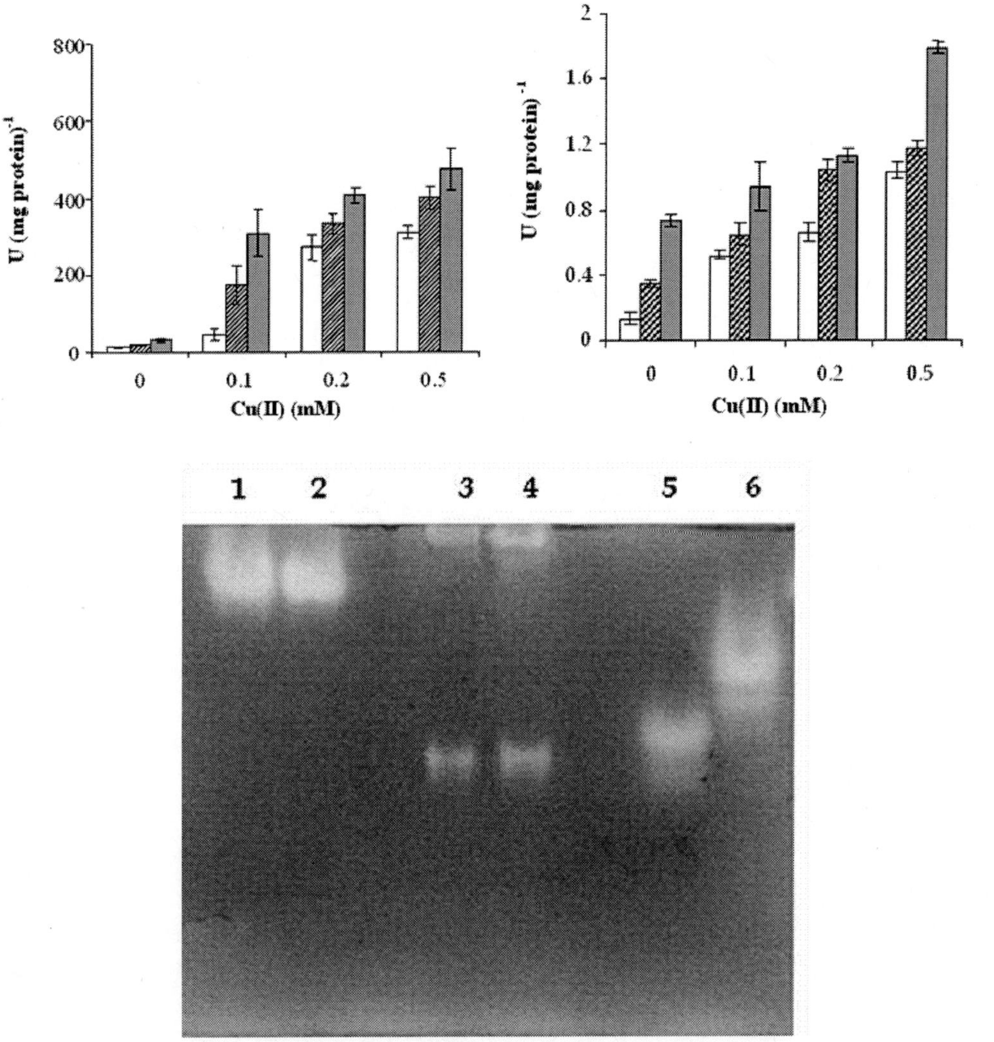

Figure 10. A: Total CAT activity B: Total SOD activity in free cell extracts of R. mucilaginosa RCL-11 grown with (0.1 to 0.5 mM) and without Cu(II) concentrations at (☐) 24,(▨) 48 and (■) 72 h of incubation. C: Activity staining for SOD in free cell extracts in 10 % native PAGE. Cell extracts incubated in the presence of Cu(II) 0.5 mM during 72 h and SOD standards, lane 1, R. mucilaginosa RCL-11 control cells; lane 2, R. mucilaginosa RCL-11 incubated with Cu(II) 0.5 mM; line 3, C. fukuyamaensis RCL-3 control cells; line 4, C. fukuyamaensis RCL-3 incubated in the Cu(II) 0.5 mM; line 5, CuZnSOD from bovine and line 6, MnSOD from E. coli.

Lamarre *et al.* (2001) found an unusual MnSOD enzyme in the *C. albicans* cytoplasm, which does not possess an amino-terminal peptide enabling mitochondrial import. These authors postulated that the switch from CuZnSOD to cytoplasmic MnSOD expression reflects the requirement for increased protection against ROS and copper toxicity. This mechanism was not found with S. Cerevisiae. These findings would explain the two bands with MnSOD activity present in the *C. fukuyamaensis* RCL-3 and the CuZnSOD absence. Then, this yeast could have a mechanism of resistance similar to *C. albicans* and different to *S. cerevisae.*

In the case of the CAT activity, it is known that *S. cerevisiae* has two CATs: catalase A, located in the peroxisome and catalase T in the cytoplasm (Jamieson 1998). *R. mucilaginosa* RCL-11, which has the capacity to keep copper inside the cell (Villegas et al. 2005), presented higher CAT activity than *C. fukuyamaensis* RCL-3. This clarify the presence of numerous peroxisomes observed in Figure 5 C when *R. mucilaginosa* RCL-11 was incubated in the presence of Cu(II) 0.5 mM.

When pre-adapted cell to copper concentrations of *R. mucilaginosa* RCL-11 were used total SOD and CAT activity were lower than the results obtained with non-adapted cells of the same strain. This results could be due to the fact that the cytoplasm of adapted cells has a larger number of proteins that sequester copper, which prevents the free ion to produce reactive oxygen species (ROS), thus decreasing oxidative stress. Consequently, adapted cells were able to grow at higher Cu(II) concentrations than non-adapted cells as was mentioned before (Villegas *et al.* 2009 b).

Cr(VI) Toxicity and Removal by Yeasts

Some works have analyzed yeasts or fungus growth at high concentration of Cr(VI). All of them indicated that the initial Cr(VI) concentration in the culture medium negatively influenced growth and the specific growth rate (Ksheminska *et al.* 2003; Dursun *et al.* 2003; Ksheminka *et al.* 2005) although specific glucose consumption rose with increasing incubation time and Cr(VI) concentration (Fig. 11) (Villegas *et. al* 2008).

Cr(VI) resistance mechanisms are not well known in yeasts. Earlier studies in some species of *Candida* and *Rhodosporidium* showed that, under aerobic conditions, chromate resistance was related to reduced ion uptake and not to biological reduction of Cr(VI) to Cr(III) species and they showed very little accumulation of chromium (Baldi *et al.* 1990; Pepi and Baldi 1992; Krauter *et al.* 1996). In contrast, more recent studies demonstrated that the yeast *Candida utilis* and a Cr(VI)-resistant yeast, isolated from tanning liquors from a leather factory and identified as *C. maltose* RR1, had ability to partially reduce Cr(VI) to Cr(III) and to accumulate chromium in the biomass (Muter *et al.* 2002; Ramirez-Ramirez *et al.* 2004). Other author assumed that alone the bioaccumulation processes is related to chromate resistance therefore they studied bioaccumulation ability of yeasts from genera: *Pichia, Candida, Debaryomyces* and *Kluyveromyces.* They reported that Cr(VI) bioaccumulation depended on incubation time but no rapid sorption phase was observed, in contrast with Cr(III). Thc proportion of the cell-accumulated to cell-bound chromium was constant (Kaszycki *et al.* 2004). These authors and other also found that Cr(VI) uptake also depended on the physiological state of the cells, incubation time, the kind of strain used, the inoculum preparation and cell density (Ksheminska *et al.* 2005; Srivastava and Thakur 2006; Fermández et. al. 2010).

Nevertheless, recent studies with yeast isolated from tannery ponds, *Lecythophora* sp. NGV-1 and *Candida* sp. NGV-9 indicated that Cr(VI) concentration decreased without changes in total extra and intracellular chromium was observed by flame atomic absorption spectroscopy (FAAS). Taking into consideration that Cr(III) is the only stable and soluble chromium component formed by Cr(VI) reduction, it could be inferred that the chromium remaining in the supernatant in this work was Cr(III). This finding may indicate that the Cr(VI) removal was due to reduction to trivalent form. However, the yeast-like isolated from the nickel-copper mine *A. pollulant* VR-8 was incubated at 0.5 mM Cr(VI) in the same conditions removed 100% of the Cr(VI) after 45 h of incubation and 68% of total chromium after 100 h of incubation. When this strain was incubated at 1 mM Cr(VI), removed 60% of Cr(VI) and only 15% of total chromium were eradicated (Fig. 12). The order of Cr(VI) removal efficiency of this strains was as follows: *A. pullulans* VR-8 > *Lecythophor*a sp. NGV-1 > *Candida* sp. NGV-9 (Villegas *et. al* 2008). Similar results were obtained with these strains under other culture conditions (Fernández *et al.* 2009).

It is important to mention that *A. pullulans* VR-8 produced exopolysaccharides (EPS) which could not be separated from the pellets by centrifugation techniques therefore these authors can not determine if chromium had accumulated in the cells or if it was adsorbed by EPS. Breierová *et al.* (2002) studied Cd(II) resistant in yeasts and confirmed that some of them synthesize extracellular polymers with the ability to adsorb important amounts of Cd(II) to protect themselves. Other authors demonstrated that EPS production by bacteria and their capacity to chelate chromium rose with increasing Cr(VI) concentrations in the culture medium (Iyer et al. 2004; 2005). On the other hand, was reported that the amount of bound heavy metals depended on the composition and amount of EPS produced by the different yeast strains (Mikes *et al.* 2005).

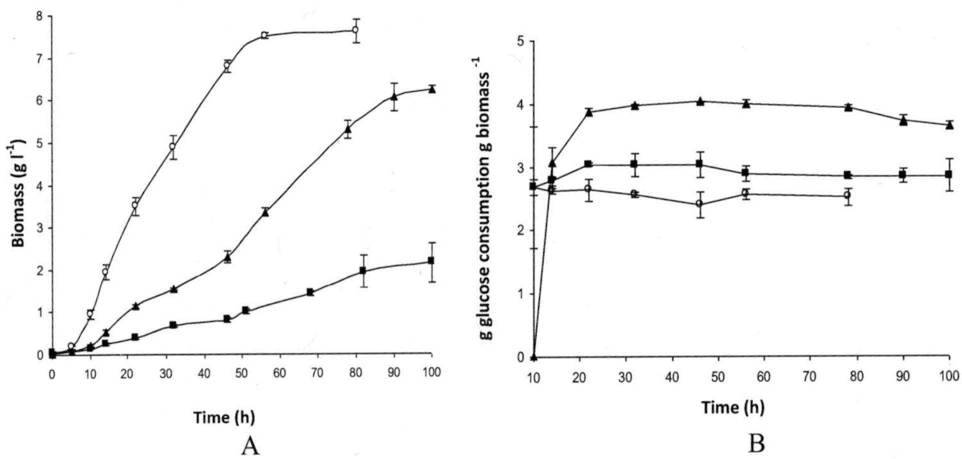

Figure 11. Effect of different Cr(VI) concentrations on A: cell growth B: specific glucose consumption in relation to biomass expressed as g glucose g biomass-1 of A. pollulans VR-8 (Villegas et. al 2008).

Specific chromate removal (defined as mg of Cr(VI) removal per gram of biomass) increased with the initial Cr(VI) concentration in the medium: *Lecythophora* sp. NGV-1 reached maximum elimination after 24 h of incubation independently of the Cr(VI) concentration and this values were constant later (Fig. 3A2). Specific Cr(VI) removal by *A. pullulans* VR-8 increased with elapsing incubation time (Villegas *et. al* 2008).

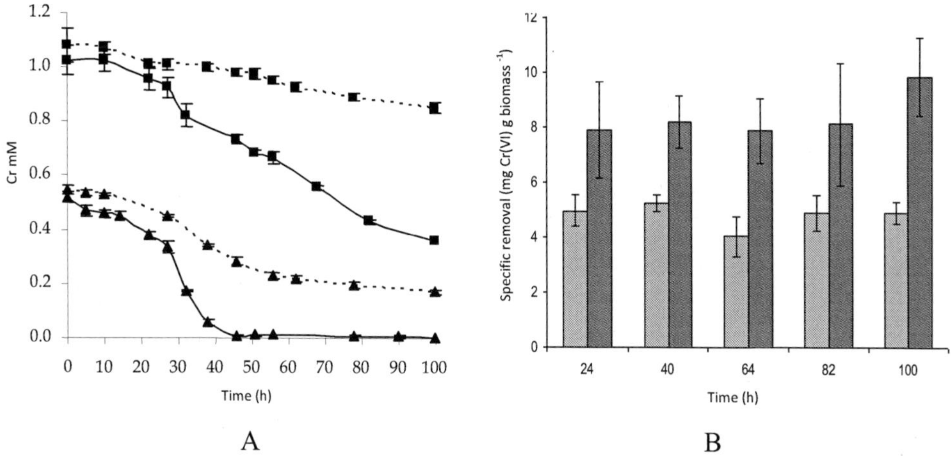

Figure 12. A: extracellular Cr(VI) concentration (__) and total Chromium (---) kinetics in the presence of 0.5 (▲) and 1 mM Cr(VI) (■); B: specific Cr(VI) removal (mg Cr(VI) g biomass-1) at 0.5 (▒) and 1 mM Cr(VI) (▧) by A. pullulans VR-8 (Villegas et. al 2008).

As was mentioned above, specific glucose consumption obtained with these strains augmented with increasing specific Cr(VI) removal, whereas other growth parameter decreased along with increasing initial Cr(VI) in the culture medium. Probably, these strains required glucose as energy source or electron donor to remove Cr(VI). So far, similar results only were reported in bacteria: Pseudomonas fluorescens and Streptomyces griseus by Wang and Xiao (1995) and Laxman and More (2002) respectively. Orozco *et al.* (2007) worked with activated sludges and showed that they are capable of removing Cr(VI) via its reduction to Cr(III) only if a suitable electron donor (such as lactose) is available.

It is known that Cr(VI) uses sulfate and phosphate transport routes to penetrate the cellular membrane due to its structural similarity to these anions. However, Cr(III) is positively charged under physiological conditions and hence the cell membranes are impermeable to this metal. Therefore, Cr(III) requires binding to an organic compound to be able to pass the plasmatic membrane (Pas *et al.* 2004; Guillén- Jimenéz *et al.* 2008).

CONCLUSION

Such investigations have indicated that yeasts which survive and flourish in heavy metals contaminated environments have developed or acquired genetic systems that allow to resist the effects of high metal ion concentrations. However, the yeast is a very heterogeneous group in terms of heavy metal resistance and accumulation. It appears that they have evolved a variety of mechanisms for metal uptake and homeostasis. These mechanisms, morphologic changes and oxidative stress response depend as much on the yeast strain under study as the growth conditions used. The metal processing capacity of yeasts can be used to concentrate, remove and recover metals from aqueous streams and enhance the efficiency of wastewater treatment processes. This variety of yeast answers or interaction with heavy metal will allow developing different bioremediation strategies. The most characteristic important is that the system is environment friendly. Further more studies are required for future applications on polluted environmental clean up.

REFERENCES

Abe F., Miura T., Nagahma T., Inoue A., Usami R. and Horokoshi K. (2001). Isolation highly copper-tolerant yeast, Cryptococcus sp., from the Japan Trench and the induction of superoxide dismutase activity by Cu2+. Biotechnol. Lett. 23: 2027-2034.

Aksu Z. and Dönmez G. (2001). Comparison of copper(II) biosorptive properties of live and treated Candida sp. J. Env. Sci. Health Part A Tox. Hazard Subst. Environ. Eng. 36: 367-81.

Amoroso M. J., Castro R. G., Carlino F. J., Romero N.C., Hill R. T. and Oliver G. (1998). Screening of heavy metal-tolerant actinomycetes isolated from the Salí River. J. Gen. Appl. Microbiol. 44: 129-132.

Arnesano F., Banci L. and Bertini E. (2001). Solution structure of the Cu(I) and apo forms of the yeast metallochaperone, AtxI. Biochem. 40: 1528-1539.

Audry, S., Schäfer, J., Blane, G., Jouanneau, J. M. (2004) Fifty year sedimentary record of heavy metal pollution (Cd, Zn, Cu, Pb) in the Lot River reservoirs (France). Environmental Pollution, 132, 413-426.

Baldi F., Vaughan A. and Olson G. (1990). Chromium(VI)-resistant yeast isolated from a sewage treatment plant receiving tannery wastes. Appl. Environ. Microbiol. 56: 913-918.

Balsalobre L., Silóniz M., Valderrama M., Benito T., Larrea M. and Peinado J. (2003). Occurrence of yeasts in municipal wastes and their behaviors in presence o cadmium copper and zinc. J. Basic. Microbiol. 43: 185-196.

Barceloux D.G. (1999a). Chromium. Clin. Toxicol., 37: 173-194.

Barceloux D.G. (1999b). Copper. Clin. Toxicol., 37: 217-230.

Bewley R.J. (1980). Effects of Heavy Metal Pollution on Oak Leaf Microorganisms. Appl Environ Microbiol. 40: 1053-1059.

Biedermann, K.A., Landolph, J.R. (19909. Role of valence state and solubility of chromium compounds on induction of cytotoxicity, mutagenesis, and anchorage independence in diploid human fibroblasts. Cancer Res. 50: 7835-7842.

Brady D. and Duncan J.R. (1994). Bioaccumulation of metal cations by Saccharomyces cerevisiae. Appl. Microbiol. Biotechnol. 41: 149-154.

Branco R., Chung A.P., Veríssimo A. and Morais P. (2005). Impact of chromium-contaminated wastewaters on the microbial community of a river. FEMS Microbiol. Ecol., 54: 35–46

Breierová E., Vajczikova I., Sasinková V., Stratilová E., Fišera M., Gregor T. and Šajbidor J. (2002). Biosorption of cadmium ions by different yeast species. Z. Naturforch. 57c: 634-639.

Clemens S., Kim E. J. Neumann D. and Schroeder J. (1999). Tolerance to toxic metals by a gene gamily of phytochelatin synthases from plants and yeast. EMBO J., 18: 3325-3333

Costa V. and Moradas-Ferreira P. (2001). Oxidative stress and signal transduction in Saccharomyces cerevisiae: insights into ageing, apoptosis and diseases. Mol Aspects Med., 22: 217-46. Review.

Debski B., Zalewski W., Gralak M. A. and Kosla T. (2004). Chromium-yeast supplementation of chicken broilers in an industrial farming system. J. Trace Elem Med Biol.18: 47-51.

Dönmez, G., Aksu Z. (2001). Bioaccumulation of copper(II) and nickel(II) by the non-adapted and adapted growing Candida sp. Water Research, 35, 1425-1434.

Dönmez, G., Aksu, Z. (1999). The effect of copper(II) ions on the growth and bioaccumulation properties of some yeasts. Process Biochemistry 35, 135-142.

Duffus J. H. (2002). Heavy metals-a meaningless term? Pure Appl. Chem., 74: 793-807.

Dursun A.Y., Uslu G., Cuci Y. and Aksu Z. (2003). Bioaccumulation of copper(II), lead(II) and chromium(VI) by growing Aspergillus niger. Process Biochem. 38: 1647-1651.

Fernández, P. M., Figueroa, L. I. C. and Fariña M. J. (2010). Critical Influence of Culture Medium and Cr(III) Quantification Protocols on the Interpretation of Cr(VI) Bioremediation by Environmental Fungal Isolates. Water, Air and Soil, 206:283-293.

Fernández, P. M., Figueroa, L. I. C. and Fariña M. J. (2010). The significance of inoculum standardization and cell density on the Cr(VI) removal by environmental yeast isolates. Water, air, soil pollut. In prenss.

Fourest E. and Roux J. C. (1992). Heavy metal biosorption by fungal mycelial by-products: mechanisms and influence of pH. Appl. Microbiol. Biotechnol., 37: 399-403.

Guillén-Jiménez, F., Morales-Barrera, L., Morales-Jiménez, J., Hernández-Rodríguez, C.H. and Cristiani-Urbina, E. (2008). Modulation of tolerance to Cr(VI) and Cr(VI) reduction by sulfate ion in a Candida yeast strain isolated from tannery astewater. J. Ind. Microbiol. Biotechnol., 35, 1277–1287.

Gulan Zetic V., Stehlik Tomas V., Grba S., Lutilsky L. and Kozlek D. (2001). Chromium uptake by Saccharomyces cerevisiae and isolation of glucose tolerance factor from yeast biomass. J. Biosci. 23: 217-223.

Halliwell B. and Gutteridge J. M (1990). The antioxidants of human extracellular fluids. Arch Biochem Biophys. 280:1-8. Review

Hassett R. and Kosman D. J. (1995). Evidence for Cu(II) reduction as a component of copper uptake by Saccharomyces cerevisiae. J. Biol. Chem., 270: 128-134.

Hernández-Saavedra, N. Y. (2003). Cu, Zn superoxide dismutase in Rhodotorula and Udeniomyces spp. Isolated from sea water: cloning and sequencing the encoding region. Yeast 20, 479-492.

Hursthouse A.S. (2001). The Relevance of Speciation in the Remediation of Soils and Sediments contaminated by Metallic Elements-An Overview and Examples from Central Scotland, UK. J. Environ. Monit., 3: 49-60.

Imlay, J. A. (2003). Pathways of oxidative damage. Annual Review of Microbiology 57, 395-418.

Ito H., Inouhe M., Tohoyama H. and Joho M. (2007). Characteristics of copper tolerance in Yarrowia lipolytica. Biometals, 20:773-80.

Iyer A., Mody K. and Jha B. (2004). Accumulation of hexavalen chromiumby an exopolysaccharide producing marine Enterobacter cloaceae. Mar. Pollut. Bull. 49: 974-977.

Iyer A., Mody K. and Jha B. (2005). Biosorption of heavy metals by a marine bacterium. Mar. Pollut. Bull., 50: 340-343.

Jamieson D. J. (1998). Oxidative stress responses of the yeast Saccharomyces cerevisiae. Yeast, 14: 1511-1527.

Kalbitz K. and Wennrich R. (1998). Mobilization of heavy metals and arsenic in polluted wetland soils and its dependence on dissolved organic matter. Sci. Total Environ., 209: 27-39.

Kaszycki P., Fedorovych D., Ksheminiska H., Babyak L., Wójcik D. and Koloczek H. (2004). Chromium accumulation by living yeast at various environmental conditions. Microbiol. Res. 159: 11-17.

Koch A. K., Peña M. M. and Thiele D. J. (1997). Copper-binding motifs in catalysis, transport, detoxification and signaling. Chem. Biol., 4: 549-560.

Krauter P, Martinelli R, Williams K, Martins S. (1996). Removal of Cr(VI) from ground water by Saccharomyces cerevisiae. Biodegradation 7: 277–86.

Ksheminska H., Fedorovych D, Babyak L., Yanovych D., Kaszycki P. and Koloczek H. (2005). Chromium(III) and (VI) tolerance and bioaccumulation in yeast: a survey of cellular chromium content in selected strains of representative genera. Process Biochem. 40: 1565-1572.

Ksheminska H., Jaglarz A., Fedorovych D, Babyak L., Yanovych D., Kaszycki P. and Koloczek H. (2003). Bioremediation of chromium by the yeast Pichia guilliermondii: toxicity and accumulation of Cr(III) and Cr(VI) and the influence of riboflavin on Cr tolerance. Microbiol. Res. 158: 59-67.

Kurtzman C. and Robnett C. (1998). Identification and phylogeny of ascomycetous yeasts from analysis of nuclear large subunit (26S) ribosomal DNA partial sequences. Antonie Van Leeuwenhoek, 73: 331–371

Labbé S, Zhu X. and Thiele D. (1997). Copper-specific transcriptional repression on yeast genes encoding critical components in the copper transport pathway. J. Biol. Chem., 272: 15951-15958.

Lamarre C., LeMay J. M., Deslauriers N. and Bourbonnais Y. (2001). Candida albicans expresses an unusual cytoplasmic manganese-containing suproxide dismutase (SOD3 gene product) upon the entry an during the stationary phase. J. Biol. Chem. 276: 43784-43791.

Laxman R. and More S. (2002). Reduction of hexavalent chromium by Streptomyces griseus. Miner. Eng., 15: 831-837.

Li Z and Yuan H (2008). Responses of Rhodotorula sp. Y11 to cadmium. Biometals.;21:613-21.

Linder M. C. (2001). Copper and genomic stability in mammals. Mutat. Res. 475: 141-152.

López-Archilla A., González A., Terron M. and Amils R. (2004). Ecological study of the fungal populations of the acidic Tinto River in southwestern Spain. Can. J. Microbiol., 50: 923-934.

Lushchak L. and Gospodaryov D. (2005). Catalases protect cellular proteins form oxidative modification in Saccharomyces cerevisiae. Cell Biol. Int., 29: 187-192.

Machado MD, Soares EV, Soares HM (2010) Removal of heavy metals using a brewer's yeast strain of Saccharomyces cerevisiae: chemical speciation as a tool in the prediction and improving of treatment efficiency of real electroplating effluents. J Hazard Mater. 180:347-53.

Malik A. (2004). Metal bioremediation through growing cells. Environ. Int., 30: 261-278.

Mannazzu I., Guerra E., Strabbioli R., Pediconi D. and Fatichenti F. (1998). The vanadate-tolerant yeast Hansenula polymorpha undergoes cellular reorganization during growth in, and recovery from, the presence of vanadate. Microbiology, 144: 2589-2597.

Marques P. A., Pinheiro H. M., Teixeira J. A. and Rosa M. F. (1999). Removal efficiency of Cu2+, Cd2+ and Pb2+ by waste brewery biomass: pH and cation association effects. Desalination, 124: 137-144.

Mikes J., Siglova M., Cejkova A., Masak J. and Jirku V. (2005). The influence of heavy metals on the production of extracellular polymer substance in the processes of heavy metal ions elimination. Water Sci. Technol. 52: 151-156.

Miura T, Abe F, Inoue A, Unsami R and Horikioshi (2002). Superoxide dismutase is involved in high tolerance to copper in the deep-sea yeast, Cryptococcus sp. N6. Biotechnol. Lett., 24, 1069–1074.

Moore M. M., Breedveld M. W. and Autor A. P. (1989). The role of carotenoids in preventing oxidative damage in the pigmented yeast Rhodotorula mucilaginosa. Arch. Biochem. Biophys., 270: 419-431.

Moore R. T. (1998). Cytology and ultrastructure of yeasts and yeastslike fungi. In: Kurtzman C.P y Fell J.W (Eds.) The yeasts. A taxonomic study, 4th edn (pp 33-44). Elsevier Science BV, Amsterdam

Muter O., Lubinya I., Millers D., Grigorjeva L., Ventinya E. and Rapoport A. (2002). Cr(VI) sorption by intact and dehydrated Candida utilis cells in the presence of other metals. Process Biochem. 38: 123-131.

Orozco A. M., Contreras E. M., Bertola N. and Zrotzky N. E. (2007). Hexavalente chormoium removal using arobic activated sludge batch systems added with powdered activated carbon. Water SA, 33: 239-244.

Pal A., Dutta S., Mukherjee P. K. and Paul K. (2005). Occurrence of heavy metal-resistance in microflora from serpentine soil of Andaman. J. Basic Microbiol., 45: 207–218.

Palmer CD, Wittbrodt PR (1991) Processes affecting the remediation of chromium-contaminated sites. Environ Health Perspect;92:25-40.

Pas M., Milacic R., Draslar K., Pollak N. and Raspor P. (2004). Uptake of chromium(III) and Chromium(VI) compounds in the yeast cell structure. BioMetals 17: 25-33.

Pepi C. and Baldi F. (1992). Modulation of chromium(VI) toxicity by organic and inorganic sulfur species in yeasts form industrial wastes. BioMetals, 5: 179-185.

Pepi C. and Baldi F. (1992). Modulation of chromium(VI) toxicity by organic and inorganic sulfur species in yeasts form industrial wastes. BioMetals 5: 179-185.

Plaper A., Jenko-Brinovec S., Premzl A., Kos J. and Raspor P. (2002). Genotoxicity of Trivalent Chromium in Bacterial Cells. Possible Effects on DNA Topology. Chem. Res. Toxicol. 15: 943-949.

Polti MA, Amoroso MJ and Abate CM. (2007). Chromium(VI) resistance and removal by actinomycete strains isolated from sediments. Chemosphere. 67:660-667.

Puig S. and Thiele D. J. (2002). Molecular mechanisms of copper uptake and distribution. Curr. Opin. Chem. Biol., 6: 171-180.

Rae T. D., Schmidt P. J., Pufahl R. A., Culotta V. C and O'Halloran R. V. (1999). Undetectable intracellular free copper: the requirement of a copper chaperone for superoxide dismutase. Science, 284: 805-808.

Ramírez-Ramírez R., Calvo-Méndez C., Ávila-Rodríguez M., Lappe P., Ulloa M., Vázquez-Juárez R. and Gutiérrez-Corono F. (2004). Cr(VI) reduction in a chromate-resistant strain of Candida maltosa isolated from the leather industry. Antonie Van Leeuwenhoek 85: 63-68.

Ramsay L. M. and Gadd G. M. (1997). Mutants of Saccharomyces cerevisiae defective in vacuolar function confirm a role for the vacuole in toxic metal ion detoxification. FEMS Microbiol. Lett. , 157: 293-298.

Raspor P., Batic M., Jamnik P., Josic D., Milacic R., Pas M., Recek M., Rezic-Dereani V. and Skrt M. (2000). The influence of chromium compounds on yeast physiology. Acta Microbiol. Immunol. Hung., 47: 143-173.

Rees E. M. and Thiele D. J. (2004). From aging to virulence: forging connections through the study of copper homeostasis in eukaryotic microorganisms. Curr. Opin. Microbiol., 7: 1-10.

Romero C., Gatti M. and Bruno E. (1999). Effects of heavy metals on microbial o water and sediment communities. World J. Microbiol. Biotechnol., 15: 161-166.

Ross I. S. and Parkin M. J. (1989). Uptake of copper by Candida utilis. Mycol. Res., 93: 33-37.

Salomons W. (1995). Environmental impact of metals derived from mining activities: Processes, predictions, prevention. J. Geochem. Exploration, 52: 5-23.

Santos V. and Linardi V. (2001). Phenol degradation by yeasts isolated from industrial effluents. J. Gen. Appl. Microbiol., 47: 213-221.

Silóniz M., Balsolobre L., Alba C., Valderrama M. and Peinado J. (2002a). Feasibility of copper uptake by the yeast Pichia guilliermondii isolated form sewage sludge. Res. Microbiol., 153: 173-180.

Silóniz M., Payo E., Callejo M., Marquina D and Peina J. (2002b). Environmental adaptation factors of two yeasts isolated from the leachate of a uranium mineral heap. FEMS Microbiol Lett., 210: 233-237.

Soares E.V., Hebbelinck K. and Soares H.M.V.M. (2003). Toxic effects caused by heavy metals in the yeast Saccharomyces cerevisiae: a comparative study. Can. J. Microbiol., 49: 336-343.

Srivastava S. and Thakur I. S. (2006). Isolation and process parameter optimization of Aspergillus sp. for removal of chromium from tannery effluent. Bioresour. Technol., 97: 1167–1173.

Teh AH, Kanamasa S, Kajiwara S, Kumasaka T. (2008). Structure of Cu/Zn superoxide dismutase from the heavy-metal-tolerant yeast Cryptococcus liquefaciens strain N6. Biochem Biophys Res Commun; 374:475-8.

Templeton D., Ariese F, Cornelis R., Danielsson L. G., Muntau H., Leeuwen H and Lobinski R. (2000). Guidelines for terms related to chemical speciation and fractionation of elements. Definitions, structural aspects, and methodological approaches. Pure Appl. Chem., 72: 1453-1470.

Turpeinen R., Kairesalo T. and Häggblom M. (2004). Microbial community structure and activity in arsenic-, chromium- and copper- contaminates soils. FEMS Microbiol. Ecol., 47: 39-50

Valls M. and Lorenzo V. (2002). Exploiting the genetic and biochemical capacities of bacteria for the remediation of heavy metal pollution. FEMS Microbiol. Rev., 26: 327-328.

Villegas L.B (2006). Estudios de la resistencia a metales pesados en levaduras aisladas de ambientes contaminados. Contribución al conocimiento de la biodiversidad. Doctoral thesis Secretary of Biochemistry, Chemistry and Farmacy. National Universit of Tucumán. Argentina.

Villegas L., Amoroso M.J. and L.I.C. de Figueroa (2004). Selection of tolerant heavy metal yeasts from different polluted sites. In: Spencer JFT & Spencer ALR (Eds.) Methods in Biotechnology, Vol 16: Environmental Biology, Methods in Protocols (pp 249-256). Humana Press Inc., Totowa, New Jersey

Villegas LB, Amoroso MJ, de Figueroa LI, and Siñeriz F. (2009b) Cu(II) removal by Rhodotorula mucilaginosa RCL-11 in sequential batch cultures. Water Sci Technol.;60(5):1225-32

Villegas LB, Amoroso MJ, de Figueroa LI. (2009a) Responses of Candida fukuyamaensis RCL-3 and Rhodotorula mucilaginosa RCL-11 to copper stress. J Basic Microbiol. 2009 Aug;49(4):395-403.

Villegas, L., Amoroso, M. J.,Figueroa, L.I.C. (2005). Copper tolerant yeasts isolated from polluted area of Argentina. Journal of Basic Microbiology 45, 381-391.

Villegas. L., Fernández. P, Amoroso. M.J.. Figueroa, L.I.C. (2008). Chromate removal by yeasts isolated from sediments of a tanning factory and a mine site in Argentina. Biometals 21, 591-600.

Volesky B. and May-Phillips H. A. (1995). Biosorption of heavy metals by Saccharomyces cerevisiae. Appl. Microbiol. Biotechnol., 42: 797-806.

Wang Y. and Xiao C. (1995). Factors affecting hexavalent chromium reduction in pure cultures of bacteria. Wat. Res., 29: 2467-2474.

Weissman Z., Berdicevsky I., Cavari B-Z. and Kornitzer D. (2000). The high copper tolerance of Candida albicans is mediated by a P –type ATPase. Proc. Natl. Acad. Sci. USA., 97: 3520-3525.

Yarrow D. (1998). Methods for the isolation, maintenance and identification of yeasts. In: Kurtzman CP & Fell JW (Eds.) The yeasts, A taxonomic study, 4th edn (pp 77-100). Elsevier Science BV, Amsterdam.

Zayed A.M. and Terry N. (2003). Chromium in the environment: factors affecting biological remediation. Plant Soil, 249: 139-156.

In: Bioremediation
Editor: Alexander C. Mason

ISBN: 978-1-61122-730-7

Chapter 7

BIOSURFACTANTS: POTENTIAL BIOMOLECULES FOR BIOREMEDIATION OF METALS

Asha A. Juwarkar**, *Mamta P. Kamat, Sanjeev K. Singh and Santosh K. Yadav

Eco-Restoration Division, National Environmental Engineering Research Institute (NEERI), Nehru Marg, Nagpur – 440020, India

ABSTRACT

Due to rapid developments and increase in mining and industrial activities, excessive levels of heavy metals are being continuously added into the soils leading to accumulation of these toxic metals into the soils. Cadmium, copper, lead, mercury, chromium, nickel and zinc are considered the most hazardous heavy metals and are included on the EPA's list of priority pollutants. The existence and fate of these metals in soil is a matter of concern not only because of their potential impact on microbial communities and plants but also because of chances for groundwater contamination and hence toxicological impact on human health. The mobility of such heavy metals, as well as their bioavailability and related eco-toxicity, strongly depend on their specific chemical forms and hence, speciation of metals is a critical factor in assessing the potential environmental impacts. Unlike organic compounds, toxic metals are not degradable in the environment, and can persist in soils for decades or even centuries. Therefore, it is highly desirable to apply suitable remedial approaches to decontaminate the polluted soil. Though, various approaches have been suggested for remediation of metal-contaminated sites, recent years have witnessed an enormous growth in the field of bioremediation technology for the restoration of contaminated sites. Biosurfactant mediated bioremediation, a new field of endeavor and innovative approach which is getting wide attention due to their low toxicity, biodegradability and better environmental compatibility. Biosurfactants are surface active molecules produced by variety of micro-organisms capable of reducing surface and interfacial tension, thereby facilitating mobilization of the metal ions by forming biosurfactant-metal complex. Metal contaminants in the environment are usually tightly bound to colloidal particles and organic matter of the soil and applications of biosurfactant for removal of available as

*Correspondence to Dr. Asha A. Juwarkar, E-mail: aa_juwarkar@neeri.res.in

well as bound forms of these contaminants are well reported. Biosurfactants have diversity in their chemical forms by the virtue of which they can be selected on the basis of their high specificity for different metals and thus metal removal efficiency. The present chapter emphasizes on various classes of biosurfactants, different possible mechanisms of interaction of biosurfactant with (cationic and anionic) metals and their application in bioremediation of metals. As far as eco-restoration of the metal contaminated site is concerned; potential of such bio-molecules should be explored to lower down the concentration of heavy metals below the limits established as a safe or acceptable by regulatory agencies.

1.0 INTRODUCTION

Over the past few decades enormous quantities of industrial pollutants have been released into the environment. Numerous industrial activities, including mining, smelting, automobile battery production, vehicle emission, land filling of industrial waste and fly ash from incineration process, accidental or deliberate release of organic and/or inorganic compounds have been responsible for release of heavy metals into the natural resources such as air, water and soil. Heavy metals are gaining attention because they are non biodegradable and tend to accumulate in the environment. The presence of heavy metals in soils lead to serious problems; (1) toxicity on biological system and (2) groundwater contamination by leaching process.

A number of remediation technologies have been applied for heavy metal contaminated soils such as soil excavation and transport of contaminated soil to hazardous waste sites for landfilling, thermal extraction for volatile metals, electrokinetics, solidification/stabilization, vitrification, chemical oxidation and soil flushing (Mulligan et al., 2001). Due to the great expense and time required for conventional techniques, alternate cost-effective remediation techniques are needed.

Recent research has focused on the development of novel technologies based on bioremediation for environmental cleanup. Now days, role of biomolecules in bioremediation is being explored as an effective tool in remediation of metal contaminated sites and their restoration, due to high affinity and selectivity for target metals. Biosurfactant, a biological chelating agent, having pronounced surface and emulsifying activity, has been reported to be an effective and eco-friendly biomolecule because of its low toxicity and biodegradability. Biosurfactants, with diverse chemical forms, can be selected on the basis of their high specificity for different metals and thus metal removal efficiency. Removal of heavy metals using biosurfactants would definitely present a more practical approach as far as bioremediation of metal contaminated soil is concerned.

2.0 METALS IN THE ENVIRONMENT

Metal contamination is being a persistent and severe environmental problem. Among metals, heavy metals are environmentally of most concern because of their toxicity even at fairly low concentrations and therefore are considered the most hazardous priority pollutants (USEPA, 1996b). Heavy metals includes both biologically essential (such as Co, Cu, Mn, Zn) and non essential (such as Cd, Pb. Hg, As) elements. The essential elements are required in low concentrations hence known as trace elements or micro elements while non essential elements are toxic to plants and animals hence known as toxic elements.

2.1. Sources of Heavy Metals

Heavy metals in aquatic and terrestrial environment originate from weathering of soil parent material and human activities. Pedogenesis, a natural process of mineral breakdown, leads to translocation of low bioavailable forms of metals through dust streams, volcanic eruptions and forest fires to more available forms. Elevated levels of As in groundwater have been linked to the As poisoning in Bangladesh and West Bengal (Kumar, 2003) is a good example of pedogenic processes.

Unlike pedogenic inputs, heavy metals added through anthropogenic activities have high bioavaibility. Heavy metals enter into the environment, anthropogenically, mainly via three routes: deposition of atmospheric particulates, disposal of metal enriched sludges and sewage effluents, and byproducts from metal mining processes. Basically, in the atmosphere, heavy metals occur in particulate form. Hence, transfer of air born particles to land or water surface by dry and wet deposition constitutes the first stage of accumulation of atmospheric heavy metals. Wong & Mak (2010) estimated heavy metal pollution of surface dust in children playgrounds build near industrial areas and traffic density regions in Hong Kong and indicated elevated concentrations of Cd, Cu, Pb and Zn, mainly due to air born particulates from fuel combustion in heavy traffic regions. Rapid industrialization has tremendously increased the chances of heavy metal contamination and is the main source of metal enhancement in soils. Industrial processes, manufacturing and improper discharge of sewage waste and effluents into the environment are leading to enrichment of the heavy metals into the environment. Indiscriminate dumping of chromium contaminated hazardous wastes in the premise of M/s Tamil Nadu Chromates and Chemicals, resulted in deterioration of nearby land and pollution of ground water in Ranipet area in Tamil Nadu (MoEF). Metal transfer from sewage sludge to soil and subsequently to groundwater represents one of the most critical long-term hazards associated with the application of these wastes to soils (McBride et al., 1997). Large land areas located in south of Tehran (approximately 6900ha) are irrigated with treated and untreated wastewater and long-term irrigation with effluent and sludge resulted in increased heavy metal contents in soil and plants (Behbahaninia et al., 2009). Environmental pollution by heavy metals is very prominent in areas of mining and old mine sites. These metals are leached out in sloppy areas, are carried by water downstream or run-off to the sea. In mining activities, water bodies are most emphatically polluted (INECAR, 2000). Wells located near mining sites have been reported to contain heavy metals at levels that exceed drinking water criteria (Peplow, 1999). Figure 1 represents the sources of pollution by anthropogenic enrichment of metal contaminants in the environment.

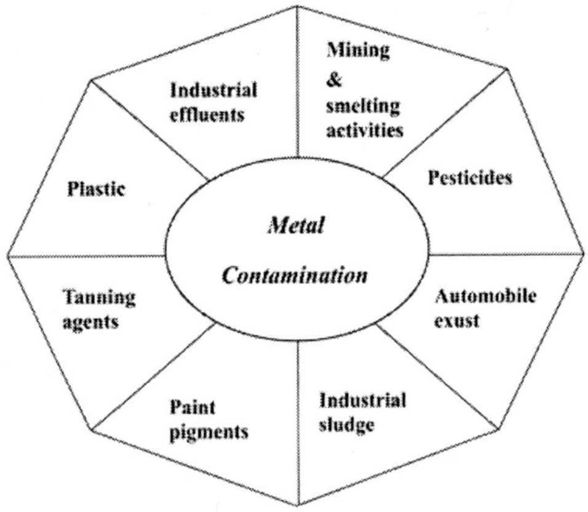

Figure 1. Environmental pollution by anthropogenic enrichment of metal contaminants.

Table 1. Heavy metal toxicity and sources of contamination

Heavy Metals	Sources	Toxicity
Cadmium	by- product of zinc (and occasionally lead) refining, electroplating, paint pigments, plastic, silver Cd batteries	Acute respiratory distress syndrome (ARDS), kidney damage, itai-itai disease, renal cancer
Lead	mining and smelting activities, automobile exhausts (by combustion of petroleum fuels treated with tetraethyl lead antiknock) and from old lead paints	Plumbism, Joint pain, loss of memory, muscle tremor, sub-encephalopathy, neurological defects, carcinogenic, hallucinations
Mercury	the degassing of the earth's crust, production of chlorine, ore processing	Stomititis, Minamata disease, Hg poisoning, numbness, blurring vision, neurological and renal disturbances
Chromium	Tanning agent, paint pigments, catalyst	Hemorrahagic diathesis, epitasis. skin ulceration, chrome holes, bronchial asthama , Throat and lung cancer
Copper	Mining activities, algicides, copper pipes	Injury to red blood cells, injury to lungs, as well as damage to liver and pancreatic functions
Nickel	Occurs naturally in the earth's crust emitted from volcanoes, oil-burning power plants, coal-burning power plants, and trash incinerators	lung cancer, nose cancer, larynx cancer and prostate cancer, Lung embolism, respiratory failure, birth defects, asthma and chronic bronchitis, skin rash, chronic bronchitis
Arsenic	earth's crust, pesticides, preserve wood	destruction of red blood cells , abnormal heart rhythm, kidney failure, skin lesion, lung cancer, death
Zinc	earth's crust, dry cell batteries, industry to make paint, rubber, dyes, wood preservatives, and ointments	stomach cramps, nausea, and vomiting, infertility, metal fume fever, skin irritation
Cobalt	to produce alloys in the manufacture of aircraft engines, magnets, grinding and cutting tools, color glass, ceramics and paints, radioacive cobalt in sterilizing medical equipment and consumer products, manufacturing plastics, and irradiating food	heart effects, liver, lung and kidney dysfunction, dermatitis
Manganese	mining activities, automobile exhaust Welding, steel factory, manganese-containing gasoline additive, fireworks, dry-cell batteries, fertilizer, paints, a medical imaging agent, cosmetics	loss of sex drive and sperm damage, nervous system effects

Source: (Lenntech, 2004; Lawrence et al., 2009; Ministry of the Environment programs and initiatives, fact sheet, 2001; ATSDR, 2010).

Among the most hazardous heavy metals, cadmium, chromium, arsenic, copper, lead, nickel, zinc and mercury are continuously being released into the environment via industrial and mining processes. The sources and toxicity of these heavy metals are depicted in Table 1. Environmental systems have a limited capacity to assimilate the inputs of the pollutants. Soil, though acts as a sink for metals, due to continuous loading, ability of the soil to retain the contaminants will significantly be reduced or even cease.

2.2 Metal Mobility and Bioavailability

Total metal concentrations do not necessarily correspond with bioavailability of metals. Bioavailability is the quantity of total metals that are available for incorporation into biota. Therefore, metal mobility and bioavailability are important parameters in determining the toxicity of the metals in the soil. Heavy metals are associated with various soil components in different ways that determine their mobility and availability.

Soil possesses an ability to bind and immobilize introduced heavy metal ions, with different forces, which keep them bound to the soil particles. Heavy metals may be bound or sorbed by particular natural substances, which may change their existing forms with ultimate increase or decrease in their mobility in the environment. In general, they may react with particular species, change oxidation states and precipitate (Dube et al., 2000).

Immobilization of heavy metals in soils and sediments is mainly due to sorption properties which are determined by physico-chemical properties of the soil such as: pH, cation exchange capacity, organic matter content, clay fraction content, and also their chemical forms and fractionation in the soil. All such properties, which collectively determine metal binding ability of the soil, are discussed below.

2.2.1 Physico-Chemical Properties of the Soil

Hydrogen ion activity, pH, is one of the major factors which govern the binding forces between soil and heavy metals. Solubility, transport, and bioavailability of metals are highly pH dependent. In acid soils (pH 4.2-6.6), the elements such as, Cd, Ni, and Zn are highly mobile, Cr is moderately mobile, and Cu and Pb practically immobile, and in neutral to alkaline soils (pH 6.7-7.8), Cr is highly mobile, Cd and Zn are moderately mobile and Ni is immobile (Fuller, 1997), which indicated that the mobility of cationic heavy metals is high at lower pH values while, the mobility of anionic heavy metals is high at higher pH values. Thus, the binding between metal cations to soils decrease with decreasing pH of the environment and vice-versa. This indicates pH is one of the important properties of the soil while determining metal binding and mobility.

The cation exchange capacity (CEC), another important soil property, is a key parameter describing the sorption and desorption of contaminants by the soil. The higher the CEC, the greater is the ability of the soil to retain heavy metals. Organic matter together with the clay, make up most of the CEC of the soil. Soil organic matter and clay mineral content have a great influence on the physical and chemical properties of soil. Being the source of negative charges in the soil, it not only contributes to mobilization and retention of cationic species but also plays an important role in forming complexes with metals. Therefore toxicity is observed only when binding sites on organic matter and clay mineral becomes saturated with cationic metals.

The type and quantity of clay determines the CEC, which increases with clay content. Clays with high montmorillonite, appear to reduce metal bioavaibility and toxicity most (Babich & Stotzky, 1977). Similarly, organic material (humic and fulvic acids), ubiquitous in most natural environments, are extremely important in controlling the mobility and bioavailability of micronutrients and pollutants due to their colloidal size and high chemical reactivity (Davies et al., 2001). Organic components of the soil constitutes high affinity for heavy metal cations because of presence of ligands or groups such as carboxylic, phenolic, carbonyl and alcoholic etc. that can form chelates with metals (Naidu & Harter, 1998). With increasing pH, the functional groups dissociate, thereby increasing the affinity of ligands ions for metal cations. The order of metal cations complexed with organics is $Cu^{2+}> Fe^{2+}> Pb^{2+}> Ni^{2+}> Co^{2+}> Mn^{2+}> Zn^{2+}$. Not only physico-chemical properties of the soil but also valancies and fractionation of the metal have broad impact on toxicity of the metal ions in the soil.

2.2.2 Valency of the Metal

The understanding of the chemical nature of metal compounds in the environment is essential because the toxicity of the metal is highly dependent upon its chemical form. It is well established that different metal species vary in their biological reactivities (Behra et al., 2002). Anionic heavy metals such as arsenic and chromium are significantly more toxic in their oxidized states compared to their reduced state. Arsenic mobility, bioavailability, and toxicity are dependent on its speciation: arsenite (AsO_3^{-3}) forms are much more toxic to biological species and are more mobile than arsenate (AsO_4^{-3}) forms (Jain & Ali, 2000). Similarly, in the environment, Cr generally exists in two forms hexavalent and trivalent, out of which hexavalent Cr is more toxic due to their carcinogenic, mutagenic, highly soluble and mobile nature as compared to trivalent Cr (Massara et al., 2007). The toxic cationic metals are usually divalent Cu^{2+}, Zn^{2+}, Cd^{2+}, Pb^{2+}. Hence, it is noteworthy that toxicity of metals strongly depends upon the forms in which they exist in the environment.

2.2.3 Fractionation of Metal in Soil

Heavy metals accumulate in soils in various geochemical forms, i.e. water-soluble, exchangeable, carbonate-associated, Fe-Mn oxide-associated, organic-associated and residual forms, which determine their bioavaibility. Water-soluble and exchangeable fractions are considered to be bioavailable while metals confined to oxide and organic fractions are released only when surrounding conditions become increasingly reducing or oxidizing respectively. Metals in the carbonate fraction are very sensitive to pH changes and they are readily leached when the pH is decreased. On the other hand, metals bound to the residual fraction are usually not expected to be released over short period of time under the conditions usually encountered in nature (Lasheen & Ammar, 2009).

Thus, it can be observed that, the properties of the soil and the state of the metals are critical factors responsible for sorption and desorption of the metals in the soil and are highly interdependent. Evaluation of such parameters is crucial while designing appropriate remediation techniques for the metals.

3.0 Conventional Remediation Approaches

Table 2. Traditional approaches for remediation of metal contaminated soils

Approaches		Advantages	Disadvantages
Isolation	An attempt has been made to isolate contamination within a designated area based on the application of semi-permeable or impermeable barriers	prevent the transport of contaminants and hence prevent further contamination of groundwater	land requirement, deterioration of the land
Excavation and offsite disposal	Contaminated material is removed and transported to permitted site	Quick and appropriate for small volumes of soil, completely removes contaminants from site	Expensive, land requirement, transportation costs, cause death hazard during the excavation, removed soil may need treatment, does not destroy the contaminants, transfers pollution to another site
Soil washing	The process, with the help of washing solutions, separates coarse soil (sand and gravel) from fine soil (silt and clay), where contaminants tend to bind and sorb.	Reduces the volume of contaminated material, washing solution may be reused after decontamination treatment	Needs the excavation of soil, generate contaminated solution, transfers contaminants to a different phase, does not destroy the contaminants,
Soil flushing	"Floods" contaminated soils with a solution that moves the contaminant to an area where they can be removed .	In situ technology that does not require excavation, relatively low invasive method	Does not destroy the contaminants, generate contaminated water, recovered fluids with the absorbed contaminants may need further treatment, incomplete removal of contaminants (strongly bounded heavy metals remain in the soil)
Stabilization/ Solidification	Reduces the risk posed by a waste by converting the contaminant into a less soluble, immobile, and toxic form	Applicable to sites with high water tables, Simple and fast method, Relatively inexpensive	does not remove the contaminants, before the treatment highly mobile form of the metal need to be converted to less mobile or immobile form, minimize potential leaching of the pollutants
Vitrification	Uses high-temperature treatments (1600–2000 °C) to volatilize and/or destroy organic and inorganic contaminants or volatile metal species (such as Hg), destroying organic pollutants by pyrolysis	Reduces the toxicity and immobilizes the contaminants , the final product can be considered to be non hazardous	Expensive and demands high energy
Electrokinetics	Removes contaminants from soil by application of an electric field to mobilize contaminants in the form of charged species	The technology mostly suitable to remediate low permeability soil or fine-grained soil that cannot readily drained	Contaminants needs to be removed from the solution with the help of other processes, Electrode corrosion, mostly applicable to saturated soils
Thermal desorption	Contaminated soil is excavated, screened, and heated to temperatures (100–600 °C) so that contaminants with boiling points in this range will vaporize and separate from the soil. The vaporized contaminants are then collected and treated by other means	Broad applicability, High removal efficiencies, No excavation required in situ, very rapid treatment time	At high temperatures contaminants may volatilize and are released from the soil into the air, wet soils ,soils high in silt and clay are harder to treat, Any heterogeneity of the soil body decreases the effectiveness of the method

In response to a growing need to address environmental contamination, many remediation technologies have been developed to treat contaminated soil. The most commonly used techniques are listed in Table 2. The combinations of one or more of these approaches are often used for treatment of a contaminated site. However, these technologies are usually expensive and soil disturbing, sometimes rendering the land useless for further activities such as plant growth. A growing public awareness and concern about environmental pollutants fostered collaboration between government and industry for the development of safe and cost-effective alternative approaches for dealing with wastes. The past two decades have seen a tremendous upsurge in the search for cost-effective and environmentally sound alternatives to traditional methods. Recently, biological-based emerging technologies are gaining wide attention due to high potential for selective removal of toxic metals. Of the technologies that have been investigated, bioremediation has emerged as the most desirable approach for the cleaning up of many environmental pollutants.

4.0 Bioremediation

Bioremediation is a process that uses microorganisms, fungi, green plants or their enzymes to return the natural environment altered by contaminants to its original condition. Bioremediation is a broad term that comprises several techniques for decontamination of contaminants. Most common bioremediation techniques are summarized below and are presented in Figure 2.

Biodegradation is the most frequently preferred bioremediation option. Micro-organisms can break down most of the compounds for their growth and/or energy. Biotransformation is a process of alteration of contaminant molecules into less or non-hazardous molecules with the help of microbes. Bioaugmentation is the practice of adding specialized microbes or their enzyme preparations to polluted matrices to accelerate transformation or stabilization of specific pollutants. Biostimulation strategy involves stimulation of the numbers and activities of natural populations, usually bacteria or fungi, so they can better break down the pollutants into harmless products. Biosorption is a process of uptake of organic and inorganic metal species through sorption to cellular surfaces or accumulation within the cells via membrane transport system. Phytoremediation strategy takes advantage of naturally occurring, specifically selected, or genetically engineered plants for environmental cleanup operations. Phytoremediation is the fastest growing area of interest among bioremediation approaches. Phytoremediation is divided into 1. Rhizofiltration is basically a water remediation technique in which metal contaminants are taken up by the plant's roots, 2. Phytoextraction refers to the use of hyperaccumulating plants to translocate the contaminants from soil to root and finally to shoot and leaves, 3. Phytotransformation which is applicable to both soil and water and involves the degradation of contaminants through plant metabolism, 4. Phytostimulation which involves the stimulation of microbial degradation through the activities of plants in the root zone, 5. Phytostabilization uses plants to reduce the migration of contaminants through the soil medium, 6. Phytovolatilization explores plants ability to transpire large amount of contaminants taken up by the plants.

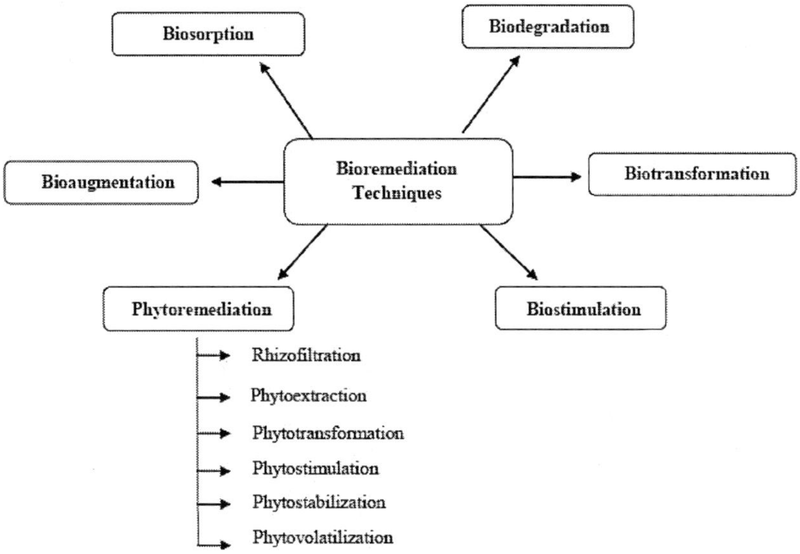

Figure 2. Frequently used bioremediation techniques.

4.1 Biomolecules in Bioremediation

Every organism synthesizes different compounds in response to adverse conditions encountered during their struggle for existence. A biomolecule is any organic molecule that is produced by a living organism (such as bacteria, fungi, algae and plants) including large polymeric molecules such as proteins, polysaccharides, as well as small molecules such as primary metabolites, secondary metabolites, and natural products. Identification and utilization of such biomolecules that can metabolize environmental pollutants has to be focused. Some of them are listed below:

- Polysaccharide: Polysaccharides are polymeric carbohydrate structures, formed of repeating units joined together by glycosidic bonds. Amine groups in polysaccharide are generally responsible for the binding of the pollutants. Cellulose and pectin are primary structural components of the plants. Application of pectin and cellulose and its derivatives for removal of metal ions such as Cd, Pb, Co (Mata et al., 2010) and Cr, Fe (Chauhan et al., 2005) have been reported. Chitin is the main structural component of the cell walls of fungi. Chitin and its derivative chitosan are reported in remediation mine impacted water (Robinson- Lora & Brennan, 2010) and removal of metal ions (Bhatnagar & Sillanpaa, 2009). Lignin is one of the primary structural components of most plants. Guo et al. (2008) reported adsorption of the heavy metal ions Pb(II), Cu(II), Cd(II), Zn(II), and Ni(II) on a lignin due to the presence of phenolic sites having affinity for metal ions.
- Enzymes: Most xenobiotics can be subjected to enzymatic bioremediation, for example, polycyclic aromatic hydrocarbons (PAHs) (Samanta, 2002), polynitrated aromatic compounds (Ramos et al., 2005), pesticides such as organochlorine insecticides (Sutherland et al., 2004, Ahuja et al., 2004), bleach-

plant effluents, synthetic dyes, polymers and wood preservatives (creosote, pentachlorophenol) (Bajpai, 2004). Many enzymes of both microbial or plant origins have been recognized to be able to transform pollutants at a detectable rate and potentially suitable to restore polluted environment. The main enzymatic classes involved in such a process are: oxidoreductases and hydrolases.

- Oxidoreductases may explicate a degradative function and transform polymeric substances into partially degraded or oxidized products that can be easily up-taken by cells. Partial oxidation of recalcitrant pollutants such as PAHs by extra cellular oxidative enzymes give rise to products of increased polarity and water solubility and thus with a higher biodegradability (Meulenberg et al., 1997). For instance, some pollutants such as 2, 4, 6-trinitrotoluene (TNT), the explosive of major environmental concern, or carbon tetrachloride (CCl_4) were significantly transformed by cellobiose dehydrogenase (Cameron & Aust, 1999; Cameron et al., 2000). Cheung & Gun (2007) reported mechanism of hexavalent chromium detoxification by *Pseudomonas maltophilia* and *Bacillus megaterium* which utilize membrane associated reductases to convert Cr^{6+} to Cr^{3+}. Laccases from fungi have been shown to be useful for the degradation of a variety of persistent environmental pollutants, including alkenes, bisphenol A, and industrial synthetic dyes (Mayer & Staples, 2002). To enhance the environmental decontamination functions of plants the laccase was introduced into tobacco plants, which was efficiently found to degrade bisphenol A. Gomare et al. (2008) reported use of lignin peroxidase for the degradation of sulphonated azo dyes.
- Hydrolases, degradative enzymes, include proteases, carbohydratases (e.g. cellulases, amylases, xylanases, etc.), esterases, phosphatases and phytases. Due to their intrinsic low substrate specificity, hydrolases may play a pivotal role in the bioremediation of several pollutants including insoluble wastes. Hydrolases responsible for the cleavage of pesticides are among the best studied groups of enzymes. Alkaline proteases are widely used in solubilization of proteinaceous, waste feathers from poultry slaughterhouses, keratinous material from poultry and recycling of polyester film etc. (Sen, 2009).

- Phytochelatins: Phytochelatins (PCs) act as chelators, and are important for heavy metal detoxification. Cobbett (2000) have demonstrated role of PCs in Cd detoxification. PCs can form complexes with Pb, Ag, and Hg in vitro (Mehra et al., 1996, Rauser, 1999). Lead-binding phytochelatins produced by *Vetiver* Grass (Andra et al., 2009) suggesting its potential usefulness for the remediation of Pb-contaminated residential sites.
- Metal binding proteins: Metallothioneins (MTs), cysteine rich proteins, are synthesized by higher organisms in response to the presence of heavy metals.

Metal ions such as Cd, Pb, Hg, Cu are sequestered by them in biologically inactive forms.

- Siderophores: Siderophores (Greek: "iron carrier") are small, high-affinity iron chelating compounds secreted by microorganisms such as bacteria, fungi and grasses (Miller & Marvin, 2008). They play important role in the complexation of toxic metals and increase their solubility. Though siderophores are highly specific for Fe^{3+} (Leong & Winkelmann, 1998), have been found to bind effectively to metals such as Mn, Cr, Mg etc. Efficiency of siderophores in removal of As from the contaminated soils has been reported by Nair et al. (2007).
- Biosurfactants: Biosurfactants are the surface-active substances synthesized by living cells. Biosurfactants are produced extracellularly or as part of the cell membrane by variety of bacteria, yeasts and fungi. Biosurfactants are capable of reducing surface and interfacial tension, and have low toxicity and biodegradability. For which it has wide range of applications in many industries such as food, cosmetic, leather, petroleum etc. and extensively used in the remediation of hydrocarbon and metal contaminated soils.

5.0 BIOSURFACTANT: A POTENTIAL BIOMOLECULE

Recent years have witnessed application of biosurfactants in bioremediation. Various reports are available on the use of biosurfactants in decontamination of hydrocarbon compounds and metal remediation. Detailed study of such biomolecule, having tremendous applicability in bioremediation, is a need of the hour. Different types and classes of biosurfactants, their structure, properties and mode of action are described in the further text.

5.1 Plant Derived Biosurfactant

A plant-derived biosurfactant, saponin, produced especially from desert plants such as *Yucca schidigera* and *Quillaja saponaria* (soapbark tree), have potential to enhance solubilization and mobilization of heavy metals in contaminated soils. Hong et al. (2002) reported utilization of saponin for effective removal of heavy metals from soils, attaining 90–100% of Cd and 85–98% of Zn removal. Application of saponin in removal of lead (89.95%), copper (81.13%) and cadmium (71.17%) was evaluated by Yuan et al. (2008). Though, very little literature on scientific research applying plant-derived biosurfactant for remediation has been found, they have various applications in mining and ore separation, in preparation of emulsions for photographic films, and extensively in cosmetics, such as lipstick and shampoo. The antifungal and antibacterial properties of saponins are important in cosmetic applications, in addition to their emollient effects.

5.2 Microbial Biosurfactants

Microbial biosurfactants are produced extracellularly or as part of the cell membrane by variety of bacteria, yeasts and fungi. Microbes synthesize them, especially during growth on water-immiscible substrates. Different types of biosurfactants produced by variety of organisms are presented in Table 3. Biosurfactants are low molecular weight organic compounds belonging to various classes. Most of these compounds are either anionic or neutral in nature. Only a few are cationic such as those containing amine groups.

In general, biosurfactants can be classified as:

- Glycolipids
- Lipopeptides
- Fatty acids neutral lipids
- Polymeric surfactants
- Particulate biosurfactants

Table 3. Different types of biosurfactants produced by variety of organisms

Types of biosurfactant	Organism	Substrates	References
Glycolipids Rhamnolipids Trehalose lipids Sophorolipids	Pseudomonas aeruginosa Rhodococcus erithropolis, Arthobacter sp Candida bombicola, Candida apicola	Oil Wastes, soybean oil, safflower oil, glycerol, distillery and whey wastes, residual sunflower frying oil, rapeseed oil	Ruggeri et al. (2009), Sadouk et al. (2008), Nitschke et al. (2005), Dubey & Juwarkar (2004), Rahman et al. (2002), Dubey & Juwarkar (2001), Babu et al. (1996), Babu (1995)
Lipopeptides Surfactin Viscosin Lichenysin Serrawettin Phospholipid	Bacillus subtilis P. fluorescens B. licheniformis Serratia marcescens Acinetobacter sp. Corynebacterium lepus	Potato peels, cassava flour wastewater, cheese whey and molasses	Das & Mukherjee (2007), Nitschke & Costa et al. (2007)
Fatty acids/neutral lipids Corynomicolic acids	Corynebacterium insidibasseosum		
Polymeric surfactants Emulsan Alasan Liposan Lipomanan	Acinetobacter calcoaceticus A. radioresistens Candida lipolytica Candida tropicalis	Agro industrial residue, ground-nut oil refinery residue	Rocha et al. (2006) Rufino et al. (2007)
Particulate biosurfactants Vesicles Whole microbial cells	A.calcoaceticus Cyanobacteria		

5.2.1 Structure of Biosurtactants

Biosurfactants are amphiphilic molecules consisting of a hydrophilic and a hydrophobic or lipophillic domain. The lipophilic portion is usually the hydrocarbon (alkyl) tail of one or more fatty acids, which may be saturated, unsaturated, hydroxylated or branched; the fatty acid is linked to the hydrophilic group by a glycosidic ester or amide bond. Due to the presence of carboxylate groups most of them are either neutral or negatively charged anionic biomolecules. Glycolipid is the largest and best-known group followed by lipopeptides. The structure of rhamnolipid, trhalolipid, sorpholipid (glycolipid type) and surfactin (lipopeptide type) are presented in Figure 3.

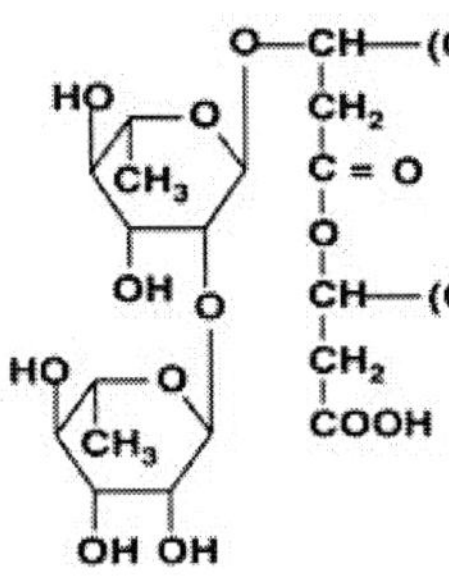

Rhamnolipid

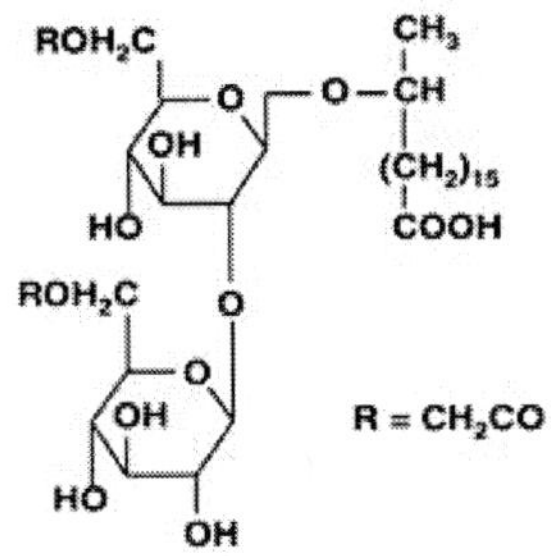

Sophorolipid

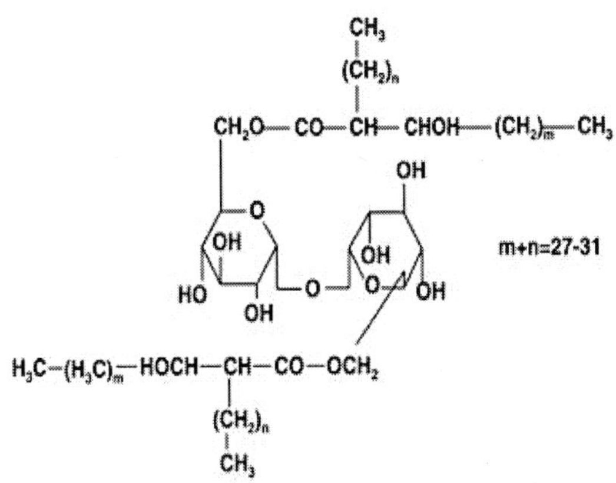

Trehalolipid

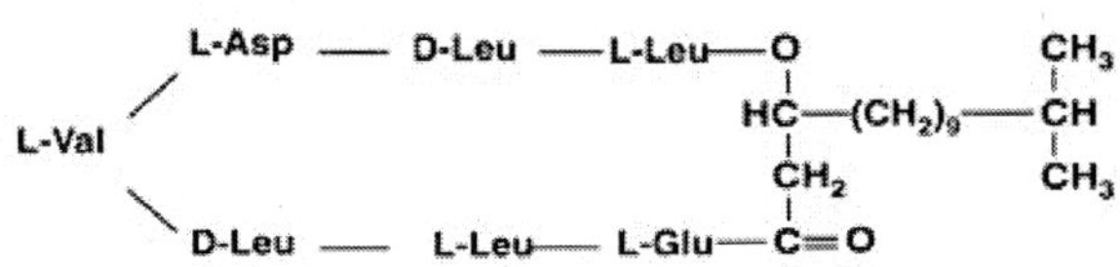

Surfactin

Figure 3. Structure of glycolipid (rhamnolipid, trehalolipid, sorpholipid) and lipopeptide (surfactin) biosurfactant.

5.2.2 Properties of Biosurfactants

Biosurfactants are attracting much interest due to their potential advantages over their synthetic counterparts in many fields spanning environmental, food, biomedical, and other industrial applications.

- Surface and interfacial tension reduction A good surfactant can lower surface tension (ST) of water from 72 to 35 mN/m and the interfacial tension (IT) water/hexadecane from 40 to 1 mN/m.Surfactants are molecules that concentrate at interfaces and decrease surface and interfacial tension. The amount of surfactant needed to obtain the lowest possible surface tension is defined as the critical micelle concentration (CMC). After CMC is reached, surface tension remains constant, and surfactants begin to form micelles. Biosurfactants are more effective and efficient and their CMC is about 10–40 times lower than that of chemical surfactants, i.e. less surfactant is necessary to get a maximum decrease in surface tension.
- Emulsification Emulsification is property of biosurfactant which primarily associated with enhanced oil recovery processes by formation of stable emulsions. Stable emulsions can be produced with a lifespan of months and years. This property is especially useful for making oil/water emulsions for cosmetics, food and hydrocarbon decontamination.
- Temperature and pH tolerance Many biosurfactants and their surface activities are not affected by environmental conditions such as temperature and pH.
- Biodegradability /biocompatibility Unlike synthetic surfactants, microbial-produced compounds are degradable and hence particularly suited for environmental applications such as bioremediation.
- Low toxicity In terms of toxicity and mutagenic properties, chemical-derived surfactants are highly toxic and mutagenic, whereas the biosurfactant was considered slightly non-toxic and non mutagenic. Very little data are available in the literature regarding the toxicity of microbial surfactants. They are generally considered as low or non-toxic products.
- Chemical diversity and specificity The chemical diversity of naturally produced biosurfactants offers a wide selection of surface-active agents with properties closely related to specific applications. Biosurfactants, being complex organic molecules with specific functional groups, are often specific in their action which would be of particular interest in detoxification of specific pollutants.
- Ability to be synthesized from renewable sources Biosurfactants can be produced from cheap raw materials which are available in large quantities; the carbon source may come from hydrocarbons, carbohydrates and/or lipids, which may be used separately or in combination with each other.

5.3 Applications of Biosurfactant

Biosurfactants have broad range of scope in environmental, industrial and biological applications. The chapter focuses on the application of biosurfactant for environmental management studies. In general, remediation of hydrocarbon contaminated soils has given much attention, however here remediation of metal contaminated soils with biosurfactant has been targeted. Other industrial and biological applications of biosurfactants have depicted in Table 4.

Table 4. Applications of biosurfactant

INDUSTRY	APPLICATIONS	ROLE Of BIOSURFACTANT
Food	Emulsification and de-emulsification Fruits and vegetables Bakery and ice cream Crystallization of sugar Cooking fat and oils	Emulsifier, solubilizer, demulsifier, suspension, wetting, foaming, defoaming, thickener, lubricating agent, improve removal of pesticides and wax coating, solubilize flavor oils, control consistency, retard staling, improve washing, reduce processing time, prevent spattering
Textiles	Preparation of fibers Dyeing and printing Finishing of textiles	Detergent and emulsifier in raw wool scoring, dispersant in viscose rayon spin, bath lubricant and antistat in spinning of hydrophobic filaments Wetting, penetration, solubilization, emulsification, dye leveling, detergency and dispersion Wetting and emulsification in finishing formulations, softening, lubricating and antistatic additives to finishes
Building and construction	Paving Concrete	Improve bond of asphalt to gravel and sand Promote air entertainment
Elastomers and plastics	Emulsion polymerization Foamed polymers Latex adhesive Plastic articles, Plastic coating and laminating	Solubilization, emulsification of monomers Introduction of air, control of cell size Promote wetting, improve bond strength Antistatic agents Wetting agents
Leather	Skins Tanning Hides Dyeing	Detergent and emulsifier in degreasing Promote wetting and penetration Emulsifiers in fat liquoring Promote wetting and penetration
Paint and protective coatings	Pigment preparation Latex paints	Dispersing and wetting of pigment during grinding Emulsification, dispersion of pigment, stabilize latex, retard
Biological	Antimicrobial agents Pharmaceuticals and therapeutics Anti-adhesive activity	Lipopolypeptides can act as antibiotics, antiviral and antitumor agents, immunomodulators or specific toxins and enzyme inhibitors, Antibacterial, antifungal, antiviral agents adhesive agents, immunomodulatory molecules, vaccines, gene therapy inhibit the adhesion of pathogenic organisms to solid surfaces or to infection sites
Agricultural	Biocontrol Phosphate fertilizers Spray application	Facilitation of biocontrol mechanisms of microbes such as parasitism, antibiosis, competition, induced systemic resistance and hypovirulence, Prevent caking during storage, Wetting, dispersing, suspending of powdered pesticides and emulsification of pesticide solutions; promote wetting, spreading and penetration of toxicant
Bioprocessing	Downstream processing	Biocatalysis in aqueous two-phase systems and microemulsions, biotransformations, recovery of intracellular products, enhanced production of extracellular enzymes and fermentation products
Cosmetic	Health and beauty products	Emulsifiers foaming agents solubilizers wetting agents, cleansers, antimicrobial agents, mediators of enzyme action

Source: (Singh et al., 2007; Kosaric, 2001).

5.3.1 Environmental Applications of Biosurfactant

Biosurfactants have been tested in environmental applications such as dispersion of oil spills, enhanced oil recovery and metal solubility.

The application of biosurfactants in the remediation of organic compounds, such as hydrocarbons, aims at increasing their bioavailability by mobilizing and removing the contaminants in a washing treatment. The application of biosurfactants in the remediation of inorganic compounds such as heavy metals, on the other hand, is targeted at chelating and removal of such ions during a washing step facilitated by the chemical interactions between the amphiphiles and the metal ions.

5.3.1.1 Bioremediation of Organic Contaminants

- Bioremediation of hydrocarbon contaminated soil Extensive research has been conducted on bioremediation of hydrocarbon contaminated soils. Oil spillage is being the main source of hydrocarbon release into the environment. The low water solubility limits their availability to microorganisms, which is a potential problem for bioremediation of hydrocarbon contaminated sites. Biosurfactants enhance the bioavailability of these hydrophobic compounds by reducing the surface and interfacial tension between two phases leading to increased solubility and subsequent biodegradation. Hydrocarbons compounds can be divided into

  - Petroleum hydrocarbons – contains Alkanes, cycloalkanes, aromatics, polycyclic aromatic hydrocarbons, asphaltenes, and resins
  - Chlorinated hydrocarbons - includes pesticides, DDT, 2, 4-D and 2, 4, 5-T, plasticizers, pentachlorophenol and polychlorinated biphenyls
  - Polycyclic or polynuclear aromatic hydrocarbons - produced during petroleum refining, coke production and wood preservation

Remediation of hydrocarbon contaminated soils and sludges have been extensively studied by many workers. Noordman et al. (2002) reported use of the biosurfactant from P. aeruginosa on hexadecane degradation. Maier and Soberon-Chavez (2000) indicated that rhamnolipid addition can enhance biodegradation of hexadecane, octadecane, n-paraffin, and phenanthrene in liquid systems, in addition to hexadecane, tetradecane, pristine, creosote and hydrocarbon mixtures in soils. Application of rhamnolipid produced by Pseudomonas sp. along with poultry litter and coir pith enhanced ex-situ bioremediation of a gasoline-contaminated soil (Rahman et al., 2002). Straube et al. (2003) evaluated the addition of P. aeruginosa, a biosurfactant-producer, with nutrients (slow-release nitrogen) and a bulking agent (ground rice hulls) to enhance bioremediation of a PAHs and pentachlorophenol (PCP) contaminated soil during land-farming.

- Bioremediation of pesticide contaminated soil Effects of biosurfactnt on remediation of pesticides, another group of contaminants, have been studied. The biosurfactant seems to bind trifluralin tightly in the micelle and releases the pesticide slowly to the aqueous phase, which could be then subjected to microbial uptake. Addition of rhamnolipid enabled biodegradation of the naphthalene to occur (Maslin & Maier, 2000). Mata-Sandoval et al. (2001)

performed biodegradation of trifluralin which was found to be enhanced in the presence rhamnolipid.

5.3.1.2 Bioremediation of Inorganic Contaminants

- Bioremediation of metal contaminated soils Metals, like many hazardous organic constituents, cannot be degraded. The existence and fate of the metals in soil is a matter of concern not only because of their potential impact on microbial communities, presents a potential health hazard for humans but also because of the potential for groundwater contamination.

5.4 Biosurfactant Assisted Remediation of Metal Contaminated Soils

As discussed earlier, metals are distributed into different fractions of the soil and removal of metals from various fractions of the soil depends on diverse physico-chemical properties of the soil. The studies revealed that biosurfactants are able to remove not only water soluble but also bound form of the metals from the soil. Juwarkar et al. (2008) have reported remediation of multi metal contaminated soils using rhamnolipid biosurfactants. Juwarkar et al., 2007 conducted column experiments to evaluate the potential of environmentally compatible rhamnolipid biosurfactant produced to remove heavy metals (Cd and Pb) from artificially contaminated soil. Washing of contaminated soil with tap water revealed that ≈2.7% of Cd and 9.8% of Pb in contaminated soil whereas washing with rhamnolipid removed 92% of Cd and 88% of Pb. Results have shown that rhamnolipid removes not only the leachable or available fraction of Cd and Pb but also the bound metals as compared to tap water which removed the mobile fraction only. Evaluation of biosurfactant for remediation of heavy metal contaminated soils has been studied for many years. Metal removal from a sandy soil contaminated with Cd and Ni was evaluated by Mulligan & Wang (2004). Removal of Cd and Ni after 20 pore volumes by rhamnolipid foam was 73.2% and 68.1% and by rhamnolipid liquid solution was 61.7% and 51.0% respectively. Rhamnolipid is the most extensively studied class of biosurfactant. Though other classes of are also evaluated for remediation of metals, rhamnolipid has been a first choice for the researchers. Batch washing experiments were conducted to determine feasibility of using Surfactin, rhamnolipids and sophorolipid biosurfactants for the removal of Zn and Cu from sediments. A single washing with rhamnolipid (0.5%) removed 65% of the copper and 18% of the zinc, sophorolipid (4%) removed 25% of the copper and 60% of the zinc. Surfactin was less effective, removing 15% of the copper and 6% of the zinc (Mulligan et al. 2001). Such a vast applications of biosurfactant in metal remediation encourage us to find possible mechanism of their interaction with metals.

5.5 Mechanism of Interaction of Biosurfactant with Metal Ions

The mechanism by which biosurfactant may remove metals has not been extensively investigated. The mechanisms of solubilization, mobilization, sorption, and emulsion formation have been postulated for hydrocarbon removal, but it is not known whether these are applicable for metals. Various mechanisms thought to be involved in the removal of metals using biosurfactants are mentioned below and are presented in Figure 4.

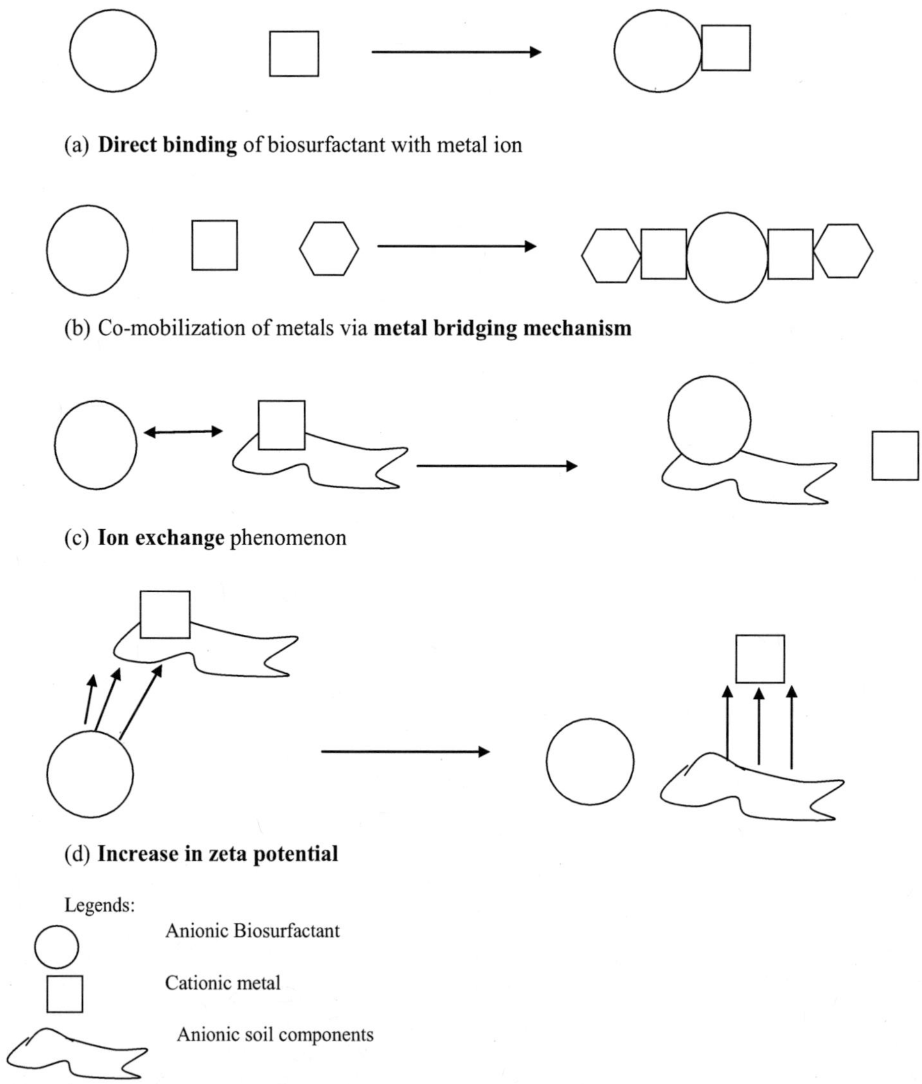

Figure 4. Various mechanisms involved in the removal of metal ions using biosurfactant.

Direct Binding

Direct binding of biosurfactant with metal ion can be a possible mechanism of interaction. Binding of metal cation with carboxylic group (COOH) of biosurfactant and must therefore is forming complexes through direct contact with the sorbed metal contaminants. The surfactant reduces the surface properties through lowering of the interfacial tension, thus decrease the force of adhesion between the metal and soil, facilitating the removal of the metal from the soil surface.

Metal Bridging

Co-mobilization of metal ions by formation of complex through metal bridging mechanism is another possible way of interaction of biosurfactant with the metal. Binding of anionic metal with anionic biosurfactant can take place with this kind of interaction. Cationic metal ions (divalent or trivalent), when present in the soil naturally or due to contamination, may bind anionic metal on one side and anionic biosurfactant on the other, thus forming a bridge between anionic metal and anionic biosurfactant. Thereby, facilitating co-mobilization of cationic metal along with the anionic metal, incorporated into the soluble micelles complexes.

Ion Exchange

Ion exchange is one of the suggested mechanisms of biosurfactant. Biosurfactant may act as an ion exchanger displacing the metal contaminant bound to soil surfaces. This may happen in case of cationic biosurfactant and cationic metal (both having same charges) or vice versa. However, negatively charged biosurfactant and positively charged metals ion exchange mechanisms are not likely in this case.

Increase in Zeta Potential

One more phenomenon that might be acting in metal removal process is that biosurfactant can increase negative zeta potential of the soil thereby mobilizing the metal contaminant through electrostatic repulsive interactions.

5.6 Interaction of Biosurfactant with Metal Ion

We have discussed different possible mechanism of interactions of biosurfactant with metal ions, however in actual system how does all these interactions lead to metal desorption and mobilization is elucidated below.

In a heterogeneous system, biosurfactant tends to interact with the phase boundary between two phases, defined as the interface. Addition of biosurfactant decreases the interfacial tension between soild surface and contaminants increasing the mobility of metals. Biosurfactants are known to form micelles at concentrations above CMC by aggregating themselves spontaneously into complex structures. A micelle is composed of a monolayer of surfactant molecules where the polar heads are oriented toward the surrounding aqueous solution and the nonpolar tails are oriented toward the hydrophobic center of the micelle. The polar head groups of micelles can bind metals. Biosurfactant e.g. anionic biosurfactant such as rhamnolipid carries a negative charge, so when the molecule encounters a cationic metal such as Cd (II) that carries a positive charge, an ionic bond, which is stronger than the metal's bond with the soil, is formed. This makes the metals more soluble in water. The micelles help to recover the metal ions from the soil surfaces and move them into aqueous solution, making them easier to recover by flushing. Diagrammatic representation proposed for the metal mobilization by biosurfactant is depicted in Figure 5. While considering metal removal, concentration of the biosurfactant is the important criteria that have to be taken into count. Exchanges of the metals from one micelle to another occur on a regular basis and this arrangement may be fairly stable at low surfactant concentrations, but as the number of micelles increases, more collisions occur between the micelles and the metal association with the micelles becomes less stable. Also at higher surfactant concentrations, the size of the micelles increases reducing the packing of the anionic heads of the micelles and thus reducing the retention of the metals on the micelles.

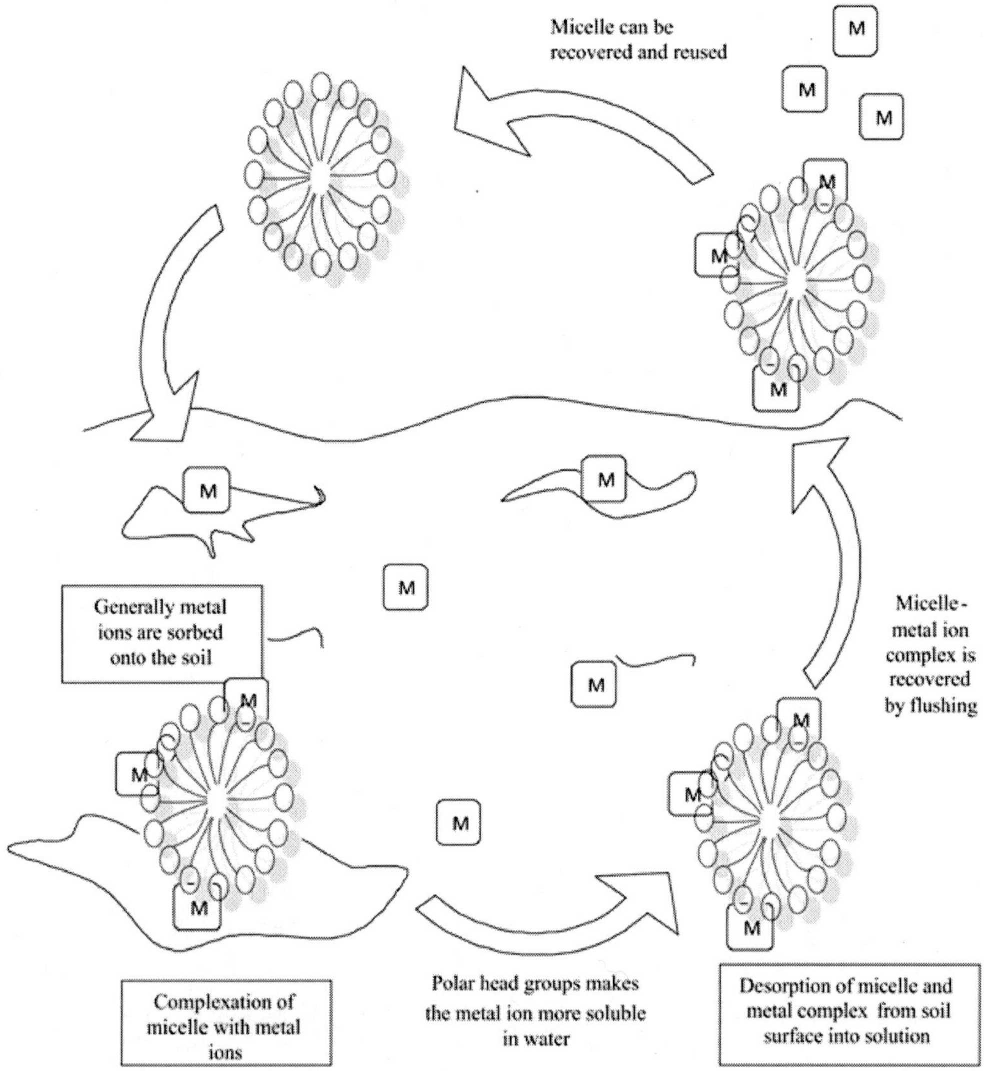

Figure 5. Interaction of biosurfactant with metal ion.

6.0 Conclusion

Remediation of metal contaminated soils is generally by physical removal to a landfill, or by *ex-situ* soil washings using synthetic chelators. None of these techniques restores the normal state of the soil and are highly destructive. Therefore, benign alternatives are needed which will be of interest for situations where the return of the soil to a healthy and productive state is of concern along with the speed and efficiency of remediation. Biosurfactants have shown their potential for remediation of metal contaminants present in the soil. Biosurfactants have high affinity for complexing metals and are capable of reducing surface and interfacial tension between the phases, thereby facilitating mobilization of the metal ions from one phase to another thus making it bioavailable in the system. Due to their biodegradability and low toxicity, they are very promising for application in remediation technologies. The chapter highlights that biosurfactant enhanced flushing can be successfully applied for the remediation of metal contaminated soil by washing the contaminated soil with biosurfactant and removing biosurfactant-metal complex by flushing. Since biosurfactant itself is biodegradable, metals from biosurfactant-metal complex can be recycled. However, further research regarding their fate and transport of the contaminants, decrease in production costs, recycling and reuse and *in-situ* production and application will be required.

Following aspects are to be strengthened in order to make a sound technology for environmental cleanup.

- Optimization of production conditions using cheaper renewable substrates could make biosurfactant production more profitable and economically feasible. Recent progress in genetic engineering is providing the driving force towards improved microbes for bioremediation. A number of genetic engineering approaches have been developed in the past several years that have proven useful in introducing the desired properties into the microorganism. Furthermore, recombinant (Wang et al., 2007) and mutant hyper-producer microbial strains (Mishra, 1997), able to grow on a wide range of cheap substrates, could produce high yield of biosurfactants and, potentially, bring the required breakthrough for their economic production. Wang et al. (2007) also reported the use of genetic engineering to obtain an *E. coli* and *P. aeruginosa* that were able to produce rhamnolipids after transposon mediated chromosome integration.
- Further studies are needed for recycling and the reuse of the biosurfactant. If the remediation is to be performed *in-situ*, production of the biosurfactants should also be *in- situ*. This would be effective with regard to technique and cost. Less labour and transport would be required and the process would be ecologically acceptable.
- To make the process more economical, biosurfactant producing organisms can be implemented on the site; however, survival of pure cultures on sites may face competition by indigenous microorganisms.
- Biosufactant mediated phytoremediation is the upcoming technology which can be effectively applied for remediation of metal contaminated soils. In genetic engineering technologies, the possibility has appeared to develop modified plants containing integrated biosurfactant producting genes. Brichkova et al., (2003, 2004) together with English researchers from the John Innes Center (Norwich), have developed rhamnolipid-carrying transgenic tobacco and arabidopsis plants that are tolerant to heavy metals and mineral oil.

REFERENCES

Agency for Toxic Substances and Disease Registry (ATSDR). (2010) *Most Viewed Toxic Substances*. http://www.atsdr.cdc.gov/.

Ahuja, S. K., Ferreira, G. M., & Moreira, A. R. (2004) *Utilization of enzymes for environmental applications*. Crit. Rev. Biotechnol., 24, 125–154.

Andra, S. S., Datta, R., Sarkar, D., Makris, K. C., Mullens, C. P., Sahi, S. V., & Bach, S. B. H. (2009) *Induction of Lead-Binding Phytochelatins in Vetiver Grass [Vetiveria zizanioides (L.)]*. J. Environ. Qual., 38, 868-877.

Babich, H., & Stotzky, G. (1977) *Reductions in toxicity of cadmium to microorganisms by clay minerals*. Appl. Environ. Microbiol., **33**, 696–705.

Babu, P. S. (1995) *Production of microbial surfactants from industrial wastes*. Ph.D. thesis. National Environmental Engineering Research Institute, Nagpur, India.

Babu, P. S., Vaidya, A. N., Bal, A. S., Kapur, R., Juwarkar, A. & Khanna, P. (1996) *Kinetics of biosurfactant production by Pseudomonas aeruginosa strain BS2 from industrial wastes*. Biotechnology Lett., 18, 263-268.

Bajpai, P. (2004) *Biological bleaching of chemical pulps*. Crit. Rev. Biotechnol., 24, 1–58.

Behbahaninia, A., Mirbagheri, S. A., Khorasani, N., Nouri, J., & Javid, A. H. (2009) *Heavy metal contamination of municipal effluent in soil and plants*. J. Food Agri. Environ., 7, 851 - 856.

Behra, R., Landwehrjohann, R., Vogel, K., Wagner, B., & Sigg, L. (2002) *Copper and zinc content of peryphyton from two rivers as a function of metal concentration*. Aquatic Sciences, 64, 300-306.

Bhatnagar, A., & Sillanpaa, M. (2009) *Applications of chitin- and chitosan-derivatives for the detoxification of water and wastewater--a short review*. Adv. Colloid Interface Sci., 152, 26–38.

Brychkova, G. G., Sorokin, A. P., & Kartel, N. A., (2004) *Bioremediation with Ecologically Safe Plants,* NATO Science Series, Series 1: Life and Behavioral Science IOS Press, 359, 147–158.

Brychkova, G. G., Sorokin, A. P., Maneshina, T. V., Kurman, P. V., Krasovskaya, L. I., Johns, J. J., & Kartel, N. A. (2003) *Development of Transgenic Plants Arabidopsis thaliana for the Purpose of Efficient Remediation of Oil Hydrocarbon-Polluted Areas, Dokl.* NAS Belarus, 47, 72–75.

Cameron, M. D., & Aust, S. D. (1999) *Degradation of chemicals by reactive radicals produced by cellobiose dehydrogenase from Phanerochaete chrysosporium*. Arch. Biochem. Biophys., 367, 115–21.

Cameron, M. D., Timofeevski, S., & Aust, S. D. (2000) *Enzymology of Phanerochaete chrysosporium with respect to the degradation of recalcitrant compounds and xenobiotics*. Appl. Microbiol. Biotechnol., 54, 751–8.

Chauhan, G. S., Singh, B., Chauhan, S., Verma, M., & Mahajan, S. (2005) *Sorption of some metal ions on cellulosic-based hydrogels*. Desalination, 181 (1-3), 217-224.

Cheung, K. H., & Gun, J. D. (2007) *Mechanism of hexavalent chromium detoxification by microorganisms and bioremediation application potential: A review*. Int. Biodet. Biodeg., 59 (1), 8-15.

Cobbett, C. S. (2000) *Update on heavy metal stress: Phytochelatins and Their Roles in Heavy Metal Detoxification*. Plant Physiol., 123, 825-832.

Das, K., & Mukherjee, A. K. (2007) *Crude petroleum-oil biodegradation efficiency of Bacillus subtilis and Pseudomonas aeruginosa strains isolated from petroleum oil contaminated soil from North-East India. Bioresour. Technol.,* 98, 1339–1345.

Davies, G., Ghabbour, E. A., & Steelink, C. (2001) *Humic Acids: Marvelous Products of Soil Chemistry.* J. Chem. Educ., 78, 1609.

Dube, A., Kowalkowski, T., Zbytniewski, R., Kosobucki, P., Cukrowska, E., & Buszewski, B. (2000) *Proceedings of the XVth International Symposium on Physico-chemical Methods of the Mixtures Separation – Ars Separatoria', Borowno n.* By dgoszcz, pp.21.

Dubey, K., & Juwarkar, A. (2001) *Distillery and curd whey waste as viable alternative to synthetic medium for biosurfactant production and pollution abatement of these wastes, World.* J. Microb. Biot., 17, 61–99.

Dubey, K., & Juwarkar, A., (2004) *Determination of genetic basis for biosurfactant production in distillery and curd whey wastes utilizing Pseudomonas aeruginosa strain BS2.* Indian J. Biotechnol., 3, 74–81.

Fuller, W. H. (1977) *Movement of selected metals, asbestos and cyanide in soil: Application to waste disposal problem.* EPA-600/2-77-020. Solid and hazardous waste research division, U.S. Environmental protection agency, Cincinnati, OH.

Gomare, S. S., Jadhav, J. P., & Govindwar, S. P. (2008) *Degradation of sulphonated azo dyes by purified lignin peroxidase from Brevibacillus laterosporus MTCC 2298.* Biotechnol. Bioprocess Eng., 13, 136-143.

Guo, X., Zhang, S., & Shan, X. Q. (2008) *Adsorption of metal ions on lignin.* J. Hazard. Mat., 151(1), 28, 134-142.

Hong, K. J., Tokunaga, S., & Kajiuchi, T. (2002) *Evaluation of remediation process with plant-derived biosurfactant for recovery of heavy metals from contaminated soils.* Chemosphere, 49, 379–387.

Institute of Environmental Conservation and Research INECAR (2000) *Position Paper Against Mining in Rapu-Rapu,* Published by INECAR, Ateneo de Naga University, Philippines (www.adnu.edu.ph/Institutes/ Inecar/pospaper1.asp).

Jain, C. K., & Ali, I. (2000) *Arsenic: Occurrence, toxicity and speciation techniques.* Water Res., 34, 4304–4312.

Juwarkar, A. A., Dubey, K. V., Nair, A., & Singh, S. K. (2008) *Bioremediation of multi-metal contaminated soil using biosurfactant – a novel approach.* Indian J. Microbiol. 48, 142–146.

Juwarkar, A. A., Nair, A., Dubey, K. V., Singh, S. K., & Devotta, S. (2007) *Biosurfactant technology for remediation of cadmium and lead contaminated soils.* Chemosphere, 68 (10), 1996-2002.

Kosaric, N. (2001) *Biosurfactants for Soil Bioremediation.* Food Technol. Biotechnol., 39 (4), 295–304.

Kumar, S. (2003) *Millions more at risk of arsenic poisoning than previously thought.* BMJ, 326 (7387), 466.

Lasheen, M. R., & Ammar, N. S. (2009) *Assessment of metals speciation in sewage sludge and stabilized sludge from different Wastewater Treatment Plants, Greater Cairo, Egypt.* J. Hazard. Mat., 164, 740–749.

Lawrence, C. J., Seigfried, T. E., Bass, H. W., & Anderson, L. K. (2009) *Predicting Chromosomal Locations of Genetically Mapped Loci in Maize Using the Morgan2McClintock Translator*. Genetics, 172, doi:10.1534/genetics.105.054155.

Lenntech Water Treatment and Air Purification (2004) *Water Treatment*, Published by Lenntech, Rotterdamseweg, Netherlands (www.excelwater.com/thp/filters/Water-Purification.htm.

Leong, S. A., & Winkelmann, G. (1998) *Molecular biology of iron transport in fungi. Met. Ions.* Biol. Syst., 35, 147–186.

Maier, R. M., & Soberon-Chavez, G. (2000) *Pseudomonas aeruginosa rhamnolipids, biosynthesis and potential applications.* Appl. Microbiol. Biotechnol., 54, 625–633.

Maslin, P., & Maier, R. M. (2000) *Rhamnolipid enhanced mineralization of phenanthrene in organic metal co-contaminated soils*. Biorem. J., 4 (2), 295-308.

Massara, H., Mulligan, C. N., & Hadjinicolaou, J. (2007) *Effect of rhamnolipids on chromium-contaminated kaolinite. Soil Sed.* Contam., 16 (1), 1–14.

Mata, Y. N., Blázquez, M. L., Ballester, A., González, F., & Munoz, J. A. (2010) *Studies on sorption, desorption, regeneration and reuse of sugar-beet pectin gels for heavy metal removal.* J. Hazard. Mat., 178 (1-3), 243–248.

Mata-Sandoval, J. C., Karns, J., & Torrents, A. (2001) *Influence of rhamnolipids and Triton X-100 on the biodegradation of three pesticides in aqueous phase and soil slurries.* J. Agric. Food Chem., 49, 3296–3303.

Mayer, A. M., & Staples, R. C. (2002) *Laccase: new functions for an old enzyme.* Phytochemistry, 60, 551–565.

McBride, M. B., Richards, B. K., Stenhuis, T., Russo, J. J., & Sauve, S. (1997) *Mobility and solubility of toxic metals and nutrients in soil fifteen years after sludge application.* Soil Sci., 162, 487–500.

Mehra, R. K., Tran, K., Scott, G.W., Mulchandani, P., & Sani, S. S. (1996) *Ag (I)-binding to phytochelatins.* J. Inorg. Biochem., 61, 125-142.

Meulenberg, R., Rijnaarts, H. H. M., Doddema, H. J., & Field, J. A. (1997). *Partially oxidized polycyclic aromatic hydrocarbons show an increased bioavailability and biodegradability*. Microbiol. Lett., 152, 45–9.

Miller & Marvin, J. (2008) *Siderophores (microbial iron chelators) and siderophore-drug conjugates (new methods for microbially selective drug delivery).* University of Notre Dame. Ministry of the Environment programs and initiatives, fact sheet. (2001). Ontario.

Mishra, K. (1997) *Improvement of biosurfactant producing strain through genetic manipulation.* Ph. D. thesis. National Environmental Engineering Research Institute, Nagpur, India.

Mulligan, C. N., Yong, R. N., & Gibbs, B. F. (2001) *Heavy metal removal from sediments by biosurfactants.* J. Hazard. Mat., 85, 111–125.

Mulligan, C. N., Yong, R. N., & Gibbs, B. F. (2001) *Remediation technologies for metal-contaminated soils and groundwater: an evaluation*. Eng. Geol., 60, 193–207.

Naidu, R., & Harter, R. D. (1998) *Effect of different organic ligands on cadmium sorption by and extractability from soils.* Soil Sci. Soc. Am. J., 62, 644-650.

Nair, A., Juwarkar, A. A., & Singh, S. K., (2007) *Production and characterization of siderophores and its application in arsenic removal from contaminated soil.* Water Air Soil Pollut., **180**, 199–212.

Nitschke, M., & Costa, S. G. V. A. O. (2007) *Biosurfactants in food industry*. Trends Food Sci. Technol., 18, 252–259.

Nitschke, M., Costa, S. G. V. A. O., Hadad, R., Goncalves, L. A., Eberlin, M. N., & Contiero, J. (2005) *Oil wastes as unconventional substrates for rhamnolipd biosurfactant production by Pseudomonas aeoginosa LBI*. Biotechnol. Progr., 21, 1562–1566.

Noordman, W. H., Wacher, J. J. J., Boer, G. L. de & Janssen, D. B. (2002) *The enhancement by biosurfactants of hexacadene degrade by Pseudomonas aeruginosa varies with substrate availability*. J. Biotechnol., 94, 195-121.

Parivesh (2005) *A Newsletter from central pollution control board*. Ministry of Environment & Forests. Hazardous Waste Management, 123-131.

Peplow, D. (1999) *Environmental Impacts of Mining in Eastern Washington*, Center for Water and Watershed Studies Fact Sheet, University of Washington, Seattle.

Rahman, K. S. M., Rahman, T. J., McClean, S., Marchant, R., & Banat, I. M. (2002) *Rhamnolipid biosurfactant production by strains of Pseudomonas aeruginosa using low-cost raw materials*. Biotechnol. Prog., 18, 1277–1281.

Ramos, J. L., González-Pérez, M. M., Caballero, A., & van Dillewijn, P. (2005) *Bioremediation of polynitrated aromatic compounds: plants and microbes put up a fight*. Curr. Opin. Biotechnol. 16, 275–281.

Rauser, W. E. (1999) *Structure and function of metal chelators produced by plants: the case for organic acids, amino acids, phytin and metallothioneins*. Cell Biochem. Biophys., 31, 19-48.

Robinson-Lora, M. A., & Brennan, R. A. (2010) *Chitin complex for the remediation of mine impacted water: geochemistry of metal removal and comparison with other common substrates*. Appl. Geochem., 25 (3), 336 – 344.

Rocha, M. V. P., Oliveira, A. H. S., Souza, M. C. M. & Gonçalves, L. R. B. (2006) *Natural cashew apple juice as fermentation medium for biosurfactant production by Acinetobacter calcoaceticus*. World J. Microbiol Biotechnol., 22, 1295-1299.

Rufino, R. D., Sarubbo, L. A., & Campos-Takaki, G. M. (2007) *Enhancement of stability of biosurfactant produced by Candida lipolytica using industrial residue as substrate*. W. J. Microbiol. Biotechnol., 23, 729–734.

Ruggeri, C., Franzetti, A., Bestetti, G., Caredda, P., La Colla, P., Pintus, M., Sergi, S., Tamburini, E. (2009) *Isolation and characterization of surface active compound-producing bacteria from hydrocarbon-contaminated environments by a high-throughput screening procedure*. Int. Biodeterior. Biodegrad., 63, 936–942.

Sadouk, Z., Hacene, H., & Tazerouti, A. (2008) *Biosurfactants production from low cost substrate and degradation of diesel oil by a Rhodococcus strain*. Oil Gas Sci. Technol.-Revue De L Institut Francais Du Petrole, 63, 747–753.

Samanta, S. K. (2002) *Polycyclic aromatic hydrocarbons: environmental pollution and bioremediation*. Trends Biotechnol., 20, 243–248.

Sen, S. (2009) *Alkaline Protease: a Tool to Clean Environment*. http://aiche.confex.com/aiche/s09/techprogram/P144084.HTM.

Singh, A., Van Hamme, J. D. & Ward, O. P. (2007) *Surfactants in microbiology and biotechnology*. Biotechnol. Adv., 25, 99–122.

Straube, W. L., Nestler, C. C., Hansen, L. D., Ringleberg, D., Pritchard, P. H., & Jones-Meehan, J. (2003) *Remediation of Polyaromatic Hydrocarbons (PAHs) through*

Landfarming with Biostimulation and Bioaugmentation. Acta. Biotechnologica., 23 (2-3), 179 – 196.

Sutherland, T. D., Weir, K. M., Coppin, C. W., Williams, M. R., Selleck, M., Russell, R. J., & Oakeshott, J. G. (2004) *Enzymatic bioremediation: from enzyme discovery to applications*. Clin. Exp. Pharmacol. Physiol., 31, 817–821.

U.S. EPA (1996b) *Report: Recent Developments for In Situ Treatment of Metal Contaminated Soils, U.S. Environmental Protection Agency, Office of Solid Waste and Emergency Response, draft.*

Wang, Q., Fang, X., Bai, B., Liang, X., Shuler, P. J., Goddard, W. A. 3rd, & Tang, Y. (2007) *Engineering Bacteria for Production of Rhamnolipid as an Agent for Enhanced Oil Recovery*. Biotechnol. Bioeng., 98, 842–853.

Wang, S., & Mulligan, C. N. (2004) *Rhamnolipid foam enhanced remediation of cadmium and nickel contaminated soil*. Water Air Soil Pollut., 157, 315–330.

Wong, J. W. C., & Mak, N. K. (2010) *Heavy Metal Pollution in Children Playgrounds in Hong Kong and Its Health Implications*. Environ. Technol., 18, 109- 115.

Yuan, X. Z., Meng, Y. T., Zeng, G.M., Fang, Y. Y., & Shi, J. G. (2008) *Evaluation of tea-derived biosurfactant on removing heavy metal ions from dilute wastewater by ion flotation. Colloid Surf. A Physico.* Eng. Aspects, 317 (1-3), 256-261.

In: Bioremediation
Editor: Alexander C. Mason

ISBN: 978-1-61122-730-7

Chapter 8

Geomicrobiology of Heritage Monuments and Artworks: Mechanisms of Biodeterioration, Bioconservation Strategies and Applied Molecular Approaches

Tikam Chand Dakal and Swaranjit Singh Cameotra*
Institute of Microbial Technology, Sector 39A, Chandigarh 160036, India

Absract

Natural stones like marbles, granite and dolomite were massively used for building construction during past centuries. High strength and durability of these stones made their uses more pronounced in constructing sculptural buildings and outstanding decorative artworks. The Taj Mahal, European Cathedrals and Roman Catacombs are few amongst them. With the introduction of different tools and their use in cutting, engraving and delicately carving of stones, the valuable and high artistic decoration was facilitated and outstanding beauty was imparted to these monuments and artifacts. Therefore, these monuments and artworks hold their own cultural and symbolic values. In the course of several centuries, from the moments they were built to the day they faced forfeiture they had seen many rise and falls of life. They bear signs of various civilizations, dominations and reign that help educating us about our past life, history, arts and literature. These heritages are our greatest natural treasures. Once decayed or devastated, it will truly become impossible for us to regain or reproduce them. During past few decades, these monuments had experienced extensive deterioration and aging. Persistent pollutants and human activity has posed a big threat to these sculptural monuments and artworks. Due to lack of appropriate knowledge of efficient repairing strategies, the previous traditional methods applied for restoration purposes turned inefficient and incompatible. Modification in the petrophysical properties of stone and altered aesthetic appearance were the characteristic features of these traditional methods. The failure of the tradition methods in restoration of physical strength and conservation of natural appearance has

* ssc@imtech.res.in

presented necessity of appropriate management tools and conservation strategies. It stimulated the researchers from all over the world to come together to elucidate, unravel and work out the issues. The global concern and consequently global involvement in characterizing the mechanism of deterioration and devising alternative remedial strategies became evident in the past few years. The aim of the studies in this field is to develop a reliable, compatible, and efficient strategies for conservation, preservation and restoration of the natural and cultural heritage of past.

This chapter detailed about the cause and mechanisms of biodeterioration, concisely summarizes the conservation strategies. Consequently, the scope of this chapter is extended to discuss the major research been carried out by some active research groups individually or in the framework of some dedicated research projects. Finally, molecular approaches and related techniques which had made the deterioration mechanisms more comprehensive are also briefly discussed.

INTRODUCTION

Since historic time when the grandeur, splendidness, magnificence and beauty of civil structures was used to measure architectural brilliance, artists and architectures used high quality, durable and spectacularly beautiful stones like marbles, granite, dolomite for the construction purposes. These civil structures include cathedrals, temples, churches, palaces and forts which are highly valued for their natural and architectural beauty. These civil structures have great significance in educating the present and future generation about the past human culture, religion, history, arts and literature. Being exposed to open environment these structures are highly susceptible to damage by atmospheric pollutants, acid rain water and invading armies of microorganisms. The general conference of UNESCO in its 17th session at Paris adopted the 'Convention Concerning the Protection of the World and Natural Heritage' and identified, listed and declared many natural and cultural sites of the world as world's heritage sites which were at the risk of deterioration. The deterioration of historic monuments and artworks is a complex process occurs as a result of microorganisms interactions with its surrounding environment. The process is often accelerates in the presence of environmental pollutants and moisture. A wide variety of microorganisms are known to be involved in the process of biodeterioration. These microorganisms using different mechanisms which ranges from simple staining and discoloration of monumental walls to more complex biofilm formation have seriously devastated and destroyed the ancient monuments and artworks which are highly valued for their beauty, magnificence and encrypted knowledge. The deterioration caused by microorganisms has resulted in undesirable and irreparable damage to these heritage buildings. With the inevitable deterioration of historic monuments and artworks, there has been a global interest in developing remedial strategies to safeguard these cultural and natural heritages. The traditional approaches to preserve these monumental structures and sculptural buildings relied mostly on the use of physical, mechanical and chemical methods but these methods faced great failure in term of their incompatibility and inefficiency in conservation and consolidation of monumental stones and artworks. Recently, the potential role of carbonatogenic bacteria in conservation and consolidation of these monumental stones and artworks has been investigated, evaluated and acknowledged. These bacteria have unique capacity of extracellular precipitation of calcium carbonate under certain experimental conditions by multiple mechanisms. This chapter details about 1) the routes and mechanisms by which microorganisms destroys, decays and deteriorates heritage buildings and artworks, 2) the different methods and remedial strategies currently been used for conservation and preservation, 3) the role of molecular techniques and bioinformatics in identification of microbial diversity, type and mechanisms of biodeterioration, and 4) future directions to and use of the conservation techniques in an efficient and productive way is also discussed.

Routes and Mechanisms to Deterioration

Microorganisms since long back are known for bringing ill health to humans, animals and plants. Besides this, these microorganisms also play a geoactive role in environment and are remarkably potent agents of biodeterioration of historic monuments and artworks. The analysis of numerous cases of biodeterioration of historic monuments revealed that damage is often associated with microbial interaction with its substrate and surrounding environment [1, 2]. The presence of microorganisms on walls of ancient historic monuments and surfaces of stoneworks and artworks and consequently their growth and activity has considerable impact of physical and aesthetical appearance of these civil structures and artworks. Most of these civil structures are made up of beautiful and highly durable stones such as marbles, granite and dolomite. These stones are highly susceptible to decay by microbes and environmental pollutants because of their porosity, surface properties and chemical compositions [3-5]. Previous research works elucidated the role of microorganisms and abiotic factors including environmental pollutants and detailed few major routes and mechanisms through which microorganisms foster deterioration [6, 7]. A detailed review of the literatures, both articles and review papers implicate that microorganisms ranged from algae and cyanobacteria to fungi, lichens, mosses and liverworts are involved in the process of biodeterioration of historic monuments and artworks [8-12].These microorganisms alone or up on interaction with abiotic factors including environmental pollutants cause staining and discoloration, accelerate bioweathering and biocorrosion, and promote salt encrustation, biofilm and patina formation resulting in irreparable and irreversible physical and aesthetical loss and damage of which some are being irremediable [9, 10, 13-20]. In this regard, five plausible mechanisms of biodeterioration that might be expected are described. Additionally a comprehensive and exemplified description of the root causes and microorganisms involved in biodeterioration is also explicated in this chapter. It is very crucial to study the root causes and mechanisms of the individual deterioration processes, as these studies will supply relevant information necessary to rationally tailor and design appropriate management tools and conservation strategies.

Role of Environmental and Abiotic Factors in Promoting Deterioration

Artist and architects during historic time used different stones (limestones, marbles, granites and dolomites etc) for the construction of splendid and magnificent monuments and for making beautiful and graceful artworks [21]. Stones employed for construction and making of these sculptural monuments and artworks were highly consolidated and durable and were obtained from naturally occurring sedimentary rocks. With the inexorable increase in industrial activity and use of automobiles, there has been a considerable increase in the environmental pollutants like oxides of nitrogen and sulfur, carbonaceous particles, suspended petroleum and diesel particulate matters [7,22-27]. The changes in the composition of environment have severe adverse impacts on the exposed natural and cultural heritage buildings. The problem is pronounced in tropical regions where the high temperature, high relative humidity and high annual rainfall favor the growth and activity of diverse communities of microorganisms [28]. Mineral matrix of the stones provides a suitable substratum and favors growth of diverse microbial community. Additionally, abiotic factors like relative humidity, temperature, wind, light and rainfall also plays a crucial role in colonization and establishment of microbial communities on the stone surfaces of monuments and artworks [1, 29]. This ability of stone to get colonized by the microorganisms is termed as bioreceptivity and is greatly influenced by stone's surface property and surrounding environment conditions including microenvironment conditions [30]. Dampening of the vertical walls of the monuments exposed to rain water and sunlight often promotes the growth of photoautotrophic bacteria like cyanobacteria and algae [31]. In the past two decades, the presence of some aggressive species of lichens like lichen *Dirina massiliensis* forma *sorediata*, *Lecanora muralis* and *Xanthoria parietina* become apparent due to drastic changes in atmospheric conditions. Persistent environmental pollutants of urban atmosphere such as oxides of sulfur and nitrogen, carbonaceous particles, fly ash, particulate matters upon settling on the surface of the monumental stones harm its aesthetic appearance [20, 32, 33]. Oxides of nitrogen and sulfur combine with the rain water and make it acidic and corrosive. Showering of such acidic rain water on the monuments has numerous deleterious effects. It is only past two-three decades, considerable efforts were made to quantify the damage caused to the sculptural buildings as a result of acid deposition. The studies regarding the spatial concentration of air pollutants and their transport mechanisms aided in understanding and estimating acid rain caused damages more accurate. Sometimes the oxides of nitrogen and sulfur may be oxidized into their corresponding acids by the humid air or moisture present on the damp stone surfaces of monuments further worsening their physical strength and durability [34-36]. Crystallization of soluble salts like sulfates and black crust formation are among the major causes of damage and devastation to the surfaces of monumental stones and artworks [37].

MONUMENTAL STONES AND THEIR BIORECEPTIVITY

Since the time historic monuments and artworks were constructed and made, they are exposed to open air. A wide variety of microbes are getting an open access to view the intricately designed wall paintings and artworks and to enjoy royal stay in historic monuments for which visitors need to pay. Microbial presence on monumental stones and artworks is often correlated with their role in biodeterioration. Microbial growth and deterioration are two different sets of phenomenon which should not be conflated. Numerous factors such as mineral matrix composition, availability of nutrients, pH, salinity, surface texture, moisture content, porosity, permeability, climatic and micro-environmental conditions remarkably influence the microbial ability to get colonize up on stone surfaces [38]. In order to explain the process of microbial growth on stones, the concept of bioreceptivity was introduced which is defined as the ability of stone to be colonized by microorganisms [30]. The ability of stones to be colonized by microbes is mainly governed by its mineralogical and surface properties together with environmental conditions while extend of colonization is mainly influenced by the surrounding environment conditions (including pollutants concentration and micro-climatic conditions) and by anthropogenic eutrophication of the atmosphere [30, 39].

Bioweathering and Stone Dissolution

Bioweathering is a process of progressive dissolution of superficial mineral surface of the monumental stones and artworks. The process is usually initiated by microbial communities secreting corrosive organic and inorganic acids and metal binding ligands. Microbial-mineral interaction serves as a good ground for studying the role of microbes in the process of biogeochemical transformations and many surface techniques such as microscopy and XRD presented evidences of alterations induced by microorganisms. This interaction represents a mechanism which microbes follow for the extraction of nutrients from the mineral surface [40-43]. The extraction of minerals as a result of stone dissolution is a way which facilitates the availability of some essential and vital nutrients such as trace-metals, phosphate, sulfate and metabolites necessary for growth and proliferation of the inhabiting microbial communities [44]. There are two mechanisms by which fungi perform stone dissolution: by forming secondary minerals and by binding metals on their cell wall or other external surfaces in a metabolically independent manner [45, 46]. The biogenic secretion of highly corrosive acids, both organic and inorganic and chelating agents by fungi and lichens on the surface of stone is among one of the mechanisms which is inadvertently involved in the promotion of bioweathering processes [38]. Though biofilm formation is noticeable on monumental stones and artworks but very less is mentioned in literature regarding their role in extraction of minerals from the stone substrates [43].

Biocorrosion: Release of Corrosive Organic and Inorganic Acids

Yet another scenario by which microorganisms can cause damage to monuments and stoneworks is biocorrosion which occurs as a consequence of biogenic secretion or release of inorganic and organic acids by microorganisms. The deterioration processes which are triggered by the release of inorganic and organic acids are respectively termed as acidolysis and complexolysis [47]. The destruction caused by secreted acids (acidolysis) is mainly associated with the chemolithotrophic bacteria which are also called as corrosion producing bacteria. Secretion of corrosive acids can be detrimental to mechanical integrity and durability of both monumental stones and other artworks. Apart from this, the carbon dioxide release as a result of cellular respiration by lichens and mosses is also a potent corrosive agent. Organic acids such as oxalic acids, citric acids etc which are being produced metabolically through cellular biochemical reactions by certain chemoorganotrophs and lichens are also known to possess strong corrosive property against work of arts.

STONE ENCRUSTATION: DEPOSITION OF CORROSIVE ORGANIC AND INORGANIC COMPOUNDS

Increasing industrial activity and automobile exhaust had severely polluted the atmospheric air and heavily loaded it with hazardous pollutants such as oxides of nitrogen and nitrogen, particulate matters, carbonaceous particles etc [26]. Oxides of both N_2 and S_2 have bioweathering effect [14, 48]. In highly polluted environments, especially those rich in oxides of nitrogen and sulfur and petroleum particulate matters, the problem of stone encrustation is most commonly evident. Observation of different stones holds the fact that SO_2 together with other dark particles (particulate matters) settles down and get deposited (dry deposition) [49] on the stone surfaces and imparts dark and yellowish color to them [50-52]. There are reports which confirms that special classes of bacteria, called corrosion-producing bacteria which includes nitrifying and sulfur-oxidizing bacteria (chemoautotrophs) upon colonizing stone surface oxidize nitrogen compounds including atmospheric ammonia (*Nitrosomonas sp. and Nitrobacter sp.*) and sulfur compunds (*Thiobacillus sp.*) into nitric and sulfuric acid respectively [14]. Being highly corrosive these acids accelerate the dissolution process of the stone surface (biocorrosion) and alter surface and textural properties. These acids react with the stone carbonates and results in the deposition of nitrate and sulfate salts. The process of sulfation (formation and deposition of sulfate salts on stones) is often accelerated by the presence of humidity and fossil fuel derived particulate matters [7, 24, 27]. There are reports which showed that both recent and archeological stones and artworks which are exposed to polluted air may survive for many centuries in relatively good conditions if sheltered from rainwater. Upon coming in contact with rain water these sulfates get dissolved and forms hydrated salts such as gypsum [53]. A significant amount of research work have been carried out to investigate the nature and origin of blackened gypsum-crust on the surface of marbles or calcareous stones [25, 54-56] and was found that the emission of carbonaceous particles (fly ash), fossil fuel particulate matters, dust and their subsequent entrapment with the gypsum (during its formation on stones) results into the formation of black and brown sulfated crust over the stone [7, 57-59]. The stones which show signs of sulfation and encrustation decay often lose their aesthetic appearance. Besides this, during these processes these sulfates may precipitates inside the porosity of the stones and upon recrystallization exert significant stress resulting in structural damage to the marble stones.

Most of the time sulfur-oxidizing bacteria are also gain benefit by the co-presence of epilithic sulfur-reducing bacteria at the base of the stone. These bacteria by the process of reduction convert the deposited sulfates to sulfides, which serves as an excellent source of energy for the sulfur-oxidizing bacteria. There are reports which support the presence of extremely and moderate halophiles on the monumental stone and artworks. Stone surfaces which are laden with the deposits of hygroscopic salts have great tendency to colonize the halophiles. These depositions are formed as a result of drying of water with high salt content on the exposed surfaces of the stone through a process known as efflorescence [60].

The experiments performed using analytical techniques for determining the chemical composition of these sulfated crusts revealed that it varies depending up on the age of the crusts [61] and particular airborne pollutants in individual areas [62, 63]. The evidences of the different salt's presence and nature of damage caused by them to stones can be characterized and determined by means of low invasive analytical techniques such as IR spectroscopy, Raman Spectroscopy etc.

SECONDARY MINERAL FORMATION: PATINA FORMATION

One of the several ways by which microorganisms accelerate deterioration is secondary mineral formation. Secondary minerals such as calcium oxalates formed by microorganisms (whewellite-$CaC_2O_4.H_2O$ and weddellite-$CaC_2O_4.2H_2O$) widely occur in nature mainly as patina on the stones substrates mainly, carbonatic lithotypes. Calcium oxalate formation on sculptural monuments and artworks is prime feature of several free living, pathogenic, plant symbiotic fungi, and lichens which stains the stone surfaces with pigments of different colors [11, 64-67]. It was hypothesized that calcium oxalate mainly came from the past treatments [68]. Some researchers rather presented a different mechanism of its formation. Calcium coming from stone substrate or atmosphere oxalic acids formed by biological (by metabolic reactions in bacteria, lichens and fungi) and chemical (by oxidation of organic compounds) means combine and precipitate calcium oxalate [69-71]. The precipitation may be induced by oxidation and UV Treatment [72]. Raman spectra analysis intended to study the biomineral products of lichen bioweathering of some stone substrates showed the presence of calcium oxalate monohydrate (by *Lecanora sulfurea* and *Aspicilia calcarea*) and dihydrate (*Dirina massiliensis* f. *sorediata*, *D. massiliensis* f. *massiliensis* and *Tephromela atra*) [73]. Earlier it was assumed that the precipitation of calcium in the form of oxalate is less common in other organisms such as algae and fungi. During recent year several studies were carried out to unravel their biogenesis in other organisms [74-76] and evidence regarding their formation by fungi were presented [70](Monte, 2003a). Some researchers attributed the formation of calcium oxalate covering on stones as a protective shield which has intrinsic property to endure adverse environment conditions and protects the surface from further weathering and decay. Only strong wind and acids are known to effects them [77].

BIOFILM FORMATION

Cyanobacteria and algae are pioneer organisms that colonize a wide variety of substrates including stone surface of ancient architectural and modern buildings [2, 9, 78-81]. Their presence is easily identifiable on the illuminated and exposed surfaces of monumental stones and artworks [2]. Cyanobacteria are often present in an intimate association with red and green algae and lichens [82, 83]. Due to their phototrophic nature, their growth on the stone is generally favored which results into the formation of patinas of various colors and incrustations [81].They are well adapted to resist adverse climatic conditions because of the presence of thick outer envelops and different phototrophic pigments [9]. The presence of water and humidity are among the major factors which facilitate their association with stone substrate and enable their fast growth as compared to other organisms [79]. Their association with stone substrate favors the formation of biofilms. Biofilms is a three-dimensional structure composed of extracellular polymeric substances within which are enclosed microorganisms. Outdoor monuments which are exposed to continuous light tend to support formation of biofilm which contain plethora of phototrophic microorganisms [84]. Biofilms are involved in facilitating attachment with the stone also regulate temperature and humidity and protecting inhabitants from desiccation and surrounding adverse abiotic conditions [85-88]. Growth of biofilm marks the initiation of petrophysical damage to the stone leading to its biodeterioration [1, 82, 84] and aesthetical damage leading to their discoloration [89-90]. Formation of biofilm is a dynamic process in which it undergoes repeated 'built-ruined-rebuilt' cycle. The composition and distribution of biofilms is mainly influenced as a consequence of spatial and temporal variation in a number of abiotic and physico-chemical factors, including micro-environment. Formation of biofilms by photosynthetic bacteria including cyanobacteria is a survival strategy that enables the inhabiting microorganisms to resist changes in environmental conditions such as extremes of temperature, drought and prolonged exposure to light [91, 92]. Other means of survival in adverse conditions, for instance, the use of stored water in the substrate [93, 94], formation of chemical compounds which provides resistance against drought [95, 96] and synthesis of protective UV shield [93, 94, 97, 98] are also known. Besides these survival strategies, cyanobacteria exhibit the property of minerals extraction and mobilization for their own nutrition [99]. Additionally, the biofilms are also known to enhance the nitrogen and phosphorous availabilities when inoculated in the soil [100]. The existence of algal population in biofilm is a good indicator of presence of other corrosion producing bacteria, for instance, sulfate-reducing bacteria (SRB) and sulfur-oxidizing bacteria (SOB). Being aerobic, algae consume oxygen in dark and create micro-environments suitable for the growth of anaerobic SRB. Algae also release oxygen during photosynthesis and can make a local spot aerobic enough for growth of sulfur-oxidizing bacteria (SOB). The biofilm is composed of microbial cells (mainly eubacteria) and extracellular polymeric substance (EPS) that facilitates the attachment of the biofilm on the solid substratum [1, 101]. The EPS acts as glue which traps atmospheric dirt and particulate matters which further accelerates the deterioration process. The thickness of biofilm regulates and prevents the oxygen to ingress deep to the base of the biofilm making the micro-environmental conditions at biofilm-base considerably anaerobic. The biofilm formation makes the stone lose its property of cohesion [1]. Stratification is also one of the most prominent features of biofilms which is controlled by plentiful of factors, including the quality of light. Low light conditions are likely to reduce the stratification and affect species diversity [102, 103] by allowing only certain species to survive [104, 105].

Biofilms shows both, detrimental and protective role, depending on the nature of substratum and microorganisms involved [106]. While biofilms and the inhabiting organisms accelerate the deterioration process [107], some communities have a more protective role [108]. The removal of this protective biofilm layer may speed up the process of deterioration by making the stones more susceptible to polluted atmosphere and to the attacks of salts [109-111].

Physical Penetration of Microbes

Physical intrusion and penetration of bacterial and fungal hyphae inside the stones is another major threat to physical and mechanical damage to monuments and artworks. The process usually gains momentum when surface conditions such as gaps, small cracks, pores have become favorable for the growth of these organisms. The physical intrusion by hyphae along the crystal plane not only destabilize the stone texture but also increase the pore size resulting into mechanical deterioration of stones and artworks [112]. Some photosynthetic microorganisms such as mosses are also known to possess rhizoids which physically intrude inside the stone but mechanical damage as a result of their intrusion is not reported in literature. Trentepohlia spp. is known to cause pitting [113] and progressive mechanical degradation of buildings [114, 115].

Microorganisms Involved in Deterioration

Bacteria

Majority of bacterial species under certain conditions are able to colonize stone surfaces of natural, architectural and modern buildings. These bacterial species are mainly related to nutritional groups: Photoautotrophs, Chemolithoautotrophs and Chemoorganotrophs. Photoautotrophic and chemolithoautotrophic bacteria identified from archaeological sites mainly constitute cyanobacteria, sulfur-oxidizing bacteria and nitrifying bacteria. Since their survival and growth rely on simpler nutritional (like inorganic minerals, atmospheric ammonia etc.) and ecological needs (like presence of light, CO_2 and water) these bacteria easily noticeable on outdoor monuments. Among these bacteria, cyanobacteria are known to posses several surviving strategies which enable them to survive under the harsh conditions such as repeated drying and rehydration occurring on exposed monument's surfaces [82, 116] and to shield themselves from the harmful ultra-violet rays by synthesizing protective pigments [82]. However, their presence sometimes becomes conspicuous on interior works of art also [90, 117-121] where these artworks are subjected to unsuitable natural or artificial illumination during visitor's hours (such as hypogean environments of Roman catacombs) [99, 122-124]. The colonization of these photosynthetic microorganisms on external surface of monuments is characterized by the formation of biofilm leading to mechanical deterioration of mineral structure (as a result of alternate shrinking and swelling cycles of colloidal biogenic slime inside the pore system), inorganic and organic acid secretion resulting in biocorrosion, undesirable unaesthetic staining of the monuments (due to the secretion of acids, pigments and metabolic bioproducts), pore enlargment (due to hyphal penetration), alteration in size of pores, their distribution and water permeability of the minerals (by deposition of surfactants) and bioweathering (due to the redox reactions and secondary mineral formation) [9, 82, 125]. Chemoorganotrophs and chemoheterotrophs bacteria found associated with deterioration process mainly comprises of sulfur-reducing bacteria and actinomycetes respectively. Population of these heterotrophic bacteria and actinomycetes abundantly found in hypogean environment where the microclimatic conditions remain stable throughout year (high relative humidity $<90\%$, constant temperature throughout year and low photon influx) [126-129]. These bacteria are mainly account for the damages caused to the indoor artworks (wall painting, frescos, stuccoes etc) and their participation in the deterioration of outdoor monuments is negligible. Some actinomycetes species such as *Geodermatophilaceae* strains [130] and also few *Micromonospor*a strains [131, 132] have been identified from monumental stones. Sometimes, actinomycetes species, mainly streptomyces (*S. julianum*) form close association with cyanobacterial partner as found in the hypogean environment of Roman catacombs. Actinomycetes also preferred to grow in intimate association with algae and chemosynthetic nitrogen fixer bacteria.

Presence of archaebacteria is though less common, but their presence become evident from few archeological sites. Two archaebacterial species were isolated and characterized from two ancient wall painting namely Catherine Chapel of the Castle Herberstein (Styria) from Austria and Roman Necropolis of Carmona from Spain. These archaeal species were then subjected to 16s rRNA sequence analysis, denaturing gradient gel electrophoresis and FT-IR analysis which established their phylogenetic relationship with *Halococcus* (halophilic) and *Natronobacterium* (alkaliphilic) [133].

The metabolic excrete of microalgae and stromatolites formed by them may have deleterious effects on the stone [134-137]. Some genera of the microalgae such as Cosmarium, Phormidium and Symploca are the major biodeteriogens of ornamental stones collected from the fountain of Bibatauin at Granada in Spain [138]. Besides this, the continuous presence of water favored the growth of some endolithic green algae and cyanobacteria and resulted into black crust formation on the Robba′s fountain statues, Ljubljana (Slovenia).

FUNGI

Fungi represent a class of chemoheterotrophic microorganims with characteristic unicellular or multicellular hyphae [139]. These are most abundant and widely distributed in nature, in both aquatic and terrestrial ecosystems. Many distinct fungal species were identified at different historical sites which are associated with the biodeterioration process. As fungi are metabolically more versatile than any other biodeteriogens in the microbial kingdom, they can colonize on wide variety of substrates including wood surfaces, surfaces of monumental stones, indoor works of arts (wall painting, frescos etc) and metal artifacts and initiate and accelerate their deterioration. Their ability to grow on variety of substrates together with the inherent capability of tolerating extremes of environmental conditions, to bring about mutualistic association with cyanobacteria and algae as lichens [45, 140] and to take up various structural, morphological and metabolic strategies further improves their versatility and adaptability [46, 141, 142]. The biodeterioration process of inorganic substrate such as carbonate by fungi completely follows a different mechanism. Since fungi are heterotrophic and hence unable to consume the inorganic carbonate substrate for their growth, but can grow unrestricted on the waste product or dead cellular debris of previous inhabiting communities and deposits of bird's excrete and decayed leaves rich in organic nitrogenous matters and aerosols [45] present on rock surface, fissures, cracks, subaerial and subsoil environment. Their capability of growing like an oligotrophs and scavenging nutrient material from atmosphere and rain water allows their growth on many hostile environments of rocks [46, 143]. The presence of fungi on the stone surface of the heritage monuments and stoneworks is often linked with the process of biodeterioration. Fungi can bring about physical and structural changes and can cause severe damage to the monumental stones and artworks in several ways. Metabolically excretion of inorganic and organic acids is one of the several ways. These metabolically generated organic acids (like oxalic acid and citric acid), siderophores, amino acids and phenolic compounds have strong chelating properties [144, 145] by which it weaken the metal-oxygen bond, increases the solubility of some metals and forms complexes with the mineral cations present on the surface matrix [146, 147]. Fungi can make their micro-environment acidic by secreting proton and organic acids and by producing carbon dioxide during cellular respiration which results in the formation of carbonic acid. Penetration of fungal hyphae along the crystal plane is known to destabilize the stone texture leading to its mechanical deterioration [112]. Another mean of mechanical damage which results due to the alternate contraction and expansion cycles of the fungal thallus, become predominant under the fluctuating environmental humidity conditions. It has been hypothesized that fungi may dissolve and burrow the mineral matrix ('rock-eating fungi' hypothesis) [148-151] resulting in the formation of fungal tunnels and may form new secondary minerals (whose chemical composition identical to host rock) within which after their death fungal tunnels are formed ('rock-building fungi' hypothesis) [152]. While the 'rock-eating fungi' hypothesis highlighted fungal role in mineral matrix dissolution, 'rock-building fungi' hypothesis explained their involvement if new mineral matrix formation. Some endolithic fungi by biochemical means cause "pitting", of the stone surface characterized by the presence of many small holes. This type of alteration is noticeable on the gate of the Cathedral of Huesca in Spain. Role of fungal species of different genera such as *Cladosporium herbarum, Aspergillus niger, Stachybotrys spp.* and *Alternaria* have been found responsible for these alterations. Use of techniques such as scanning electron microscopy has revealed the presence of 'etch marks' which is a form of biochemical weathering [11, 153, 154]. Fungi can stimulate transformation and weathering the mineral matrix of stone surface and results in the formation of secondary minerals (or neogenesis) such as calcium oxalate by combining the precipitated calcium carbonate with the secreted oxalic acid [155, 156]. Calcium oxalate formed by fungi exhibit a variety of crystalline forms (tetragonal, bipyramidal, plate-like, rhombohedral or needles) [157]. Formation of oxalate film on carrara marble from Pisa tower is attributed to the growth of fungal species *Sporotrichum* [158]. Calcium oxalate formed on carbonatic rocks may again convert into calcium carbonate and cause cementation of the pre-existing limestones [159]. Fungi may produce oxalates of other metals such as Cd, Co, Cu, Mg, Mn, Sr, Zn, Ni and Pb by interacting with a variety of different metals and metal-bearing substrates [46, 64, 160-162]. The formation of toxic metal oxalates is considered as a mechanism which enables fungi to survive and withstand high concentrations of these toxic metals [163, 164].

Fungal Diagenesis

Fungal diagenesis can be describes as a complex process of biochemical [141] and biomechanical alterations [45, 155] of carbonate substrates induced by fungus [165] resulting in the formation of different secondary minerals. Carbonate stones mainly consisted of limestone or dolomite serves as an unusual niche for the fungal communities. These substrates are highly susceptible to fungi and their interaction promotes extensive microbial diagenesis of these substrates. Extensive diagenesis results in the dissolution (or bioweathering) of the carbonate substrate due to exuded organic acids (oxalic, citric and malic acid) and transformation of the original minerals of the substrate with new one as consequences of fungal biomineralization process [162, 165].

LICHENS

Lichens grow as a visible film on the stone surface and due to their macroscopic structure, their presence on stones is visibly evident. Lichens represent the symbionts of fungi (mainly ascomycete) and algae (mainly green algae) or fungi and cyanobacteria (less common) [19, 166, 167]. Lichens existence as facultative or obligate symbioses with algal or cyanobacterial partners is a mechanism enabling lichens to grow in almost all terrestrial environments. It is estimated that lichens cover 6% of the earth land surface [168]. They participate in retention and distribution of nutrients (such as carbon and nitrogen) and trace metals, soil formation and rock weathering. As compared to other microorganisms except fungi, lichens can efficiently withstand climatic variations such as extreme temperature and desiccation which enables them to grow in wide variety of habitats, of which some of them may be inhospitable for other forms of lives [169, 170]. Lichens are the pioneer organisms which inhabits the exposed stone surfaces showing abundance of organic nitrogen rich matters such as excretes of birds (crows and pigeons). They are thought to be one of the major contributors of biogeophysical and biogeochemical deterioration of the monumental stone and artworks. Lichens have capability of accumulating many toxic metals such as lead, copper, and some other elements of environmental concern, including radionuclides, in significantly high quantity [171]. They can also form metal oxalates [66, 172], for instance, copper oxalate (Moolooite) which is formed with lichen thallus by lichens growing of copper-sulfide bearing rocks [174, 175]. The CO_2 produced during cellular respiration is converted into carbonic acid (a potent weathering agent) inside the lichen thallus [19, 176] which brings about solubilization of calcium and magnesium carbonates in calcareous stones [177]. Presence of some specialized structures like hyphae (crustose lichens) and rhizoids (foliose and fructicose lichens) aid lichens in gaining attachment with stone substrate and penetrating into the pores, cracks and fissures of the stones leading to structural and biomechanical damage [11, 146, 172, 178]. Their metabolic activities are often associated with the release of highly corrosive organic carboxylic acids (like oxalic acid etc) and metal-chelating compounds by which they form complexes with the mineral cations of the stone substrates [11, 179]. Nitrogen fixation by them is also known to improve the bioweathering potential [43]. Previously documented in literature were lichen acids, which represents a group of polyphenolic compounds such as anthraquinones [11] having polar moiety that causes metal cations chelation and promote the process of bioweathering. Death of lichen leave behind on stones, a pitting corrosion with etches mark formed either their metabolic activity and by incorporation of mineral fragments into thallus. The alternate contraction and expansion of lichen thallus due to desiccation and rehydration result in peeling and lifting up of superficial mineral grains [19, 180-182]. Their presence was also suspected for the formation of oxalate layer on patinas [11, 19].

Fungi within lichens are known to be involved in the secretion of several hundred compounds commonly known as lichen substance. These lichen substances include simple aliphatic organic acids, aromatic polyphenol compounds (such as depsides, depsidones, depsones and carotenoids) [40, 41] and chelating agents (such as norstictic, psoromic, iso-usnic, and usnic acid) [40]. Lichen substances are believed to have an important role in the extraction of nutrients from the mineral matrix of stone substrate [40, 41]. There are few studies which demonstrated the process of lichen-stone mineral interaction and established that the secretion of some lichen substances such as usnic acid, zeroin, and leucotylin (in *Lecanora muralis*), divaricatic and usnic acid (in *Ophioparma ventosa*), parietina and rhizocarpic acid (in *Xanthoria elegans*) and thamnolic acid (in *Ophioparma ventosa* and *Pertusaria corallina*) is not associated with any sort of biodeterioration [20, 183].

Mosses and Liverworts

Mosses and liverworts are bryophytes which prefer to grow on the surface shows high abundance of humus deposits. Dead algal cells, soil and atmospheric particles constitutes major portion of humus deposits. Mosses and liverworts are photosynthetic organisms having characteristic pigments but they lack vascular tissues meant for nutrient transport. Their presence is noticeable mostly with algae communities in tropical environment and damped surfaces of historic monuments. The damage caused by them (usually green-grey patches) is mainly associated with aesthetic appearance with few other types of damages are also reported. The carbonic acid produced by them as a result of cellular respiration can cause acid damage to stone over an extended period of time leading to their biocorrosion. They are also associated with extraction of mineral cations from the stones [63]. There are literatures that demonstrated the ability of mosses (*Grimmia pulvinata* (Hedw.) Sm.) to uptake calcium and established their role in biodeterioration, though their role in biodeterioration is still assumed very negligible [12]. Regardless of the presence of distinct structures called rhizoids, their role in causing biomechanical damage to stone surface by physical intrusion has not been reported yet. As a consequence of the death on mosses, the stone surface of monuments often get enriched with humus content which serves as good substrate for successive establishment of higher plants.

Remedial Measures and Strategies to Safeguard Monuments and Artworks

With the inevitable increase in the atmospheric pollution and emergence of species, the deterioration of sculptural monuments and artworks became evident. For the past few decades it has been a global concern and had necessitated and demanded global involvement in searching innovative bioconservation strategies for monuments and artworks preservation. Since some years now great research has been inspired and directed toward the development of new methodologies for the bioconservation, in a technical sense. [184-190]. The choice of an appropriate conservation and restoration strategies partly depends up on the nature of the deterioration and partly on the petrophysics properties of the substrate and surrounding environmental conditions. Plentiful of physical, mechanical and chemical methods [191, 192] were devised and tested for the cleaning and restoration purposes but all yielded poor and unsatisfactory results in regard to preservation, cementation and impermeabilization of the deteriorated monuments and works of art [193, 194]. Few methods rather induced alterations in the treated stones in term of change in surface texture, loss of physical strength and aesthetic appearance [13] [194]. Majority of the traditional methods which offered pore plugging ware characterized by their unfruitfulness and inefficiency [195] and their incompatibility resulted in irreversible, irreparable or even irremediable damage to the stone surface leading to weathering, discoloration and loss of aesthetic beauty of the treated stone structures [196]. The noticeable incompatibility of the conventional methods has stimulated the research for developing compatible, effective and eco-friendly conservation treatments in recent years [197, 198]. Alternative techniques were investigated based on their ability to conserve and consolidate the porous stones by products with similar nature. These techniques are mainly based on the methods with which it is possible to induce the formation of new mineral phase, such as calcite. Recently, alternative methods based on exploitation of microbial capability of calcium carbonate precipitation for cleaning procedures were experimented and their efficacy evaluations were investigated. The microbial participation in monuments and artworks conservation showed propitious results in past few years [193, 195, 199] and their carbonatogenic property of these microbes has allowed us to ascertain its potential in bioconservation.

TRADITIONAL METHODS OF CONSERVATION: USE OF PHYSICAL, MECHANICAL AND CHMEICAL METHODS

Conservation and consolidation of sculptural monuments and artistic works is a big challenge nowadays. During previous years, particular attention had been focused on using physical, mechanical and chemical methods for formulating conservation regime. Majority of these tested methods, with few exceptions [184, 188], has proven inefficient and seemed unlikely to yield successful results. While physical methods of conservation included irradiation of affected material with ultra-violet, gamma and X-rays, mechanical methods make use of certain tools which by manual rubbing cleans the surface of the stones or artifacts [192]. Rinsing with low pressure water and steam also offered cleaning. Chemical methods employed use of inorganic, organic, quaternary ammonium compounds, phenols and urea derivatives for cleaning purposes [192]. Application of fumigants and use of few compounds which has shown selective protective activity against microorganisms such as biocides (Biotin N) [200], algicides (such as benzalkonium chloride) [201, 202], fungicides and insecticides have been widely reported in literatures [192]. During previous years, advances in the field of applied biotechnology, in particular, the use of carbonatogenic potential of microbes in conservation and restoration regimes has attracted the researcher's attention and a great number of studies were carried out in this field. Though, it has become obvious that conversation and restoration strategies mainly relies on microbial ability of calcium carbonate precipitation but the potential and contribution of other remedial strategies cannot be under estimated. Recently, the exploitation of laser induced cleaning techniques has made the restoration process more promising [184].

Current Method of Conservation: Use of Carbonatogenic Bacteria

The microbial colonization on sculptural monuments and artworks has always been supposed detrimental or destructive and therefore their role in conservation is often disregarded. Recently, the exploitation of microorganism in bioconservation and restoration of monuments and stone work of arts was evaluated. These microorganisms are called as 'carbonatogenic bacteria' or 'calcifying microbes, because of their ability of extracellular precipitation of calcium carbonate under certain environmental conditions. These microorganisms though not related to a single phylogenetic group and are widely disturbed among microorganisms. Because of their capability of calcium carbonate precipitation, most of these microorganisms are able to confer protection to sculptural monuments and stone works. These microorganisms mainly belong to different genera within the phylum *Firmicutes, Proteobacteria and Actinomycetes* [203, 204]. Besides this, precipitation of calcium carbonate by few other microorganisms such as sulfate reducing bacteria, cyanobacteria [205, 206] and fungi [207] is also well documented in literature but their bioconservation potential is still to be evaluated and confirmed.

Dedicated Research Projects

The general conference of UNESCO at its 17th session at Paris in 1972 adopted the 'Convention Concerning the Protection of the World and Natural Heritage'. The principle aim of thus programme was to identify, catalogue and conserve the heritage sites of natural, cultural and architectural importance. Under this programme many historic buildings and heritage sites of the world which were invariably ruined or deteriorated by microbes and environmental pollutants were recognized and listed as world's cultural heritages. To deeply understand the mechanisms of deterioration and the methodologies for safeguarding the monuments and artworks, a new phase of projects were launched in few countries especially EU countries. Besides, some dedicated research group had stepped forward to carry research in this field to unravel the mechanisms of the deterioration, to ascertain the role or microbe-environment interaction in biodeterioration and to develop remedial methodologies for preservation of the heritage buildings from the detrimental effects of micro-organism and air borne pollutants. The research is been funded either by EU or other government funding agencies. The necessary information regarding the contributions of some major EU funded projects in bioconservation and of heritage sites is retrieved from European Union (EU) commission website, IST website and respective project websites (http://ec.europa.eu/research/environment/index_en.cfm?pg=projects&area=all; http://www.ist-world.org/) and the research activity of the selected research groups in the field of geomicrobiology of historic monuments and artworks is concisely summarized in the table I.

MOLECULAR STRATEGIES AND RELATED APPROACHES IN GEOMICROBIOLOGICAL RESEARCH

Geomicrobiological research in the field of biodeterioration of historic monuments and artworks could not be accomplished without the application of molecular and other related techniques. Strategies ranging from the use of conventional culturing methods to use of molecular techniques and other highly sophisticated instruments have enabled a better understanding of the cause and mechanisms of deterioration and aided in designing appropriate conservational regimes. Traditional techniques which involved the use of culture media for culturing and identification of microorganisms associated with deterioration have faced great defeat in many ways. The traditional culturing based methods often yielded inaccurate results. Due to the co-presence of other inhibitory and predatory species of fungi and other organisms in environmental samples, the identification and detection of few organisms such as cyanobacteria is severely affected. Because of fungal activity, the cyanobacterial species which are capable of growing in dry environment significantly lost during culture. It is also felt that a single method cannot be applied for the identification and characterization of all species in environmental sample. The method is also not suitable for those microorganisms which cannot be cultured (non-culturable microorganisms).

With the advent of molecular methods and other detection techniques, the research in the field of identification of potent biodeteriogens and mechanisms of biodeterioration of historic monuments and artworks has reached to level of high sophistication. The techniques such as scanning electron microscopy, denaturing DNA gel electrophoresis, 16S-rRNA sequencing, ITS sequencing, ARADA, ARISA and bioinformatics tools has enabled us not only in identification and characterization of the microorganism but also in establishing its phylogenetic relationship. The techniques such as scanning electron microscopy, confocal laser scanning microscopy, IR and Raman spectroscopy has been widely used in identification of type of deterioration cause to monuments and artworks by the microorganisms, in analysis of composition of deposited and encrusted salt on stone surface, in analysis of secondary minerals of biological origin, and in studying composition of biofilms.

CONCLUSION

The process of biodeterioration is mainly accomplished through a variety of known mechanisms, if not constraint, may lead to disastrous damage to historic monuments and artworks. In the era of globalization and industrialization, the conservation and protection of historic monuments and artworks from the detrimental effects of polluted air and microbes is a challenging task. A detailed study of literature in this field has extended our understanding in the geomicrobiology of historic monuments and artworks. Based on our knowledge and understanding of the routes and mechanisms of biodeterioration, we would like to suggest few points in order to make the conservation strategies more efficient and effective. Human activity, artificial ventilation and illumination for a long time, have been the major cause for the introduction of new microorganisms (or pollutants) which are not the natural inhabitants of that place [29, 237]. The definition and demarcation of the area to be conserved or protected can be a one of the many major management tools. This delineation and demarcation will reduce the risk of biodeterioration caused due to anthropological reasons. Any type of physical abrasive cleaning of stone structure should be avoided while the poultice and laser-based cleaning procedures as described in literatures should be encouraged [184]. With respect to use of chemical methods for cleaning, while few chemicals showed detrimental effects after treatment other resulted in better cleaning and showed promising results. The recurrence of microbial growth on monumental surfaces and those of artworks confirmed the inefficacy of several methods previously applied during conservation and consolidation processes. Of many reasons, which resulted in recurrence of microbial growth, the lack of appropriate future monitoring protocols was the major one. Therefore, it becomes very essential for the bioconservators to periodically take care and look after the monuments and artworks after treatment process to ensure that it do not show any signs of microbial growth and activity.

REFERENCES

[1] Saiz-Jimenez, C. (1999). Biogeochemistry of Weathering Processes in Monuments. Geomicrobiology Journal, 16, 27-37.

[2] Lamenti, G. Tiano, P. & Tomaselli, L. (2006). Biodeterioration of ornamental marble statues in the Boboli Gardens (Florence, Italy). Journal of Applied Phycology, 12, 427–433.

[3] Figueirdedo, C; Folha, R; Maurício, A; Alves, C & Aires-Barros, L. Pore structure and durability of Portuguese limestones: a case study. In : Smith, BJ; Gomez-Heras, M; Viles, HA & Cassar, J editors. Limestone in the Built Environment: Present-Day Challenges for the Preservation of the Past. London: Geological Society; 2010; 157-170.

[4] Rescic, S; Fratini, F & Tiano, P. On-site evaluation of the 'mechanical' properties of Maastricht limestone and their relationship with the physical characteristics. In : Smith, BJ; Gomez-Heras, M; Viles, HA & Cassar, J editors. Limestone in the Built Environment: Present-Day Challenges for the Preservation of the Past. London: Geological Society; 2010; 157-170.

[5] Beck, K &Al-Mukhtar, M. Weathering effects in an urban environment: a case study of tuffeau. In: Smith, BJ; Gomez-Heras, M; Viles, HA & Cassar, J editors. Limestone in the Built Environment: Present-Day Challenges for the Preservation of the Past. London: Geological Society; 2010; 157-170.

[6] Wakefield, R.D. & Jones, M.S. (1998). An introduction to stone colonizing micro-organisms and biodeterioration of building stone. Quaterly Journal of Engineering Geology and Hydrogeology, 31, 301-313.

[7] Cultrone, G. Arizzi, A. Sebastián, E. & Rodriguez-Navarro, C. (2008). Sulfation of calcitic and dolomitic lime mortars in the presence of diesel particulate matter. Environ. Geol., 56, 741-752.

[8] Cuzman, O.A. Ventura, S. Sili, C. Mascalchi, C. Turchetti, T. D' Acqui, L. P. & Tiano, P. (2010). Biodiversity of phototrophic biofilms dwelling on monumental fountains. Microb Ecol, 60, 81-95.

[9] Crispim, C. A. & Gaylarde, C. C. (2005). Cyanobacteria and Biodeterioration of Cultural Heritage: A Review. Microb Ecol, 49, 1-9.

[10] Gadd, G. M. (2010). Metals, minerals and microbes: geomicrobiology and bioremediation. Microbiology, 156, 609–643.

[11] Adamo, P. & Violante, P. (2000). Weathering of rocks and neogenesis of minerals associated with lichen activity. Appl Clay Sci., 16, 229–256.

[12] Altieri, a. & Ricci, S. (1997). Calcium uptake in mosses and its role in stone biodeterioration. International Biodeterioration & Biodegradation, 40, 201-204.

[13] Nuhoglu, Y. Oguz, E. Uslu, H. Ozbek, A. Ipekoglu, B. & Ocak, I. (2006). The accelerating effects of the microorganisms on biodeterioration of stone monuments under air pollution and continental cold climatic conditions in Erzurum, Turkey. Science of Total Environment, 364, 272–283.

[14] Mansch, R. & Bock, E. (1998). Biodeterioration of natural stone with special reference to nitrifying bacteria. Biodegradation, 9, 47–64.

[15] Uroz, S. Calvaruso, C. Turpault, M. P. & Frey-Klett, P. (2009). Mineral weathering by bacteria: ecology, actors and mechanisms. Trends in Microbiology, 17, 378-387.

[16] Grvić, M. V. L. & Vukojević, J. B. (2009). Role of fungi in biodeterioration process of stones in historic buildings. Proc. Nat. Sci, Matica Srpska Novi Sad, 116, 245-251.

[17] Burford, E.P. Kierans, M. Gadd, G. M. (2003). Fungal involvement in bioweathering and biotransformation of rocks and minerals. Mineral. Mag. 67, 1127–1155.

[18] Seaward, M.R.D. (2003). Lichens, agents of monumental destruction. Microbiology Today, 30, 110-112.

[19] Oksanen, I. (2006). Ecological and biotechnological aspects of lichens. Appl Microbiol Biotechnol, 73, 723–734.

[20] Cicek, A. Aslan, A. Yazıcı, K. & Koparal, A. S. (2009). Effects of environmental conditions on historical buildings: lichens and NOx gases. Environ Monit Assess, 154, 187–195.

[21] Gauri, K.L., & Bandyopadhyay, J.K., (1999). Carbonate Stone: Chemical Behaviour, Durability, and Conservation. New York, NY: Wiley-Interscience.

[22] Del-Monte, M. Sabbioni, C. & Vittori, O. (1981) Airborne carbon particles and marble deterioration. Atmos Environ, 15, 645–652.

[23] Moropoulou, A. Bisbikou, K. Van Grieken, R. Zezza, F. & Macri, F. (1998). Origin and growth of weathering crusts on ancient marbles in industrial atmosphere. Atmos. Environ., 32, 967–982.

[24] Simão, J. Ruiz-Agudo, E. & Rodriguez-Navarro, C. (2006). Effects of particulate matter from gasoline and diesel vehicle exhaust emissions on silicate stones sulfation. Atmos Environ, 40, 6905– 6917.

[25] Sabbioni, C. (1995). Contribution of atmospheric deposition to the formation of damage layers. Water, Air and Soil Pollution, 63, 305–316.

[26] Zappia, G. Sabbioni, C. Pauri, M. G. & Gobbi, G. (1992). Effect on SO2-riched atmosphere on ancient and modern building materials. Mater Engg, 3, 445–458.

[27] Rodriguez-Navarro, C. & Sebastian, E. (1996). Role of particulate matter form vehicle exhaust on porous building stones (limestone) sulfation. Sci Total Environ, 187, 79–91.

[28] Kumar, R. & Kumar, A. V. (1999). Biodeterioration of Stone in Tropical Environments – An Overview, The Getty Conservation Institute, USA.

[29] Saiz Jimenez, C. (1995). Deposition of anthropogenic compounds on monuments and their effect on airborne microorganisms. Aerobiologia, 11, 161–175.

[30] Guillitte, O. (1995). Bioreceptivity: a new concept for building ecology studies. The Science of the Total Environment, 167, 215-220.

[31] Barberousse, H. Lombardo, R.J. Tell, G. & Couté, A. (2006). Factors involved in the colonisation of building façades by algae and cyanobacteria in France. Biofouling, 22, 69-77.

[32] Hamilton, R. S. Revitt, D. M. Vincent, K. J. & Butlin, R. N. (1995). Sulphur and nitrogen particulate pollutant deposition on to building surfaces. Sci of The Total Environ, 167, 57–66.

[33] Grossi, CM; Brimblecombe, P; Bonazza, A; Sabbioni, C & Zamagni, J. Sulfate and carbon compounds in black crusts from the Cathedral of Milan and tower of London. In: Gomez-Heras & Vazquez-Calvo editiors. Heritage, Weathering and Conservation. London: Taylor and Francic Group; 2006; 441-446.

[34] Johansson, L. G. Lindqvist, O. & Mangio R. E. (1988). Corrosion of calcareous stones in humid air containing SO2 and NO2. Durabil Build Material, 5, 439–449.

[35] Johansson, L.G. (1989). Synergistic effects of air pollutants on the atmospheric corrosion of metals and calcareous stones. Marine Chemistry, 30, 113-122.

[36] Charola, AE & Ware, R. Acid deposition and the deterioration of stone: a brief review of a broad topic. In: Siegesmund, S; Weiss, T & Vollbrecht, A editors. Natural stone, weathering phenomena, conservation strategies and case studies. London: Geological Society; 2002; 329–345.

[37] Vallet, J. M. Gosselin, C. Bromblet, P. Rolland, O. Vergès-Belmin, V. & Kloppmann, W. (2006). Origin of salts in stone monument degradation using sulphur and oxygen isotopes: First results of the Bourges cathedral (France). Journal of Geochemical Exploration, 88, 358-362.

[38] Warscheid, T. & Braams, J. (2000). Biodeterioration of stone: a review. Int Biodet Biodeg, 46, 343-368.

[39] Prieto, B. & Silva, B. (2005). Estimation of the potential bioreceptivity of granitic rocks from their intrinsic properties. Int Biodeterior Biodegradation, 56, 206–215.

[40] Huneck, S. & Yoshimura, I. (1996). Identification of lichen substance. Berlin: Springer-Verlag.

[41] Purvis, W. (2000). Lichens. Washington, DC: Smithsonian Inst. Press.

[42] Neaman, A. Chorover, J. & Brantley, S. L. (2005) Implication of the evolution of organic acid moieties for basalt weathering over ecological time. Am. J. Sci. 305, 147–185.

[43] Seneviratne, G. & Indrasena, I. K. (2006). Nitrogen fixation in lichens is important for improved rock weathering. J. Biosci, 31, 639–643.

[44] Gadd, G. M. and Sayer, J. A. (2000). Fungal transformations of metals and metalloids. In: Lovley, DR editor. Environmental Microbe-Metal Interactions. Washington, DC: American Society of Microbiology; 2000; 237–256.

[45] Sterflinger, K. (2000). Fungi as geologic agents. Geomicrobiology Journal. 17, 97–124.

[46] Gadd, G. M. (2007). Geomycology: biogeochemical transformations of rocks, minerals, metals and radionuclides by fungi, bioweathering and bioremediation. Mycological Research III, 3-49.

[47] Gaylarde, C. Silva, M. R. & Warscheid, T. (2003). Microbial impact on building materials: an overview. Materials and Structures / Matdriaux et Constructions, 36, 342-352.

[48] Olaru, M. Aflori, M. Simionescu, B. Doroftei, F. & Stratulat, L. (2010). Effect of SO2 dry deposition on porous dolomitic limestones. Materials, 3, 216-231.

[49] Haneef, S. J. Johnson, J. B. Dickinson, C. Thompson, G. E. & Wood, G. C. (1992). Effect of dry deposition of NOx and SO2 gaseous pollutants on the degradation of calcareous building stones. Atmos.Environ. 26A, 2963.

[50] Brimblecombe, P. & Grossi, C. M. (2005). Aesthetic thresholds and blackening of stone buildings. Science of The Total Environ, 349, 175-189.

[51] Grossi, C. M. Brimblecombe, P. Esbert, R. M. & Alonso, F. J. (2007). Color changes in architectural limestones from pollution and cleaning. Color Research and Application, 32, 320-331.

[52] Grossi, C M & Brimblecombe, P. (2007). Effect of long-term changes in air pollution and climate on the decay and blackening of European stone buildings. Geological Society, London, Special Publication, 271, 117-130.

[53] Böke, H. Gokturk, E. H. Caner-Saltik, E. N. & Demirci S. (1999). Effect of airborne particle on SO2-calcite reaction. Appl Surf Sci, 140, 70–82.

[54] Moroni, B. Pitzurra, L. & Poli, L. (2004). Microbial growth and air pollutants in the corrosion of carbonate building stone: results of laboratory and outdoor experimental tests. Environmental Geology, 46, 436–447.

[55] Jimenez De Haro, M C; Justo, A; Durán, A; Sigüenza, M B; Pérez-Rodriguez, JL & Bueno, J. Crusts formed on different building materials from the cathedral of Seville, Spain. In: Saiz-Jimenez, C editor. Air Pollution and Cultural Heritage. Leiden: Balkema; 2004; 79–84.

[56] Sabbioni, C; Bonazza, A; Zamagni, J; Gedni, N; Grossi, CM & Brimblecombe, P. The Tower of London: a case study on stone damage in an urban area. In: Saiz-Jimenez, C editor. Air Pollution and Cultural Heritage. Leiden: Balkema; 2004; 57-62.

[57] McAlister, J. J. Smith, B. J. & Török, A. (2006). Element partitioning and potential mobility within surface dusts on buildings in a polluted urban environment, Budapest. Atmospheric Environment, 40, 6780–8790.

[58] McAlister, J. J. Smith, B. J. & Török, A. (2008).Transition metals and water-soluble ions in deposits on a building and potential catalysis of stone decay. Atmospheric Environment, 42, 7657–7668.

[59] Grossi, C.M. Esbert, R. M. Díaz-Pache, F. & Alonso, F. J. (2003). Soiling of building stones in urban environments. Build Environ, 38, 147–159.

[60] Rodriguez-Navarro, C & Dohene, E. (1999). Salt weathering: influence of evaporation rate, supersaturation and crystallization pattern. Earth Surf Proc Land, 24, 191–209.

[61] Toniolo, L Carlotta, M. & Bugini, R. (2009). Black layers on historical architecture. Environ Sci Pollut Res. 16, 218–226.

[62] Saiz-Jimenez, C., (1993). Deposition of airborne organic pollutants on historical buildings. Atmospheric Environment, 27B, 77–85.

[63] Saiz-Jimenez, C. Biodeterioration of stone in historic buildings and monuments. In: Llewellyn GC; Dashek WW & O' Rear CE, editors. Biodeterioration research 4: mycotoxins, wood decay, plant stress, biocorrosion,and general biodeterioration. New York: Plenum; 1994; 587-603.

[64] Gadd, G.M. (1999). Fungal production of citric and oxalic acid: importance in metal speciation, physiology and biogeochemical processes. Adv. Microb. Physiol. 41, 47-92.

[65] Gharieb, M. M. Sayer, J. A. & Gadd, G. M. (1998). Solubilization of natural gypsum (CaSO4.2H2O) and the formation of calcium oxalate by Aspergillus niger and Serpula himantioides. Mycol. Res. 102, 825-830.

[66] Adamo, P. Vingiani, S. & Violante, P. (2002). Lichen–rock interactions and bioformation of minerals. Dev Soil Sci, 28B, 377–391.

[67] Pavía, S. & Caro, S. (2006). Origin of Films on Monumental Stone. Studies in Conservation, 51, 177-188.

[68] Lazzarini, L. & Salvadori, O. (1989). A reassessment of the formation of a patina called scialbatura. Studies in Conservation, 34, 20-26.

[69] Cunningham, J.E. & Kuiack, C. (1992). Production of citric and oxalic acids and solubilization of calcium phosphate by Penicillium bilaii. Applied and Environmental Microbiology, 58, 1451-1458.

[70] Monte, M. (2003a). Biogenesis of oxalate patinas on marble specimens in fungal culture. Aerobiologia, 19, 271–275.

[71] Ruitger, G.J.G., Van de Vondervoort, P.J.I. & Visser, J. (1999). Oxalic acid production by Aspergillus niger: an oxalate-non-producing mutant produces citric acid at pH 5 and in the presence of manganese. Microbiology, 145, 2569-2576.

[72] Cariati, F. Rampazzi, L. Toniolo, L. & Pozzi, A. (2000). Calcium oxalate films on stone surface: experimental assessment of the chemical formation. Studies in Conservation, 45, 180–189.

[73] Edwards, H.G.M. Seaward, M. R. D. Attwood, S. J. Little, S. J. de Oliveira, L. F. C. & Tretiach, M. (2003). FT-Raman spectroscopy of lichens on dolomitic rocks: an assessment of metal oxalate formation. Analyst. 128, 1218–1221.

[74] Monte, M. et al. (2000) Damage caused by microorganisms on marble samples. In: Sequeira, CAC editor. Microbial Corrosion n. 29. Proceedings of the fourth intern. Portugal: ESC workshop, Portugal-1999; 2000; 339–355

[75] Monte, M. (2001) Effect of Fungal Growth on Marble Samples: Formation of an Oxalate Deposit. In Proceedings of the 11th workshop "Eurocare – Euromarble EU 496, Vienna. Vienna: Centre Kartause Mauerbach, Vienna-2000; 2001; 103–107.

[76] Monte, M. Experimental evidence of oxalate formation by fungal strains on marble samples. In: Saiz-Jimenez, C editor. Molecular Biology and Cultural Heritage. Lisse, The Netherlands: Swets and Zeitlinger B.V.; 2003b; 263-266.

[77] Bonaventura, M. P. D. Del Gallo, M. Cacchio, P. Ercole, C. & Lepidi, A. (1999). Microbial formation of oxalate films on monument surfaces: bioprotection or biodeterioration? Geomicrobiology Journal, 16, 55-64.

[78] Ortega-Calvo, J.J. et al. (1991) Biodeterioration of building stones by cyanobacteria. Int Biodeterior. 28, 165–186.

[79] Tiano, P. Accolla, P. & Tomaselli, L. (1995). Phototrophic biodeteriogens on lithoid surfaces: an ecological study. Microb Ecol, 29, 299–309.

[80] Cecchi, G. Pantani, L. Raimondi, V. Tomaselli, L. Lamenti, G. Tiano, P. & Chiari, R. (2000). Fluorescence lidar technique for remote sensing of stone monuments. J Cult Herit, 1, 29–36.

[81] Tomaselli, L. Lamenti, G. Bosco, M. & Tiano, P. (2000a). Biodiversity of photosynthetic microorganisms dwelling on stone monuments. International Biodeterioration and Biodegradation. 46, 251–258.

[82] Crispim, C. A. Gaylarde, P. M. & Gaylarde, C. C. (2003). Algal and cyanobacterial biofilms on calcareous historic buildings. Curr Microbiol, 46, 79–82.

[83] Ariño, X. Ortega-Calvo, J.J. Gómez-Bolea, A. Saiz-Jimenez, C. (1995). Lichen colonization of the Roman pavement at Baelo Claudia (Cadiz, Spain): biodeterioration vs. bioprotection. Sci Total Environ 167:353–363.

[84] Ramírez, M. Hernández-Mariné, M. Novelo, E. & Roldán, M. (2010). Cyanobacteria-containing biofilms from a Mayan monument in Palenque, Mexico. Biofouling, 26, 399-409.

[85] Albertano, P; Moscone, D; Palleschi, G; Hermosin, B; Saiz-Jimenez, C; Sanchez-Moral, S; Hernandez-Marine, M; Urzi, C; Groth, I; Schroeckh, V; Saarela, M; Mattila-Sandholm, T; Gallon, JR; Graziottin, F; Bisconti, F; & Giuliani, R. Cyanobacteria attack rocks (CATS): control and preventive strategies to avoid damage caused by cyanobacteria and associated microorganisms in Roman hypogean monuments. In: Saiz-Jimenez, C editor. Molecular Biology and Cultural Heritage. Balkemla, Lisse: 2003; 151 –162.

[86] Decho, A.W. (2000). Microbial biofilms in intertidal systems: an overview. Cont. Shelf Res., 20, 1257– 1273.

[87] Tamaru, Y. Takani, Y. Yoshida, T & Sakamoto, T. (2005). Crucial Role of Extracellular Polysaccharides in Desiccation and Freezing Tolerance in the Terrestrial Cyanobacterium Nostoc commune. Applied and Environment Microbiology, 71, 7327-7333.

[88] Ophir, T. & Gutnick, D. L. (1994). A role for exopolysaccharides in the protection of microorganisms from desiccation. Appl Environ Microbiol, 60, 740–745.

[89] Pietrini, AM; Ricci, S; Bartolini, M & Giuliani, MR. (1985). A reddish colour alteration caused by algae on stoneworks. Preliminary studies. In: Felix, G editor. Proceedings of the 5th International Congress on Deterioration and Conservation of Stone. Lausanne, Switzerland: Presses Polytechniques Romandes; 1985; 653–662.

[90] Hernández-Mariné, M. Clavero, E. & Roldán, M. (2003). Why there is such luxurious growth in the hypogean environments. Arch Hydrobiol Suppl Algol Stud, 109, 229–239.

[91] Gorbushina, A. A. & Broughton, W. J. (2009). Microbiology of the atmosphere-rock interface: how biological interactions and physical stresses modulate a sophisticated microbial ecosystem. Annu. Rev. Microbiol, 63, 431–450.

[92] Novelo, E. & Ramírez, M. (2006). Algae and cyanobacterial diversity and distribution patterns on mayan buildings in Palenque, Chiapas. Cyanobacterial diversity and ecology on historic monuments in Latin America. Rev Latinoam Microbiol, 48, 188–195.

[93] Gorbushina, AA & Krumbein, WE. Subaerial Rock dwelling fungal communities: diversity of life styles and colony structure. In: Seckbach, J editor. Journey to Diverse Microbial Worlds. Dordrecht, The Netherlands: Kluwer Acadmic Publishers; 2000a; 317-334.

[94] Gorbushina, AA & Krumbein, WE. Subaerial microbial mats and their effects on soil and rock. In: Riding, RE & Awramik, SM editors. Microbial Sediments. Berlin, Germany: Springer Verlag; 2000b; 161–170.
[95] Potts, M. (1999). Mechanisms of desiccation tolerance in cyanobacteria. Eur. J. Phycol., 34, 319–328.
[96] Wynn-Williams DD. Cyanobacteria in deserts – life at the limit? In: Whitton, BA & Potts, M editors. The ecology of cyanobacteria. Dordrecht: Kluwer Academic Publishers; 2000; 341–366.
[97] Castenholz, RW& Garcia-Pichel, F. Cyanobacterial responses to UV-radiation. In: Whitton BA & Potts, M editors. The ecology of cyanobacteria. The Netherlands: Kluwer Academic Publisher; 2000; 591–611.
[98] Fleming, E.D. & Castenholz, R.W. (2007). Effects of periodic desiccation on the synthesis of the UV-screening compound, scytonemin, in cyanobacteria. Environ Microbiol, 9, 1448–1455.
[99] Albertano, P. Methodological Approaches to the Study of Stone Alteration caused by Cyanobacterial Biofilms in Hypogean Environments. In: Koestler RJ; Koestler, VR; Charola, AE & Nieto-Fernandez, FE editors. Art, Biology and Conservation: Biodeterioration of Works of Art. New York: The Metropolitan Museum of Art; 2003; 302-315.
[100] Seneviratne, G. & Jayasinghearachchi, H.S. (2005). A rhizobial biofilm with nitrogenase activity alters nutrient availability in a soil. Soil Biol Biochem, 37, 1975–1978.
[101] Perry IV, T. D. Klepac-Ceraj, V. Zhang, X. V. McNamara, C. J. Polz, M. F. Martin, S. T. Berke, N. & Mitchell, R. (2005). Binding of Harvested Bacterial Exopolymers to the Surface of Calcite. Environ. Sci. Technol., 39, 8770–8775.
[102] Albertano, P. & Kovacik, L. (1996). Light and temperature responses of terrestrial sciaphilos strains of Leptolyngbya sp. in cross-gradient cultures. Arch Hydrobiol Suppl Algol Stud, 83, 17–28.
[103] Hernández-Mariné, M. Roldán, M. Clavero, E. Canals, A. & Ariño, X. (2001). Phototrophic biofilm morphology in dim light. The case of the Puigmolto sinkhole. Nova Hedwigia, 123, 235–251.
[104] Horat ,T. Neu, T. R. & Bachofen, R. (2006). An endolithic microbial community in dolomite rock in Central Switzerland: characterization by reflection spectroscopy, pigments analyses, scanning electron microscopy, and laser scanning microscopy. Microb Ecol, 51, 353–364.
[105] Walker, J.J. & Pace N. R. (2007). Endolithic microbial ecosystems. Ann Rev of Microbiol, 61, 331–347.
[106] Hoppert, M. Flies, C. Pohl, W. Gunzl, B. & Schneider, J. (2004). Colonization strategies of lithobiontic microorganisms on carbonate rocks. Environ Geol, 46, 421–428.
[107] Schiavon, N. Biodeterioration of calcareous and granitic building stones in urban environments. In: Siegemund, S; Weiss, T & Vollbrecht, A editors. Natural Stone, Weathering Phenomena, Conservation Strategies and Case Studies. London: Geological Society; 2002; 195–205.
[108] Carter, N. E. A. & Viles, H. A. (2004). Lichen hotspots: raised rock temperatures beneath Verrucaria nigrescens on limestone. Geomorphology, 62, 1–16

[109] Goudie, A.S., & Viles, H.A. (1997). Salt Weathering Hazards. Chichester: Wiley and Sons.

[110] Fitzner, B; Heinrichs, K & La Bouchardiere, D. Limestone weathering of historical monuments in Cairo, Egypt. In: Siegemund, S; Weiss, T & Vollbrecht, A editors. Natural Stone, Weathering Phenomena, Conservation Strategies and Case Studies. London: Geological Society; 2002; 217-239.

[111] Török, A. Leithalkalk-type limestones in Hungary: an overview of lithologies and weathering features. In: Prikryl, R. & Siegl, P. editors. Architectural and Sculptural Stone in Cultural Landscape. Prague: Karolinum Press; 2004; 157-172.

[112] Mohammadi, P. & Krumbein, W. E. (2008). Biodeterioration of ancient stone materials from the Persepolis monuments (Iran). Aerobiologia, 24, 27–33.

[113] Gaylarde, P. M. Englert, G. Ortega-Morales, B. O. Gaylarde, C. C. (2006). Lichen-like colonies of pure Trentepohlia on limestone monuments. Int Biodeterior Biodegr, 58, 119–123.

[114] Wakefield, R. Jones, M. Wilson, M. Young, M. Nicholson, K. & Urguhart C. (1996). Investigations of decayed sandstone colonized by a species of Trentepohlia. Aerobiologia, 12, 19–25.

[115] Noguerol-Seoane, A. & Rifon-Lastra, A. (1997). Epilithic phycoflora on monuments. A survey of San Esteban de Ribas de Sil Monastery (Ourense, NW Spain). Cryptogam Algol, 18, 351–361.

[116] Ortega-Calvo, J. J. Hernández-Marine, M. & Saiz-Jimenez, C. (1991). Biodeterioration of building materials by cyanobacteria and algae. Int Biodeterior, 28, 165–185.

[117] Hoffmann, L. Caves and other low-light environments: aerophytic photoautotrophic microorganisms. In: Bitton, G editor. Encyclopedia of Environmental Microbiology. New York: John Wiley; 2002; 835-843.

[118] Lefèvre, M. (1974). La maladie verte de Lascaux. Studies in Conservation, 19, 126–156.

[119] Ortega-Calvo, JJ; Hernandez-Marine, M & Saiz-Jimenez, C. Cyanobacteria and algae on historic buildings and monuments. In: Garg, KL; Garg, N & Mukerji, KG editors. Recent advances in biodeterioration and biodegradation, Calcutta, India: Naya Prokash; 1993a; 173-203.

[120] Ortega-Calvo, J.J. Sanchez-Castillo, P. M. Hernandez-Marine, M. & Saiz-Jimenez, C. (1993b). Isolation and characterization of epilithic chlorophytes and cyanobacteria from two Spanish cathedrals (Salamanca and Toledo). Nova Hedwigia, 57, 239 – 253.

[121] Roldán, M. Thomas, F. Castel, S. Quesada, A. & Hernández-Marine, M. (2004). Noninvasive pigment identification in single cells from living phototrophic biofilms by confocal imaging spectrofluorometry. Appl Environ Microbiol, 70, 3745–3750.

[122] Albertano, P. (1991). Effects of monochromatic lights on four species of Leptolyngbya. Archives für Hydrobiologie, Algological Studies. 64, 199–214.

[123] Albertano, P. & Bruno, L. The importance of light in the conservation of hypogean monuments In: Saiz-Jimenez, C. editor. Molecular Biology and Cultural Heritage. Balkema, Lisse: 2003; 171-177.

[124] Ariño, X. Hernandez-Marine, M. & Saiz-Jimenez, C. (1997). Colonization of Roman tombs by calcifying cyanobacteria. Phycologia, 36, 366–373.

[125] Gaylarde, C.C. & Morton, L.H.G. (1999). Deteriogenic biofilms on buildings and their control: a review. Biofouling. 14, 59–74.

[126] Groth, I. & Saiz-Jimenez, C. (1999). Actinomycetes in hypogean environments. Geomicrobiology Journal, 16, 1-8.

[127] Pantazidou, A; & Theoulakis, P. Cyanophytes and associated flora at the neoclassical Palace of Sts George and Michael in Corfu (Greece). Aspects of cleaning procedures. In: Moropoulou, A. Zezza, F. Kollias, E & Papachistodoulou, I editors. Proceedings of the 4th International Symposium on the Conservation of Monuments in the Mediterranean Basin. Rhodes: Technical Chamber of Greece; 1997; 355-368.

[128] Urzì, C. Interactions of some microbial communities in the biodeterioration of marble and limestone. In: Guerrero, R & Pedros-Alio, C editors. Trends in Microbial Ecology. Barcelona: Spanish Society of Microbiology; 1993; 667–672.

[129] Albertano, P. & Urzì, C. (1999). Structural interactions among epilithic cyanobacteria and heterotrophic microorganisms in roman hypogea. Microb Ecol., 38, 244–252.

[130] Urzì, C. Brusetti, L. Salamone, P. Sorlini, C. Stackebrandt, E. Daffonchio, D. (2001). Biodiversity of Geodermatophilaceae isolated from altered stones and monuments in the Mediterranean basin. Environmental Microbiology, 3, 471-479.

[131] Zezza, F; Urzì C; Moropolou T; Macrì F & Zagari M. Indagini microanalitiche e microbiologiche di patine e croste presenti su pietre calcaree e marmi esposti all'aerosol marino e all'inquinamento atmosferico. In: Biscontin, G & Driussi, G editors. Proceedings of XI Convegno Scienze e Beni culturali- La Pulitura dell'Architettura. Libreria Progetto Editoriale Padova; 1995; 293-303.

[132] Urzì, C. & Realini, M. (1998). Colour changes of Noto's calcareous sandstone as related to its colonisation by microorganisms. International Biodeterioration and Biodegradation. 42, 45-54.

[133] Piñar, G. Saiz-Jimenez, C. Schabereiter-Gurtner, C. Blanco-Varela, M. T. Lubitz, W. Rölleke, S. (2001). Archaeal communities in two disparate deteriorated ancient wall paintings: detection, identification and temporal monitoring by denaturing gradient gel electrophoresis. FEMS Microbiol Ecol, 37, 45–54.

[134] Sand, W. (1997). Microbial mechanisms of deterioration of inorganic substrates - A general mechanistic overview. Internat. Biodet. Biodeg., 40, 183-190.

[135] Golubic, S. The relationship between blue-green algae and carbonate deposits. In: Carr NG, Whitton BA, editors. The biology of the blue-green algae. Oxford: Blackwell Scientic Publ; 1973; 434-472.

[136] Golubic, S. (1987). The role of diatoms in stromatolite growth from modern freshwater settings. Journal of Phycology. 23, 195–201.

[137] Borowitzka, MA. Carbonate calcification in algae: initiation and control. In: Mann, S; Webb, J & Williams, RJP Editors. Biomineralization, Chemical and Biochemical Perspectives. Berlin: VHC Verlag; 1989; 63-94.

[138] Zurita, Y. P. Cultrone, G. Castillo, P. S. Sebastián, E. & Bolívar, F. C. (2005). Microalgae associated with deteriorated stonework of the fountain of Bibatauín in Granada, Spain. Int Biodeterior Biodegradation, 55, 55–61.

[139] Hawksworth, D.L. (2001). The magnitude of fungal diversity: the 1.5 million species estimate revisited. Mycological Research, 105, 1422–1432.

[140] Gorbushina, A. Krumbein, W. E. Panina, L. Soukharjevsky, S. Wollenzien, U. & Hamann, K.H. (1993). On the role of black fungi in colour change and biodeterioration of antique marbles. Geomicrobiology Journal, 11, 205–221.

[141] Burford, E. P. Fomina, M. & Gadd, G. M. (2003a). Fungal involvement in bioweathering and biotransformation of rocks and minerals. Mineral Mag, 67, 1127–1155.

[142] Gadd, G. M. (2004). Microbial influence on metal mobility and application for bioremediation. Geoderma, 122, 109–119.

[143] Wainwright, M. Tasnee, A. A. & Barakah, F. (1993). A review of the role of oligotrophic microorganisms in biodeterioration. International Biodeterioration and Biodegradation, 31, 1–13.

[144] Burgstaller, W. & Schinner, F. (1993). Leaching of metals with fungi. J Biotechnol, 27, 91–116.

[145] Ortega-Morales, O. Guezennec, J. Hernandez-Duque, G. Gaylarde, C.C. & Gaylarde, P.M. (2000). Phototrophic biofilms on ancient Mayan buildings in Yucatan, Mexico. Curr Microbiol, 40, 81–85.

[146] Banfield, J. F. Barker, W. W. Welch, S. A. & Taunton, A. (1999). Biological impact on mineral dissolution: application of the lichen model to understanding mineral weathering in the rhizosphere. Proc Natl Acad Sci USA, 96, 3404–3411.

[147] Harley A.D. & Gilkes R. J. (2000). Factors influencing the release of plant nutrient elements from silicate rock powders: a geochemical overview. Nutrient Cycling in Agroecosystems, 56, 11–36.

[148] Jongmans, A. G. van Breemen, N. Lundstrom, U. S. van Hees, P. A. W. Finlay, R. D. Srinivasan, M. Unestam, T. Giesler, R. Melkerud, P.A. & Olsson, M. (1997). Rock-eating fungi. Nature, 389, 682–683.

[149] Landeweert, R. Hoffland, E. Finlay, R. D. Kuyper, T. W. & Van Breemen, N. (2001). Linking plants to rocks: ectomycorrhizal fungi mobilize nutrients from minerals. Trends Ecol Evol, 16, 248–254.

[150] Golubic, S. Radtke, G. & Le Campion-Alsumard, T. (2005). Endolithic fungi in marine ecosystems. Trends Microbiol, 13, 229–235.

[151] Cockell, C. S. & Herrera, A. (2008). Why are some microorganisms boring? Trends Microbiol, 16, 101–106.

[152] Fomina, M. Hillier, & Charnock, S. J. (2005). Role of oxalic acid overexcretion in transformations of toxic metal minerals by Beauveria caledonica . Appl. Environ. Microbiol. 71, 371-381.

[153] Callot, G. Maurette, M. Pottier, L. & Dubois A. (1987). Biogenic etching of microfractures in amorphous and crystalline silicates. Nature, 328, 147–149.

[154] Lee, M.R. & Parsons, I. (1999). Biomechanical and biochemical weathering of lichen-encrusted granite: textural controls on organic-mineral interactions and deposition of silica-rich layers. Chem. Geol, 161, 385– 397.

[155] Burford, E. P. Kierans, M. & Gadd, G. M. (2003b). Geomycology: fungal growth in mineral substrata. Mycologist, 17, 98–107.

[156] Adeyemi, A. O. & Gadd, G. M. (2005). Fungal degradation of calcium, lead and silicon-bearing minerals. Biometals, 18, 269–281.

[157] Arnott, HJ. Calcium oxalate in fungi. In: Khan, SR editor. Calcium Oxalate in Biological Systems. Boca Raton, FL: CRC Press; 1995; 73-111.

[158] Monte, M. (2003c). Oxalate film formation on marble specimens caused by fungus. Journal of Cultural Heritage, 4, 255-258.

[159] Verrecchia, EP; Braissant, O & Cailleau, G. The oxalate carbonate pathway in soil carbon storage: the role of fungi and oxalotrophic bacteria. In: Gadd, GM editor. In Fungi in Biogeochemical Cycles. Cambridge: Cambridge University Press; 2006; 289-310.
[160] White, C. Sayer, J. A. & Gadd, G. M. (1997). Microbial solubilization and immobilization of toxic metals: key biogeochemical processes for treatment of contamination. FEMS Microbiol Rev, 20, 503–516.
[161] Sayer, J. A. Cotter-Howells, J. D. Watson, C. Hillier, S. & Gadd, G. M. (1999). Lead mineral transformation by fungi. Curr Biol, 9, 691–694.
[162] Sayer, J. A. & Gadd, G. M. (1997). Solubilization and transformation of insoluble metal compounds to insoluble metal oxalates by Aspergillus niger. Mycol Res, 101, 653–661.
[163] Gadd, G. M. (1993a). Interactions of fungi with toxic metals. New Phytol, 124, 25–60.
[164] Jarosz-Wilkołazka, A. & Gadd, G. M. (2003). Oxalate production by wood-rotting fungi growing in toxic metal-amended medium. Chemosphere, 52, 541–547.
[165] Kolo, K. Keppens, E. Preat, A. & Claeys, P. (2007). Experimental observations on fungal diagenesis of carbonate substrates. J Geophys Res, 112, 1–20.
[166] Hawksworth, D. L. Kirk, P. M. Sutton, B. C. & Pegler, D.N. (1995). Dictionary of the fungi. CAB, Wallingford.
[167] Piervittori, R. Salvadori, O. & Isocrono, D. (2004). Literature on lichens and biodeterioration of stonework. IV. Lichenologist, 36, 145–157.
[168] Haas, JR & Purvis, OW. Lichen biogeochemistry. In: Gadd GM editor. Fungi in biogeochemical cycles. Cambridge: Cambridge University Press; 2006; 344–376.
[169] Kranner, I. Cram, W. J. Zorn, M. Wornik, S. Yoshimura, I. Stabentheiner, E. Pfeifhofer, H. W. (2005). Antioxidants and photoprotection in a lichen as compared with its isolated symbiotic partners. Proc Natl Acad Sci USA, 102, 3141–3146.
[170] Vrablikova, H. McEvoy, M. Solhaug, K. A. Bartak, M. Gauslaa, Y. (2006). Annual variation in photoacclimation and photoprotection of the photobiont in the foliose lichen Xanthoria parietina. Journal of Photochemistry and Photobiology B: Biology, 83, 151–162.
[171] Purvis, O. W. & Pawlik-Skowronska, B. Lichens and metals. In Stress in Yeasts and Filamentous Fungi. Amsterdam: Elsevier; 2008; 175-200.
[172] Chen, J, Blume, H. P. & Beyer, L. (2000). Weathering of rocks induced by lichen colonization-A review. Catena, 39, 121– 146.
[173] Purvis O. W. (1996). Interactions of lichens with metals. Science Progress, 79, 283–309.
[174] Purvis, O. W. & Halls, C. (1996). A review of lichens in metal-enriched environments. Lichenologist, 28, 571–601.
[175] Drewello, R. & Weissmann, R. (1997). Microbially influenced corrosion of glass. Appl Microbiol Biotechnol, 47, 337–346.
[176] Tiano, P. (2002). Biodegradation of cultural heritage: decay, mechanisms and control methods. Seminar article, New University of Lisbon, Department of Conservation and Restoration, 7–12 January, 2002 (http://www.arcchip.cz/w09/w09_tiano.pdf).
[177] Favero-Longo, S.E. Castelli, D. Salvadori, O. Belluso, E. Piervittori, R. (2005). Pedogenetic action of the lichens Lecidea atrobrunnea, Rhizocarpon geographicum gr. and Sporastatia testudinea on serpentinized ultramafic rocks in an alpine environment. Int Biodeterior Biodegrad, 56, 17–27.

[178] Kiurski, J. S. Ranogajec, J. G. Ulhelji, A. L. Radeka, M. M. & Bokorov, M.T. (2005). Evaluation of the effect of lichens on ceramic roofing tiles by scanning electron microscopy and energy-dispersive spectroscopy analyses. Scanning, 27, 113–119.

[179] de los Rios, A. Wierzchos, J. Sancho, L. G. & Ascaso, C. (2004). Exploring the physiological state of continental Antarctic endolithic microorganisms by microscopy. FEMS Microbiol. Ecol, 50, 143–52.

[180] Gaylarde, CC & Morton, LHG. Biodeterioration of mineral materials. In: Bitton G editor. Encyclopedia of environmental microbiology. New York: Wiley; 2002; 516–527.

[181] Seaward, M. R. D. (1988). Lichen damage to ancient monuments: A case study. Lichenologist 10, 291–295.

[182] Bjelland, T. & Thorseth, I. H. (2002). Comparative studies of the lichen–rock interface of four lichens in Vingen, western Norway. Chemical Geology. 192, 81-98.

[183] Altieri, A. & Ricci, S. (1997). Calcium uptake in mosses and its role in stone biodeterioration. International Biodeterioration and Degradation. 40, 201-204.

[184] Siano, S. & Salimbeni, R. (2009). Advances in Laser Cleaning of Artwork and Objects of Historical Interest: The Optimized Pulse Duration Approach. Accounts of Chemical Research, 43, 739-750

[185] Ranalli, G. Alfano, G. Belli, C. Lustrato, G. Colombini, M. P, Bonaduce, I. Zanardini, E. Abbruscato, P. Cappitelli, F. & Sorlini, C. (2005). Biotechnology applied to cultural heritage: biorestoration of frescoes using viable bacterial cells and enzymes. J Appl Microbiol. 98, 73–83.

[186] Da Silva Nunes-Halldorson, V. & Duran, N. (2003). Bioluminescent bacteria: lux genes as environmental biosensors. Braz. J. Microbiol, 34, 91–96.

[187] Slaton, D. & Normandin, K.C. (2005). Masonry cleaning technologies. J. Architect. Conserv, 11, 7–31.

[188] Favaro, M. Tomasin, P. Ossola, F. & Vigato, P. A. (2008). A novel approach to consolidation of historical limestone: the calcium alkoxides. Appl. Organometal. Chem, 22, 698–704.

[189] González-Muñoz, M.T. Rodriguez-Navarro, C. Jimenez-Lopez, C. Rodriguez-Gallego, M. (2008). Method and product for protecting and reinforcing construction and ornamental materials, WO2008009771.

[190] Barbesi, C. Galizzi, A. Mastromei, G. Rossi, M. Tamburini, E. & Perito, B. (2007). Bacillus subtilis gene cluster involved in calcium carbonate biomineralization. Journal of Bacteriology, 189, 228–235.

[191] Salvadori, O. (2003). The control of biodeterioration. Coalition, 6, 16–20.

[192] Fernandes, P. (2006). Applied microbiology and biotechnology in the conservation of stone cultural heritage materials. Appl Microbiol Biotechnol, 73, 291–296.

[193] Cappitelli, F. Toniolo, L. Sansonetti, A. Gulotta, D. Ranalli, G. Zanardini, E. &Sorlini, C. (2007a). Advantages of using microbial technology over traditional chemical technology in removal of black crusts from stone surfaces of historical monuments. Appl Environ Microb, 73, 5671–5675.

[194] González-Muñoz, M. T. (2008) Bacterial biomineralization applied to the protection-consolidation of ornamental stone: current development and perspectives. Coalition, 15, 12–18.

[195] Tiano, P. Biagiotti, L. & Mastromei, G. (1999). Bacterial bio-mediated calcite precipitation for monumental stones conservation: methods of evaluation. J Microbiol Meth, 36, 139–145.

[196] Scheerer, S. Ortega-Morales, O. & Gaylarde, C. (2009). Microbial Deterioration of Stone Monuments-An Updated Overview. Advances in Applied Microbiology, 66, 97-139.

[197] Hansen, E. Doehne, E. Fidler, J. Larson, J. Martin, B. Matteini, M. Rodriguez-Navarro, C. Sebastián Pardo, E. Price, C. de Tagle, A. Teutonico, J. M. & Weiss, N. (2003). A review of selected inorganic consolidants and protective treatments for porous calcareous materials. Reviews in Conservation, 4, 13–25.

[198] De Muynck, W. Cox, K. Belie, N. D. & Verstraete, W. (2008). Bacterial carbonate precipitation as an alternative surface treatment for concrete. Construction Build. Mater, 22, 875–885.

[199] Castanier, S; Le Métayer-Levrel, G; Orial, G; Loubiere, JF; & Perthuisot, JP. Bacterial carbonatogenesis and applications to preservation and restoration of historic property. In: Ciferri, O; Tiano, P & Mastromei, G editors. Of microbes and art: the role of microbial communities in the degradation and protection of cultural heritage. New York: Phenum; 2000; 201-216.

[200] Polo, A. Cappitelli, F. Brusetti, L. Principi, P. Villa, F. Giacomucci, L. Ranalli, G. & Sorlini, C. (2010). Feasibility of removing surface deposits on stone using biological and chemical remediation methods. Microb Ecol, 60, 1–14.

[201] Capitelli, F. Fermo, P. Vechi, R. Piazzalunga, A. Valli, G. Zanardini, E. & Sorlini, C. (2009). Chemical–physical and microbiological measurements for indoor air qualita` assessment at the Ca'Granda Historical Archive, Milan (Italy). Water, Air, and Soil Pollution, 201, 109–120.

[202] Ranalli, G; Belli, C; Baracchini, C; Caponi, G; Pacini, P; Zanardini, E & Sorlini, C. Deterioration and bioremediation of frescoes: A case-study. In: Saiz-Jimenez, C. editor. Molecular Biology and Cultural Heritage. Lisse, The Netherlands: A.A. Balkema Publishers; 2003; 243-246.

[203] Jroundi, F. Fernández-Vivas, A. Rodriguez-Navarro, C. Bedmar, E. J. González-Muñoz, M. T. (2010). Bioconservation of deteriorated monumental calcarenite stone and identification of bacteria with carbonatogenic activity. Microb Ecol, 60, 39-54.

[204] Sprocati, A.R. Alisi, C. Tasso, F. Vedovato, E. Barbabietola, N. Cremisini, C. (2008). A microbial survey of the Etruscan Mercareccia Tomb (Italy): Contribution of microorganisms to deterioration and restoration. In 9th International Conference on NDT of Art, Jerusalem Israel.

[205] Wright, D.T. (1999). The role of sulphate-reducing bacteria and cyanobacteria in dolomite formation in distal ephemeral lakes of the Coorong region, South Australia. Sedimentary Geology, 126, 147–157.

[206] Warthmann, R. van Lith, Y. Vasconcelos, C. McKenzie, J. A. & Karpoff, A. M. (2000). Bacterially induced dolomite precipitation in anoxic culture experiments. Geol, 28, 1091–1094.

[207] Masaphy, S. Zabari, L. Pastrana, J. & Dultz, S. (2009). Role of fungal mycelium in the formation of carbonate concretions in growing media—An investigation by SEM and synchrotron-based X-ray tomographic microscopy. Geomicrobiology Journal, 26, 442–450.

[208] Zammit, G. Psaila, P. & Albertano, P. (2008). An investigation into biodeterioration caused by microbial communities colonizing artworks in three Maltese Palaeo-Christian Catacombs. In 9th International Conference on NDT of Art, Jerusalem Israel, 25-30 May 2008.

[209] Miller, A. Z. Laiz, L. Dionísio, A. Macedo, M. F. & Saiz-Jimenez, C. (2009). Growth of phototrophic biofilms from limestone monuments under laboratory conditions. Int Biodeterior Biodegradation, 63, 860-867.

[210] Gómez-Alarcón, G. Cilleros, B. Flores, M. & Lorenzo, J. (1995a) Microbial communities and alteration processes in monuments at Alcala de Henares, Spain. Science of The Total Environment, 167, 231-239.

[211] Tomaselli, L; Tiano, P & Lamenti, G. Occurrence and fluctuation in photosynthetic biocoenoses dwelling on stone monuments. In: Ciferri, O; Tiano, P & Mastromei, G editors. Of Microbes and Art – The Role of Microbial Communities in the Degradation and Protection of Cultural Heritage. New York: Kluwer; 2000b; 63-76.

[212] Caneva, G; Nugari, MP; Ricci, S & Salvadori, O. Pitting of marble Roman monuments and the related microflora. In: Delgado, J; Henriques, F & Telmo, F editors. Proceedings of the 7th International Congress on Deterioration and Conservation of Stone. Lisbon: Laborato´ rio Nacional de Engenharia Civil; 1992; 521-530.

[213] Gómez-Alarcón, G. Muñoz, M. L. Ariño, X & Ortega-Calvo, J. J. (1995b). Microbial communities in weathered sandstones: the case of Carrascosa del Campo church, Spain. Science of The Total Environment, 167, 249-254.

[214] Gómez-Alarcón, G. Muñoz, M. L. & Flores, M. (1995c). Excretion of organic acids by fungal strains isolated from decayed sandstone. International Biodeterioration and Degradation, 34, 169-180.

[215] de la Torre, M. A. Gomez-Alarcon, G. Melgarejo, P. & Saiz-Jimenez, C. (1991). Fungi in weathered sandstone from Salamanca cathedral, Spain. Science of The Total Environment, 107, 159-168.

[216] Ortega-Calvo, J. J. Ariño, X. Hernandez-Marine, M. & Saiz-Jimenez, C. (1995). Factors affecting the weathering and colonization of monuments by phototrophic microorganisms. Sci Total Environ, 167, 329–341.

[217] Berner, M. Wanner, G. & Lubitz, W. (1997). A comparative study of the fungal flora present in medieval wall paintings in the chapel of the castle Herberstein and in the parish church of St Georgen in Styria, Austria. Internation Biodeterioration and Degradation, 40, 53-61.

[218] Ciferri, O. (1999). Microbial degradation of paintings. Applied and Environmental Microbiology, 65, 879-885.

[219] Piñar, G. Ripka, K. Weber, J. & Sterflinger, K. (2009). The micro-biota of a sub-surface monument the medieval chapel of St. Virgil (Vienna, Austria). International Biodeterioration and Degradation, 63, 851-859.

[220] Ariño, X. & Saiz-Jimenez, C. (1997). Deterioration of the Elephant Tomb (Necropolis of Carmona, Seville, Spain). International Biodeterioration and Degradation, 40, 233-239.

[221] Bolívar Galiano F.C. & Sánchez Castillo, P. M. (1997). Biomineralization processes in the fountains of Alhambra, Granada, Spain. Int Biodet Biodegr, 40, 205–215.

[222] Ascaso, C; Garciá del Cura, MA & De Los Ríos, A. Microbial biofilms on carbonate rocks from a quarry and monuments in Novelda (Alicante, Spain). In: St Clair, LL &

Seaward, MRD editors. Biodeterioration of stone surfaces. Lichen and biofilms as weathering agents of rocks and cultural heritage. Dordrecht, Netherlands: Kluwer; 2004; 79-98.

[223] Leite Magalhães, S & Sequeira Braga, MA. Biological colonization features on a granite monument from Braga (NW, Portugal). In: V. Fassina, V editor. Proceedings of the 9th international congress on deterioration and conservation of stone. Amsterdam: Elsevier; 2000; 79-98.

[224] Sarró, M. I. García, A. M. Rivalta, V. M. Moreno, D. A. & Arroyo, I. (2006). Biodeterioration of the Lions Fountain at the Alhambra Palace, Granada (Spain). Building and Environment, 41, 1811-1820.

[225] Ortega-Calvo, J. J. Hernandez-Marine, M. & Saiz-Jimenez, C. (2003). Biodeterioration of building materials by cyanobacteria and algae. International Biodeterioration, 28, 165-185.

[226] Bellinzoni, A. M. Caneva, G. & Ricci, S. (2003). Ecological trends in travertine colonization by pioneer algae and plant communities. Int Biodeterior Biodegradation, 51, 203–210.

[227] Flores, M. Lorenzo, J. & Gómez-Alarcón, G. (1997). Algae and bacteria on historic monuments at Alcala de Henares, Spain. Int Biodeterior Biodegradation, 40, 241–246.

[228] Cappitelli, F. Principi, P. Pedrazzani, R. Toniolo, L. & Sorlini, C.http://www.sciencedirect.com/science?_ob=ArticleURL&_udi=B6V78-4P7FCSY-2&_user=529620&_coverDate=10%2F15%2F2007&_rdoc=1&_fmt=high&_orig=search&_sort=d&_docanchor=&view=c&_acct=C000026599&_version=1&_urlVersion=0&_userid=529620&md5=9da641e215130c9fd89bc935ef36486d - aff1 (2007b). Bacterial and fungal deterioration of the Milan Cathedral marble treated with protective synthetic resins. Science of The Total Environment, 385,172-181.

[229] Macedo, M. F. Miller, A. Z. Dionísio, A. & Saiz-Jimenez, C. (2009). Biodiversity of cyanobacteria and green algae on monuments in the Mediterranean Basin: an overview. Microbiology, 155, 3476–3490.

[230] Anagnostidis, K. Economou-Amilli, A. & Roussomoustakaki, M. (1983). Epilithic and chasmolithic microflora (Cyanophyta, Bacillariophyta) from marbles of the Parthenon (Acropolis-Athens, Greece). Nova Hedwigia, 38, 227–287.

[231] Kramar, S. & MirTič, B. (2008). Characterization of black crusts of Robba´s fountain statues, Ljubljana (Slovenia). RMZ – Materials and Geoenvironment, 55, 490–504.

[232] Saarela, M. Alakomi, H. L. Suihko, M. L. Maunuksela, L. Raaska, L & Mattila-Sandholm, T. (2004). Heterotrophic microorganisms in air and biofilm samples from Roman catacombs with a special emphasis on actinobacteria and fungi. Int. Biodeterior. Biodegrad. 1, 27–37.

[233] Miller, A. Z. Laiz, L. Gonzalez, J. M. Dionísio, A. Macedo, M. F. & Saiz-Jimenez, C. (2008). Reproducing stone monument photosynthetic-based colonization under laboratory conditions. Sci Total Environ, 405, 278–285.

[234] Laiz, L. Miller, A. Z. Jurado, V. Akatova, E. Sanchez-Moral, S. Gonzalez, J. M. Dionísio, A Macedo, M. F. & Saiz-Jimenez, C. (2009). Isolation of five Rubrobacter strains from biodeteriorated monuments. Naturwissenschaften, 96, 71-79.

[235] Altieri, A; Pietrini, AM; Ricci, S & Roccardi, A. The temples of the archaeological area of Paestum (Italy): a case study on biodeterioration. In: Fassina, V editor. Proceedings

gof the 9th International Congress on Deterioration and Conservation of Stone. Amsterdam: Elsevier; 2000; 433-443.

[236] Saiz-Jimenez, C. Hermosin, B. Ortega-Calvo, J. J. & Gómez- Alarcón, G. (1991). Applications of analytical pyrolysis to the study of stony cultural properties. J Anal Appl Pyrol, 20, 239–521.

[237] Dupont, J. Claire Jacquet, C. Dennetière, B. Lacoste, S. Bousta, F. Orial, G. Cruaud, C. Couloux, A. & Roquebert, M-F. (2007). Invasion of the French Paleolithic painted cave of Lascaux by members of the Fusarium solani species complex. Mycologia, 99, 526–533.

[238] Miller, AZ & Macedo, MF(2006). Mapping and characterization of a green biofilm inside of Vilar de Frades Church (Portugal). In: Fort, R; de Buergo, MA; Gomez-Heras, M & Vazquez-Calvo, C editors. Heritage, Weathering and Conservation. London: Taylor & Francis; 2006; 329-335.

Glossary

Bioreceptivity: The ability of a stone to be colonized by microorganisms is termed as its bioreceptivity. This property of stone is synergistically controlled and influenced by the numerous factors like mineral composition, nutrient availability, pH, salinity, surface texture, moisture content, porosity, permeability, climatic and micro-environmental conditions.

Biodeterioration: Biodeterioration can be defined as undesirable alterations brought about by synergistic interaction of microorganisms with its surrounding environment which leads to loss of in physical and mechanical strength, durability and aesthetic appearance of monumental stones and artworks.

Bioconservation: A total of all remedial and safeguarding measures from prevention of microbial colonization on stones to their protection, preservation and consolidation and future look after collectively termed as bioconservation (with respect to historic monuments and artworks)

Biofilm: Biofilm is a three-dimensional structure consisted of extracellular polymeric substances (EPS) secreted by cyanobacteria and related species attached to stone outer surface and provide favorable niche for growth of eubacteria and other microbial communities by regulating temperature and humidity.

Fungal Tunnels: According to 'rock-building fungi' hypothesis, fungal tunnels are distinct signatures of previous biogeochemical activity of fungi which are formed in the superficial patina, crust and secondarily formed mycogenic mineral matrix after their death or degradation of their hyphae.

Pitting and Etching: Pitting and etching are marks formed as a result of microtopographic changes brought about by the metabolically secreted protons, corrosive acids and metabolites and by the incorporation of mineral fragments into the thallus by fungi and lichens leading to dissolution of superficial mineral matrix of stones.

Salt Encrustation: A biogeochemical process of deposition of environmental pollutants such as petroleum particulate matters, dirt and fly ash (dry deposition), precipitated or dried salts and acids, of both environmental and biological origin, on the surface of monumental stones and artworks is called salt encrustation.

Ecological Succession: Ecological succession is a gradual, constant and more or less predictable natural process initiated by formation of new, unoccupied habitats by some form of life which bring about orderly changes in composition and structure of ecological communities over time.

Fungal Diagenesis: Fungal diagenesis can be describes as a complex process of biochemical and biomechanical alterations of carbonate substrates induced by fungi resulting in the formation of different secondary minerals.

In: Bioremediation
Editor: Alexander C. Mason

ISBN: 978-1-61122-730-7

Chapter 9

MICROBIAL SURFACE ACTIVE AGENTS AS AGROCHEMICALS

Anil Kumar Singh and Swaranjit Singh Cameotra*
Institute of Microbial Technology, Chandigarh, India

ABSTRACT

Surface active agent of microbial origin has engrossed great deal of expectation as it exhibit wide applications in different fields under environment friendly regime. Microbial surfactants have proven its metal in field of agriculture, bioprocessing, biomedical, cosmetics, environmental, food and petroleum industries. Some specific application like stabilization of nanoparticals has been greatly appreciated in recent past. Most of these applications have been highlighted in several reviews and chapters. However, discussion of biosurfactant in context to agriculture is rare. Microbial surfactant and some microbe producing surfactant have excellent potential to topple petro-chemical based agrochemicals. Microbial surfactant score much higher as compared to their synthetic counterpart as they are biodegradable, less toxic, have specific mode of action and can be produced from cheap substrates. Biosurfactant can serve as bio control agent, agriculture adjuvant and growth enhancers. The ability to beneficially influence phytoremediation and induce systematic resistance in plants has opened new frontiers for their application in agriculture sector. Surfactant producing microbes can be applied as biopesticide and most potent candidate for this is *Bacillus* sp. which is renowned for vast array of biologically active molecules potentially inhibitory for phyto-pathogen growth. Their spore-forming ability also makes these bacteria some of the best candidates for developing efficient microbial surfactant based bio-pesticide products from a technological point of view. On Commercial gains, use of Novel recombinant varieties of biosurfactant producing microorganisms, which can grow on a wide range of cheap substrates and produce biosurfactant at high yields, along with efficient downstream processing methods, can potentially bring the required breakthrough in the biosurfactant production.

* Corresponding author. Tel. No: 0172-6665233, Fax No.: 0172-2690632/585, E-mail address: ssc@imtech.res.in

1. INTRODUCTION

Human civilization has not only gifted man with comfort but also added new challenges for him to protect its environment from undesirable substances that play with the health of biosphere. Although most of the contaminants are anthropogenic, yet, sole efforts of human to confiscate them effectively from the environment have proven to be insufficient. Fortunately, nature's most primitive, smartest and smallest organisms have proven their ability to decontaminate almost all the materials, classified as contaminants, present in their vicinity by the phenomenon known as **Bioremediation**. Bioremediation utilize living forms particularly microorganism to reduce, eliminate, contain or transform to benign products contaminants present in soils, sediments, water or air. Although bioremediation has been in existence since the beginning of recorded history it has gained all inclusive acclamation in recent times only because of increase in environmental contamination and concern for the environment. Bioremediation, in present scenario, has become one of the most promising avenues of biotechnology which is being applied for environmental remediation and management.

But as it is well accomplished that "*the prevention is better than cure*", environmentalist all over the world unanimously recommend minimizing the use of detrimental compounds or replacing them, as far as possible, with environmental friendly natural compounds. In itinerary of this view in early 1990s US Environmental Protection Agency (http://www.epa.gov/) introduce the concept of "green chemistry" to promote chemicals technologies that reduce or eliminate the use or generation of hazardous substances in design, manufacture and use of chemicals products. Replacing petrochemical based product with eco-friendly greener product is presently revolving around products like Bio ethanol, Biopolymers, Bio-solvents and Bio-lubricants etc. With surfeit of substitution available for variety of synthetic products, replacing petro chemically derived surfactant with natural green bio based surfactant is often overlooked. Surfactants because of its unique property to reduce interfacial tension between two fluids, finds wide application. There extensive uses and low biodegradability cumulate to create environmental nuisance particularly for aquatic life forms.

The potential candidate that can replace chemical surfactants and considered to be most viable candidate for environmental friendly management practice are ***"microbial surface active compound"*** more commonly referred as **Biosurfactant**. Biosurfactants are amphiphilic compounds produced by living organisms, mostly by microbial cells, containing hydrophobic and hydrophilic moieties that confer the ability to accumulate between fluid phases, thus reducing surface and interfacial tension at the surface and interface respectively.

Surface active molecules producing microorganism are ubiquitous, and their presence has been reported from every imaginable environment worldwide, including soil, marine environments, wastewater, hydroponic systems, human lungs tissue and plant surface. Microorganism can produce biosurfactant on water soluble substrate or hydrophobic substrate or on both. Biosurfactant produced by microbe can be released extracellular into surrounding environment or may be localized on the surface. In later case biosurfactant remain associated with cell membrane and organism itself acts as surface active agent. Biosurfactant are secondary metabolites produced during late log phase and stationary phase. Importance of biosurfactant in microbial world can be reflected by high occurrence of surfactant producing microbe as well as diversity of produced surface active compound. Activity of biosurfactant primarily depends on its specific structure and production characteristic. Even subtle changes in structure can cause major alteration in biological activity. It plays essential roles for survival of biosurfactant producing microorganisms through nutrient transport or microbe-host interactions or by acting as biocide agents (D'aes et al., 2009).

Table 1. Application of microbial surfactants in various industries

Industry	Application	Role of Biosurfactant
Agriculture	Biocontrol	Facilitation of bio control mechanisms of microbes such as parasitism, antibiotics, competition, induced systemic resistance and hypovirulance
Bioprocessing	Downstreaming processing	Biocatalyst in aqueous two phase systems and micro emulsion; biotransformation; recovery of intracellular products; enhanced production of extracellular enzymes and fermentation products
Biomedical	Pharmaceuticals and therapetics	Antibacterial, antifungal antiviral agents; anti adhesive agents, immunomodulators molecules; vaccines and gene therapy.
Biological	Microbiological	Physiological behavior such as cell mobility, cell communication, nutrient accession , cell-cell competition, plant and animal pathogenesis
Cosmetics	Health and beauty products	Emulsifiers, foaming agents ,solubilizers, wetting agents, cleansers, antimicrobial agent
Environmental	Bioremediation; Soil flushing	Emulsification of hydrocarbon ; lowering of interfacial tension ; metal sesquestration dispersion ;foaming agent
Food	Emulsification and de-emulsifing	Emulsifier; solubilizer; demulsifier; suspension, wetting, defoaming, thickener, lubricating agent
Petroleum	Enhanced oil recovery ; de-emulsification	Improving oil drainage into well bore; stimulating release of oil entraped by capillaries; reduction of oil viscosity and oil pour point; lowering of interfacial tension.
Nanotechnlogy	Stabilizer	Stabilize s nanoparticles in the liquid phase

On industrial point of view, application of biosurfactant has engrossed great deal of attention as it finds wide application in diverse fields like agriculture, bioprocessing, biomedical, cosmetics, environmental, food and petroleum industries (table 1). Functional properties namely emulsification, wetting, foaming, cleansing, phase separation, surface active, reduction in viscosity of hydrophobic substance, functional at extreme temperatures, pH, salinity and biodegradability has been extensively exploited for its diverse application. Use of biosurfactant in recent past has been effectively extended to nanotechnology. Natural biosurfactant, rhamnolipid, was employed to stabilize silver nanoparticles in liquid phase. The silver nanoparticles were initially synthesized using $NaBH_4$ as reducing agent in rhamnolipid reverse micelles and was extracted from the micellar solution to be dispersed in heptanes. Silver nanoparticals were reported to be stable for at least two month without addition of passivator (Xie et al., 2006). Recently rhamnolipid has been demonstrated as an efficient capping agent for synthesis of uniform Zn-S nanoparticles (Narayanan et al., 2010). Similarly Surfactin, a lipopeptide biosurfactant, was used to stabilize the formation of gold nanoparticles (Reddy et al., 2009). Kasture et al., 2007 reported Sophorolipids, a class of Glycolipids as capping agent for Co nanoparticle which made nanoparticles stable and water re-dispersible. Sophorolipids have also been reported as effective reducing and capping agent in synthesis silver nanoparticles (Kasture et al., 2008).

With plethora of application known for biosurfactant, we in this chapter highlight application of microbial surfactants in agriculture sector. This chapter not only discusses feasibility of replacing synthetic surfactant with microbial one but also bring forward additional application of microbial surfactant that crowns them as environmental friendly agrochemicals. Here beneficial interactions between biosurfactant producing microbe and plant are also discussed as it can have noteworthy influence on agriculture. Furthermore, phytoremediation of organic pollutants and metals, facilitated by biosurfactant and biosurfactant producing bacteria is also included as it adds new dimension to plant mediate environment management.

2. Surfactant in Agriculture Sector

Surfactants are vital group of chemicals that knowingly or unknowing influence our daily life. They find application not only in household works but also in industries. The rapid development of chemical industries that began in 1960s led to development of wide variety of petroleum –based chemical surfactants. At present surfactant industry is dominated by several types: alkylbenzene sulfonates, alcohol ethoxylates and ethersulfates. These are the major components of laundry detergents, household and personal care products and account for over half of all the surfactants used. Interest in increasing performance in these areas has also led to research into mixed surfactant systems. The world surfactant market is somewhat about 8.600 kt volume with a value of about D 10.500 million, each about D 3.300 million for Western Europe, North America and Asia-Pacific (Fernandez et al., 2005).

Surfactant as agro-chemical has quite momentous place in agriculture sector. At present Chemicals surfactants are exploited in wide variety of agricultural applications which ranges from crop protecting agents to growth enhancers. Application of surfactant as agro chemical can be dated to dawn of agriculture. One of the earliest applications of surfactant as agrochemicals was in form of agricultural adjuvant. Adjuvants are substances added to pesticide formulation to enhance performance of pesticide. Adjuvants in case of herbicide enhance the penetration into plants by ensuring adequate spray coverage and keeping the herbicide in contact with plant tissues, or by increasing rates of foliar and/or stomatal penetration.

Noteworthy efforts for making surfactant momentous part of agro-chemical were taken by Tomlinson and Faithfull 1980. During their studies on the control of big vein of hydroponically grown lettuce, which is caused by a virus vectored by zoospores of *Olpidium brassicae*, they observed that disease control was achieved following the addition of a benzimidazole fungicide (Bavistin, BASF) to the recirculating nutrient solution. It was subsequently discovered that the inert ingredients in the fungicide formulation exhibited lytic activity against zoospores of *Olpidium*. It was suspected that the inert ingredients contained a synthetic surfactant. Further studies demonstrated the lytic effects of various surfactants, including anionic (Manoxol O/T, Marasperse CB, Sodium lauryl sulfate), nonionic (Agral, Ethylan CPX, Spreadite, Triton X100), and cationic (Cetrimide, Deciquam 222, Hyamine 1622) types, to zoospores of *O. brassicae*.

Surfactants are believed to influence rumen microbial communities by escalating enzyme release from microbial cells due to increased cell permeability and also by increasing enzyme substrate affinity. Chemical surfactant Sorbitan trioleate SOLFA-850 was found to increase digestion of rice straw in cow rumens by increasing protease, amylase, carboxymethylcellulase and xylanase activity (Lee et al, 2004). A similar effect was observed in a separate study when Tween -80 was added alongside the ionophore moensin to synergistically improve rumen fermentation condition (Wang et al, 2004). Thus opening path where chemical surfactant can replace antibiotic as growth enhancers in animal feed. However application of synthetic surfactant in agriculture is vilified by environment protecting groups as they are associated with water and farmland contamination. Surfactants that can degrade faster in natural condition and do not form any harmful intermediate during degradation are considered most appropriate for such applications. Biosurfactant fulfil above mentioned criteria and thus considered greener substitute for petrochemical based surfactant.

3. Advantage of Microbial Surfactants Over Synthetic Surfactants

Microbial surfactant has several advantages associated with it as compared to their synthetic petrochemical derived counterparts. Apart from being environmental friendly biosurfactant engrosses several advantages not associated with synthetic surfactants. Some of which are mention below:-

a) **Low toxicity:** Biosurfactant are generally considered as low or non-toxic products hence considered suitable for pharmaceutical, cosmetic and food uses. There are very few reports demonstrating toxicity level of biosurfactant. A report suggested that a synthetic anionic surfactant (Corexit) displayed an LC50 (concentration lethal to 50% of test species) against *Photobacterium phosphoreum* ten times lower than rhamnolipids, demonstrating the higher toxicity of the chemical-derived surfactant. It was also reported that biosurfactant showed higher EC50 (effective concentration to decrease 50% of test population) values than synthetic dispersants (Poremba, et al., 1991). A biosurfactant from *Pseudomonas aeruginosa* was compared with a synthetic surfactant (Marlon A-350) widely used in the industry, in terms of toxicity and mutagenic properties. Both assays indicated higher toxicity and mutagenic effect of the chemical-derived surfactant, whereas the biosurfactant was considered non-toxic and nonmutagenic (Flasz at al., 1998).

b) **Biodegradability, Biocompatible and Digestibility**: Rate of degradation of biosurfactant is relatively faster than synthetic surfactant thus making it environmentally acceptable. While comparing the toxicity of six biosurfactants, four synthetic surfactants and two commercial dispersants, it was found that biosurfactant degraded faster, except for a synthetic Sucrose-stearate that showed structure homology to Glycolipids and was degraded more rapidly than the biogenic Glycolipids (Poremba et al., 1991). Moreover their biocompatibility and easy digestibility make it better aspirant for cosmetic, food, agricultural and pharmaceuticals applications.
c) **Specificity**: Microbial surfactants are complex organic molecules possess specific functional group which imparts specificity in their action. This makes them good candidate particularly in detoxification of specific pollutants, de-emulsification of industrial emulsion, pharmaceuticals, agricultural, cosmetic and food applications.
d) **Production from cheap Substrates**: Inexpensive raw materials like industrial and agricultural wastes that are easily available in large quantity can be used in production of microbial surfactants thus reducing not only production cost but also pollution associated with these wastes.
e) **Effectiveness at extremes abiotic conditions:** Biosurfactant are even functional at extreme temperatures, pH and salinity thus making them very much appropriate for application at ruthless conditions like detoxification of pollutants at extreme abiotic environments.

4. Diversity of Natural Surfactants

Naturally occurring surfactants are diverse and widely distributed in living forms. Their occurrence has been reported in all five kingdoms of living beings i.e. Monera, Protista, Fungi, Plantae and Animalia (Table 2). Phospholipids an important part of plasma membrane is also an amphiphilic molecule.

Generally surfactants of microbial origin is considered as Biosurfactant and often considered as better source of natural surfactant as they require less space, time, as well as less energy for production. Moreover, structure of microbial surfactant can be modified by genetic and metabolic manipulation of microbe (Dogan et al., 2006; Tahzibi et al., 2004; Ohno, et al., 1995) or changing growth media or growth conditions (Morita et al., 2007; Wei et al., 2005; Hu and Ju, 2001).

Table 2. List of few surfactants present in five kingdoms of living beings

Kingdom		Organisms	Surfactant	References
Prokaryote	Monera	*Pseudomonas aeruginosa*	Rhamnolipids	
		Bacillus subtilis	Surfactin/iturin/fengycin	Muthusamy et al.,
		Acinetobacter calcoaceticus	Emulsan	2008
Eukaryote	Protista	*Akashiwo sanguinea*	Mycosporine-like amino acids	Jessup et al., 2009
	Fungi	*Ustilago maydis*	Mannosylerythritol lipids	Hewald et al.,2005
		Candida bombicola	Sophorose lipids	
	Plantae	*Saponaria officinalis L.*	Triterpenoid Saponins	
		Lonicera japonica Thunb.	Triterpenoid Saponin loniceroside C	Kwak et al.,2003
	Animalia	*Equus caballus*	Latherin	McDonald et al,2009

Biosurfactant has been classified variously depending on different features of microbial surfactants.

a) **Residual Charge:** Depending upon Residual charge surfactant has been divided into cationic, anionic, non -ionic and amphoteric. However, cationic microbial surfactant has not been reported till date as surfactant with positive residual charge is contemplated to be toxic for organisms.
b) **Source of Origin:** Biosurfactant can be of Animal, Plant or Microbial origin. More over microbial surfactant can be further be grouped into yeast, fungal or bacterial derived.
c) **Molecular Weight:** Depending upon molecular weight microbial surface active compounds has been divided in to low molecular mass, which efficiently lower surface and interfacial tension, and high-molecular-mass polymers, which are more effective as emulsion stabilizing agents. (Rosenberg and Ron, 1999)
d) **Chemical Composition:** Gives better view of diversity present in the world of biosurfactants, as presented in table 3 while chemical structure of some common biosurfactant is presented in figure 1.

Table 3. Major Biosurfactant class and microorganism involved in its production

Surfactant class	Composition	Microorganisms
Glycolipids	Carbohydrates linked to long chain aliphatic acids or hydroxyaliphatic acid by means of ether or ester group	
❑ Rhamnolipids		*Pseudomonas aeruginosa*
❑ Trehalose lipids		*Rhodococcus erithropolis*
❑ Sophrolipids		*Candida bombicola*
❑ Mannosylerythritol lipids		*C.apicola,C. antartica*
Lipopeptides /Lipoproteins	Lipid attached to polypeptide chain by	
❑ Surfactin/Iturin/Fengycin		*Bacillus subtilis*
❑ Viscocin		*P.fluorescens*
❑ lichenysin		*B.licheniformis*
❑ Serrawettin		*Serratia marcescens*
❑ **Phospholipids & Fatty acids and neutral lipids**	Component of cell membrane; Phosphate group attached to long fatty acid chain;	*Acinetobacter sp.* *Corynebacterium lepus* *Corynebacterium insidibasseosum*
Polymeric surfactants	Polysaccharide-protein complex;	
❑ Emulsan		*Acinetobacter calcoaceticus*
❑ Alasan		*A.radioresistens*
❑ Liposan		*C.lipolytica*
❑ Lipomanam		*C.tropicalis*
Particulate Biosurfactants	Extracellular membrane vesicles composed primarily of proteins,phospholipids and lipopolysaccharides; whole cell	
❑ Vesicles		*Acinetobacter calcoaceticus*
❑ Whole microbial Cells		*Cyanobacteria*

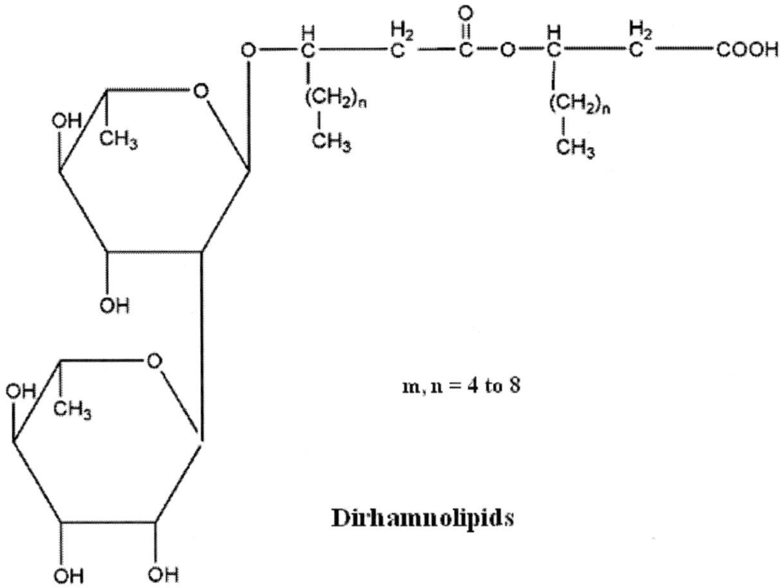

Dirhamnolipids

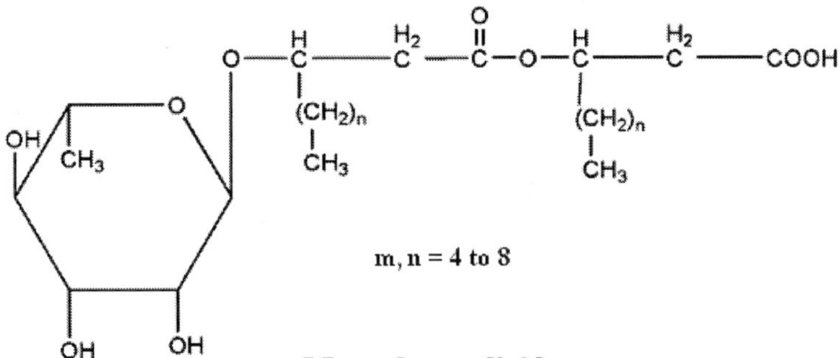

Monorhamnolipids

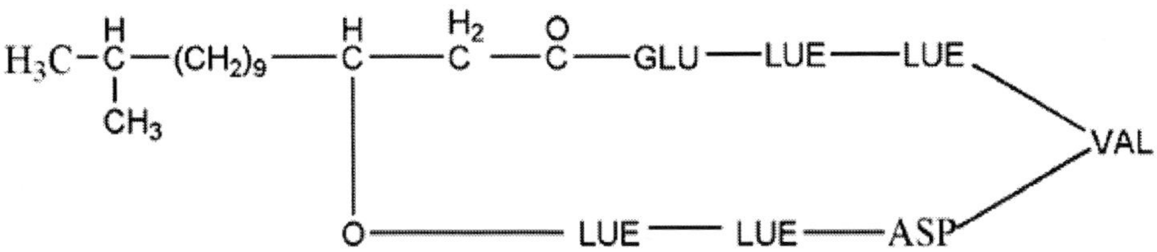

Surfactin

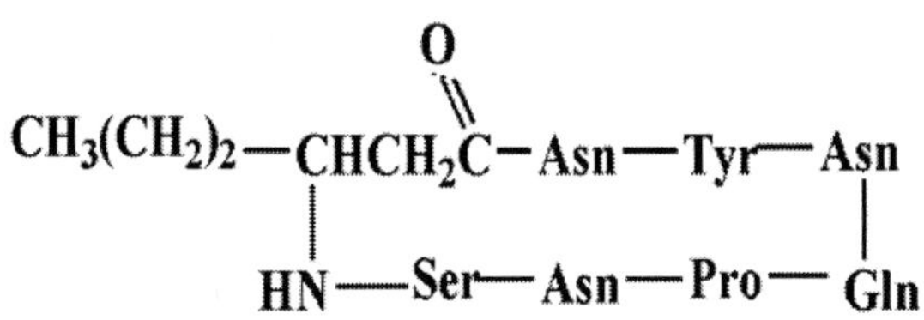

Iturin

Figure 1. (Continued).

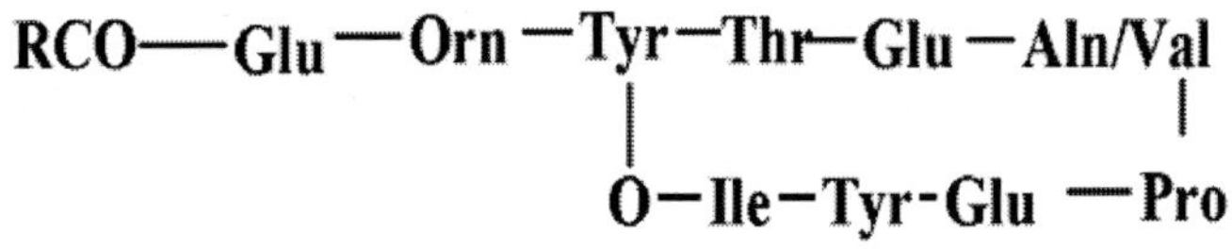

Fungycin

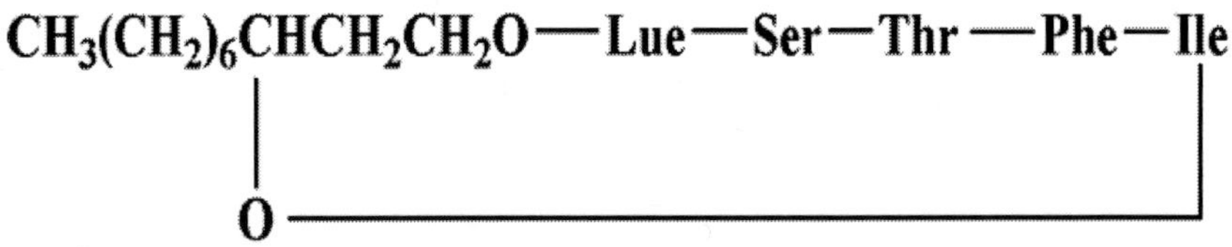

Serrawatin

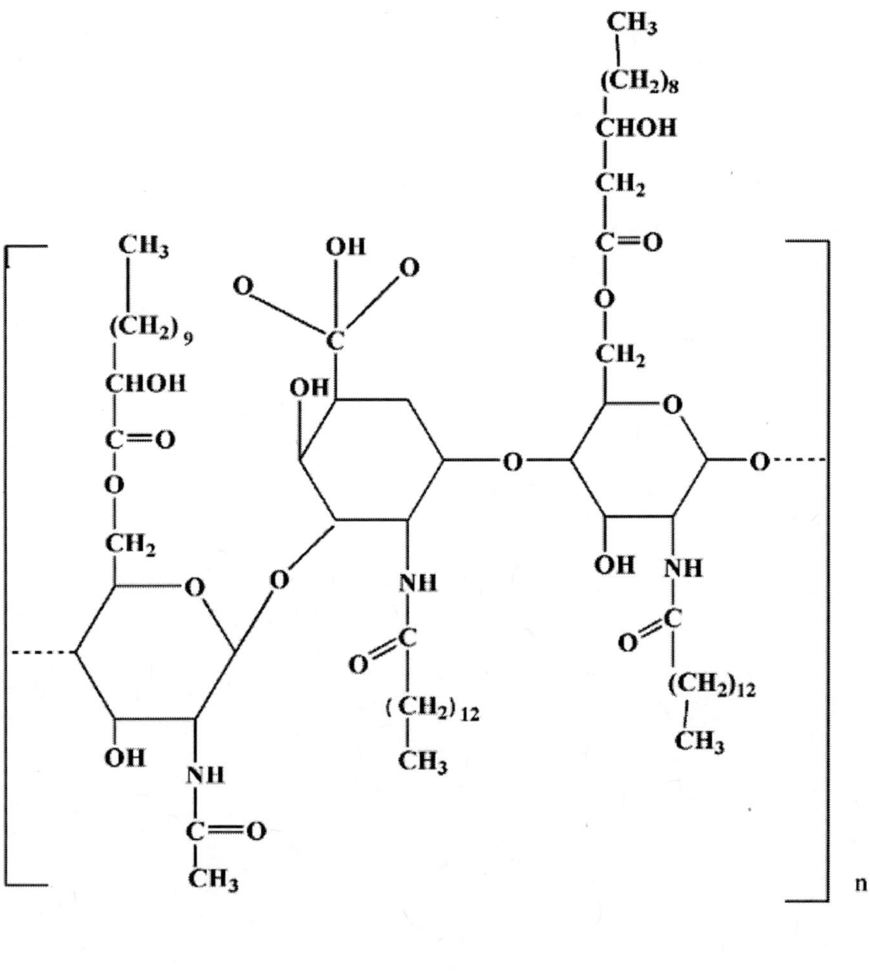

Emulsan

Figure 1. (Continued).

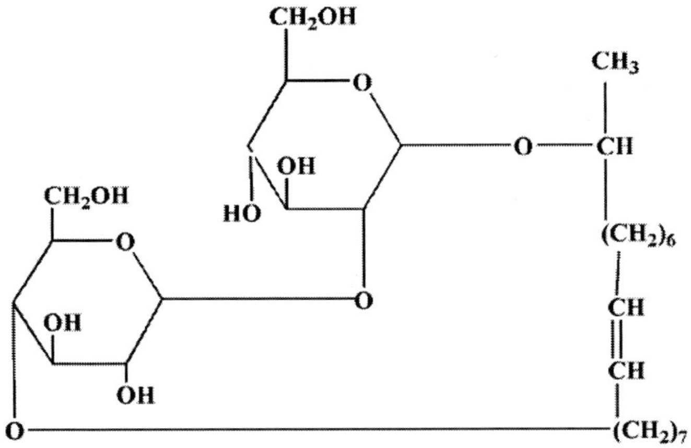

Lactonic Sophorolipids

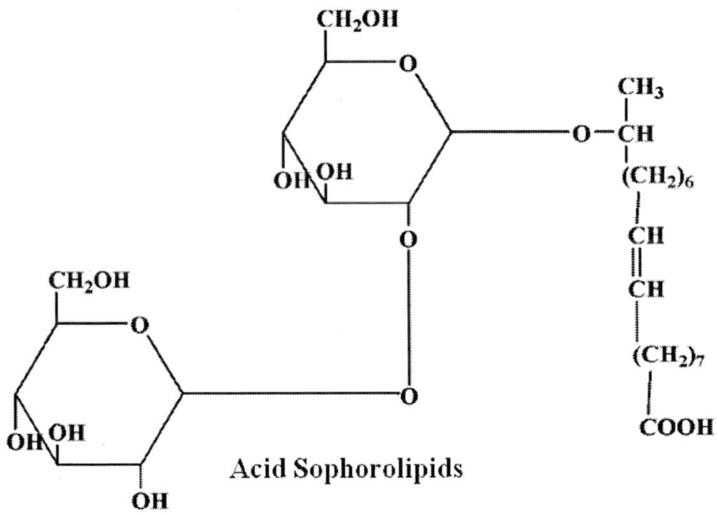

Acid Sophorolipids

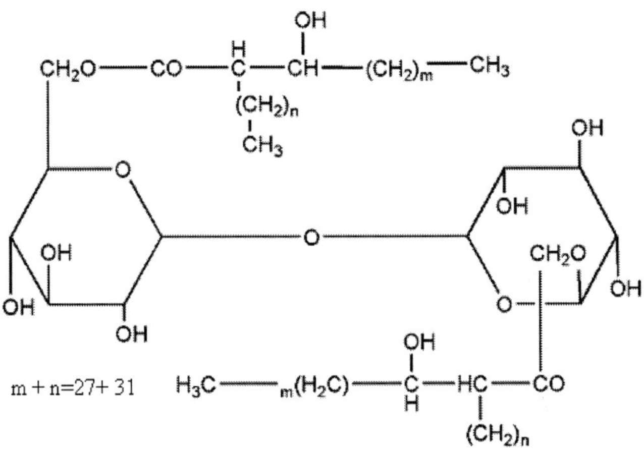

Trehalose dimycolates

Figure 1. (Continued).

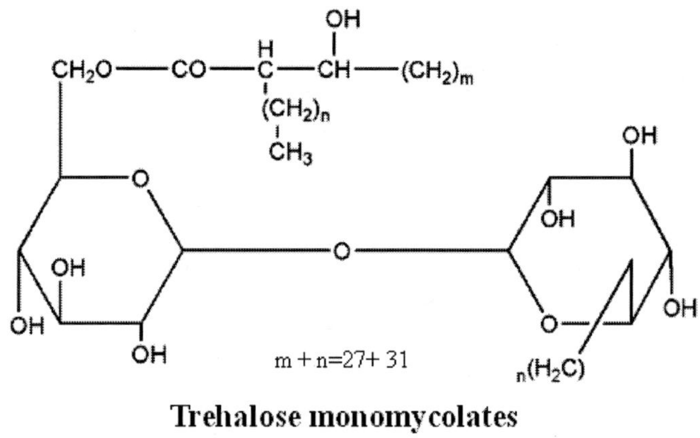

Figure 1. Chemical structure of some common microbial surfactants.

5. Application of Microbial Surfactants in Agriculture

Microbes and plants have coexisted for millions of years on earth. This long association have resulted in development of beneficial as well as detrimental relationship among them. Hot spots of such interaction such as rhizosphere and phylosphere are inhabited by biosurfactant producing microorganisms. Studies have showed that in general, numbers of biosurfactant producers on plant surface are quite high even though variable, ranging from 10^3 to 10^5 colony forming units per gram fresh roots for colonizing fluorescent *Pseudomonas*. Nielsen and colleagues (2002) found that about 60% of fluorescent *Pseudomonas* isolates from the sugar beet rhizosphere in a sandy soil produced biosurfactants, whereas only 6% of sugar beet rhizosphere isolates from a loamy soil were able to produce biosurfactants. Screening of the rhizosphere of different black pepper fields in Vietnam showed that approximately 1.3% of cultivable pseudomonades produced biosurfactant (Tran et al., 2008). Survey for biosurfactant producing fluorescent pseudomonades in a wheat rhizosphere from different agricultural soils yielded on average 1–5% of biosurfactant producers (De Souza et al., 2003). Presence of biosurfactant producing bacteria in vicinity of vigorous plant indicates its potential to serve agriculture in eco-friendly manner. Microbial surfactant has knack to play significant role in agriculture as *(a) Biological control agent (b) Agriculture Adjuvant (c) Growth enhancer (d) Inducer of systematic resistance in plant and (e) Enhancer of Phytoremediation.*

(5a.) Biological Control Agent

Biosurfactants can act as broad spectrum cytolytic agents, causing destruction of membranes and subsequent death of bacteria, fungi, oomycetes and viruses. Microbial surfactants due to their amphiphilic nature interact with biological membranes causing disruption of amphiphilic lipid bilayers and proving detrimental for target organism. Usually, biosurfactant are most toxic towards specific groups of organisms, showing less significant activity against others. These differences in activity can be attributed to structural properties of both surfactant and membrane, which will determine whether or not the surfactant can sufficiently interact with the target membrane to cause deleterious effects. From agriculture perspective biosurfactant mediated antibiosis can lead to beneficial effects and can serve as potent eco friendly biological control agent.

Biosurfactant as promising biological control agent was shown by Stanghellini and Miller (1997). They demonstrated that Rhamnolipids can disrupt zoospore membrane and cause lysis of zoospores of several oomycetes plant pathogens. Since then, several reports have described the possible use of rhamnolipids and rhamnolipid producing bacteria to suppress zoospore producing pathogens (Nielsen et al., 2006; De Jonghe et al., 2005; Kim et al., 2000). At present the powerful antimicrobial effects of various microbial surfactants for controlling variety of important plant pathogens has been well recognized (De Souza et al., 2003, Kondoh et al., 2001).

Control of Zoospores by Rhamnolipids is patented by US Pat. No. 5767,090, which claims application of rhamnolipid for controlling zoosporic plant pathogens (Stanghellini, et al., 1998) .Rhamnolipids biosurfactant, is currently being promoted in marked as Zonix™ Biofungicide by Jeneil Biosurfactant Company. Zonix™ Biofungicide act as contact biofungicide to prevent and control plant pathogenic fungi in horticulture, turf and agriculture settings. It can be applied at concentration between 85 and 125 ppm using conventional spray fog or drenching equipment or applied via chemigation or by hydroponics. It can also be used to treat seed transplants, bulbs and cutting (http://epa.gov/opp00001/ biopesticides/ ingredients/tech_docs/brad_110029.pdf) Application of microbial surfactant has been extended to bio pesticide in US patent: 20050266036, that relates to microbial biosurfactant and biosurfactant producing microbes to be used as agents for controlling pests. This patent reports rhamnolipid as effective bio control agent (Biopesticide) against incest such as ants, aphids, thrips, whiteflies, scales, lice, cockroaches, termites, houseflies, mosquitoes, etc; arachnids such as mites, spiders, ticks; nematodes; mollusks; amoeba; parasites; and algae (Awada et al., 2005). In one of the reports cyclic lipopeptides from *Bacillus subtilis* was found to have mosquito larvicidal and pupicidal activity (Geetha and Manonmani, 2007).

Biosurfactant has also been applied to control inevitable algal blooms. An algal bloom is a rapid increase or accumulation in the population of algae in an aquatic system. The significant adverse impacts of Harmful algal blooms (HABs) involve risks to human health, loss of natural or cultured seafood resources, impairment of tourism or recreational activities and damage to non-commercial marine resources and wildlife (GEOHAB, 2001), which seriously constrain the sustainable development of coastal areas. According to a conservative estimate, about 2000 cases of human poisoning are reported each year (Zingone and Enevoldsen, 2000) and the economic loss from a serious HAB event is usually over millions of US dollars (GEOHAB, 2001).

Most HAB organisms have the capability of phototaxis and usually move to the surface during the daytime. During this period it is anticipated that mitigation of HABs by surfactants will be efficient. Generally it is believed that surfactants are efficient for HAB mitigation in conditions of high cell density with little damage to other aquatic organisms. In this context surface active agent of microbial origin has also been investigated for potential to control HAB. *Sophorolipids* was found to be effective aganist *Cochlodinium polykrikoides* (Gymnodiniaceae) and *Alexandrium tamarense* (Dinophyceae) (Sun et al., 2004). Combination of sophorolipid and loess showed synergistic effects both on the removal efficiencies and on the mitigation cost. The combination was reported to be effective against

Cochlodinium polykrikoides (Gymnodiniaceae) and *Alexandrium tamarense* (Dinophyceae) (Sun et al., 2004). Use of sophorolipid–yellow clay mixture in feild experimrnt was found to be more efficient in mitigating the *Cochlodinium* bloom (95% removal efficiency after 30 min) than yellow clay alone (79% after 30 min) (Lee et al.,2008). Rhamnolipids have shown algicidal effects on harmful algal bloom caused by *Heterosigma akashiwo* (Raphidiace) and *Protocentrum denatum* (Dinophyceae) at concentration of 0.1 to 10.0 mg/L (Wang et al., 2005). Recently, a rhamnolipid biosurfactant produced by the bacterium *P. aeruginosa* was shown to have agicidal effect on marine as well as freshwater HAB species. HAB species included *Alexandrium minutum* (Dinophycaee), *Karenia brevis* (Dinophyceae), *Pseudonitzschia sp.* (Bacillariophyceae) in marine ecosystems and *Gonyostomum semen* (Raphidophyceae) and *Microcystis aeruginosa* (Cyanophyecae) in freshwater. Authors suggest that biosurfactant may be used to either control initial algal blooms without causing negative side effect to the ecosystem or to provoke lethal effects when necessary (Gustafsson et al ,2009)

(5b.) Growth Enhancers

Growth enhancers hasten growth of crop or animals, thus providing farmer a fast and better economic return for his labour and investment. Burden on agriculture sector to sustain ever increasing population has made crop or animal growth enhancer an auspicious part of agrochemical. Growth enhancers generally act by making nutrient/ micronutrient easily available to target organism or by increase enzymatic activity so as hasten digestion in case of animals.

Generally antibiotics are used in animal feed as growth promoters. The use of antibiotics as feed additives has been a hallmark of modern animal husbandry, but this widespread practice is not without criticism. Alternative, chemical surfactants are being explored as replacements for antibiotics as growth enhancers in animal feed due to their impacts on rumen microbial communities. Lee et al. (2004) found that the surfactant sorbitan trioleate SOLFA-850 increased digestion of rice straw in cow rumens by increasing protease, amylase, carboxymethylcellulase and xylanase activity. In addition, increased NH_3-N *in ruminal fluid* was detected and the total number of viable bacteria and anaerobic fungi increased with the surfactant-supplemented diet. A similar effect was observed in a separate study when Tween-80 was added alongside the ionophore monensin to synergistically improve rumen fermentation conditions (Wang et al., 2004) Surfactants are believed to increase enzyme-substrate affinity and to increase enzyme release from microbial cells due to increased permeability. However application of microbial surface active compounds with similar prospect has not been reported to best of our knowledge.

Natural surfactant phospholipids which are major component of cell membrane have been shown to play important role in animal nutrition. Phospholipids possess both hydrophobic and hydrophilic part as result has surfactant like properties. They are natural constituents of bile and also occur in feeds. These phospholipids help in fat digestion by their emulsifying properties and in nutrient absorption by the formation of micelle structures in the gastrointestinal tract. Xing et al., 2004, showed that growth performance of nursery pigs may be improved by lysolecithin supplementation to diets with added lard and by encapsulation of lard through spray drying.

Use of Phospholipids based stabilizer for processed animal feeds and feed ingredients is being marketed by Kemin AgriFoods North America under trade name of LYSOFORTE. LYSOFORTE contains lecithins enriched in lysophospholipids and it assists in fat absorption by facilitating the emulsification of fat in feed. With better fat absorption, it can increase energy utilization and assist in absorption of other fat soluble nutrients and vitamins (http://www.kemin.com/agrifoods/north-america/production-efficiency/lysoforte).

Several biosurfactant producing bacteria like *Pseudomonas fluorescens*, *Acinetobacter calcoaceticus Burkholderia cenocepacia, Bacillus subtilis, Serratia marcescens* are known to enhance plant growth via direct or indirect mechanisms and are included under plant growth promoting rhizobacteria (PGPR). The mechanisms by which PGPR promote plant growth are not fully understood, but are thought to include: **(i)** the ability to produce or change the concentration of the plant hormones indoleacetic acid (IAA; Sachdev et al., 2009), gibberellic acid (Kang et. al., 2009: Joo et. al., 2009), cytokinins (Tien et al., 1979), and ethylene (Glick et al., 1998); **(ii)** asymbiotic N_2 fixation(Kennedy et al., 1997); **(iii)** antagonism against phytopathogenic microorganisms by production of siderophores (Díaz de Villegas et al., 2002), chitinases (Kamil et al., 2007), antibiotics (Arguelles-Arias et al., 2009) and cyanide (Flaishman et al., 1996) and **(iv)** solubilization of mineral phosphates and other nutrients (Kuklinsky-Sobral et al., 2004; Caballero-Mellado et al., 2007).

Studies revealing exact role of biosurfactant in plant growth promotion is not well elucidated. Studies suggest that biosurfactant can enhances plant growth by activities like antagonism against phytopathogenic microorganisms, solubilising nutrients, inducing systematic plant resistance and altering gene expression. One of recent studies shows increased zinc uptake by *Brassica napus* var. Pinnacle roots in ^{65}Zn-spiked ice-cold solutions in presence of rhamnolipid, as compared with $ZnSO_4$ alone. Rhamnolipid appeared to facilitate Zn absorption via a non-metabolically mediated pathway. Synchrotron μ-X-ray Fluorescence (XRF) and μ-X-ray Absorption Spectroscopy (XAS) studies showed that Zn was present in roots as Zn-phytate-like compounds when roots were treated with Zn-free solutions, $ZnSO_4$ or Zn-EDTA. With rhamnolipid application, Zn was primarily found in roots as the Zn-rhamnolipid complex. In case of durum wheat (*Triticum durum* L. cv. Balcali-2000) and bread wheat (*Triticum aestivum* L. cv. BDME-10) rhamnolipid application increased dry matter production and Zn concentrations in shoot (Stacey et al., 2008). Aforesaid ability of Rhamnolipids elucidates its possible application in fertilizer to enhance absorption of micronutrient metals by plant roots.

(5c.) Agriculture Adjuvants

Adjuvant in general is a broad term that describes any non pesticide material added to pesticide spray mixture or pesticide product to enhance pesticide activity. Usually pesticides such as insecticides, insect repellents, fungicides, bactericides, herbicides as well as plant growth regulators are formulated in such manner that it can easily combined with water and applied in the form of a liquid or a semi-solid dispersion. As water molecule has a high surface tension and this leads to formation of water droplets when sprayed on a waxy plant surfaces. This results in pesticides not being able to spread evenly and leaving an irregular coverage consequently higher use of pesticides. A spray drop must be able to wet the foliage and spread out or cover an area of the plant for the pesticide to perform its pest control function. The adjuvant acts here by reducing the surface tension of the water on the surface of the spray drop and by reducing the interfacial tension between the spray drop and surface of the leaf. Good adjutants increase the formulation's penetration ability through the cuticular wax, cell walls and/or stomatal openings.

Adjuvants can be divided into two general types: **(1) Formulation adjuvants** and **(2) Spray adjuvants.** The first type consists of adjuvants, which are part of the formulation, while the second types of adjuvants are added along with the formulated product to the water in the tank of the spray equipment before application on the fields (Krogh et al., 2003). Agriculture adjuvants comprises of a large and heterogeneous group of substances. However major part is contributed by Petrochemical based surfactants, especially Non-ionic and Anionic surfactants (Underwood, 1992; Foy, 1996). Low biodegradability or persistence nature of petrochemical based surfactants results in creation of nuisance for environment. The long-term fates of most adjuvants in soils and elsewhere in the environment are largely unknown because of the lack of long-term monitoring data. The major environmental problem related to these compounds is the ability to enhance the mobility of other pollutants. Reports suggest that it can interact with proteins and membrane phospholipids by modifying their structure and permeability at cellular levels. Moreover, these surfactants are known to have detrimental effect on earthworms, insects, fish, crustaceans, phytoplankton, and soil micro-organisms in dose dependent manner (Krogh et al., 2003).

Concern for environment has increased interest in developing renewable/sustainable agriculture adjuvant that is benign to the human health and environment. Surfactant from *Bacillus subtilis* RB14-C was found to have synergistic effect with chemical pesticide, flutoianil against *Rhizoctonia solani,* casual agent of damping off plant. Integration of RB14-C culture broth with flutolanil reduced the amount of flutolanil used to one-fourth of that of the single use of flutolanil, with the same efficacy of reducing disease occurrence. This adjuvant like activity was primarly due to Iturin A as surfactin production reduced in presence of pesticide and also Iturin- A has suppressive action against plant pathogens (Kondoh et al., 2001).

(5d.) Induction of Systematic Resistance

Interaction of plant and pathogen can result in either infection or development of resistance. In later case set of localized responses is elicited in and around the infected host cells that renders the cells less susceptible to subsequent infection in distal tissue. This is long lasting phenomenon and termed as Systematic Acquired Resistance (SAR) (Durrant and Dong, 2004). Some non pathogenic microbe particularly rhizobacteria can also reduce susceptibility to diseases through the stimulation of inducible plant defence mechanisms, a phenomenon termed as Induced Systematic Resistance(ISR) (Van Loon and Bakker, 2006). Analogues to pathogen induced SAR, protection afforded through ISR is non- specific regarding the nature of the infection agent (Ongena and Jacques, 2008). ISR is effective against root- infecting pathogens as well as reduces damage from pathogens that are exclusively active on foliage, flowers or fruits. ISR is considered suitable for controlling disease caused by fungi, bacteria and viruses. It can also be applied successfully as a strategy in management of nematode and insect pest in several crops. Even though the ISR do not confer total protection against pathogen infection, ISR-based bio control strategies are considered to be promising because the phenomenon is long-lasting and not conducive to the development of pathogen resistance.

Some of biosurfactant producing bacteria like *Bacillus* and *Pseudomonas* species are well established for their ability to Induce Systematic Resistant in plants. Biosurfactants particularly belonging to class lipopeptides has been reported to be involved in ISR. Fengycin and surfactin-biosurfactant belonging to class lipopeptides can interact with plant cells as a bacterial determinant for turning on an immune response through the stimulation of the induced systemic resistance phenomenon (Ongena and Jacques, 2008). In bean (*phaseolus vulgaris*) pure fengycins and surfactin provide a significant ISR mediated protective effect against plant pathogen *Botrytis cinerea*, which was very much similar to one induced by living cells of *Bacillus subtilis* strain S499. Moreover, experiments conducted on bean and tomato plants by this group showed that overexpression of both surfactin and fengycin biosynthetic genes in the naturally poor producer *Bacillus subtilis* strain 168 was associated with a significant increase in the potential of the derivatives to induce resistance. In tomato cells, key enzymes of the lipoxygenase pathway appeared to be activated in resistant plants following induction by lipopeptide overproducers (Ongena et al., 2007).

During their *in vitro* study Jourdan et al, 2009 showed that surfactin was main inducer of early defence responses in tobacco cell culture while the effect of iturin and fengycin was very limited. Addition of surfactin but not fengycin or iturin in the micromolar range to tobacco cell suspensions induced defence-related early events such as extracellular medium alkalinization coupled with ion fluxes and reactive oxygen species production. Surfactin also stimulated the defence enzymes phenylalanine ammonia lyase and lipoxygenase and modified the pattern of phenolics produced by the elicited cells without causing any significant cell death or any marked toxicity. The most potent elicitors were surfactin homologues with longer fatty acid chains (C14 or C15), while surfactins with C12 and C13 fatty acid chains were markedly less active. Ramification of fatty acid chains also improved eliciting capacity of surfactin, while linearization or methylation of the peptide part led to a decrease in its activity. As these structural traits contribute to surfactant character and to membrane-destabilising potential of this molecules

Jourdan et al, 2009 postulated that surfactin does not act by binding to a specific receptor in the plant cell, but forms irreversible pore in a way sufficient to induce disturbance or transient channelling in the plasma membrane that can, in turn, activate a biochemical cascade of molecular events leading to defensive responses.

Biosurfactant of *Pseudomonas* origin like Rhamnolipids and Cyclic lipopeptides are also able to induce resistant in plants. Rhamnolipids was found to trigger defence responses in grapevine cell cultures and confer resistance to *Botrytis cinerea* on *in vitro* grapevine plantlets. Both mono- and dirhamnolipids was reported as effective elicitors, whereas L-rhamnose was not capable of inducing any defence-related effects in grapevine cells (Varnier et al., 2009). These findings by Varnier and colleagues point to a role for the amphiphilic nature of the molecule and thus the importance of plant membrane interactions for the elicitor activity of rhamnolipids. When tomato (*Lycopersicon esculentum*) roots were treated with purified Massetolide A (lipopeptide of *P. fluorescens* SS101 origin) the leaves showed enhanced resistance to infection by *Phytophthora infestans* (Tran et al., 2007). Exact mechanism involved in Massetolide A induced resistant is not known but it is plausible that Massetolide A may influence plant plasma membrane to Induce Systematic Resistance.

Collectively, these studies show that biosurfactant from some non pathogenic *Bacillus* and *Pseudomonas* species constitute a novel class of microbial-associated molecular patters that can be perceived by plant to mount defence responses. Exploiting these potential beneficial effects would require more insight in this phenomenon. It is not clear whether induced resistance events are intentional ecological processes that benefit bacteria and/or plants. Ratio and amounts of the different biosurfactant produced by *Bacillus* and *Pseudomonas* species are influenced by the environment and more specifically by the host plant. Accordingly, it is possible that plants somehow decide which biosurfactant particularly belonging to class lipopeptide is most interesting to its own benefit and stimulate its production (Ongena and Jacques, 2008;

Jourdan et al., 2009). On the other hand, bacteria could also take advantage of a more resistant plant, since it will not be easily damaged or killed, therefore constituting a more reliable host. To our best knowledge, a potential action of these surface active compounds of microbial origins as plant defence inducing compounds has not been investigated until now and would provide new insights in ISR mechanisms of these bacteria and associated plant responses.

(5e.) Biosurfactant in Phytoremediation

Phytoremediation uses living plants for *in situ* risk reduction and/or removal of pollutants from polluted soil, water, sediments and air. Phytoremediation is often considered as potentially clean, energy efficient, inexpensive and aesthically pleasing method of remediating sites with low to moderate levels of contamination. Phytoremediation involves array of processes by green plant for remediating contaminated sites:

i. **Phytoextraction** involves uptake and concentration of substances from the environment into the plant biomass.
ii. **Phytovolatilization** is removal of substances from soil or water with release into the air, sometimes as a result of phytotransformation to more volatile and/ or less polluting substances. **(iii) Phytostabilitzation** is reducing the mobility of substances in the environment, for example by limiting the leaching substances from the soil.
iii. **Rhizofiltration** uses plant roots to remove toxic substances from polluted waters as a result pollutants remain absorbed in or adsorbed to the roots of the plant.
iv. **Phytortransformation** is chemical modification of environmental substances as a direct result of plant metabolism, often resulting in their inactivation, degradation (phytodegradation) or immobilization (phytostabilization).

Phytoremediation has shown its potential to detoxify both metal as well as organic pollutants (Shoji et al., 2008; Vervaeke et al., 2003; Ma et al., 2001). For phytoremediation of heavy metals and radio nuclides various plant species such as tobacco, sunflower, mustard, maize and sand rocket are being used based on ability to absorb or hyper accumulate the metal pollutants (Hooda 2007; Meagher, 2000). Organic pollutants like biphenyls, polycyclic aromatic compounds, nitro aromatic or liner halogenated hydrocarbons has been mineralized using poplar trees, willow, alfalafa and different grass varieties. Since natural distribution of plant that can efficiently remediate polluted site is limited, use of plant beneficial microbes and generation of transgenic popular for the purpose is seen as present and future prospect in this field.

As mention earlier that microbe in vicinity of plant can impart beneficial or harmful effect on plant thus influencing their various physiological functions. Realizing the importance of plant –microbe association, particularly bacteria, in phytoremediation has been studied in past decade with great inquisitiveness. The efficacy of phytoremediation in association with plant growth promoting /degradative bacteria has been demonstrated under laboratory conditions and it is expected that during the next five to ten years its use will become widespread (Glick, 2010).

Importance of biosurfactant and biosurfactant producing bacteria in phytoremediation has also been accredited only in recent past. Biosurfactant producing bacteria, apart from supporting growth of plant under hostile condition prevalent at contaminate site, facilitate bioavailability of relatively insoluble compound like polycyclic aromatic hydrocarbon, halogenated hydrocarbons and metals, thus facilitating the decontamination. In a hydroponic experiment application of rhamnolipid enhanced the uptake of polycyclic aromatic hydrocarbons (PAHs) by ryegrass roots. With increasing concentration of rhamnolipids, the PAH content in ryegrass roots initially increased and then decreased, while the PAH content in ryegrass shoots did not change (Zhu and Zhang, 2008). Zhu and Zang reported that increase of permeability of ryegrass root cells with the increase of rhamnolipid concentration may have lead to the initial enhancement of PAH content in ryegrass roots and the decrease of PAH adsorption onto the root surface with further increase of rhamnolipids may have led to the decrease of PAH content in ryegrass roots. (Zhu and Zhang, 2008). Introduction of rhamnolipid along with arbuscular mycorrhizal fungi and aromatic hydrocarbon degrading bacteria into phytoremediation system of alfalfa increased phytoremediation by 251.83% (Zhang et al., 2010). Moreover aforesaid multi process system removed most of the high molecular weight polycyclic aromatic hydrocarbon likes fluoranthene, pyrene and benzo[α] pyrene. Application of biosurfcatnt increased bioavailability of polycyclic aromatic hydrocarbon while arbuscular mycorrhizal fungal accelerated plant growth and increase PAHs removal from contaminated sites. In another report, biosurfactant-producing *Bacillus* sp. J119 isolated from heavy metal contaminated soils was found to reduce cadmium toxicity in tomato, maize and rape. Cd treatment in presence of live bacterial inoculation of *Bacillus* sp. J119 did not significantly decreased growth of tomato, maize and rape plants. In the Cd-added soil, above-ground biomass and root dry weights of tomatoes were increased by 24 and 59 %, respectively, in live bacterial inoculation compared to dead bacterial inoculation control (Sheng et al, 2008).

Typically, transgenic plants exhibit new or improved phenotypes that are engineered by the overexpression and/or introduction of genes from other organisms such as bacteria or mammals (Van, 2008). This approach has been in applied effectively in plant like *Brassica juncea* for phytoremediation of heavy metals from soil *Helianthus anus* (Dushenkov et al., 1995) and *Chenopodium amaranticolor* (Eapen et al., 2003) for rhizofiltration of uranium.

Inspiration that surfactant secreting plant may accelerate breakdown of environmental pollutant by increasing its bioavailability, led researcher to created transgenic plant harbouring microbial surfactant gene. *Pseudomonas aeruginosa* biosurfactant gene *rhlA* and *rhlB* which encodes for rhamnosyltransferase and participate in rhamnolipid biosynthesis were introduced in *Nicotiana tobacum* and *Arabidopsis thaliana.* Transgenic plants were able to secrete glycolipids in the root zone. Plant also had good concentration of rhamnolipid in the tissues as determined by anthrone regent at λ=290 nm. Biodegradation study of oil hydrocarbon by transgenic plant containing either the *rhlA* or *rhlB* genes produced 25 %-34 % reduction in oil over 45 days. Presence of both *rhlA* and *rhlB* genes in plant reduced the oil by 18-30% over the same period showing efficient detoxification ability. However, non-transformed plants produced no oil reduction. (www.isb.vt.edu/articles/apr0403.htm). Genetically modified tobacco plant containing rhamnolipid genes *(rhlA* and *rhlB*) showed high aluminium tolerance of 200 mM in ionite soil (Brichkova et al., 2007). This protective effect of rhamnolipid may be attributed to its high chelating ability with aluminium resulting in formation of nontoxic aluminium complex. Apart from phytoremediation benefits of aforesaid transgenic plant, according to Kartel's team that has done remarkable work in this area, by combining rhamnolipid production ability with lectin expression may reduce pesticide application in agriculture (www.isb.vt.edu/articles/apr0403.htm).

6. Biosurfactant Producer as Bio-Pesticide

Using biosurfactant producing microorganism as bio-pesticide is a good alternative for replacing synthetic / petrochemical based counterparts. With steadily increase global demand for bio-pesticide, future for it seems to be bright and worth mentioning. Biosurfactant / biosurfactant producers as bio- control agents can be used solely or may be applied together with chemicals, either in rotation to reduce the possible development of pathogen resistance or in an integrated pest management strategy with the goal of minimizing the use of synthetic pesticides. Most potent candidate for this purpose is *Bacillus* sp. Their ability to produce vast array of biologically active molecules potentially inhibitory for phytopathogen along with spore forming ability makes these bacteria one of the best candidates for developing efficient bio-pesticide products from a technological point of view.

7. Overcoming Economic Barriers

Successful commercialization of every biotechnology product depends largely on its bioprocess economics. Like any other biotechnological product, biosurfactant too suffer ruthlessness of high production economy thus, hampering its extensive commercialization and uses in different felid. Undisputedly diverse application of biosurfactant makes it '*Jack of all trade'* but economical hurdle compel it to be *'master of none.'*

Nowadays it is recognized that a broader strategic approach is needed to improve the overall process performance and to alleviate capacity bottle-necks in downstream purification of microbial surfactants. To make biosurfactant commercially viable, the use of cheaper substrates, optimizing production conditions coupled with novel and efficient multistep downstream processing methods and the use of hyper producing microbial strains (recombinant/mutant) has engrossed the attention of several investigators. Improvement in production technology of biosurfactant has already enabled 10-20 fold increase in productivity although further significant improvements are required (Muthuswamy et al., 2008).

A variety of cheap raw materials, including industrial waste, urban waste and agro-based raw materials have been reported to support biosurfactant production (Seghal et al., 2010; Kiran et al.,2010; Yilmaz et al., 2009 ;Raza, et al., 2007). Use of waste material for production not only reduces production cost but also lessen the environmental nuisances attached to waste disposable.

For most of biotechnological products the downstream processing costs account for ~ 60% of the total production costs. Thus economical and efficient downstream steps can make significant saving to production process. Reports suggest that use of modern approaches like foam fractionation, membrane technique or adsorption process in biosurfactant downstream process can be beneficial (Sen and Swaminathan, 2005; Chen et al., 2006; Monteiro et al., 2009; Qiao and Shao, 2009; Sarachat et al., 2010). Novel recombinant varieties of biosurfactant producing microorganisms, which can grow on a wide range of cheap substrates and produce biosurfactant at high yields along with efficient downstream processing methods, can potentially bring the required breakthrough in the biosurfactant production.

Although, a large number of biosurfactant producers have been reported in the literature, biosurfactant research, particularly related to production enhancement and economics has been confined mostly to a few genera of microorganisms such as *Bacillus* sp., *Pseudomonas* sp. and *Candida* sp. A large group of biosurfactant producers belonging to the generas *Rhodococus, Gordonia, Torulopsis* and *Acinetobacter* have not been exploited properly for the economical production of biosurfactants. Recently, novel rhamnolipids producing bacteria (*Acinetobacter calcoaceticus, Enterobacter asburiae, Enterobacter hormaechei, Pantoea stewartii, Pseudomonas aeruginosa*) were isolated thus keeping up the expectation of finding more from different genus (Rooney et al., 2009) that can be profitably exploited for biosurfactant production.

8. Conclusion

Surfactant undisputedly is an important class of chemical as it finds diverse applications in different fields which range from household to large industries. Surfactant also find momentous place among cosmic army of agrochemicals. Nuisance associated with their extensive application and fondness for eco friendly greener product in recent past has led to augmentation of liking for microbial surfactants. Applications of natural surfactant particularly microbial surfactant in agriculture sector scores much higher than synthetic counterpart. Microbial surfactant has shown potentials to serve agriculture not only as 'surface active agent' but also as biological control agent, induction of systematic resistance, growth enhancer, agriculture adjuvant and agents for enhanced phytoremediation. Agricultural applications of microbial surfactants have been demonstrated in laboratory scale or in green house condition while field successes are still to be appreciated. Some of the uses like induction of systematic resistance in plant are boon of recent researches, thus in infancy stage. More work is needed in this context to have better understanding. More over the knowledge of interaction between biosurfactant producers and plant is important for insight into pathogenesis as well as into plant growth promoting and disease suppressing effects.

Application of biosurfactant producers in agricultures fields need intense investigation for production of other plant advantageous compound like Phytohormones , extracellular proteins in relation to biosurfactant production under natural conditions. Integration of microbial surfactant gene with plant system by Kartel's team has paved new path, which has potentials to bring new credits in biosurfactant research. Work at present mainly revolves around microbial surfactants belonging to few genera namely *Bacillus* sp., *Pseudomonas* sp. and *Candida* sp. Much remains to be revealed about physicochemical effects of lesser studied Biosurfactants and their biological importance for plant. Laurels of any biotechnology product is accomplished only when it reaches commoner and for this future research of biosurfactant should be more focused on economization of the biosurfactant production process, particularly through development of novel recombinant hyper producers along with cost-effective bioprocesses.

REFERENCES

Arguelles-Arias, A., Ongena, M., Halimi, B., Lara, Y., Brans, A., Joris, B., Fickers, P. (2009). *Bacillus amyloliquefaciens* GA1 as a source of potent antibiotics and other secondary metabolites for biocontrol of plant pathogens. *Microb. Cell. Fact.* 8: 63, 1475-2859.

Awada, S. M., Spendlove, R. S., Awada, M.(2005). *Microbial Biosurfactants as agents for controlling pests* 2005 Application Number: 11/141669 Publication Date:12/01/2005

Bais, H.P., Fall, R., Vivanco, J.M. (2004). Biocontrol of *Bacillus subtilis* against infection of *Arabidopsis* roots by *Pseudomonas syringae* is facilitated by biofilm formation and surfactin production. *Plant Physiol.* 134(1), 307-319.

Boyette, C.D., Walker, H.L., Abbas, H.K. (2002). Biological control of kudzu (Pueraria lobata); with an isolate of Myrothecium verrucaria. *Biocontrol Sci. Technol.* 12, 75–82.

Brichkova, G.G., Shishlova, A.M., Maneshina, T.V., Kartel', N.A. (2007). The tolerance of tobacco genetically modified plants to aluminium. *Tsitol Genet.* 41(3), 23-28.

Caballero-Mellado, J., Onofre-Lemus, J., Estrada-de Los Santos, P., Martínez-Aguilar L. (2007). The tomato rhizosphere, an environment rich in nitrogen-fixing *Burkholderia* species with capabilities of interest for agriculture and bioremediation. *Appl. Environ. Microbiol.* 73(16), 5308-5319.

Chen, C., Baker, S.C., Darton, R.C. (2006). Continuous production of biosurfactant with foam fractionation. *J. Chem. Technol. Biotechnol.* 81, 1915–1922.

D'aes, J., Maeyer, K., Pauwelyn, E., Höfte, M. (2010). Biosurfactants in plant–*Pseudomonas* interactions and their importance to biocontrol. *Environmental Microbiology Reports* (2010) 2(3), 359–372.

De Souza, J.T., De Boer, M., De Waard, P., Van Beek, T.A., Raaijmakers, J.M. (2003). Biochemical, genetic, and zoosporicidal properties of cyclic lipopeptide surfactants produced by *Pseudomonas fluorescens*. *Appl. Environ. Microbiol.* 69(12), 7161-7172.

De Jonghe, K., De Dobbelaere, I., Sarrazyn, R., Hofte, M. (2005). Control of *Phytophthora cryptogea* in the hydroponic forcing of witloof chicory with the rhamnolipid based biosurfactant formulation PRO1. *Plant Pathol.* 54, 219–226.

Díaz de Villegas, M.E., Villa, P., Frías, A. (2002). Evaluation of the siderophores production by Pseudomonas aeruginosa PSS. Rev. Latinoam. *Microbiol.* 44(3-4), 112-117.

Dogan, I., Pagilla, K. R., Webster, D. A., Stark, B. C. (2006). Expression of Vitreoscilla haemoglobin in *Gordonia amarae* enhances biosurfactant production. *J. Ind. Microbiol. Biotechnol.* 33, 693–700.

Dushenkov, V., Kumar, P.B.A.N., Motto, H., Ruskin, I. (1995). Rhizofiltration: The use of plants to remove heavy metals from aqueous streams. *Environ. Sci. Technol.*, 29, 1239-1245.

Durrant, W.E., Dong, X. (2004). Systemic acquired resistance. *Annu. Rev. Phytopathol.* 42, 185–209.

Eapen, S., Suseelan K.N., Tivarekar S., Kotwal S.A., Mitra R. (2003). Potential for rhizofiltration of uranium using hairy root cultures of *Brassica juncea* and *Chenopodium amaranticolor*. *Environ. Res.,* 91(2), 127-133.

Fernandez, A. M., Held, U., Willing, A., Breuer, W. H. (2005). New green surfactants for emulsion polymerization. *Progress in Organic Coatings* 53, 246–255.

Flaishman, M.A., Eyal, Z., Zilberstein, A., Voisard, C., Hass, D. (1996). Suppression of *Septoria tritici* blotch and leaf rust of wheat by recombinant cyanide-producing strains of *Pseudomonas putida. Mol. Plant-Microbe Interact.* 9, 642-645.

Flasz, A., Rocha, C. A., Mosquera, B. and Sajo, C. (1998). A comparative study of the toxicity of a synthetic surfactant and one produced by *Pseudomonas aeruginosa* ATCC 55925. *Med. Sci. Res.* 26, 181–185.

Foy, C.L., 1996. Adjuvants—current technology and trends. In: Foy, C.L., Pritchard, D.W. (Eds.), *Pesticide Formulation and Adjuvant Technology*. CRC Press, Boca Raton, FL, pp. 323–352.

GEOHAB 2001. Global ecology and oceanography of harmful algal blooms, In: Glibert, P., Pitcher, G. (Eds.), *Science Plan. SCOR and IOC*, Baltimore and Paris, 87 pp.

Geetha, I., Manonmani, A.M. (2008) .Mosquito pupicidal toxin production by *Bacillus subtilis* subsp. *Subtilis. Biological Control* 44, 242–247.

Glick R. B., (2010). Using soil bacteria to facilitate phytoremediation. *Biotechnology Advances* 28, 367–374.

Glick, B.R., Penrose, D.M., Li, J. (1998). A model for the lowering of plant ethylene concentrations by plant growth-promoting bacteria. *J. Theor. Biol.* 190, 63-68.

Gronwald, J.W., Plaisance, K.L., Ide, D.A., Wyse. D.L. (2002). Assessment of *Pseudomonas syringae* pv. tagetis as a biocontrol agent for Canada thistle. *Weed Sci.*50, 397–404.

Gustafsson, S., Hultberg, M., Figueroa, R.I., Rengefors K. (2009). On the control of HAB species using low biosurfactant concentrations. *Harmful Algae* 8, 857–863.

Hooda V. (2007). Phytoremediation of toxic metals from soil and waste water. *J. Environ. Biol.* Apr; 28(2 Suppl),367-376.

Hu, Y., Ju, L. (2001).Sophorolipid production from different lipid precursors observed with LC-MS. *Enzyme and Microbial Technology* 29, 593–601.

Jourdan, E., Henry, G., Duby, F., Dommes, J., Barthelemy, J.P., Thonart, P., and Ongena, M. (2009). Insights into the defense-related events occurring in plant cells following perception of surfactin-type lipopeptide from *Bacillus subtilis*. *Mol Plant Microbe Interact* 22, 456–468.

Joo, G.J., Kang, S.,M., Hamayun, M., Kim, S.K., Na, C.I., Shin, D.H., Lee, I.J. (2009). *Burkholderia* sp. KCTC 11096BP as a newly isolated gibberellin producing bacterium. *J. Microbiol.* 47(2), 167-171.

Kang, S.M., Joo, G.J., Hamayun, M., Shin, D.H., Kim, H.Y., Hong, J.K., Lee, I.J. (2009). Gibberellin production and phosphate solubilization by newly isolated strain of *Acinetobacter calcoaceticus* and its effect on plant growth. *Biotechnol. Lett.* 31(2), 277-281.

Kamil, Z., Rizk, M, Saleh, M., Moustafa, S. (2007). Isolation and Identification of Rhizosphere Soil Chitinolytic Bacteria and their Potential in Antifungal Biocontrol *Global Journal of Molecular Sciences* 2 (2): 57-66.

Kasture, M., Singh, S., Patel, P., Joy, P.A., Prabhune, A.A., Ramana, C.V., Prasad, B.L. (2007) .Multiutility sophorolipids as nanoparticle capping agents: synthesis of stable and water dispersible Co nanoparticles. *Langmuir*. 23(23), 11409-11412.

Kasture, M., Patel, P., Prabhune, A., Ramana, C.V., Kulkarni, A.A., Prasad,B.L. (2008).Synthesis of silver nanoparticles by sophorolipids: Effect of temperature and sophorolipid structure on the size of particles. *J. Chem. Sci.*, Vol. 120, No. 6, 515–520.

Kennedy, I.R., Pereg-Gerk, L.L., Wood, C., Deaker, R., Gilchrist, K., Katupitiya, S. (1997). Biological nitrogen fixation in non-leguminous field crops: Facilitating the evolution of an effective association between *Azospirillum* and wheat. *Plant Soil* 194, 65-79.

Kim, B.S., Lee, J.Y., and Hwang, B.K. (2000). *In vivo* control and *in vitro* antifungal activity of rhamnolipid B, a glycolipid antibiotic, against *Phytophthora capsici* and *Colletotrichum orbiculare*. *Pest Manag. Sci.* 56, 1029–1035.

Kiran, G.S., Thomas, T.A., Selvin, J. (2010).Production of a new glycolipid biosurfactant from marine *Nocardiopsis lucentensis* MSA04 in solid-state cultivation. *Colloids Surf B Biointerfaces*. 78(1), 8-16.

Kondoh, M., Hirai, M., Shoda, M. (2001). Integrated biological and chemical control of damping-off caused by Rhizoctonia solani using Bacillus subtilis RB14-C and flutolanil. *J. Biosci. Bioeng*.91(2),173-177.

Krogh, K.A., Halling-Sørensen, B., Mogensen, B.B., Vejrup, K.V. (2003). Environmental properties and effects of nonionic surfactant adjuvants in pesticides: a review. *Chemosphere*. 50(7), 871-901.

Kuklinsky-Sobral, J., Araújo, W.L., Mendes. R., Geraldi. I.O., Pizzirani-Kleiner, A.A., Azevedo, J.L. (2004). Isolation and characterization of soybean-associated bacteria and their potential for plant growth promotion. *Environ. Microbiol.* 6(12), 1244-1251.

Lee, S.S., Kim, H.S., Moon, Y.H., Choi, N.J., Ha. J.K. (2004). The effects of a non-ionic surfactant on the fermentation characteristics, microbial growth, enzyme activity and digestibility in the rumen of cows. *Anim. Feed Sci. Technol.* 115, 37–50.

Lee Y.J., Choi J.K., Kim E.K., Youn S.H., Yang E.J. (2008). Field experiments on mitigation of harmful algal blooms using a Sophorolipid—Yellow clay mixture and effects on marine plankton . *Harmful Algae* 7(2) , 154-162.

Ma, L.Q., Komar, K.M., Tu, C., Zhang, W., Cai, Y., Kennelley, E.D. (2001). A fern that hyperaccumulates arsenic. *Nature* 409(6820), 579.

Meagher, R.B. (2000). Phytoremediation of toxic elemental and organic pollutants. *Curr. Opin. Plant Biol.* 3(2), 153-162.

Morita, T., Konishi, M., Fukuoka, T., Imura, T., Kitamoto, D. (2007). Physiological differences in the formation of the glycolipid biosurfactants, mannosylerythritol lipids,between Pseudozyma antarctica and Pseudozyma aphidis. *Appl. Microbiol. Biotechnol.* 74(2), 307-315.

Monteiro, A.S., Coutinho, J.O., Júnior, A.C., Rosa, C.A., Siqueira, E.P., Santos, V.L. (2009). Characterization of new biosurfactant produced by *Trichosporon montevideense* CLOA 72 isolated from dairy industry effluents. *J. Basic Microbiol.* 49(6), 553-563.

Muthusamy, K., Gopalakrishnan, S., Kochupappy, T. R., Sivachidambaram, P. (2008). Biosurfactants: Properties, commercial, production and application. *Current Science* 94, 736-747.

Narayanan, J., Ramji, R., Sahu, H., Gautam, P. (2010). Synthesis, stabilisation and characterisation of rhamnolipid-capped ZnS nanoparticles in aqueous medium. *IET Nanobiotechnol.* 4(2), 29-34.

Nielsen, C.J., Ferrin, D.M., and Stanghellini, M.E. (2006). Efficacy of biosurfactants in the management of *Phytophthora capsici* on pepper in recirculating hydroponic systems. *Can. J. Plant Pathol.* 28, 450–460.

Ohno, A., Ano, T., Shoda, M. (1995). Production of a lipopeptide antibiotic, surfactin, by recombinant Bacillus subtilis in solid-state fermentation. *Biotechnol. Bioeng.* 47, 209–214.

Ongena, M., Jacques, P. (2008). Bacillus lipopeptides: versatile weapons for plant disease biocontrol. *Trends Microbiol.*16 (3), 115-125.

Ongena, M., Jourdan, E., Adam, A., Paquot, M., Brans, A., Joris, B., Arpigny, J.L., Thonart P.(2007). Surfactin and fengycin lipopeptides of *Bacillus subtilis* as elicitors of induced systemic resistance in plants. *Environ. Microbiol.* 9(4), 1084-1090.

Perneel, M., D'Hondt, L., De Maeyer, K., Adiobo, A., Rabaey, K., and Hofte, M. (2008). Phenazines and biosurfactants interact in the biological control of soil-borne diseasescaused by *Pythium* spp. *Environ Microbiol* 10, 778–788.

Poremba, K., Gunkel, W., Lang, S., Wagner, F. (1991).Toxicity testing of synthetic and biogenic surfactants on marine microorganisms. *Environ. Toxicol. Water Qual.* 6, 157–163.

Qiao, N., Shao, Z., (2009) .Isolation and characterization of a novel biosurfactant produced by hydrocarbon-degrading bacterium *Alcanivorax dieselolei* B-5. *J. Appl. Microbiol. Aug* 6.1207-1216.

Raza, Z.A., Rehman, A., Khan, M.S., Khalid, Z.M. (2007). Improved production of biosurfactant by a *Pseudomonas aeruginosa* mutant using vegetable oil refinery wastes. *Biodegradation* 18(1), 115-121.

Reddy, A.S., Chen, C.Y., Chen, C.C., Jean, J.S., Fan, C.W., Chen, H.R., Wang, J.C., Nimje, V.R. (2009). Synthesis of gold nanoparticles via an environmentally benign route using a biosurfactant. *J. Nanosci. Nanotechnol.* 9(11), 6693-6699.

Rooney, A.P., Price, N.P., Ray K.J., Kuo TM., (2009). Isolation and characterization of rhamnolipid-producing bacterial strains from a biodiesel facility. *FEMS Microbiol. Lett.* 295(1), 82-87.

Rosenberg, E., and Ron, E.Z. (1999). High- and low-molecular-mass microbial surfactants. *Appl Microbiol Biotechnol 52*, 154-162.

Sarachat, T., Pornsunthorntawee, O., Chavadej, S., Rujiravanit. (2010). Purification and concentration of a rhamnolipid biosurfactant produced by *Pseudomonas aeruginosa* SP4 using foam fractionation. *Bioresource Technology* 101 (1), 324-330.

Sachdev, D.P., Chaudhari, H.G., Kasture, V.M., Dhavale, D.D., Chopade, B.A. (2009). Isolation and characterization of indole acetic acid (IAA) producing Klebsiella pneumonia strains from rhizosphere of wheat (Triticum aestivum) and their effect on plant growth. *Indian J Exp Biol.* 47(12), 993-1000.

Seghal, K. G., Anto, T. T., Selvin, J., Sabarathnam, B., Lipton, A.P. (2010). Optimization and characterization of a new lipopeptide biosurfactant produced by marine *Brevibacterium aureum* MSA13 in solid state culture. *Bioresour Technol.* 101(7), 2389-2396.

Sen, R. and Swaminathan, T. (2005). Characterization of concentration and purification parameters and operating conditions for the small-scale recovery of surfactin. *Process Biochem.* 40, 2953–2958.

Sheng, X., He, L., Wang, Q., Ye, H., Jiang, C. (2008). Effects of inoculation of biosurfactant-producing *Bacillus* sp. J119 on plant growth and cadmium uptake in a cadmium-amended soil. *J. Hazard Mater.* 155(1-2), 17-22.

Shoji, R., Yajima, R., Yano, Y. (2008). Arsenic speciation for the phytoremediation by the Chinese brake fern, Pteris vittata. *J. Environ. Sci.* (China). 20(12), 1463-1468.

Stacey, S.P., McLaughlin, M.J., Cakmak, I., Hettiarachchi, G.M., Scheckel, K.G., Karkkainen, M. (2008). Root uptake of lipophilic zinc-rhamnolipid complexes. *J. Agric. Food Chem.* 56(6), 2112-2117.

Stanghellini, M. E. ,Miller, R. M., Rasmussen, S.L., Kim, D. H.,Zhang, Y.(1998). *Microbially produced rhamnolipids (biosurfactants) for the control of plant pathogenic zoosporic fungi US Patent* 5,767,090 , June 1998.

Stanghellini, M.E., Miller, R.M. (1997) Biosurfactants. Their identity and potential efficacy in the biological control of zoosporic plant pathogens. *Plant Dis.* 81, 4–12.

Sun, X., Han, K.N., Choi, J.K., Kim, E.K. (2004). Screening of surfactants for harmful algal blooms mitigation. *Marine Pollution Bulletin* 48, 937–945.

Sun, X.X., Choi, J.K., Kim, E.K. (2004). A preliminary study on the mechanism of harmful algal bloom mitigation by use of sophorolipid treatment. *J. Exp. Mar. Biol. Ecol.* 304, 35–49.

Sun, X.X., Lee, Y.J., Choi, J.K., Kim, E.K. (2004). Synergistic effect of sophorolipid and loess combination in harmful algal blooms mitigation. *Mar. Pollut. Bull.* 48(9-10), 863-872.

Tahzibi, A., Kamal, F., Assadi, M. M., (2004). Improved Production of Rhamnolipids by a *Pseudomonas aeruginosa* Mutant. *Iranian Biomedical Journal* 8 (1), 25-31.

Thrane, C., Nielsen, T.H., Nielsen, M.N., Sorensen, J., Olsson, S. (2000) Viscosinamide-producing *Pseudomonas fluorescens* DR54 exerts a biocontrol effect on *Pythium ultimum* in sugar beet rhizosphere. *FEMS Microbiol Ecol* 33, 139–146.

Thrane, C., Nielsen, M.N., Sorensen, J.,Olsson, S.(2001). *Pseudomonas fluorescens* DR54 reduces sclerotia formation, biomass development, and disease incidence of *Rhizoctonia solani* causing damping-off in sugar beet. *Microb. Ecol.* 42, 438–445.

Tien, T.M., Gaskins, M.H., Hubbell, D.H. (1979). Plant Growth Substances Produced by *Azospirillum brasilense* and Their Effect on the Growth of Pearl Millet (*Pennisetum americanum* L.). *Appl. Environ. Microbiol.*37 (5), 1016-1024.

Tomlinson, J. A., Faithfull, E. M. (1980). Effects of fungicides and surfactants on the zoospores of *Olpidium brassicae*. *Ann. Appl. Biol.* 93, 13-19.

Tran, H., Ficke, A., Asiimwe, T., Hofte, M., and Raaijmakers, J.M. (2007). Role of the cyclic lipopeptide massetolide A in biological control of *Phytophthora infestans* and in colonization of tomato plants by *Pseudomonas fluorescens*. *New Phytol* 175, 731–742.

Tran, H., Kruijt, M., Raaijmakers, J.M. (2008). Diversity and activity of biosurfactant-producing Pseudomonas in the rhizosphere of black pepper in Vietnam. *J. Appl. Microbiol.* 104(3), 839-851.

Underwood, A.K., (1992). Concerns within the pesticide industry relating to spray adjuvants. In: Foy, C.L. (Ed.), *Adjuvants for Agrochemicals*. CRC Press, Boca Raton, FL, pp. 489–501.

Varnier, A.L., Sanchez, L., Vatsa, P., Boudesocque, L., Garcia-Brugger, A., Rabenoelina, F., *et al.* (2009). Bacterial rhamnolipids are novel MAMPs conferring resistance to *Botrytis cinerea* in grapevine. *Plant Cell Environ.* 32, 178–193.

Van Loon, L.C. and Bakker, P.A.H.M. (2006) Induced systemic resistance as a mechanism of disease suppression by rhizobacteria. In PGPR: *Biocontrol and Biofertilization* (Siddiqui, Z.A., ed.), pp. 39–66, Springer

Van Aken B. (2008).Transgenic plants for phytoremediation: helping nature to clean up environmental pollution. *Trends Biotechnol.* 26(5), 225-227.

Vervaeke, P., Luyssaert, S., Mertens, J., Meers, E., Tack, F., M., Lust, N. (2003). Phytoremediation prospects of willow stands on contaminated sediment: a field trial. *Environ Pollut.* 126(2), 275-282.

Wang, X., Xiulin Wang, Liangyu Gonga, Shengkang Liang a, Xiurong Hana, Chenjian Zhua, Yanbin Li .(2005). Algicidal activity of rhamnolipid biosurfactants produced by *Pseudomonas aeruginosa. Harmful Algae,* 4, 433–443.

Wang, Y., Alexander, T.W., McAllister, T.A. (2004). In vitro effects of Monensin and Tween 80 on ruminal fermentation of barley grain: barley silage-based diets for beef cattle. *Animal Feed Sci. Technol*, 16,197–209.

Wei, Y., Choub, C., Chang, J. (2005). Rhamnolipid production by indigenous *Pseudomonas aeruginosa* J4 originating from petrochemical wastewater. *Biochemical Engineering Journal* 27, 146–154.

Xie, Y., Ye, R., Liu, H. (2004). Synthesis of silver nanoparticles in reverse micelles stabilized by natural biosurfactant Colloids and Surfaces A: *Physicochem. Eng. Aspects* 279 (2006) 175–178.

Xing, J. J., Heugten, V., Li, D. F., Touchette, K. J., Coalson J. A., Odgaard R. L., Odle J.(2004). Effects of emulsification, fat encapsulation, and pelleting on weanling pig performance and nutrient digestibility. *J. Anim. Sci.* 82, 2601-2609.

Yilmaz, F., Ergene, A., Yalçin, E., Tan, S. (2009). Production and characterization of biosurfactants produced by microorganisms isolated from milk factory wastewaters. *Environ Technol.* 30(13), 1397-1404.

Zhang, J., Yin, R., Lin, X., Liu, W., Chen, R., Li, X. (2010). Interactive Effect of biosurfcant and Microoragnism to enhance Phytoremediation for removal of aged aromatic Hydrocarbon from contaminated Soil. *Journal of Health Science* 56(3), 257-266.

Zhu, L., Zhang, M. (2008). Effect of rhamnolipids on the uptake of PAHs by ryegrass. *Environ. Pollut.* 156, 46-52.

Zingone, A., Enevoldsen, H.O. (2000). The diversity of harmful algal blooms: a challenge for science and management. *Ocean Coast. Manage.* 43, 725–748.

In: Bioremediation
Editor: Alexander C. Mason

ISBN: 978-1-61122-730-7

Chapter 10

BIOREMEDIATION OF INDUSTRIAL WASTEWATERS CONTAMINATED WITH HAZARDOUS COMPOUNDS

Vicky Dritsa and Fotis Rigas
National Technical University of Athens,
School of Chemical Engineering, Athens, Greece

ABSTRACT

Intensive industrialization generates by-products or residues, which are usually hazardous (toxic and/or persistent) to the environment and to humans. The most hazardous pollutants include heavy metals, polyaromatic hydrocarbons (PAHs), polychlorinated biphenyls (PCBs), pesticides and synthetic dyes. For more than a century industrialized countries contaminate systematically soil, water and air with such materials. Yet, it was only after the 1980s that remedial actions were undertaken and the laisser-faire approach was set aside. With regard to soil and water contamination, the general term "site remediation" is usually used. This term is not limited to the cleaning of contaminants in either a solid matrix (mainly soil) or in liquid media (groundwater or surface water), because in some cases there is no clean-up at all, and in fact the contaminants are contained or immobilized in place. Therefore, the term "remediation" should be seen as having broader meaning than "cleanup".

Biological treatment technologies have played an important role in the degradation of hazardous environmental pollutants. Bioremediation is a diverse clean-up technology, cost-effective and environmentally friendly compared to the traditional commonly used physico-chemical techniques. The term "soft technology" is sometimes used for biotechnology to distinguish from the most drastic and environmentally affecting physico-chemical techniques. Bioremediation involves the use of various microbial organisms for eliminating environmental pollutants. It may be a permanent solution and may even lead to complete mineralization of organic pollutants, leaving the ecosystem intact. The most important factors for the success of this technology are the availability of suitable microbes in the contaminated medium, the type and concentration of pollutants and the environmental conditions.

This chapter reviews the recent bioremediation research on a range of pollutants in wastewaters produced by industries such as alcohol distilleries, olive mills, and paper and pulp mills industries. Treatment and reuse of effluents present significant challenges due to the characteristics of the wastewater such as high chemical oxygen demand (COD), high content of persistent pollutants and dark colour. Many species have been isolated and characterized as pollutant-degrading microorganisms. Biodegradation and biosorption have been extensively studied using pure or mixed cultures of fungi and bacteria. The evaluation of the proper microorganism(s) is normally the first step for a biological treatment of pollutants. An extended review of the microorganisms, emphasized on bacteria and fungi that have been tested for the removal of the specific pollutants in the industrial wastewater is given in this chapter.

Industrial Pollution

Water and soil pollution can be traced to waste management practices, accidental or incidental leakage, spills and improper disposal. The nature of waste being generated has changed significantly with time. The rapid expansion and technological improvement in industrial fields has led to an increasing amount and complexity of toxic waste effluents. Moreover, the presence of these compounds in drinking or irrigation water or in cultivated land represents a significant environmental hazard but also a health hazard. Environmental pollution can harm the flora and fauna of affected habitats, resulting in the uptake and accumulation of toxic chemicals in food chains and causing serious health problems and/or genetic defects in humans (Rigas and Dritsa, 2007). The increasing public awareness and the stricter regulations established by governments have reduced chronic industrial pollution over recent years but major accidents still occur and a considerable number of known polluted sites exist worldwide.

Physicochemical Methods

Various physicochemical methods are available to treat and manage wastes, either alone or in combination. Uneconomical physicochemical methods due to non-sustainable contaminants are continuously being replaced by newer more efficient methods. Biological processes appear to be a potentially economical, energy efficient and environmentally friendly approach. Bioremediation represents a diverse clean-up strategy. Bioremediation is defined as the elimination, attenuation or transformation of polluting or contaminating substances by the use of biological processes (Lynch and Maffat, 2008). Bioremediation uses microorganisms to treat contaminants by degrading organic compounds to less toxic materials such as carbon dioxide, metals, water and inorganic salts. Most important factors that should be taken into consideration in water and soil remediation processes are (Farhadian et al., 2008):

- Sources and concentration pollutants
- Physical, chemical and microbiological characteristics of contaminated water (pH, BOD, COD, etc.)
- Chemistry and mechanics of soil at contaminated site
- Energy, time and personnel needs
- Hydrogeology and hydrology of contaminated site

- Environmental conditions (pH, temperature, pressure, presence of electron acceptors, nutrient sources)
- Biodegradabiltity of contaminats
- Transport and dispersion of pollutants
- Physical and chemical properties of pollutants (adsorption, solubility, volatility)
- Method of detection, determination and monitoring pollutant.

BIOLOGICAL METHODS

Microorganisms (usually bacteria, fungi and in some cases, algae) play an important role in biological processes. They are ubiquitous and have the physiological and metabolic capabilities for pollutant removal. In microbial metabolism, the typical electron acceptors are oxygen, nitrate, ferrum (III), sulfate and carbon dioxide. In the case of aerobic respiration, oxygen is utilised and the others are involved in anaerobic respiration. Microorganisms use the contaminant as a carbon and energy source via aerobic respiration, anaerobic respiration, fermentation and cometabolism. Other nutrients necessary for their growth and survival such as nitrogen, phosphorus, sulfur and trace elements are added to the medium. Not only medium composition but also environmental conditions such as pH, temperature, salinity, and pressure play a vital role. Also, mixed cultures of bacteria or fungi have been able to metabolise a variety of organic compounds (Farhadian et al., 2008). Biodegradation rates can be further improved by raising bioavailability (for instance by the addition of surfactants that improve the accessibility of these substrates to microbial attack) and genetically modified microorganisms (Quan et al., 2004).

Anaerobic Treatment

Anaerobic treatment of waste is a complex biological process involving several groups of microorganisms (Cha and Noike, 1997). In general, complex wastes are stabilized in three basic steps: hydrolysis, acid fermentation and methanogenesis. Anaerobic digestion is accomplished by a complex ecosystem in which diverse groups of microorganisms interact in a symbiotic, synergistic, competitive and antagonistic association. The main groups of bacteria are three. The first group comprises fermentative bacteria, which perform hydrolysis and acidogenesis. Acetogenic bacteria constitute the second group and are responsible for breaking down the products of the acidification step to form acetate. The third group involves methanogenic bacteria that convert acetate or carbon dioxide and hydrogen into methane (Acharya et al., 2008). Anaerobic digestion can convert a significant portion of COD to biogas, which may be used as in-plant fuel. The steps in anaerobic digestion process have been described as follows (Mata-Alvarez, 2003):

1) Hydrolysis of organic polymers, carbohydrates, lipids and proteins to sugars, fatty acids and amines, respectively by hydrolytic and/or acidogenic bacteria.
2) Fermentation of monomeric organic substances to hydrogen or formate, carbon dioxide, volatile fatty acids, ethanol and ketones by fermentative bacteria.

3) Oxidation of reduced compounds to hydrogen, carbon dioxide and acetate by obligatory hydrogen-producing acetogens.
4) Acetogenic respiration of bicarbonate compounds by homoacetogenic bacteria.
5) Oxidation of reduced compounds (alcohols, butyric and propionic acids) to carbon dioxide and acetate by sulphate-reducindg bacteria and nitrate-reducing bacteria in the presence of sulphates and nitrates.
6) Acetate oxidation to carbon dioxide and oxidation of hydrogen by sulphate - reducindg bacteria and nitrate-reducing bacteria.
7) Conversion of acetic acid by acetoclastic methanogenic bacteria and methanogenic respiration of carbon dioxide by hydrogenofil methanogenic bacteria.

Aerobic Treatment

A huge number of bacterial and fungal genera and in some cases algae possess the capability to degrade organic pollutants. Enzymatic key reactions of aerobic biodegradation are oxidations catalyzed by oxygenases and peroxidases.

Bacteria are the predominant degraders of organopollutants, able to use a huge number of natural and xenobiotic compounds as carbon sources and electron donors for the generation of energy. Bacterial dehalogenases catalyse the cleavage of carbon-halogen bonds, which is the critical step in aerobic mineralization of many halogenated compounds.

Fungi, especially ligninolytic fungi, produce extracellular enzymes with very low substrate specificity. The three main lignin-modifying enzymes (LMEs) are lignin peroxidase, manganese dependent peroxidase, and laccase. White-rot fungi contain all three enzymes and are therefore able to breakdown and mineralize several environmental pollutants into non-toxic forms. This takes place through the generation of radical species that cause the complete biodegradation of lignin polymers. The main reactions that are catalyzed by the ligninolytic enzymes include depolymerization, demethoxylation, decarboxylation, hydroxylation and aromatic ring cleavage. Many of these reactions result in oxygen activation, creating radicals that perpetuate oxidation of the organopollutants (Reddy and Mathew, 2001).

Algae play an important role in the conversion of solar energy into chemical energy through their photosynthetic activities. Photosynthetic conversion can be expected to be an efficient and alternative process in several industrial fields such as environmental protection. Algae possess biodegradative capabilities on aromatic compounds, ranging from simple monocyclic to more complex polycyclic pollutants (Semple et al., 1999)

Wastewaters discharged by the industries are one of the major causes of environmental contamination and pollution. Synthetic dyes, pesticides, PCBs are of great environmental concern because of their abundance and their toxic, carcinogenic and teratogenic effects on animal and humans. Microbial degradation and decolourisation of industrial wastes is an environment-friendly and cost- competitive, alternative to chemical decomposition processes of wastes minerilization. This chapter discusses the strategies implemented for the treatment of representative wastewaters produced by alcohol distilleries, olive mills and pulp and paper mills, emphasizing on the biological methods. Various groups of microorganisms, which have the potential to decolourise and degrade toxic organic compounds from the selected industrial effluents, are discussed.

ALCOHOL INDUSTRIES

A large and increasing wastewater is produced globally by alcohol industries. Alcohol industries are divided in three categories: brewing, distilling and wine manufacture. The manufacture of alcohol in distilleries consists of four mains steps: feed preparation, fermentation, distillation and packaging. On an average 8–15 L of effluent is generated for every liter of alcohol produced (Saha et al., 2005). Distilleries generate spentwash as wastewater. Sources of wastewater are stillage, fermenter and condenser cooling water, and fermenter wastewater. The liquid residues in alcohol production are liquor, sugar cane washing water from condensers and from washing the equipment except for the residual water (Pant and Alholeya, 2007). The production and the characteristics of wastewaters are highly variable and dependent on the raw material used for the fermentation. Alcohol industry usually uses molasses as a raw material for the fermentation, which generates high volumes of high strength wastewater.

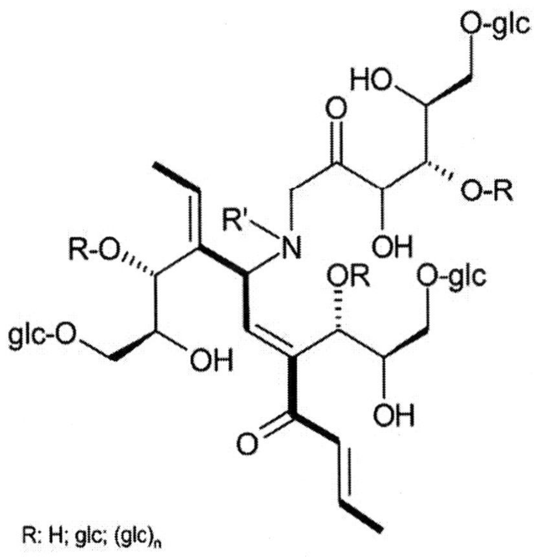

Figure 1. Basic melanoidin structure formed from carbohydrates and amino acid (Cammerer et al., 2002).

The common characteristics of spent wash are generally acidic, high in chemical oxygen demand (COD), unpleasant odour and colour and may contain phenolic compounds (Strong and Burgess, 2008). Nitrogen, potassium, phosphates, calcium, sulphates are also contained in high amounts. Spentwash has a dark colour due to the presence of melanoids from Mailard reaction of sugars with proteins and are highly recalcitrant and toxic to many microorganisms. A wide range of reactions including cyclizations, dehydrations, retroaldolizations, rearrangements isomerizations and condensations take place for the formation of nitrogenous melanoidins. The basic melanoidin structure formed from carbohydrates and amines is shown in figure 1. It is thought to have similar properties to humic acids (Naik et al., 2008). Although it is a widely distributed biopolymer in nature, it is hardly decomposed by microorganisms. Melanoidins or related formation products can occur in different processes of beverage manufacture such as heat concentrated juices and musts, beers or wines (Kroh, 1994). Melanoidins are considered as the most aggressive by-products generated by ethanol-production industry. Apart from melanoidins, caramel from overheated sugars, various sugar decomposition compounds, anthocyanines, tannins and other xenobiotic compounds are present in the wastewater, which are resistant to degradation (Pandey et al., 2003). The disposal of wastewater into the environment is hazardous and poses a serious threat to the quality of the water in the groundwater and natural water bodies. The highly coloured effluents block out the sunlight in rivers and streams, reducing the oxygenation of the water by photosynthensis that is detrimental to aquatic life. Additionally, the heavy pollution load leads to eutrophication of the water sources. Its disposal into soil causes inhibition of seed germination and depletion of vegetation by acidified soil.

TREATMENT TECHNOLOGIES

Several treatment technologies have been applied for the removal of the recalcitrant compounds and reduction of the pollution load of distillery effluents, in order to be discharged into environment. Physicochemical and biological methods have been used for the treatment of highly polluted effluents.

Physicochemical Methods

Physicochemical methods that have been practiced for the colour removal of distillery effluents include adsorption, coagulation and flocculation, oxidation, radiation and membrane processes. Adsorption has been widely employed for the removal of colour and specific organic compounds. The most common adsorbent is activated carbon, an extremely good adsorber of organic chemicals. Almost complete decolourization and over 90% COD and BOD removal was observed in packed bed studies using a commercial activated charcoal (Chandra and Pandey, 2000). Another efficient absorbent is chitosan, a biopolymer used as anion-exchanger with 98% colour removal and 99% COD removal (Lalov et al., 2000).

Coagulation and flocculation are suitable processes for the removal of suspended colloids that do not settle naturally. Cationic coagulants provide positive electric charges to reduce the negative charge. The addition of coagulating and flocculating chemicals results in the settling of suspended solids to remove fine solids that will not settle naturally. The process uses the addition of specialized chemicals to wastewater in order to convert fine suspended solids into flocs, which can be removed by filtration. Ferric chloride ($FeCl_3$), ferric sulfate, aluminium sulfate and calcium oxide are some of the inorganic compounds that were used for the flocculation of melanoidins (Migo et al., 1993). The use of a commercial oganic anionic polyelectrolyte (Percol 47), in combination with ferrous sulfate and lime, resulted in 99% reduction in colour and 87 and 92% reduction in COD and BOD, respectively (Mandal et al., 2003).

Oxidation is a destructive process that converts organic compounds to carbon dioxide and water (mineralization). Especially, advanced oxidation processes (AOP) are based on the formation of free radicals, especially hydroxyl radicals (Beltran et al., 1993). This is accomplished, for instance, by the addition of ozone and hydrogen peroxide, as combined oxidants. Oxidation by ozone alone achieved 80% decolourisation for biologically treated spentwash (Pena et al., 2003). The formation of free radicals using TiO_2 and solar radiation is a very effective destructive oxidation process, which leads to complete mineralization of the effluent (Kulkarni, 1998). Fenton's technology is also based on the formation of free radicals obtained by the reaction of hydrogen peroxide and iron salts (Hadavifar et al., 2009). The efficiency varied from 18 to 97% for the photo-Fenton process and 5-47% for the conventional Fenton process.

Membrane-based separation processes, like nanofiltration (NF) and reverse osmosis (RO), have been applied to purify the distillery wastewater from colour and contaminants. Complete COD and colour removal were achieved this way (Nataraj et al., 2006). Radiation technology was also applied and involved a combination of electron–beam and coagulation purification of molasses distillery slops (Pikaev, 2002).

Anaerobic Treatment

Biological treatment processes to treat wastewater from distilleries include anaerobic digestion and aerobic treatment. Both are effective methods for highly polluted wastewater and are used to treat alcohol industries effluents. In the last decades, there is an increasing interest in utilizing microbial activity by pure bacteria and fungi for the decolourisation and mineralization of wastewaters instead of mixed cultures.

Bioremediation of distillery spent wash by anaerobic digestion is an attractive primary treatment due to its reputation as a low cost, environment friendly and socio economically acceptable technology. Although anaerobic digestion has the ability to treat the high organic loads in distillery wastewaters, the treatment process can be inhibited by the presence of phenolic compounds, salt ions and sulphides.

At present, the anaerobic biological treatment of distillery effluents is an effective technology for the removal of 90% COD in an effluent stream (Pant and Aldoheya, 2007). However, the performance and treatment efficiency of anaerobic process can be influenced, both, by inoculum source and feed pre-treatment. Various types of anaerobic processes have been developed for different industrial wastes. Anaerobic systems can be operated as single-phase, where one reactor for the microorganisms is used to digest organic matter or two-phase systems that separate the acidogenic and methanogenic organisms in two different reactors (Mohana et al., 2007). Biphasic systems are more appropriate for high strength wastewater. Anaerobic lagoons and continuous stirred tank reactors (CSTR) are the simplest choice for anaerobic treatment of distillery waste. Treatment of molasses wastewater in CSRT in a single as well as in biphasic operations has resulted in 80-90% reduction after 10-15 days (Pathade, 2003). Although, anaerobic lagooning is advantageous due to the low operational costs and energy requirements, the ground water contamination cannot be prevented (Bories et al., 2005). In the field of distillery waste management, the application of two pilot–scale lagoons in series, resulted in 82-92% BOD removal (Rao, 1972). Anaerobic sequencing batch reactor (ASBR) has not been extensively applied. Winery wastewaters were treated in ASBR and the removal efficiency of COD reached 98% (Ruiz et al., 2002). Anaerobic fixed film bioreactors have a biofilm support for the sustainment of the biomass in the bioreactor, thus leading to a sufficient biodegradation. Various materials have been used as support media, such as glass beads, sand polyurethane foam, nylon fibers and various other plastics.

A comparative study of low cost materials for the treatment of distillery spentwash using anaerobic fixed film bioreactors was performed. Cocconut coir was the supporting material compared to charcoal and nylon fibers (Acharya et al., 2008) Upflow anaerobic sludge blanket bioreactors (UASB) have been extensively used in the treatment of distillery effluents. The success of the UASB is based on the formation of active and settlable granules (Fang et al., 1994). Uzal et al., (2003) investigated an anaerobic treatment of whisky distillery spentwash in two-stage UASB and the system resulted in an efficient COD reduction (95%).

Anaerobic fluidized bed reactors utilize small fluidized media particles to induce extensive cell immobilization, thereby achieving a high reactor biomass hold-up and a long mean cell residence time (Sowmeyan and Swaminathan, 2008). It is an effective technology for high strength wastewaters as it favours the transport of microbial cells from the bulk to the surface and as a result it enhances the contact between microorganisms and substrate. In the anaerobic fluidised bed bioreactor for the treatment of spentwash, the carrier particles were perlite, which allows low energy requirements and provides a good surface for biomass attachment. The system achieved 84% COD removal.

Aerobic Treatment

Bacteria Treatment

Naturally occurring microbial activities have been the starting point for all biotechnological applications. Various bacteria have been isolated and proved to be capable of decolourisation and bioremediation of anaerobically treated distillery spent wash. A detailed list of bacteria tried by different researchers for decolourization of distillery effluent is given in Table 1.

The first study on the biodegradation of distillery spent wash by six bacterial strains (gram-positive named I to VI) was conducted by Kumar and Viswanathan (1991). The products of the biodegradation were found to be biomass, carbon dioxide, and volatile acids. These strains were able to reduce the COD of distillery effluent by 80% after 4-5 days.

Two aerobic cultures LA1 and D-2 decolourised (36.5% and 32.5%) and reduced COD (41% and 39%) on a diluted spentwash under optimized conditions in eight days (Kumar et al., 1997a). An aerobic culture isolated from soil samples of a distillery plant, L-2 achieved 31% decolourisation and 57% COD reduction after 7 days incubation under specific conditions. The strain L-2 belonged to Genus *Lactobacilli* (Kumar et al., 1997b). A *bacillus* strain isolated from soil samples was selected among various microorganisms by Nakajima-Kame and co-workers (1999) for the decolourisation of molasses wastewater under thermophilic and anaerobic conditions. The strain decolourised 35.5% of molasses pigment within 20 days at 55°C under anaerobic conditions, but no decolourisation activity was observed under aerobic conditions. Gel filtration chromatography revealed that the decolourisation of molasses pigment by the isolated strain, which most closely resembled to *B. smithii*, was accompanied by a decrease in small and large molecules.

Table 1. Bacteria employed for treatment of distillery wastewaters

Bacteria	COD removal (%)	Color removal (%)	References
Lactobacillus hilgardii		40	Ohmomo et al., 1988
Gram-positive named I toVI		80	Kumar and Viswanathan, 1991
Lactobacillus L-A1	41	37	Kumar et al., 1997a
Lactobacillus D-2	39	33	
Lactobacillus D-2	57	31	Kumar et al., 1997b
Bacillus sp.		36	Nakajima-Kambe et al., 1999
Aeromonas formicans	57	55	Jain et al., 2000
Pseudomonas fluorescence		76	Dahiya et al., 2001a
TA2		67	Asthana et al., 2001
TA4		64	
Mixed (TA2+TA4)		75	
Oscillatoria boryana		60	Kalavathi et al., 2001
Oscillatoria sp.		96	Patel et al., 2001
Lyngbya sp.		81	
Synechocystis sp.		26	
Pseudomonas putida	44	60	Ghosh et al., 2002
Aeromonas strain	44		
Bacillus cereus	81	75	Jain et al., 2002
Acetogenic bacteria			Cibis et al., 2002
Acetogenic bacteria strain No. BP103	77	76	Sirianuntapiboon et al., 2004b
Pseudomonas, Enterobacter, Stenotrophomonas, Aeromonas, Acinetobacter, Klebsiella	44		Ghosh et al., 2004
Bacillus thuringiensis, Bacillus brevis, Bacillus sp. (MTCC6506)	36-64	29-51	Kumar and Chandra, 2006
Microbacterium hydrocarbonoxydans *Achromobacter xylosoxidans* *Bacillus (subtilis, megaterium, anthracis, licheniformis, cereus, thuringiensis)* *Achromobacter sp.,Staphylococcus Epidermidis, Pseudomonas migulae* *Alcaligens faecalis*	86	76	Chaturvedi et al., 2006
Pseudomonas aerugenosa *Stenotrophomonas maltophila* *Proteus mirobilis*	67	51	Mohana et al., 2007
Bacillus species	90		Krzywonos et al., 2008
Alcaligenes faecalis	67		Chandra et al., 2009
Bacillus cereus	63		
Mixed	74		

Cell immobilisation was used for the treatment of distillery effluent. Calcium alginate immobilised cells of the strain *Lactobacillus hilgardi* were able to decolourise 40% of melanoidin solution (Ohmomo et al., 1988). Immobilised cells of *Pseudomonas fluorescens* isolated from soil samples contaminated with molasses, on cellulose carrier decolourised molasses wastewater (MWW) samples up to 76% under non-sterile conditions in four days at 30°C. Cellulose carrier coated with collagen was found to be the most effective carrier, producing the highest decolourisation activity of 94% in a 4-day process. This carrier could be reused with 50% of the decolourisation activity retained until the seventh day (Dahiya et al., 2001a). Three bacterial strains, which were isolated from the activated sludge of a distillery waste water treatment plant, *Xanthomonas fragairae*, *B. megaterium* and *B. Cereus* were capable to remove COD and colour in a range of 55-68% and 38-58%, respectively (Jain, 2002). *B. cereus* resulted in maximum COD (81%) and colour (75%) reduction in free form. The reduction efficiencies increased marginally with immobilization.

Acetogenic bacteria are capable of oxidation decomposition of melanoidins, removing low molecular weight compounds in untreated molasses spentwash and almost all the low and high molecular weight compounds in anaerobically treated molasses spentwash. The biodegradation of potato slops by acetogenic bacteria reaching 77% COD removal was reported (Cibis et al., 2002). Acetogenic bacteria strain No.BP103 could decolourize 73.5% of molasses pigments in molasses wastewater supplemented with nutrients (glucose, yeast extract and basal mineral salts), whereas the decolourisation of this strain was decreased to only 9.75% in the absence of nutrients (Sirianuntapiboon et al., 2004a). The decolourisation activity might be due to a sugar oxidase.

Bacterial strains were isolated from the soils of abandoned sites of distillery effluent discharge and were characterised for their ability of reducing the chemical oxygen demand (COD) of the spent wash. The strains were identified as *Pseudomonas, Enterobacter*, *Stenotrophomonas*, *Aeromonas*, *Acinetobacter* and *Klebsiella*. The extent of COD reduction collectively by the six strains (44%) was equal to that reduced individually by *Aeromonas*, *Acinetobacter*, *Pseudomonas* and *Enterobacter* (Ghosh et al. 2004).

Several nitrifying bacteria were screened for detoxification of distillery effluent and *Nitrosococcus oceanus* was noticed to be a potential candidate for detoxification. The detoxified distillery effluent was used in rice plant culture (Arora et al., 1992).

Marine cyanobacteria are considered to be ideal for treatment of distillery effluent. A marine cyanobacterium *Oscillatoria boryana* was reported to degrade melanoidin due to the production of H_2O_2, hydroxyl, perhydroxyl and active oxygen radicals, resulting in the decolourization of the effluent (Kalavathi et al., 2001). The organism decolourised pure melanoidn pigment by 75% and crude pigment in the distillery effluent by 60% in 30 days. A reduction in colour by 96%, 81% and 26% of distillery effluent through bioflocculation by *Oscillatoria sp.*, *Lyngbya sp.* and *Synechocystis sp.*, respectively, was reported by Patel et al. (2001).

Fifteen culturable bacteria were isolated from rhizosphere soil of *Phragmites australis* and identified as *Microbacterium hydrocarbonoxydans*, *Achromobacter xylosoxidans*, *Bacillus subtilis*, *B. megaterius*, *B. anthracis*, *B. licheniformis*, A. *xylosoxidans*, *Achromobacter sp.*, *B. thuringiensis*, *B. licheniformis*, *B. subtilis*, *Staphylococcus epidermidis*, *Pseudomonas migulae*, *Alcaligens faecalis* and *B. cereus*. All the fifteen isolates grew on effluent supplemented medium as a sole carbon source resulting in the reduction of the levels of distillery pollutants and their colour by 76% and 85–86% BOD and COD reduction of the effluent within 30 days (Chaturvedi et al., 2006).

Mixed cultures have been used for molasses wastewater decolourisation. It seems to be a promising technique according to the fact that catabolic activities of microoganisms in mixed culture complement each other and the interactions lead to complete mineralization of the effluent (Adikane et al., 2006, Kumar and Chandra, 2006). Adikane and his co-workers (2006) reported that 69% decolourisation of molasses spent wash was achieved by using soil samples as inoculum instead of isolated microorganisms. The decolourisation of four synthetic melanoidins by three isolates, *Bacillus thuringiensis*, *Bacillus brevis*, and *Bacillus sp.* was up to 22%, 27.4%, and 27.4% colour removal, respectively.

The bacterial consortium comprising of *Pseudomonas aeruginosa*, *Stenotrophomonas maltophila* and *Proteus mirabilis* achieved its maximum molasses decolourisation (67%) and 51% COD reduction within 72 h in the presence of 0.5% glucose (Mohana et al., 2007). Krzywonos and co-workers (2008) studied the effect of temperature on the extent of aerobic batch biodegradation of potato stillage with a mixed culture of bacteria of the genus *Bacillus*. Over the temperature range of 20–63°C, the removal efficiency was very high, by about 90% after 80h (with an extent of COD reduction following solids separation that varied between 77.57% and 89.14% after 125 h). No COD reduction was observed at 65°C.

Two aerobic bacteria, namely *Alcaligenes faecalis* and *Bacillus cereus,* were capable to degrade and utilize sucrose–glutamic acid Maillard products (SGMPs) as carbon, nitrogen and energy source. The mixed bacterial culture was found more effective compared to axenic culture of *Alcaligenes faecalis* and *Bacillus cereus* decolourizing 73.79%, 66.80% and 62.56% SGMPs, respectively. From the LC-MS–MS and other spectrophotometric analysis, it was revealed that most of the SGMPs detected in control were diminished from bacteria treated samples (Chandra et al., 2009).

Fungal Treatment

Fungi play an important role in the degradation of organic compounds and various basiodiomycetes and ascomycetes have been used in the decolourisation of natural and synthetic melanoidins in connection with colour reduction from distilleries effluents. The fungi attack organic compounds to obtain carbon and nitrogen for nutrition, purifying the effluent and reducing its BOD and COD. Additionally, the produced fungal biomass is protein rich and may be used for animal feed. Fungi have the capacity to adapt to severe environments and have lower sensitivity in temperature, pH, nutrients and aeration. Several fungi that have been investigated for their ability to decolourise molasses are shown in Table 2.

One of the most studied fungi is *Aspergillus*. Specifically *Aspergillus fumigatus*, *Aspergillus niger*, *Aspergillus niveus* achieved an average of 69–75% decolourisation along with 70–90% COD reduction (Ohmomo et al., 1987; Miranda et al., 1996; Jimnez et al., 2003, Shayegan et al., 2004, Angayarkanni et al., 2003, Mohammad et al., 2006). It was suggested that the decolourisation by the fungi takes place due to the destruction of chromophore molecules and partially sorption to mycelium. A longer aeration period causes the adsorbed colour molecules to be released as a result of endogenous respiration and cell death, hence reducing decolourization efficiency (Mohana et al., 2007). The effect of various organic and inorganic nitrogen sources was studied on the degradation and decolourisation of melanoidins by *Aspergillus fumigatus*. Results determined that the supplementation of organic nitrogen source affected harmfully the degradation of melanoidins, as addition of organic nitrogen sources such as yeast extract, peptone, beef extract and tryptone brought down the level of decolourisation, whereas inorganic nitrogen sources such as sodium nitrate and ammonium nitrate reduced the decolourisation values (Ohmomo et al., 1987).

Table 2. Fungi employed for treatment of distillery wastewaters

Fungi	COD removal (%)	Colour removal (%)	References
Coriolus sp. No. 20		80	Watanebe et al., 1982
Coriolus versicolour		80	Aoshima et al., 1985
Aspergillus fumigatus *Aspergillus niger* *Aspergillus niveus*	70–90	69–75	Ohmomo et al., 1987 Miranda et al., 1996
Phanerochaete chrysosporium		85	Fahy et al., 1997
Trametes versicolor	75	80	Benito et al., 1997
Phanerochaete chrysosporium *Coriolus versicolor*	73 70	54 72	Kumar et al., 1998
Geotrichum candidum, Coriolus versicolour, Phanerochaete chrysosporium, Mycelia sterilia		53	FitzGibbon et al., 1998
Geotrichum candidum		80	Kim and Shoda, 1999
Coriolus hirsutus		80	Miyata et al., 2000
Coriolus hirsutus		45	Fujita et al., 2000
Aspergillus niger			Petrucioli et al., 2000
Trametes sp	62	73	Gonzalez et al., 2000
Penicillium decumbens	51	41	Jimenez et al., 2003
Flavodon flavus		80	Rughukumar et al.,2004
Coriolus versicolor	49	63	Kahraman and Yesilada, 2003
Funalia trogii	62	57	
Phanerochaete chrysosporium	57	37	
Pleurotus pulmonarius	34	43	
Citeromyces sp.		69	Sirianuntapiboon et al., 2004a
Aspergillus niger		80	Patil et al., 2003
Phanerochaete chrysosporium		75	Vahabzadeh et al., 2004
C. versicolour	53	53	Chopra et al., 2004
Phanerochaete chrysosporium		55	Guimaraes et al., 2005
P. chrysosporium		82	Thakkar et al., 2006
Marine basidiomycetes NIOCC		100	D'Souza et al., 2006
Phanerochaete chrysosporium	83	86	Hossain, S.M., 2007
Trametes pubescens, *P. cinnabarinus*		86 69	Strong, 2010

White rot fungi has recently received considerable attention for their bioremediation potential, as they have been shown to be effective at degrading many hazardous environmental pollutants, through their non-specific, extracellular enzymatic system. The major enzymes associated with lignin-degrading ability are laccase, manganese peroxidase and lignin peroxidase (Rigas and Dritsa, 2007). The biological decolourisation of wastewaters containing melanoidins depends on pH, temperature, concentration of nutrients, oxygen and inoculum size, while the enzymatic system responsible for the degradation of melanoidins consists mainly sugar oxidases and peroxidases as sorbose oxidase, glucose oxidase, manganese dependent (MnP) and independent peroxidases (MIP). The first strain that was tested for its ability to decolourise a melanoidin solution was *Coriolus sp.* No 20. The decrease in decolourisation was about 80% under optimal conditions. This decolourisation occurred with an intracellular enzyme, which was prepared from an extract of integrated mycelia and required aeration and glucose and sorbose. In the absence of sugars, no decolourisation was observed (Watanabe et al., 1982). Melanoidins were suggested to be decolourised by the active oxygen (O_2, H_2O_2) species produced by the reactions catalyzed by enzymes probably oxidases.

Later, *Coriolus versicolour* Ps4a decolourized the effluent by decomposition of melanoidins and not by the partial transformation of chromophores and achieved 80% decolourisation under optimum conditions. The decolourizing activity was attributed to an intracellular enzyme, which is induced in the presence of melanoidin pigment (Aoshima et al., 1985). In another study, it was observed that *Coriolus hirsutus* pellets grown in a melanoidin-containing medium showed decolourising activities due to the production of extracellular H_2O_2 and the peroxidases, manganese-independent peroxidase (MIP)and manganese peroxidase (MnP) (Miyata et al., 1998). Treatment of biodigested distillery wastewater by *C. versicolour* has also been investigated by Chopra et al. (2004). The fungus was able to reduce COD and colour up to 53% in 8 days, respectively.

Phanerochaete chrysosporium and *Trametes versicolour* are the most widely studied white rot fungi. *P. chrysosporium* JAG-40 resulted in 80% decolourisation of diluted synthetic melanoidin (Dahiya et al., 2001b). In another research, extensive decolourisation (75%) of molasses wastewater (MWW) resulted after five days of incubation. The colour removal ability was found to be correlated to the activity of ligninolytic enzyme system, lignin peroxidase (LiP) and manganese peroxidase (MnP) (Vahabzadeh et al., 2004). Batch aerobic studies of the biological process showed that the *Phanerochaete chrysosporium* removed 83% COD, 86% BOD and 88 % colour of distillery wastewater, respectively (Hossain, 2007). The bioremediation potential of *Phanerochaete chrysosporium* strains NCIM 1073, NCIM 1106 and NCIM 1197 to decolourise molasses in solid and liquid molasses media was also studied. Under submerged cultivation conditions, two strains could decolourise molasses up to 82%. Under stationary cultivation conditions, none of the strains could decolourise molasses (Thakkar et al., 2006). The capacity of *Trametes sp.* I-62 to detoxify Cuban distillery wastewater reached 73.3% decolourisation and 61.7% COD reduction after seven days of fungal treatment, when 20% (v/v) of distillery vinasses was added to the culture medium (Gonzalez et al., 2000). The fungus resulted in 82% decolourisation of 12.5% anaerobically–aerobically treated effluent. Of this, 90% colour was removed biologically and the rest by adsorption on the mycelium (Benito et al., 1997). *T. versicolour* produces an extracellular enzyme identified which is involved in mineralization of melanoidins. The induction of laccases by molasses wastewaters and molasses melanoidins in the basidiomycetous fungus *Trametes sp.* I-62 was observed by Gonzalez et al. (2008). The time course of effluent decolourisation and laccase activity in the culture supernatant of the fungus were correlated.

Four fungi *Geotrichum candidum*, *Coriolus versicolour*, *Phanerochaete chrysosporium* and *Mycelia sterilia* were studied for their decolourizing ability of molasses wastewater. *G. candidum* and *P. chrysosporium* growth rates increased in the presence of increasing concentrations up to 50% v/v of molasses spent water (MSW). However, the growth of *M. sterilia* and *C. versicolour* was inhibited at spent wash concentrations above 5% (v/v). Increasing the concentration of MSW from 6.25% (v/v) to 12.5% (v/v) increased the decolourizing ability of each fungus, except for *M. sterilia*. *C. versicolour* exhibited 53% colour reduction) after 10 days growth in 12.5% (v/v) MSW (FitzGibbon et al., 1998). Full-strength wastewater was treated using shake-flask monocultures of four white rot fungi *Trametes pubescens* MB 89, *Ceriporiopsis subvermispora, Pycnoporus cinnabarinus* or *Phanerochaete chrysosporium*. The greatest COD removal efficiencies in the suspended solids containing wastewater were achieved by *Trametes pubescens* (74%) and *Phanerochaete chrysosporium* (64%). *Trametes pubescens* MB 89 and *P. cinnabarinus* displayed the greatest removal of phenolic compounds (86 and 69 %, respectively) (Strong, 2010).

Kahraman and Yesilada (2003) studied the effect of cotton stalk on decolourizing and COD removing capability of four fungi *Coriolus versicolour*, Funalia trogii, *Phanerochaete chrysosporium* and *Pleurotus pulmonarius*. Cotton stalk stimulated the decolourisation activity and was used as an attachment place and a source of nutrients. *C. versicolour* could effectively decolourise 48% of 30% diluted vinasse without any additional carbon source and increased to 71% with the addition of cotton stalks.

Immobilized fungal isolate of *A. niger* UM2 resulted in a 72% decolourisation of diluted synthetic melanoidin and 80% decolourisation of 50% biodigested effluent (Patil et al., 2003). Similarly, a thermophilic strain of *A. fumigatus* G-2-6 decolourized 75% of melanoidin pigment solution at 45°C (Ohmomo et al., 1987). The dye decolourizing fungus *G. candidum* Dec 1 immobilized on polyurethane foam resulted in 80% removal of colour in diluted molasses solution (Kim and Shoda, 1999).

Immobilised fungus *Flavodon flavus*, a white-rot basidiomycete fungus isolated from a marine habitat, was immobilised in polyurethane foam. The immobilised fungus could effectively decolourise molasses spent water (MSW) by 73% after 7 days of incubation. The concentration of benzo(a)pyrene, a polycyclic aromatic hydrocarbon (PAH) in the MSW decreased by 68% by day 5 on treatment with the fungus. A possible mechanism of decolourisation of MSW is via the action of glucose oxidase accompanied by production of hydrogen peroxide that may ultimately act as a bleaching agent (Raghukumar et al., 2003).

Apart from fungi, yeasts have been used for the treatment of molasses waste waters. The yeast *Citeromyces* WR-43-6 resulted in high and stable removal efficiency in both colour intensity and organic matter. The removal efficiencies for diluted spentwash were 75% for colour intensity and 76% for BOD (Sirianuntapiboon et al., 2004a). Yeast *Saccharomyces cerevisiae* had the ability to utilize non-reducing sugars and also recalcitrant material resulting in decolourisation of effluent. Colour removal reached 83% by the addition of glucose or maltose (Rajor et al, 2002).

Algal Treatment

The feasibility of combining microalgae, *Chlorella vulgaris* and macrophyte *Lemna minuscule* for bioremediation of wastewater from alcohol distilleries was investigated. This combination resulted in 61% COD reduction and 52% colour reduction. The microalgal treatment led to removal of organic matter and further treatment with macrophytes removed other organic matter, colour and precipitated the microalgae (Valderrama et al., 2002).

OLIVE OIL INDUSTRIES

Olive mill wastewater (OMW) generated by the olive oil extraction industries is a great pollutant due to its organic load, especially in the Mediterranean countries that produce more than 98% of the world's olive oil. Olive oil extraction involves different processes such as olive washing, grinding, beating and the extraction itself, that constitutes the basic stage of the whole process. Traditionally, three-phase or two-phase continuous centrifugation systems are used for olive oil production. Due to the utilisation of large amounts of water in the process, large amounts of wastewater or sludge are produced (Niaounakis and Halvadakis, 2006). The composition of OMW varies according to the vegetation water, olive oil extraction process and storage time. Olive mill wastewater is one of the most complex agro-industrial effluents because of its high organic carbon content, lipids and aromatic compounds and sugars. OMW pollution is attributed to the recalcitrance of phenolic fraction to microbial degradation. Phenolic compounds are responsible for several biological effects, including phytotoxicity and antibiosis (Asses et al., 2008). Two major phenolic categories are observed, phenolic monomers (flavoinoids, not autoxidated tannins) and dark coloured phenolic polymers. Phenolic compound are strictly dependent on the processing system used for olive oil production. Hydroxytyrosol is the main natural polyphenolic compound in OMW, characterised by its bio-antioxidant activity. In contrast, tyrosol (Figure 2) and its derivatives are substantially resistant to air/oxygen, bacterial and enzymatic degradation, and are of a highly polluting nature.

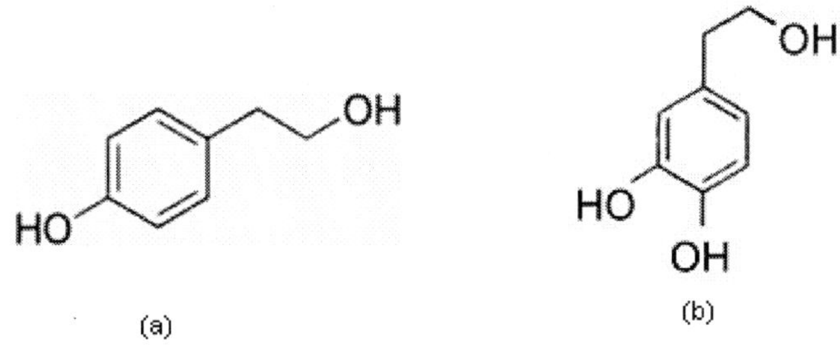

Figure 2. Structure of tyrosol (a) and hydroxytyrosol (b) in olive mill wastes.

Fructose, mannose, glucose, saccharose, and traces of sucrose and some pentose are present in OMW depending on the variety of olives, the climatic conditions during growth and the extraction methods used. It is characterized by a dark brownish colour and a strong odour. They also contain amounts of oleanolic and maslinic acids and inorganic compounds such as iron, potassium, phosphorus, calcium and magnesium. The improper disposal of

OMW to the environment is prohibited due to its toxicity to microorganisms and its potential threat to surface and groundwater. It may affect thephysical and chemical properties soil, such as porosity and pH. The high concentration of darkly coloured polyphenols in OMW can colour streams and rivers. This change in colour is attributed to the oxidation and subsequent polymerization of tannins giving darkly coloured polyphenols, which are difficult to remove from the effluent. In addition, the high concentration of reduced sugars can stimulate microbial respiration, lowering dissolved oxygen concentrations, while the high phosphorus content can lead to eutrophication (Niaounakis and Halvadakis, 2006). The presence of phytotoxic phenolics generally prohibits the use of untreated olive mill wastewaters for irrigation purposes in agricultural production.

TREATMENT TECHNOLOGIES

Phsysicochemical Methods

Several methods have been used for the treatment of olive mill effluents but it is a complicated phenomenon due to the seasonal and geographical diffuse nature of olive oil production (Mcnamara et al., 2008). The most important physicochemical processes include oxidation (ozonation, Fenton's reagent, electrochemical oxidation and others), adsorption, flocculation/ coagulation, ultrafiltration and reverse osmosis (Mameri et al., 2000). Adsorption of phenols and organics of OMW is mainly due to hydrophobic interactions. Adsorption on activated carbon reached 94% removal of phenols in less than 4 h (Azzama et al., 2005). Coagulation with materials such as lime, alum, ferric chloride and ferrous sulphate has also been employed in OMW treatment. The application of quicklime allowed the reduction of almost 37% of the initial COD, and approximately 88% and 71% of the colour and phenolic content of the waste, respectively (Jaouani et al., 2005).

Chemical oxidation, utilizing ozone and advanced oxidation technologies (e.g. ozone and hydrogen peroxide or UV radiation), is another possible way to reduce COD and polyphenol content. Ozone reacts with the phenolic compounds of wastewater and at the same time reduces the alkalinity of the media for further biological processing (Beltran et al., 1999). Advanced oxidation processes (AOP) use a combination of strong oxidizing agents (e.g H_2O_2, O3) with catalysts (e.g. transition metal ions) and irradiation, including Fenton's reagent, a combination of UV radiation and hydrogen peroxide as well as photo-Fenton (Beltran et al., 2001). The main mechanism of AOPs function is the generation of highly reactive free radicals. Beteta and co-workers (2009) studied the treatment of an actual industrial waste with three advanced oxidation processes: conductive-diamond electrooxidation (CDEO), ozonation and Fenton oxidation. A complete mineralization was achieved by CDEO. Likewise, both ozonation and Fenton oxidation were able to treat the wastes, but they obtained very different results in terms of efficiency and mineralization.

Filtration through membranes of different pore size and materials is increasingly being used by the table olive processing industry to reuse, concentrate, or eliminate wastewater. The membrane technology for the treatment of olive mill wastewater (OMW) includes microfiltration, ultrafiltration, nanofiltration and reverse osmosis. A membrane technique to treat olive oil mill washing water was achieved by using different commercial ultrafiltration membranes: one organic membrane (PCI) and two ceramic (Ceraver) membranes. The organic PCI membrane could reduce pollution due to organic matter by decreasing the value of the chemical oxygen demand by about 90%. Moreover, the nature of the ultrafine pore membrane appeared to be an important parameter which may strongly increase or decrease the capacity of the membrane (Mameri et al., 2000). Nanofiltration membranes prepared from selected types of poly(amidesulfonamide) have been proven to be effective in removing oil from OMW. With reverse osmosis a good water quality is obtained that allows reutilization of the water for industrial production processes. Combinations of ultrafiltration and nanofiltration (NF) or reverse osmosis (RO) were found to be very efficient for the treatment and fractionation OMW (Paraskeva et al., 2007). The final effluent contained very low content of organic compounds and dissolved ionic salts. Phenols present in the OMW were removed to an extent exceeding 95% of the initial value following the nanofiltration step. Better efficiency was achieved applying reverse osmosis after ultrafiltation.

Anaerobic Treatment

Anaerobic Bacteria

Biological processes employ the use of microorganisms for the degradation of pollutants. The microbial treatment of olive mill wastewater can positively affect its composition, solving toxicity problems. The selection of the appropriate microorganism is critical and the strain should have the ability to degrade organics. Anaerobic digestion is used for the removal of high concentrations of organic matter. Methane and carbon dioxide are the final products of the transformation of the organic compounds in OMW. A significant advantage of anaerobic processes over aerobic is the generation of methane which may be used as an energy source for other processes (Mcnamara et al., 2008). Anaerobic bacteria are employed in anaerobic biodegradation. Fermentative bacteria, (perform hydrolysis and acidogenesis), acetogenic bacteria (acidification to form acetate) and methanogenic bacteria (perform methanogenesis) constitute the three groups for anaerobic process. Anaerobic biodegradation is affected by temperature, retention time, pH, H_2, partial pressure, chemical composition of wastewater and the presence of toxic compounds. Anaerobic biodegradation can take place under psychrophilic (<20°C), mesophilic (25–40°C), or thermophilic (50–65°C) conditions which are usually applied due to the higher loading rates and to greater destruction of pathogens (Niaounakis and Halvadakis, 2006). Several researchers have investigated the biological disposal of OMW by anaerobic digestion. The major problem for methanogenic bacteria is the presence of toxic compounds (phenolics and lipids). One approach to the problem was the dilution of olive mill wastewater in order to reduce the concentration of the compounds with the inhibitory effects. A reduction in COD from 80-85% was obtained in an anaerobic process (Aveni, 1984).

Various types of bioreactors have been used for the treatment of OMW such as up-flow anaerobic sludge blanket (UASB) reactor, anaerobic baffled reactor (ABR), continuous-flow stirred tank reactor (CSTR), anaerobic contact reactor, anaerobic filter reactor (up-flow and down-flow), expanded or fluidized bed reactor, combinations of various reactors. The initial trials on the biotreatment of OMW were conducted using continuous-flow stirred tank reactor (CSTR) (anaerobic contact reactor). One of the first researchers who used UASB reactor for the anaerobic treatment of OMW was Aveni (1985). The formation of recalcitrant phenolics such as condensed tannins and the presence of long-chain fatty acids which inhibited microbial growth in an UASB reactor were reported by Zouari and Ellouz, 1996b. The surface active/detergent activity of long chain fatty acids (LCFAs) present in OMW (oleic and linoleic acid) can cause microbial cell lysis, and acetate-utilizing methanogens used during the methanogenesis are particularly vulnerable pathogens (Niaounakis and Halvadakis, 2006). The key factor to prevent inhibition of methanogenesis is the degradation of unsaturated fatty acids bysaturation followed by b-oxidation (Beccari et al., 1998).

The anaerobic treatability of OMW was investigated in a UASB reactor under different operating conditions such as pH, feed strength, and hydraulic retention time (Ubay and Ozturk, 1997). COD removal reached about 75% under the optimum conditions and methane conversion rate was $0.35m^3$ per kg COD.

The ABR reactor has been successfully used in treating diluted OMW. Khabbaz and co-workers (2004) reported stable and efficient COD removal. Maximum COD removal was achieved at an influent COD concentration of 3 g/l.

Combinations of bioreactors have been employed for the treatment of olive mill wastewaters. The supernatant and the sludge, when using a fixed-bed type for the supernatant and a plug-flow type for the sludge, from settled OMW was investigated by Georgacakis and Dalis (1993). COD reduction was determined at 94.02%. An up-flow digester combined with a fixed-bed-type reactor was an effective treatment for reducing the organic load of total raw OMW.

In a study, a two-stage pilot plant, an upflow-sludge bed and an anaerobic filter operated in a series achieved 91% COD removal. The use of an anerobic filter and a UASB reactor was applied for the treatment of wash waters from the purification of virgin olive oil. The COD removal efficiency reached 89% (Borja-Padilla et al., 1996). AquatecOLIVIA technology — for the treatment of OMW (DE19829673, 2000) in a two-phase anaerobic treatment , consisted of an initial step of acidification and a second step of methanogenesis using moving-bed reactor with hollow cylindrical supports acting as a carrier material (Niaounakis and Halvadakis, 2006). The purification ranged from 92 to 97% and the production of biogas had a methane content of 60–65%.

Aerobic Treatment

Aerobic Bacteria

Various studies have used different bacteria in aerobic conditions for the biodegradation of olive mill wastewaters. A detailed list of bacteria tried by different researchers for the treatment of olive mill effluent is given in Table 2. Aerobic bacteria have been tested primarily for their ability to remove phytotoxic compounds (monoaromatic or simple phenolics) from olive mill effluents. The strain *Azotobacter chroococcum* was the first bacteria which grew on substrates containing aromatic compounds and could metabolize them (Hardisson et al., 1969, Moreno et al. 1990).

A different strain *A. vinelandii* could remove 50% COD. It was also shown that a complete removal of the phytotoxicity of OMW was achieved by the enrichement of the medium with culures isolated from industrial and municipal wastes and agricultural soil (Zouari and Ellouz, 1996a). The former strain has been used as an inoculum for aerobic biological treatment aiming at detoxifying OMW by other researchers (Ehaliotis et al., 1999, Piperidou et al., 2000). They achieved greater than 90% removal of phytotoxic compounds.

Bacillus pumilus, was able to reduce 50% of the phenolic content of OMW. A complete degradation of protocatechuic and caffeic acid was observed but it had no effect on tyrosol (Ramos-Cormezana et al., 1996).

A gram positive bacteria, *Arthrobacter sp.*, transformed completely tyrosol to 4-hydroxyphenylacetic acid (Knupp et al., 1996).

The effectiveness of two aerobic bacterial strains, *Ralstonia sp.* and *Pseudomonas putida* was tested to degrade monocyclic aromatic acids in Italian OMW (DiGioia et al., 2001). The two strains grew and biodegraded extensively a synthetic mixture of nine monocyclic aromatic acids commonly found at high concentrations in OMW. A total of 10 bacterial strains were isolated from a compost of corn treated with olive mill wastewaters (OMW) and selected by their capacity to synthesize exopolysaccharides (EPS). The strain which was able to produce the highest EPS yield was chosen to perform an assay for testing its putative detoxifying activity and it showed to reduce more than half the toxic capacity of the OMW (Aguilera et al., 2008). Some of the bacteria that have been used for the treatment of olive mill wastewaters are shown in Table 3.

Table 3. Bacteria employed for treatment of olive mill wastewaters

Bacteria	COD removal (%)	Reference
Azotobacter chroococcum		Hardisson et al., 1969 Moreno E. et al. 1990
Azotobacter vinelandii	50	Zouari and Ellouz, 1996a
Azotobacter vinelandii	90	Ehaliotis C.C., 1999 Piperidou C.I., 2000
Bacillus pumilus	50	Ramos-Cormezana et al., 1996
Arthrobacter sp		Knupp et al., 1996
Ralstonia sp. Pseudomonas putida		DiGioia et al., 2001b
Bacterial stain From compost of corn	50	Aguilera et al., 2008

Fungi

Fungi may also have a role in OMW biodegradation. The strong potential of fungi for degradation of organic pollutants is related to their ability to grow in environmental conditions unfavorable for bacteria (e.g., low pH). Interest in bioremediation led scientists to evaluate the ability of white rot fungi to degrade phenolics in OMW due to their non-specific extracellular enzymes. In addition, reduction of COD and colour has been achieved. Recent studies have therefore focused on species such as *Pleurotus ostreatus*, *Lentinula edodes*, *Coriolus versicolour*, *Phanerocahete sp. Lentinula edodes* (shiitake mushroom) was able to decolourise an OMW, removing 75% COD and 66% phenolics (Vinciguerra et al., 1995). Tyrosol was converted by cell-free preparation of the fungus into a dimeric tetracyclic ketone. In another study, the same fungus reached around 88% removal of COD and phenolics and a significant decolourisation was observed (D'Annibale et al., 1998). The study revealed that the presence of laccase and manganese-peroxidase (MnP) were responsible for the depolymerisation of high molecular weight fraction and phenol removal.

Phanerochaete chrsysosporium is a fungus that has been extensively used as a model organism to metabolize a wide range of diverse environmental pollutants. Lignin peroxidase (LiP) and manganese peroxidase (MnP) were the major enzymes in the degradation of phenolic compounds in OMW (Sayadi and Ellouz, 1992). *P. chrysosporium* decreased the COD about 50%. The comparison between *P. chrysosporium* and *Pl. ostreatus* showed that *P. chrysosporium* isolate was more efficient in decolourizing and detoxifying OMW, reducing more than 50% of the colour and phenols (Kissi et al., 2001). A different strain of *Phanerochaete*, P. *flavido-alba*, attained 90% decrease in the OMW phenolic content under nitrogen-limited media. The enzymes involved in OMW decolourisation process were the manganese-peroxidase (MnP) and the phenol oxidase (laccase), whereas lignin peroxidase (LiP) was not detected in the growth environment (Hamman et al. 1999).

The ability of sixteen strains of six *Pleurotus* species to degrade and decolourise OMW was investigated in solid cultures (Flouri et al., 1996). The capability of *Pleurotus ostreatus* to degrade phenols of OMW was investigated in different conditions such as in sterilized and thermally processed (at 100°C) wastewater, with and without dilution. *P. ostreatus* removed phenols from the culture medium, under all different conditions that were examined. The degradation of phenols reached up to 78.3% for the sterilized and 50% diluted OMW, 66.7% and 64.7% for the thermally processed OMW, with and without dilution, respectively (Fountoulakis et al., 2002). The experiments were conducted in agar plates containing different concentrations of OMW. For all strains tested, decolourisation proceeded more slowly than radial growth. Among the six Pleurotus species, a strain of *P. ostreatus* and *P. cornucopiae* were the most efficient, whereas two strains of *P. cystidiosus* showed no decolourisation. The edible mushroom, *Pleurotus ostreatus*, has been shown to degrade phenolic compounds from OMW. Laccase plays an important role in the oxidation of phenolic compounds and aromatic amines, by reducing molecular oxygen to water (Tomati et al., 1991). Laccase is able to catalyse the oxidation of polyphenols, aromatic amines and a series of other compounds. In a typical laccase reaction, as it is shown in Fig. 3, a phenolic substrate is subjected to a one-electron oxidation giving rise to an aryloxy radical. This active species can be converted to a quinone in the second stage of oxidation. The quinine as well as the free radical product undergo non-enzymatic coupling reactions leading to polymerisation (Minussi et al., 2002).

In another study (Tsioulpas et al., 2002), several *Pleurotus spp.* strains were tested to remove phenolic compounds from OMW, with respect to their laccase activity. All strains tested in this work were able to grow in OMW without any addition of nutrients and a high laccase activity was measured in the growth medium, while 69–76% of the initial phenolic compounds were removed. The black colour of OMW became yellow-brown and brighter as the strains grew. A decrease of the phytotoxicity was noticed in OMW treated with some *Pleurotus spp.* strains.

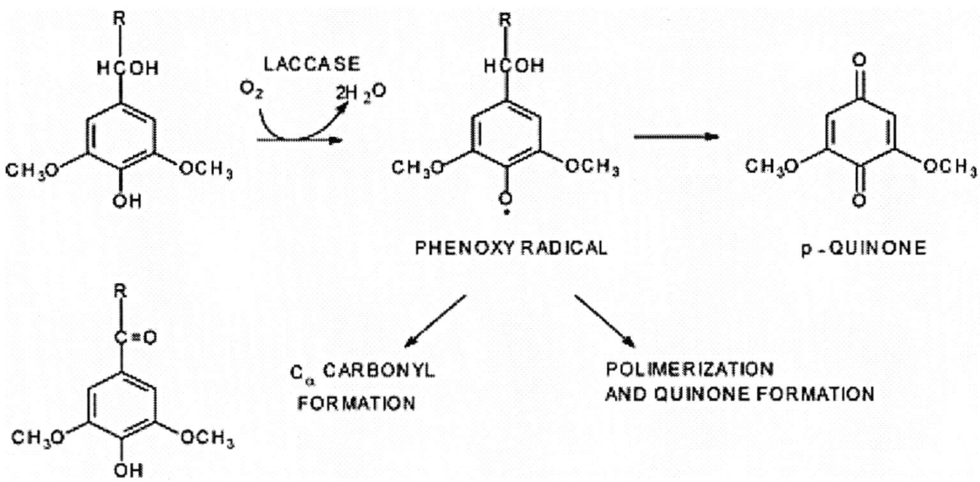

Figure 3. Phenols oxidation catalysed by laccase (Minussi et al., 2002).

Funalia trogii was used for the phenol removal and decolourisation of OMW in static and agitated cultures (Yesilada O. et al., 1995, and 1998). The white rot fungus *F. trogii* showed 38% colour removal, 77 % phenol removal and 40% COD reduction. The white rot fungus *Coriolus versicolour* removed 80% phenol and 50% colour without any additional nutrients (Yesilada and Fiskin, 1996). Laccase was secreted by the white fungi, *C. versicolour* and *F. trogii*, grown on media with diluted OMW and vinasse. Addition of spent cotton stalks enhanced the laccase activity with a maximum after 12 days of cultivation (Yesilada O. et al., 1998). During growth in optimum conditions, *C. versicolour* removed approximately 63% COD, 90% phenol and 65% colour within 6 days. *F. trogii* removed approximately 70% COD, 93% phenol and 81% colour of the OMW used.

High biodegradation yields were also obtained by fungi immobilized in calcium alginate gels (Yesilada O. et al., 1998). It has been well established that various *Aspergillus sp.* are able to degrade phenols and remove the colour in a variety of wastewaters, probably through their capacity to excrete various extracellular enzymes (Borja et al., 1990, 1995, Miranda et al., 1996, Hamdi et al., 1991). *Aspergillus niger* and *A. terreus* reduced COD and phenol concentration greater than 60% (Martinez Nieto et al., 1993). Free cells of *A. niger* were grown on OMW supplemented with rock phosphate in a 5 L air-lift bioreactor for 8 days at 30°C. The fungus grew well, attaining a maximum removal of COD by 64%. A 60% reduction in total sugars was also achieved, but no alteration was observed in phenol levels (Cereti et al., 2004).

Stored-black OMW are not readily biodegradable compared to the fresh because darkly coloured phenolic compounds are formed through autooxidation and polymeristaion of phenolic compounds and tannins. *Geotrichum candidum* grown on fresh OMW reduced COD and colour by 50 and 75%. On the other hand, no decolourisation was observed on the stored black OMW and the reduction of COD was about 25% (Assas et al., 2000, and 2002).

The effectiveness of different types of fungi, namely *Geotrichum sp.*, *Aspergillus sp.*, and *Candida tropicalis* for the biodegradation of OMW was studied (Fadil et al., 2003). They were selected for their tolerance to the polyphenols and the initial concentrations of COD varied. The reduction of COD (55-63%) and phenolics (44-52%) was similar for the three species and significant decolourisation was evident. Two white rot fungus (*Phanerochaete chrysosporium* and *Geotrichum candidum*) and two species of *Aspergillus* (*A. niger* and *A. terreus*) were also compared for removal of phenols in OMW. The order of the removal efficiency of total phenols was *P. chrysosporium*>*A. niger*> *A. terreus*, while no removal was observed for *G. candidum* (Garcia Garcia et al., 2000).

Yeasts have used for the bioremediation of OMW and were effective at COD reduction and phenol removal. Immobilized cells of *Candida tropicalis* in calcium alginate beads resulted in 69.7, 69.2, and 53.3% reduction of COD, monophenols and polyphenols, respectively (Ettayebi et al., 2003). Three yeasts *Candida krusei*, *Saccharomyces chevalierie*, and *Saccharomyces rouxii* were selected on their ability to tolerate the polyphenols on OMW, reducing BOD_5 at a range 40–50% (Gharsallah, 1993). *Yarrowia lipolytica* demonstrated the greatest potential to reduce COD on OMW in shaker-flasks, among five yeasts (Scioli and Felice, 1993). In another study, *Trichosporon cutaneum* (Chtourou et al., 2004), showed a significant reduction of COD and phenol which exceeded 80% of the original loading after eight days of treatment of the OMW extract. Some of the fungi that have been used for the treatment of olive mill wastewaters are shown in Table 4.

Table 4. Fungi employed for treatment of olive mill wastewaters

Fungi	COD removal (%)	Colour removal (%)	Phenolics removal (%)	Reference
Phanerochaete chrsysosporium	50		50	Sayadi and Ellouz, 1992 Kiss et al., 2001
Aspergillus niger, *A. terreus*	>60		>60	Martinez Nieto et al., 1993
Lentinula edodes	75		66	Vinciguerra et al.,1995
Funalia trogii	38	40	77	Yesilada et al., 1995
Coriolus versicolour	50		80	Yesilada and Fiskin, 1996
Lentinula edodes	88		88	D'Annibale et al., 1998
C. versicolou	63	65	90	Yesilada et al., 1998
r F. trogii	70	81	93	
Phanerochaete flavido-alba	90			Hamman et al. ,1999
Geotrichum candidum	50	75		Assas N. et al., 2000
Phanerochaete chrysosporium, Geotrichum candidum, Aspergillus niger, A. terreus			*P. chrysosporium*> *A. niger*> *A. terreus*	Garcia Garcia et al., 2000
Pleurotus species			69–76	Tsioulpas et al., 2002
Pleurotus ostreatus			78	Fountoulakis et al., 2002
Geotrichum sp., Aspergillus sp., Candida tropicalis	55-63		44-52	Fadil K. et al.,2003
A. niger	64			Cereti et al., 2004

Algae

Chlorella pyrenoidosa and *Scenedesmus obliquus*, two microalgae, reduced the amounts of certain components, essentially sugars and salts, thereby diminishing the pollutant effect of this residue (Sanchez-Villasclaras et al., 1996). *Ankistrodesmus braunii* and *Scenedesmus quadricauda*, two phenol resistant microalgae, showed a limited reduction of phenol content after 5 days of treatment. In contrast, the co- culture of the two algae, grown in the dark, degraded over 50% of the low molecular weight phenols contained in OMW into other non-identified aromatic compounds (Pinto et al., 2003).

PULP AND PAPER MILL INDUSTRIES

The wood puplping and the production of paper generate large amounts of wastewaters. On an average 330 m^3 of effluents are generated for a ton of product. Pulping process is the initial stage for paper making and utilizes large amounts of water, resulting in the largest source of pollution. The main steps in pulp and paper manufacturing are raw material preparation, such as wood debarking and chip making, pulp manufacturing, pulp bleaching, paper manufacturing and fiber recycling. Among the processes, chemical pulping generates a high-strength wastewater (Pokhrel and Viraraghavan, 2004). The quantity and quality of the wastewaters discharged depend on the type of process, type of wood materials, process technology, management practices, internal recirculation of the effluent and the used amount of water. The effluent generated at the pulping stage (black liquor) contains compounds originating from the raw material of the pulp such as cellulose, hemicellulose, lignin (Fig. 4), resin acids, fatty acids, tannins and phenols.

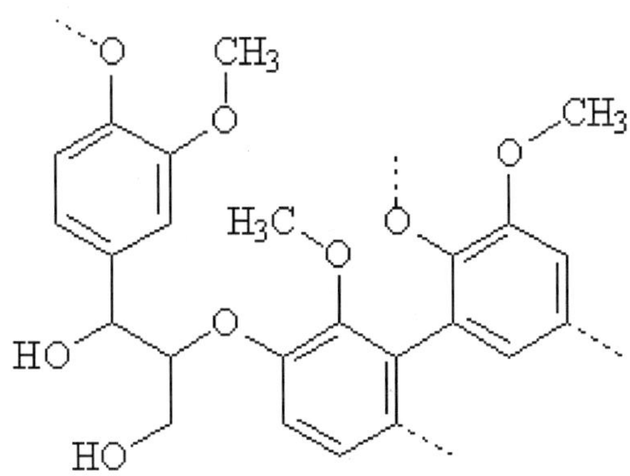

Figure 4. Simplified structure of lignin, a three-dimensional polyaromatic polymer in woody plants.

These organic compounds are also responsible for giving the effluent its characteristic dark brown colour and toxicity (Singhal and Thakur, 2009). The dark brown colour is due to the formation of lignin degradation products during the processing of lignocellulosics from paper and pulp manufacture. Wastewaters also contain the chemicals added during the manufacturing process. Sulfur compounds and nitrogen oxides are emitted to the air, and chlorinated and organic compounds, nutrients, and metals are discharged to the wastewaters (Rintala and Puhakka, 1994). About 500 different chlorinated organic compounds have been identified in paper mill effluent including chloroform, chlorate, resin acids, chlorinated hydrocarbons, phenols, catechols, guaiacols, furans, dioxins, syringols, vanillins, etc. (Savant et al., 2006). The high chemical diversity of these pollutants causes a variety of carcinogenic, endocrinic and mutagenic effects on fishes. The undiluted effluents are toxic to aquatic organisms and exhibit a strong mutagenic effect. Furthermore, some compounds in the effluents are resistant to biodegradation and can bioaccumulate in the aquatic food chain. Serious soil chemistry changes have been caused by the irrigation with pulp and paper mill effluents. The various elements introduced via pulpmill wastewater irrigation differ not only in their effects on crop growth and soil properties but also with regard to their relative mobility in the soil profile (Howe and Michael, 1998).

TREATMENT TECHNOLOGIES

Physicochemical Methods

Although physical and chemical methods are available for the treatment of pulp and paper mill effluent, biological processes are more desirable. Physicochemical wastewater treatment typically includes sedimentation/floatation, screening, adsorption, coagulation, oxidation, ozonation, reverse osmosis, ultra-filtartion and nano-filtration technologies. Sedimentation removes suspended matters in wastewaters such as bark particles, fiber debris, filler and coating materials. Sedimentation contributed greater than 80% removal of suspended solids in paper mills in UK (Thompson et al., 2001). In another study, air flotation improved wastewater treatment and was able to remove 95% of the total suspended solids (TSS) (Wenta and Hartmen, 2002). Coagulation–flocculation is one of the most used water effluent treatments. It employes a cationic metal as a coagulant agent that usually promotes water hydrolysis and the formation of hydrophobic hydroxide compounds with different charges, depending on the solution pH. Several researchers have applied chitosan as a coagulant alone or associated with iron and aluminum salts (Pinnoti et al., 1997, Bratskaya et al., 2004). Coagulation–flocculation using $FeCl_3$ as a coagulant agent eliminated several impurities from water efficiently, allowing the replacement of aluminum salts, which may be hazardous to health (Rodriguesa et al., 2008). Removal of TC, colour and turbidity of up to 88, 90 and 98%, respectively, were observed in a combined bleached chemithermomechanical/thermomechanical (BCTMP/TMP) effluent (Stephenson and Duff, 1996). Activated carbon adsorption has been successfully implemented as a unit process in both potable and wastewater treatment. It has been found that chlorinated phenols can be strongly adsorbed by activated carbon. Synthetic and natural allophanic compounds were used to evaluate the phenolic compounds and colour removal capacity from a biological treated softwood bleached Kraft mill effluent. Phenolic compounds and colour removal reached 90 and 75%, respectively (Diez et al., 1999).

In a study, pulp mill effluent containing toxic chemicals was treated by different advanced oxidation processes (AOPs). The treatments consisted of hydrogen peroxide, Fenton's reagent (H_2O_2/Fe^{2+}), UV, UV/ H_2O_2, photo-Fenton (UV/ H_2O_2/Fe^{2+}), ozonation and peroxone (ozone/ H_2O_2) in laboratory-scale reactors for colour, total organic carbon (TOC) and adsorbable organic halogens (AOX) removals from the pulp mill effluent. Fenton's reagent utilizing H_2O_2/Fe^{2+} resulted in the highest colour, TOC and AOX removals under acidic conditions when compared with the other AOPs tested. Approximately, 88% TOC, 85% colour and 89% AOX removals were obtained by the Fenton's reagent. Photo-Fenton process yielded comparable TOC (85%), colour (82%) and AOX (93%) removals within 5 min due to oxidations by UV light in addition to the Fenton's reagent. Hydrogen peroxide treatment alone resulted in very low TOC (5.1%), AOX (34%) and colour (24%) removals, which were not satisfactory (Catalkaya and Kargi, 2007).

Membrane filtration is being adopted increasingly in the paper mills effluents. Application of membrane technology is one such option which can improve the recycled water quality by removal of heavy metal contaminates, COD, TDS, AOX (Adsorbable organic halides) and colour. Ultrafiltration in combination with air flotation achieved the removal efficiencies in terms of conductivity, TOC, colour, TSS, 40%, 65%, 90%, and 100%, respectively (De Pinho et al., 2000). It was reported that 88% and 89% removal of BOD, and COD, respectively, was achieved by reverse osmosis (RO) (Dube et al. 2000).

Biological Treatment

Anaerobic Treatment

Anaerobic filter, upflow sludge blanket (UASB), fluidized bed, anaerobic lagoon and anaerobic contact reactors are anaerobic processes that are commonly used to treat pulp and paper mill effluents. Among all anaerobic systems, the UASB reactor has become the most popular reactor system for the treatment of paper mill effluents. The first UASB reactors were installed for the treatment of paper and board mill effluents (Habets and Knelissen, 1985). These processes are very sensitive to environmental conditions and, therefore, strict controls of pH, temperature, volatile acids, nutrients and toxicity are required for their effective operation.

In a study, black liquor and bleach effluent from pulp and paper mill were treated anaerobically to reduce their high chemical oxygen demand (COD) and adsorbable organic halide (AOX) contents. After the addition of glucose, black liquor and bleach effluent produced only 80 and 76% methane with COD reductions of 71 and 76%, respectively (Ali and Sreekrishnan, 2000). A bench scale upflow anaerobic sludge blanket reactor (UASB) was evaluated for the treatment of diluted black liquor from a Kraft pulp plant. The UASB reactor was stable, easy to operate with low maintenance and operational costs. The average chemical oxygen demand removal efficiency during the entire experiment was 80% (Buzzini and Pires, 2002). Upflow anaerobic sludge blanket reactor was installed to replace the conventional anaerobic lagoon treating bagasse wash wastewater from agro-based pulp and paper mill. In the plant 80–85% COD reduction was achieved with biogas production (Chinnaraj and Rao, 2006). Another upflow anaerobic sludge blanket reactor showed over 70% of COD removal efficiency for the wastewater from a liner paper factory. A considerable amount of alkalinity was produced in the reactor when sulphate was reduced to hydrogen sulphide by sulphate reducing bacteria, SRB (Kim et al., 2003). Perez et al. (1998) evaluated two anaerobic systems (anaerobic filters and fluidized bed) in laboratory-scale reactors and reported that 81.5% organic removal efficiency was obtained in the case of fluidised bed with porous packing and 50% removal was obtained in the case of anaerobic filters on corrugated plastic tubes.

It has been also shown the feasibility of anaerobic treatment contaminated with Kraft black liquors and other pulp and paper industry wastewaters in methanogenic, thermophilic fluidised bed reactors (Poggi-Varaldo et al., 1996). Raizer-Neto et al. (1989) studied the decrease in chlorinated organics in bleaching effluents in a continuously fed fluidized bed reactor. Polyurethane (PU) foam cubes were used as the support material. They found that COD and AOX removal efficiency were primarily affected by AOX loading rate, whereas BOD removal efficiencies were affected at much higher AOX concentrations. The microbial community in the sludge blanket of a bench scale upflow anaerobic sludge blanket (UASB) reactor with partial recirculation of the effluent, treating wastewater from an unbleached pulp plant, was evaluated. The average COD removal efficiency was 80–86% without recirculation and 75–78% with recirculation (Buzzini et al., 2006).

The behaviour of twelve selected chlorinated compounds, phenols, catechols, guaiacols and vanillins during the high-rate anaerobic treatment of segregated kraft mill bleach plant effluents was studied. Eleven of the twelve detected chlorinated compounds were reduced by 80%. The behaviour of the chlorinated compounds was affected by the chlorine position and degree of chlorination (Parker et al., 1993).

The thermophilic (55°C) anaerobic removal of chlorinated phenolic compounds from softwood bleaching effluents using 4 different types of anaerobic processes—an upflow anaerobic sludge blanket (UASB) digester, a UASB digester enriched with sulfate, a UASB digester with recirculation and a fixed bed reactor with recirculation was tested (Lepisto and Rintala, 1994). In all processes, most of the chlorinated phenols, catechols, guaiacols and hydroquinones were either eliminated or reduced at a range from 80 to 95%. All digesters removed 30–70% of the COD and 25–67% of the AOX. The alkyl dehalogenators include physiologically diverse groups of strict anaerobes and facultative anaerobes, such as methanogens species of *Clostridium*, *Acetobacterium woodii*, *Shewanella* and proteobacteria such as *Desulfitobacterium*, etc. Three mechanisms have been suggested by which anaerobic bacteria utilize chlorinated organic compounds, as a source of carbon and energy, a cometabolite and an electron acceptor (Savant et al., 2006).

Aerobic Treatment

Bacteria

Bacterial strains reduce colour of kraft mill effluent up to a limited extent but they are quite effective in reduction of recalcitrant organic phenolics and derivatives formed during degradation. Various bacteria can decompose monomericlignin substructure models, but only a few strains are able to attack lignin derivatives obtained from different pulping processes (Bajpai and Bajpai, 1994). Several bacteria have been tested for their ability to degrade resin acids such as *Bacillus* species, *E. coli* , *Pseudomonas*, *Flavobacterium* species, *Anthrobacter sphinomonas*, *Zooglea*, *Commamonas*, *Mortierella*, *Chaetomium cochliolidae*, *Corticum sasaki* and *Fomes annosus* (Liss et al., 1997).

The capability of three bacteria, *Pseudomonas aeruginosa*, *P. aeruginosa* and *Bacillus megaterium* was exploited to reduce the BOD and COD level of pulp and paper mill effluents up to permissible levels. A concomitant reduction in TDS (total dissolve solid), AOX and colour (76%) was also observed (Tiku et al., 2010).

Four different bacterial strains were isolated from pulp and paper mill sludge, in which one alkalotolerant isolate (LP1) having higher capability to remove colour and lignin (69% and 53%, respectively), was identified as *Bacillus* sp. Enzymes involved in the process of decolourisation of effluent were found to be xylanase and manganese peroxidase (Mishra and Thakur, 2010). *Bacillus cereus* and *Serratia marcescens*, two pentachlorophenol (PCP)-

degrading bacterial strains were used for the treatment of pulp and paper mill effluent (Chandra et al., 2008). These two bacterial strains effectively reduced colour (45-52%), lignin (30-42%), BOD (40-70%), COD (50-60%), total phenol (32-40%) and PCP (85-90%) within 168 h of incubation. However, mixed culture treatment showed the highest reduction in colour (62%), lignin (54%), BOD (70%), COD (90%), total phenol (90%) and PCP (100%).

Three lignin-degrading bacterial strains, *Paenibacillus sp.*, *Aneurinibacillus aneuvinilyticus* and *Bacillus sp.* were used for the treatment of pulp and paper mill effluent. All three bacterial strains effectively reduced colour (39-61%), lignin (28-53%), biochemical oxygen demand (BOD) (65-82%), chemical oxygen demand (COD) (52-78%) and total phenol (64-77%) within six days of incubation. However, the highest reduction was achieved by *Bacillus sp.* (Raj et al., 2007). In another study, the three former strains proved capable to effectively degrade the kraft lignin. The mixed culture of these three bacteria showed the highest reduction of colour by 69%, lignin by 40% and total substrate by 50% compared to the pure cultures of bacteria (43-65% colour removal). Complete degradation of lignin was concluded from HPLC analysis (Chandra et al., 2007).

Pseudomonas fluorescens obtained from a bacterial consortium degraded 4-chlorophenol through an oxidative route. The bacterial strain was applied for the treatment of pulp and paper mill effluent in sequential bioreactor. The strain significantly reduced colour (75%), phenol (66%), COD (79%) and lignin (45%) and increased chloride content (92%) after 15 days in bioreactor (Chauhan and Thakur, 2002).

Batch and continuous studies were conducted on the treatment of black liquor from a kraft pulp and paper mill by a bacterial strain of *Aeromonas formicans*. The strain was able to remove 71% and 78% of COD and lignin, respectively, while the colour removal efficiency was around 86% in batch cultures. Similar removal activities were obtained in continuous bioreactor (Gupta et al., 2001).

Two resin-acid-degrading bacteria, *Pseudomonas abietaniphila* BKME-9 and *Zoogloea resiniphila* DhA-35, were tested in laboratory sequencing batch reactors (SBRs) for their ability to enhance resin acid removal. Both bacteria enhanced resin acid removal but not removal of total organic carbon (TOC) by either pH-shocked or starved activated sludge (Yu and Mohn, 2001).

Eighteen bacterial strains able to use chlorophenols as the sole carbon source were isolated from water samples from a river receiving pulp mill discharges. Five of them were *Pseudomonas fluorescens*, two *Alcaligenes sp.*, one *Citrobacter freundii*, one *Aeromonas sp.* and nine unidentified Gram negative bacilli (Martinez et al., 2000).

Ancylobacter aquaticus A7, *Pseudomonas* P1, and *Methylobacterium* CP13, strains indigenous to a bleached kraft-mill effluents treatment system, were tested for growth on chlorinated acetic acids and alcohols and for adsorbable organic halogen (AOX) reduction in batch cultures of sterile bleached kraft-mill effluents from three sources. Mixed inocula of all three species combined achieved highest removal AOX (Fulthorpe and Allen, 1995).

Many mesophilic bacteria have been isolated and characterized for their ability to degrade resin acids. *Pseudomonas*, IpA-1 and IpA-2, were capable of growing on isopimaric acid as the sole carbon source and electron donor and showed good resin acid removal capacities (Wilson et al. 1996).

Pseudomonas aeruginosa was capable of reducing Kraft mill effluent colour by 26-54% or more under aerobic conditions (Blair and Davis, 1980).

Reddy et al. (2005) treated the pulp and paper mill effluent by using thermophilic microorganisms in batch systems. Maximum removal efficiency of 55.2% was achieved at 40ºC. Ruiz-Ordaz et al. (2001) investigated phenol biodegradation using repeated batch culture of *Candida tropicalis* in a multistage bubble column. The phenol removal efficiency of 98.7% was achieved. Some of the bacteria that have been used for treatment of pulp and paper mill wastewaters are shown in Table 5.

Table 5. Bacteria employed for treatment of pulp and paper mill wastewaters

Bacteria	COD removal (%)	Colour removal (%)	Phenolics removal (%)	Reference
Pseudomonas aeruginosa		26-54		Blair and Davis, 1980
Ancylobacter aquaticus Pseudomonas Methylobacterium				Martinez et al., 2000
Aeromonas formicans		86		Gupta et al., 2001
Candida tropicalis			98	Ruiz-Ordaz et al., 2001
Pseudomonas fluorescens	79	76	75	Chauhan and Thakur,2002
Paenibacillus sp. Aneurinibacillus aneuvinilyticus Bacillus sp.	52-78	39-61	64-77	Raj et al.,2007
Bacillus cereus Serratia marcescens	50-60	45-52%	85-90	Chandra et al., 2008
Bacillus sp		69		Mishra and Thakur, 2010
Pseudomonas aeruginosa P. aeruginosa Bacillus megaterium		76		Tiku et al., 2010

Fungi

Fungi, especially the white rots, have been shown to be the most efficient degradersof lignin. White rot fungi are filamentous fungi that inhabit the wood of dead or dying trees. One of the most efficient lignin degrading and decolourizing white rot fungus is *P. chrysosporium*, a well-studied microorganism. *Phanerochaete chrysosporium* was successfully employed for the decolourisation of pulp and paper mill effluent and removal of undesirable pollutants. Lignin metabolism in *P. chrysosporium* is a secondary metabolic event and is triggered by carbon-sulfur or nitrogen limitation, even in the absence of lignin (Kirk et al., 1976, Eaton et al., 1980).

In a study, an evaluation of the effects of a tertiary treatment by fungi (*Pleurotus sajor-caju*, *Trametes versicolour* and *Phanerochaete chrysosporium* and *Rhizopus oryzae*) on individual organic compounds of a *Eucalyptus globulus* bleached kraft pulp and paper mill final effluent after secondary treatment (final effluent). The tertiary treatment with the fungi was performed in batch reactors, which were inoculated with separate species and monitored throughout the incubation period. The four fungi species showed an adequate capacity to remove organic compounds and colour. *R. oryzae* was able to remove 99% of organic compounds. *P. sajor-caju*, *T. versicolour* and *P. chrysosporium* were able to remove 97%, 92% and 99% of organic compounds, respectively. Lignin derivatives such as vanilic acid, syringic acid, guaiacol, syringol and phloroglucinol were totally removed by *R. oryzae* (Rocha-Santos et al., 2010).

The potential of white rot fungi, namely, *Polyporus versicolour* FRI 165, *Polyporus hirsutus* FRI 534, *Polyporus hirsutus* FRI 2862, *Phanerochaete chrysosporium* BKMF 1767 and *Sporotrichum pulverulentum* for the reduction of colour of pretreated pulp and paper mill effluents under sterilised and unsterilised conditions. Maximum decolourisation (70%) of the effluent was obtained with pH adjusted to 4.5 at 40°C, along with more than 50% reduction in biological oxygen demand (BOD) and chemical oxygen demand (Dev Mittar et al., 1992).

Eight isolates could be cultured to test for thermotolerance and to screen for the presence of ligninolytic enzymes using *Phanerochaete chrysosporium* as a reference. Three species, *Daedaleopsis sp.*, *Schizophyllum commune* PT, and *S. commune* SL, were able to grow above 40°C. All fungal cell suspensions tested decolourized wastewater from the pulping process and wastewater from the pulping process combined with that from the paper recycling process. Laccase activities were detected in the decolourized effluents and all fungi tested reduced the COD by 59-83% (Prasongsuk et al., 2009).

Two basidiomycetous fungi (*Merulius aureus* and *Phlebia sp.* and an unidentified genus) and a deuteromycetous fungus (*Fusarium sambucinum*) were isolated from soils polluted with effluents of a pulp and paper mill. The isolates were immobilized on nylon mesh and the consortium was used for bioremediation of pulp and paper mill effluent in a continuously aerated bench-top bioreactor. The treatment resulted in 79% colour reduction, 79.0% reduction of lignin and 90% COD removal (Malaviya and Rathore, 2007).

Pleurotus sajor-caju secretes phenol-oxidases that enable the use of recalcitrant compounds as substrates. Cultures of the fungus resulted in about 42% colour reduction. The polyphenol content was also reduced by 59% by the 13th day of culture (Munari et al., 2007).

A marine fungal isolate, NIOCC2a cultured from decaying mangrove wood. The fungus also decolourized several synthetic dyes. Laccase was the most dominant lignin-degrading enzyme. Industrial effluents and synthetic dyes added to the growing culture of this fungus were decolourized to a great extent (D'Souza et al., 2006).

Aspergillus foetidus was studied for its ability to remove colour, COD and lignin from bagasse-based pulp and paper mill wastewater (Sumathi and Phatak, 1999). Nearly 90–95% of the total colour was removed, and the simultaneous reduction in colour and lignin level indicated a strong correlation between the decolourisation and ligninolytic processes.

The lignin-degrading abilities of five white rot fungi, *Phanerochaete chrysosporium*, *Pleurotus ostreatus*, *Lentinus edodes*, *Trametes versicolour* and S22, on the supporting media were evaluated. The fungi were grown on a porous plastic media, were individually used to treat black liquor from a pulp and paper mill. Over 71% of lignin and 48% of chemical oxygen demand (COD) were removed from the wastewater (Wua et al., 2005).

Purified cellobiose dehydrogenase from two different fungal sources was assessed for its ability to remove and/or reduce colour from pulp mill bleach plant effluent. Cellobiose dehydrogenase purified from *Phanerochaete chrysosporium* was shown to prefer acidic conditions and was consequently used to treat the acid effluent stream discharged from a pulp mill bleach plant, while an analogous enzyme originating from *Humicola insolens* preferred alkaline conditions, and was applied to the effluent discharged from the caustic sewer of the bleach plant. Up to 50% of the effluent colour was removed within 4 days when treated under optimised conditions (Wingate et al., 2005). Under optimal conditions using cell free system, total colour and COD (chemical oxygen demand) decreased by almost 82% and 75%, respectively, after cultivation for 5 days. Simultaneously, over 80% of total aromatic compounds were degraded (Yusef et al., 1999).

Gliocladium virens, a saprophytic soil fungus decolourised paper and pulp mill effluents by 42% due to the production of hemicellulase, lignin peroxidase, manganese peroxidase and laccase (Murugesan, 2003).

Trametes versicolour was used in the form of pellets in aerated reactors (fluidized, stirred and air-pulsed reactors). Reductions in colour and aromatic compounds of 70-80% and in COD of 60% were achieved. From the results, a relationship between laccase production and toxicity reduction was revealed (Font et al., 2003).

The potential of *Pleurotus ostreatoroseus* for use in the treatment of the first alkali extraction stage (E1) effluent of a kraft bleach plant was studied. The best results were obtained when the effluent was diluted and enriched with glucose. The removal of colour, of total phenols and of lignin/chlorolignin was 85%, 82% and 72%, respectively (Santos et al., 2002).

In a rotating biological contactor (RBC) reactor effluent decolourisation was studied using *Coriolus versicolour*, a white-rot fungus and *Rhizomucor pusillus* strain RM7, a mucoralean fungus. Decolourisation by both fungi was directly proportional to initial colour intensities. It was shown that decolourisation of 53 to 73% could be attained using a hydraulic retention time of 23 h. With *R. pusillus*, 55% of AOX were removed compared to 40% by *C. versicolour* (Van Driessel and Christov, 2001). Decolourization and phenol reduction of paper mill effluent by the fungus, *Aspergillus fumigatus* isolated from paper mill effluent were studied. It was observed 87.5% removal of colour and 51% reduction of phenol in the effluent (Jeyaramraja et al., 2001).

Aspergillus foetidus was studied for its ability to remove colour, COD (chemical oxygen demand) and lignin from dissolved bagasse lignin, raw black and alkali stage liquors in nutrient medium. Nearly 90-95% of the total initial colour was removed from growth medium. Simultaneous reduction in colour and lignin content of the samples indicate strong correlation between the decolourisation and lignolytic processes (Sumathi and Phatak, 1999)

The bagasse-based paper mill effluent employed for a decolourization study by the white-rot fungus *Trametes versicolour*. A high degree of decolourization (40%) by the native effluent microflora was noted on the second day in effluent supplemented with ammonium nitrate and glucose but uninoculated with the fungus. A maximum colour removal of the order of 60% was achieved on the fourth day of culture in effluent inoculated with *T. versicolour* after addition of ammonium nitrate and glucose cosubstrate (Modi et al., 1998).

A comparative study investigated the biodegradation of lignin and reduction of BOD, COD and colour in the pulp and paper mill effluents by natural flora initially present in the effluents and the fungal culture *Pleurotus ostreatus*. The inoculated fungal culture showed better results with 76.8% COD reduction, 60.0% BOD reduction and about 80% colour removal (Choudhury et al., 1998).

The reduction of colour and adsorbable organic halides (AOX) from the kraft bleach plant effluents was investigated using a laboratory-scale fungal bioreactor, containing calcium-alginate-entrapped white-rot fungus *Trametes versicolour*. A colour reduction percentage of 69% (± 6%) and AOX reduction percentage of 58% (± 6%) were achieved (Pallerla and Chambers, 1997).

Table 6. Fungi employed for treatment of pulp and paper mill wastewaters

Fungi	COD removal (%)	Colour removal (%)	Phenolics removal (%)	Reference
Lower fungi	48	60		Marchand, 1978
Coriolus versicolour		80		Livernoche et al., 1983
Polyporus versicolour *Polyporus hirsutus* *Polyporus hirsutus* *Phanerochaetec hrysosporium* *Sporotrichum pulverulentum*	50	70		Dev Mittar et al., 1992
Trametes versicolour.	69	92		Mehna et al., 1995
Trametes versicolour		63-75		Pallerla and Chambers, 1997
Pleurotus ostreatus	77	80		Choudhury et al., 1998
Trametes versicolour		60		Modi et al., 1998
Phanerochaete chrysosporium	75	82		Yusef et al., 1999
Aspergillus foetidus		90–95%		Sumathi and Phatak, 1999
Aspergillus fumigatus		88	51	Jeyaramraja et al., 2001
Coriolus versicolour *Rhizomucor pusillus l*		53 - 73		Van Driessel and Christov, 2001
Pleurotus ostreatoroseus		85	82	Santos et al.,2002
Trametes versicolour	60	70-80		Font et al., 2003
Phanerochaete chrysosporium		50		Wingate et al., 2005
Phanerochaete chrysosporium *Pleurotus ostreatus* *Lentinus edodes* *Trametes versicolour* S22	48			Wua, et al., 2005
marine fungi NIOCC2a				D'Souza et al., 2006
Merulius aureus *Phlebia sp.* *Fusarium sambucinum*	90	79		Malaviya and Rathore, 2007
Pleurotus sajor-caju		42	59	Munari et al., 2007
Daedaleopsis sp., *Schizophyllum commune* *S. commune*	59-83			Prasongsuk et al., 2009
Pleurotus sajor- caju *Trametes versicolour* *Phanerochaete chrysosporium* *Rhizopus oryzae*		97 92 99 99		Rocha-Santos et al., 2010

Different effluent streams of a small pulp mill utilizing agro-residues were characterized for their pollution load and were decolourized with the white rot fungus *Trametes versicolour*. The maximum colour reduction of 92% with a chemical oxygen demand reduction of 69% was obtained (Mehna et al., 1995). Laccase protein fractions secreted by *T. versicolour* could directly mineralize the organochlorine from toxic chlorophenolics commonly found in these effluents. The results indicated that each of the fungal laccase proteins tested can rapidly partially dechlorinate a number of toxic polychlorinated phenols and guaiacols. As toxic chlorophenolic waste treatment enzymes, *T. versicolour* laccases appear to offer a variety of advantages over the secreted peroxidase of *P.chrysosporium* because no H_20_2 addition was needed. Additionally, these laccases were much more stable, were constitutively produced in large quantity and were more effective in dechlorination (Roy-Arcand and Archibald, 1991).

Coriolus versicolour was selected among sixteen strains of white rot fungi. It was grown in liquid culture and removed over 60% of the colour of the effluent within six days in the presence of sucrose. Treatment of the same effluent with this fungus, immobilized in beads of calcium alginate gel, resulted in 80% decolourisation after three days in the presence of sucrose. Recycled beads could remove colour efficiently and repeatedly in the presence of air but not under anaerobic conditions (Livernoche et al., 1983).

Among the microorganisms isolated from a biological treatment of pulp and paper mill sulfate waste waters, *Pseudomonas putida*, *Nocardia corallina*, *Torula sp.* and a strain of lower fungi degrade the alkali lignins, the only source of carbon in the medium. Their growths led to a decrease of COD, a disappearing of phenolic compounds and the production of extracellular enzymes such as laccase and phenoloxidase. The fungus showed a COD reduction of about 60% and 48% colour removal (Marchand, 1978). Some of the fungi that have been used for treatment of pulp and paper mill wastewaters are shown in Table 6.

Algae

Many investigators have suggested that algae can replace fungi to remove colour and AOX. *Microcystis sp.*, have been reported to decolourize diluted bleach Kraft mill effluents. Both pure and mixed algal cultures were found to be capable of removing up to 70% of the colour within 2 months of incubation. Colour removal by algae was due to metabolic transformation of coloured molecules and not adsorption to biomass (Lee et al., 1978). A mixed algal culture including *Chlorella*, *Chlamydomonas*, *Microcystis*, achieved a 70% AOX reduction, while colour was reduced by 80% in 30 days under continuous lighting conditions. The main colour removal mechanism was metabolism rather than adsorption (Dilek et al., 1999).

Algae were tested to treat t a wood-based pulp and paper industry wastewater under different lighting and initial wastewater strength conditions. Algae were found to remove up to 58% of COD, 84% of colour and 80% of AOX from pulp and paper industry wastewaters. It was also revealed that the main mechanism of colour and organics removal from pulping effluents was partly metabolism and partly metabolic conversion of coloured and chlorinated molecules to non-coloured and non-chlorinated molecules (Tarlan et al., 2002).

CONCLUSION

Biological processes, aerobic and anaerobic treatment systems are feasible to treat wastewater from all types of distilleries, olive mills and pulp and paper mills. The combination using an anaerobic process followed by an aerobic treatment system under optimum conditions can make use of the advantages of both the treatment processes, resulting in high removal of toxic pollutants. Effective bioremediation resulting in significant reductions in COD, phenolics and colour allows safe and economical disposal of industrial wastewater onto land or into surface waters. Investigation of three representative industries confirmed the usefulness and the potential of various microorganisms for the treatment of industrial effluents.

REFERENCES

Acharya, B.K., Mohana, S., Madamwar, D.*Biores. Technol.* 2008, 99, 4621-4626.

Adikane, H.V., Dange, M.N., Selvakumari, K. *Biores. Technol.* 2006, 97, 2131-2135.

Aguilera, M., Quesada, M.T., del Águila, V.G., Morillo, J.A., Rivadeneyra, M.A., Ramos-Cormenzana, A., Monteoliva-Sánchez, M. *Biores. Technol.* 2008, 99, 5640-5644

Ali, M., Sreekrishnan, T.R. *Process Biochem.* 2000, 36, 25-29.

Angayarkanni, J., Palaniswamy, M., Swaminathan, K. *B. Environ. Contam. Tox.* 2003, 70, 268-277.

Aoshima, I., Tozawa, Y., Ohmomo, S., Ueda, K. *Agric. Biol. Chem.* 1985, 49, 2041–2045.

Arora, M., Sharma, D.K., Behera, B.K. *Resour., Conserv. Recy.* 1992, 6, 347–353.

Assas, N., Ayed, L., Marouani, L., Hamdi, M. *Process Biochem.* 2002, 38, 361–365.

Assas, N., Marouani, L., Hamdi, M. *Bioprocess Eng.* 2000, 22, 503–507.

Asses, N. L., Ayed, H. Bouallagui, I. Ben Rejeb, M. Gargouri, M. Hamdi, *Biores. Technol.* 2009, 100, 2182–2188

Aveni, A. *Water Res.* 1985, 19 (5), 667–669.

Aveni, A. Biogas recovery from olive oil mill waste water by anaerobic digestion. In Ferrero, M.P., Ferranti, G.L. and Naveau, H (eds), Anaerobic Digestion and Carbohydrate Hydrolysis of Waste, Commission of the European Communities, Luxembourg. Elsevier Applied Science Publishers, New York, USA, 1984, 489–491.

Azzama M. O. J.; Kamal, I., Nehal, Abu-Laila, I. *J. Environ. Sci. Health, Part A* 2005, 39, 269 - 280

Bajpai, P., Bajpai, P.K. *J. Biotechnol.* 1994, 33, 211-220.

Beccari, M., Majone, M., Torrisi, L. *Water Sci. Technol.* 1998, 38, 53–60.

Beltran, F.J., Alvarez, P.M., Rodriguez, E.M., García-Arayaa, J.F., Rivas, J. *Biotechnol. Prog.* 2001, 17, 462-467.

Beltran, F.J., Garcia-Araya, J.F., Frades, J., Alvarez, P., Gimeno, O. *Water Res.* 1999, 33, 723-732.

Beltran, F.J., Encinara, J.M., García-Arayaa, J.F. *Water Res.*1993, 27, 1023-1032.

Benito, G.G., Miranda, M.P., De los Santos, D.R.. *Biores. Technol.* 1997, 61, 33–37.

Beteta, A., Canizares, P., Rodrigo, M.A., Rodriguez, L., Saez, C. *J. Hazard. Mater.* 2009, 168, 358-63.

Blair, J.E., Davis, L.T. Process for decolourizing pulp and paper mill wastewater. US Patent 4, 199444, 1980.

Borja, R., Martin, A., Alonso, V., Garcia, I., Banks, C.J. *Water Res.* 1995, 29, 489–95.

Borja, R., Martin, A., Fiestas, J.A., Maestro, R. *Grasas Aceites* 1990, 41, 397–403.

Borja-Padilla, R., Alba-Mendoza, J., Banks, C.J. *Process Biochem.* 2006, 31, 25-30S

Bories, Y., Sire, Y., Colin, T. *Wa. Sci. Technol.* 2005, 51, 129–136.

Bratskaya, S., Schwarz, S., Chervonetsky, D. *Water Res.* 2004, 38, 2955–2961.

Buzzini, A.P., Sakamoto, I.K., Varesche, M.B.,. Pires, E.C. *Process Biochem.* 2006, 41, 168-176

Buzzini, A.P., Pires, E.C. *Process Biochem.* 2002, 38, 707-713.

Cammerer, B., Jalyschkov, V., Kroh, L.W. *Int. Congress Series* 2002, 1245, 269–273.

Catalkaya, E.C., Kargi, F. *J. Haz. Mat.* 2007, 139, 244–253.

Cereti, C.F., Rossini, F., Federici, F., Quaratino, D., Vassilev, N., Fenice, M. *Biores. Technol.* 2004, 91, 135–140.

Cha, G.C., Noike, T. *Wat. Sci. Tech.* 1997,36, 247-253.

Chandra, R., Pandey, P.K. Indian Journal of Environmental Protection 2000, 21, 134–137.

Chandra, R., Raj, A., Purohit, H.J., Kapley, A. *Chemosphere* 2007, 67, 839-842.

Chandra, R., Bharagava, R.N., Rai, V., Kumar, S., Singh, D. *Biores. Technol.* 2009, 100, 6665-6668.

Chaturvedi, S., Chandra, R., Rai, V. *Ecol. Engin.* 2006, 27, 202-207.

Chauhan, N., Thakur, I.S. *Pollution Res.* 2002, 21, 429-434.

Chinnaraj, S., Venkoba Rao, G.V. *Biomass and Bioenergy* 2006, 30, 273-277

Chopra, P., Singh, D., Verma, V., Puniya, A.K. *Ind. J. Microbiol.* 2004, 44, 197-200.

Choudhury, S., Sahoo, N., Manthan, M., Rohella, R.S. *Journal of Industrial Pollution Control* 1998, 14, 1-13.

Chtourou, M., Ammar, E., Nasri, M., Medhioub, K. *J. Chem. Technol. Biotechnol.* 2004, 79, 869–878.

Cibis, E., Kent, C.A., Krzywonos, M., Garncarek, Z., Garncarek, B., Miskiewicz, T. *Biores. Technol.* 2002, 85, 57– 61.

Dahiya, J., Singh, D., Nigam, P. *Biores. technol.* 2001a, 78, 111-114.

Dahiya, J., Singh, D.,Nigam, P. *Biores. Technol.* 2001b, 78, 95-98.

De Pinho M.N, Minhlma M., Rosa .M.J.,Taborda F. *Pulp Paper Mag. Can.* 2000, 101, 50-54

Dev Mittar, Khanna1, P.K., Marwaha, S.S., Kennedy, J.F. *J. Chem. Technol. Biotechnol.* 1992, 53, 81-92.

Diez, M.C., Mora, M.L., Videla, S. *Water Res.* 1999, 33, 125-130.

Dilek, F.B., Taplamacioglu, H.M., Tarlan, E. *Appl. Environ.Microbiol.* 1999, 52, 585-591.

DiGioia, D., Fava, F., Bertin, L., Marchetti, L. *App. Microbiol. Biotechnol.* 2001b, 55, 619–626.

D'Annibale, A., Crestini, C., Vinciguerra, V., Sermanni, G. *J.Biotechnol.* 1998, 61, 209–218.

D'Souza, D.T., Tiwari, R., Sah, A.K., Raghukumar, C. *Enz. Microb. Technol.* 2006, 38, 504-511.

Dube, M., McLean, R., MacLatchy, D., Savage, P. *Pulp Pap Can. Mag.* 2000, 101, 42– 5.

Eaton, D., Chang, H.M., Kirk, T.K. *Tappi* 1980, 63, 103-106.

Ehaliotis, C., Papadopoulou, K., Kotsou, M., Mari, I., Balis, C. *FEMS Microbiol. Ecol.* 1999, 30, 301–311.

Ettayebi, K., Errachidi, F., Jamai, L., Tahri-Jouti, M.A., Sendide, K., Ettayebi, M. *Microbiol. Lett.* 2003, 223, 215–219.

Fadil, K., Chahlaoui, A., Ouahbi, A., Zaid, A., Borja, R. *Int. Biodeter. Biodegr.* 2003, 51, 37–41.

Fahy, V., FitzGibbon, F.J., McMullan, G., Singh, D., and Marchant, R. *Biotechnol. Lett.* 1997, 19, 97-99.

Fang, H.H.P., Chui, H.K., Li, Y.Y. *Wa. Sci. Technol.* 1994, 30, 87-96.

Farhadian, M., Duchez, D., Vachelard, C., Larroche, C. *Water Res.* 2008, 42, 1325 – 1341.

FitzGibbon, F., Singh, D., McMullan, G., Marchant, R. *Process Biochem.* 1998, 33, 799–803.

Flouri, F., Sotirchos, D., Ioannidou, S., Balis, C. *Int. Biodeter. Biodegrad.* 1996, 38, 189-192.

Font, X., Caminal, G., Gabarrell, X., Romero, S., Vicent, M.T. *J. Chem. Technol. Biotechnol.* 2003, 78, 548-554.

Fulthorpe, R.R., Allen, D.G. *App. Microbiol. Biotechnol.* 1995, 42,782-789.

Fountoulakis, M.S., Dokianakis, S.N., Kornaros, M.E., Aggelis, G.G., Lyberatos, G. *Water Res.* 2002, 36, 4735-4744

Garcia Garcia, I., Jimenez Pena, P.R., Bonilla Vaceslada, J.L., Martin Martin, A., Martin Santos, M.A., Ramos Gomez, E. *Process Biochem.* 2000, 35, 751–758.

Georgacakis, D., Dalis, D. *Biores. Technol.* 1993, 46, 221–226.

Gharsallah, N. *Environ. Technol.* 1993, 14, 391-395.

Ghosh, M., Verma, S.C., Mengoni, A., Tripathi, A.K. *J. App. Microbiol.* 2004, 96, 1278-1286.

Gonzalez, T., Terrón, M.C., Yagüe, S., Junca, H., Carbajo, J.M., Zapico, E.J., Silva, R., González, A.E. *Res. Microbiol.* 2008, 159, 103-109.

Gonzalez, T., Terron, M.C., Yague, S., Zapico, E., Galletti, G.C., Gonzalez, A.E. *Rapid Comm. Mass Spect.* 2000, 14, 1417–1424.

Gupta, V.K., Minocha, A.K., Jain, N. *J. Chem. Technol. Biotechnol.* 2001, 76, 547-552.

Habets, L.H.A., Knelissen, H.J. (1985) Wat. Sci. Tech. 1985, 17, 61-75.

Hadavifar, M., Zinatizadeh, A.A., Younesi, H., Galehdar, M. *Asia—Pacific J. Chem. Engin.* 2009, 5, 454-464.

Hamdi, M., Bou-Hamed, H., Ellouz, R. *Appl. Microbiol. Biotechnol.* 1991, 36, 285–288.

Hamman, O.B., Rubia, T., Martınez, J. *Environ. Toxicol. Chem.* 1999, 18, 2410–2415.

Hardisson, C., Sala, J.M., Stainer, R.Y. *J. Genet. Microbiol.* 1969, 59, 1–11.

Hossain S.M. *Ind. J. Environ. Prot.* 2007, 27,362-366.

Howe, J., Michael, R.W. *Environ. Pollut.* 1998, 105, 129– 135.

Jain, N., Minocha, A.K., Verma, C.L. *Ind. J. Exp. Bot.* 2002, 40, 101-105.

Jaouani, A., Vanthournhout, M., Penninckx, M.J. Environ. Technol. 2005, 26,633-641.

Jeyaramraja, P.R., Anthony, T., Rajendŕan, A., Rajakumar, K. *Pollution Res.* 2001, 20, 309-312.

Jimenez, A.M., Borja, R., Martin, A. *Process Biochem.* 2003, 38, 1275-1284.

Kahraman, S. Yesilada, O. *Folia Microbiol.* 2003, 48, 525-528.

Kalavathi, D.F., Uma, L., Subramanian, G. 2001. *Enz. Microb. Technol.* 2001, 29, 246–251.

Khabbaz, M.S., Vossoughi, M., Shakeri, M. performance of an anaerobic baffled reactor fro olive mill wastewater treatment. In: Proceedings of the 39th Central Canadian Symposium on Water Quality Research, 2004, Burlington, Ontario. Canada.

Kim, Y.-H., Han, K., Lee, W. *Process Biochem.* 2003, 38, 925-931

Kim, S.J., Shoda, M. *J. Biosc. Bioeng.* 1999, 88, 586–589.

Kirk, T.K., Connors, W.J., Zeikus, J.G. *Appl. Environ. Microbiol.* 1976, 32, 192-194.

Kissi, M., Mountadar, M., Assobhei, O., Gargiulo, E., Palmieri, G., Giardina, P., Sannia, G. *Appl. Microbiol. Biotechnol.* 2001, 57, 221–226.

Knupp, G., Rucker, G., Ramos-Cormenzana, A., Garrido Hoyos, S., Neugebauer, M., Ossenkop, T. *J Agric. Food Chem.* 1996, 48, 1941–1948.

Kroh, L.W. *Food Chem.* 1994, 51, 373–379.

Krzywonos, M., Cibis, E., Miśkiewicz, T., Kent, C.A. *Biores. Technol.* 2008, 99, 7816-7824.

Kulkarni, A.K. *Ind. Chem. Engin.r* 1998, 40, 169–172.

Kumar, S. Viswanathan, L. *Enz. Microb. Tech.* 1991, 13, 179-187.

Kumar, V., Wati, L., Nigam, P., Banat, I.M., McMullan G., Singh, D., Marchant, R. *Microbios* 1997a, 89, 81-90.

Kumar, V., Wati L., F. FitzGibbon, P. Nigam, I.M. Banat, D. Singh, Marchant, R. *Biotechnol. Lett.* 1997b, 19, 311-313.

Kumar, V., Wati L.; Nigam P.; Banat I.M.; Yadav B.S.; Singh D.; Marchant R. *Process Biochem.* 1998, 33, 83-88.

Kumar, P., Chandra, R. *Biores. Technol.* 2006, 97, 2096-2102.

Lalov, I.G., Guerginov, I.I., Krysteva, M.A., Fartsov, K. *Water Res.* 2000, 34, 1503-1506.

Lee, E.G., Mueller, J.C., Walden, C.C. *TAPPI* 1978, 61, 59–62.

Lepisto R., Rintala, J. *Water Sci. Technol.* 1994, 29, 373–380.

Liss, S.N., Bicho, P.A., Saddler, J.N. *Can. J. Microbiol.* 1997, 43, 599-611.

Livernoche, D., Jurasek, L., Desrochers, M., Dorica, J., Veliky, I.A. *Biotechnol. Bioeng.* 1983, 25, 2055-2065.

Lynch, J.M., Moffat A.J. *Ann Appl Biol*, 2005, 146, 217–221.

Malaviya, P., Rathore, V.S. *Biores. Technol.* 2007, 98, 3647-3651.

Mameri, N., Halet, F., Drouich, M., Grib, H., Lounici, H., Pausse, A., Piron, D.,

Belhoucine, D. *Can. J. Chem. Engin.* 2000, 78, 590–595.

Mandal, A., Ojha, K., Ghosh, D.N. *Ind. Chem. Engin.* 2003, 45, 264–267.

Marchand, M. *Rev. Ecol. Biol. Sol.* 1978, 15 321-331.

Martinez, M., Baeza, J., Freer, J., Rodriguez, J. *Toxicol. Environ.l Chem.* 2000, 77, 159-170.

Martinez Nieto, L., Garrido Hoyos, S.E., Camacho Rubio, F., Garcia-Pareja, M.P., Ramos - Cormenzana, A. *Biores. Technol.* 1993, 43, 215–219.

Mata –Alvarez, J. Fundamentals in anaerobic digestion process. In J. Mata-Alvarez (ed), Biomethanization of the organic fraction of municipal solid wastes, IWA publishing, London, 2003, 1-20.

Mcnamara, C.J., Anastasiou, C.C., Flaherly, V.O., Mitchell, R. *Int. Bodeter. Biodegrad.* 2008, 61, 122-134.

Mehna, A., Bajpai, P., Bajpai, P.K. *Enz. Microb. Technol. 1995*, 17, 18-22. .

Migo, V., Veronica P., Del Rosario, E.J., Matsumura, M. *J.fermen. bioengin.* 1993, 83, 287-291.

Minussi R.C., Pastore, G.M., Bollag, J.M. *Trends Food Sci. Technol. 2002*, 13, 205-216.

Miranda, P.M., Benito, G.G., Cristobal, N.S., Nieto, C.H. *Biores. Technol.* 1996, 57, 229-235.

Miyata, N., Iwahori, K., Fujita, M. Journal of Fermentation and Bioengineering 1998, 85, 550–553.

Mishra, M., Thakur, I.S. *Biodegradation* 2010, 1-12.

Modi, D.R., Chandra, H., Garg, S.K. *Biores. Technol.* 1998, 66, 79-81.

Mohammad, P., Azarmidokht, H., Fatollah, M., Mahboubeh, B. *Int. Bodeter. Biodegrad.* 2006, 57, 195-199.

Mohana, S. C., Desai, D., Madamwar, D. *Biores. Technol.* 2007, 98, 333-339.

Moreno, E., Quevedo-Sarmiento, J., Ramos-Cormenzana, A. In P.N. Cheremisinoff (ed), Antibacterial activity of washwater from olive mills, Encyclopaedie of Environmental Control Technology, 1990, Gulf publications, 731–736.

Munari, F.M., Gaio, T.A., Dillon, A.J.P. *Biocat. Biotrans.* 2007, 25, 24-28.

Murugesan, K. Indian J. *Exper. Biol.* 2003, 41, 1239-1248.

Naik, N.M., Jagadeesh, K.S., Alagawadi, A.R. *Ind. J. Microbiol.* 2008, 48, 41-48.
Nakajima-Kame, T., Shimomura, M., Nomura, N., Chanpornpong, T., Nakahava, T. *J. Biosci. Bioengin.* 1999, 87, 119-121.
Niaounakis, M., Halvadakis, C.P. Olive processing waste management: literature review and patent survey, second ed. Elsevier, 2006, Amsterdam.
Nataraj, S.K., Hosamani, K. M., Aminabhavi, T.M. *Water Res.* 2006, 40, 2349-2356.
Ohmomo, S., Daengsabha, W., Yoshikawa, H., Yui, M., Nozaki, K., Nakajima T., Nakamura, I. *Agric. Biol. Chem.* 1988a, 57, 2429-2435.
Ohmomo, S., Kaneko, Y., Sirianuntapiboon, S., Atthasampunna, P., Nakamura, I. *Agric. Biol. Chem.* 1987, 51, 3339–3346.
Pallerla, S., Chambers, R.P. *Biores. Technol.* 1997, 60, 1-8.
Pandey, R. A., Malhotra, S., Tankhiwale, A., Pande, S., Pathe, P.P., Kaul, S.N. *Int. J. Environ. Studies* 2003, 60, 263 – 275.
Pant, D., Adholeya, A. *Biores. Technol.* 2007, 98, 2321–2334.
Paraskeva, C.A., Papadakis, V.G., Tsarouchi, E., Kanellopoulou, D.G., Koutsoukos, P.G. *Desalination* 2007, 213, 218-229.
Parker, W. J. Farquhar, G. J., Hall, E. R. *Environ. Sci. Technol. 1993*, 27, 1783–1789.
Patel, A., Pawar, R., Mishra, S., Tewari, A. *Ind. J. Environ. Prot.* 2001, 21, 1118–1121.
Pathade, G.R. A review of current technologies for distillery wastewater
treatment. In P.K. Goel (ed.), Advances in Industrial Wastewater Treatment, ABD Publishers, 2003, Jaipur, India, 180-239.
Patil, P.U., Kapadnis, B.P., Dhamankar, V.S.*All India Distiller's Association (AIDA) Newsletter* 2003, 53–56.
Pena, M., Coca, M., Gonzalez, R., Rioja, R., Garcıa, M.T. *Chemosphere* 2003, 51, 893–900.
Perez, M., Romero L.I., Sales, D. *Water Res.* 1998, 2, 559 –564.
Pikaev, A.K. *Radiat. Phys. Chem.*, 2002, 65, 515–526.
Pinotti, A., Bevilacqua, A., Zaritzky, N. *J. Food Eng.* 1997, 32, 69-81.
Pinto, G., Pollio, A., Previtera, L., Stanzione, M., Temussi, F. *Biotechnol. Lett.* 2003, 25, 1657–1659.
Piperidou, C.I., Chaidou, C.I., Stalikas, D., Soulti, K., Pilidis, G.A., Balis,
C., *J. Agricult. Food Chem.* 2000, 48, 1941–1948.
Poggi-Varaldo, H.M., Estrada-Vazquez, C., Fernandez-Villagomez, G.,
Esparza-Garcia, F. Pretreatment of black liquor spills effluent. In Proceedings of the Industrial Waste Conference, West Lafayette,USA, 1996;51, 651–61.
Pokhrel, D., Viraraghavan, T. *Sci. Total Environ.t* 2004, 333, 37– 58.
Prasongsuk, S., Lotrakul, P., Imai, T., Punnapayak, H. *ScienceAsia* 2009, 37-41.
Quan, X., Shi, H., Liu, H., Wang, J., Qian, Y. *Process Biochem.* 2004, 39, 1701-1707.
Rajor, A., Singh, R., Mathur, R.P. *Ind. J. Environ. Prot.* 2002, 22, 1241-1252.
Raghukumar, C., Mohandass, C., Kamat S., Shailaja, M.S. *Enz. Microb.Technol.* 2003, 35, 197-202.
Rao, S.B. *Water Res.* 1972, 6, 1275-1282.
Raj, A., Reddy, M.M.K., Chandra, R.S. *J. Chem. Technol. Biotechnol* 2007, 82, 399-406.
Ramos-Cormenzana, A., Juarez-Jimenez, B., Garcıa-Pareja, M.P.International Biodeterioration and Biodegradation 1996, 38, 283–290
Raizer-Neto, E., Pichon, M., Rouger, J. Decreasing chlorinated organics in bleaching effluents in an anaerobic fixed bed bioreactor. In Proceedings of 4th International

Biotechnology Conference in Pulp and Paper Industry, Raleigh, North Carolina, 1989, 279–287.
Reddy, P., Pillay V.L., Kunammeni, A., Singh, A.K.S. *Water SA* 2005, 31, 575-580.
Reddy, C.A., Mathew, Z. Bioremediation potential of white rot fungi. In G.M. Gadd (ed.), Fungi in Bioremediatio, Cambridge University Press, 2001, 52-78.
Rigas, F., Dritsa, V. Fungal Remediation of Contaminated Sites. In R.W. Gore (ed), Environmental Reasearch at the leading edge, Nova publishers, 2007, 293-331.
Rintala, J.A., Puhakka, J.A. *Bioresour. Technol.* 1994, 47, 1– 18.
Rocha-Santos, T., Ferreira, F., Silva, L., Freitas, A.C., Pereira, R., Diniz, M., Castro, L., Duarte, A.C. *Environ. Sci. Poll. Res.* 2010, 17, 866-874.
Rodriguesa A.C., Boroskia, M., Shimadaa, N.S., Juliana Garcia Carla, Jorge Nozaki, a and Noboru Hioka Journal of Photochemistry and Photobiology A: Chemistry 2008, 194, 1-10
Roy-Arcand, L., Archibald, F.S. *Enz. Microb. Technol.* 1991, 13, 194-203.
Ruiz, C., Torrijos, M., Sousbie, P., Lebrato, J., Martinez, R., Delgenes, J.P.
Water Sci. Technol. 2002, 45, 219-224.
Ruiz-Ordaz, N., Ruiz-Lagunez, J.C., Castanon- Gonzalez, J.H., Hernandez-Manzano, E., Urbina, C., Galindez-Mayer, J. Reviste Latinoamericana de Microbiologia 2001, 43, 19-25.
Saha, N.K., Balakrishnan, M., Batra, V.S. *Res. Conserv. Recycl.*2005, 43, 163–174.
Sanchez-Villasclaras, S., Martinez Sancho, M.E., Espejo Caballero, M.T., Delgado Perez, A. *Int. Bioter. Biodegrad.* 1996, 38, 245-247.
Santos, A.Z., Tavares, C.R.G., Gomes-da-Costa, S.M. Braz. *J. Chem. Engin.* 2002, 19, 371-375.
Savanta, D.V., Abdul-Rahmanb, R., Ranade, D.R. *Biores. Technol.* 2006, 97, 1092-1104.
Sayadi, S., Ellouz, R. *Appl. Microbiol. Biotechnol.* 1992, 37, 813–817.
Semple, K.T., Cain, R.B., Schmidt, S. *FEMS Microbiol. Lett.,* 1999, 170, 291–300.
Shayegan, J., Pazouki, M., Afshari, A. *Process Biochem.* 2004, 40, 1323-1329.
Strong, P.J., Burgess, J.E. *Biores. Technol.* 2008, 99, 6134-6142.
Sirianuntapiboon, S., Zohsalam, P., Ohmomo, S. *Process Biochem.* 2004a, 39, 917–924.
Sirianuntapiboon, S., Phothilangka, P., Ohmomo, S. *Biores. Technol.* 2004b, 92, 31-39.
Singhal A, Thakur IS. *J. Hazard. Mater.* 2009,171,619-25.
Sowmeyan, R., Swaminathan, G. *Biores. Technol.* 2008, 99, 3877–3880.
Stephenson, R.J., Duff, S.J.B. *Water Res.* 1996, 30, 781-792.
Strong, P. *World J. Microbiol. Biotechnol.* 2010, 26, 133–144.
Sumathi, S. Phatak, V. *Environ. Technol.* 1999, 20, 93–98.
Tarlan, E., Dilek, F.B., Yetis, U. *Biores. Technol.* 2002, 84, 1-5.
Thakkar, A.P., Dhamankar, V.S., Kapadnis, B.P. *Biores. Technol.* 2006, 97, 1377-1381.
Thompson G, Swain J, Kay M, Forster C.F. *Bioresour. Technol.* 2001, 77, 275–286.
Tiku, D.K., Kumar, A., Chaturvedi, R., Makhijani, S.D., Manoharan, A., Kumar, R. *Biodeter. Biodegr.* 2010, 64, 173-183
Tomati, U., Galli, E., Di Lena, G., Buffone, R. *Agrochimica* 1991, 35, 275–279.
Tsioulpas, A., Dimou, D., Iconomou, D., Aggelis, G. *Biores. Technol.* 2002, 84, 251–257.
Ubay, G., Ozturk, I. *Water Sci. Technol.* 1997, 36, 287–294.
Uzal, N., Gokcay, C.F., Demirer, G.N. *Process Biochem.* 2003 39, 279-286.

Vahabzadeh, F., Mehranian, M., Saatari, A.R., W. *J. Microbiol. Biotechnol.* 2004, 20, 859-864.

Valderrama, L.T., Del Campo, C.M., Rodriguez, C.M., Bashan, L.E., Bashan, Y. *Water Res.* 2002, 36, 4185-4192.

Van Driessel, B., Christov, L. *J. Biosci. Bioeng.* 2001, 92, 271-276.

Vinciguerra, V., D'Annibale, A., Gacs-Baitz, E., Delle Monache, G. *J. Mol. Catalysis B* 1997, 3, 213–220.

Watanabe, Y., Sugi, R. Tanaka, Y., Hayashida, S. *Agric. Biol. Chem.* 1982, 46, 1623-1630.

Wenta, B., Hartmen, B. *Pulp Pap. Can. Mag.* 2002, 76, 43 –47.

Wilson, A.E.J., Moore, E.R.B., Mohn, W. *Appl. Environ. Microbiol. 1996*, 62, 3146-3151.

Wingate, K.G., Stuthridgel, T., Shawn, D., Mansfield, D. *Biotechnol. Bioengin.* 2005, 90, 95–106.

Wing-Hong Chan, Sai-Cheong Tsao, Journal of Applied Polymer Science, Volume 87, Issue 11, pages 1803–1810, 2003

Wua, J., Xiao Y.-Z., Yua, H.-Q. *Biores. Technol.* 2005, 96 1357-1363.

Yesilada, O., Sik, S., Sam, M. *World J. Microbiol. Biotechnol.* 1998, 14, 37–42

Yesilada, O., Fiskin, K. *J. Biol.* 1996, 20, 73–79.

Yesilada, O., Fiskin, K., Yesilada, E. *Environ. Technol.* 1995, 16, 95–100.

Yu, Z., Mohn, W.W. *Water Res.* 2001, 35, 883-890.

Yusef, H.H., Ghanem, K.M., El-Kassas, H.Y. *Fresen. Environ. Bull.* 1999, 8, 668-677.

Zouari, N., Ellouz, R. *J.Ind. Microbiol.* 1996a, 16, 155–162.

Zouari, N., Ellouz, R. *J. Chem. Technol. Biotechnol.* 1996b, 66, 414–420.

In: Bioremediation
Editor: Alexander C. Mason
ISBN: 978-1-61122-730-7

Chapter 11

SACCHAROMYCES CEREVISIAE'S THREE B-S: BAKERY, BREWERY AND BIOREMEDIATION

Ileana Cornelia Farcasanu [*] ***and Mihaela Matache*** [†]
University of Bucharest, Faculty of Chemistry, Sos. Panduri 90-92, Bucharest, Romania

ABSTRACT

As strict regulations compelled industries to develop environmental friendly, low-cost and efficient treatment techniques, the biotechnological approaches to clean contaminated environments received more and more attention in the recent years. *Saccharomyces cerevisiae* is a microorganism mainly used as baker's yeast and for some type of fermentation, but its potential as bioremediation effector is becoming more evident. *S. cerevisiae* is an excellent model system to study the biosorption, the uptake and the accumulation of various chemicals, including pollutants. This simple organism can grow on low-cost media, can be manipulated easily, and is not pathogenic. Its genome has been entirely sequenced and many cellular processes have been identified at molecular level. The high biosorption capacity as well as the possibility to combine biosorption with continuous metabolic uptake render *S. cerevisiae* one of the promising bioremediator candidates. This review focuses on the applicabiliy of *S. cerevisiae* cells to remove or harvest from contaminated sites an array of chemicals, from metal ions to complex organics. Bioremediation using dead biomass or growing cells, biosorptive or bioaccumulating strains, strains isolated from contaminated sites or engineered strains is discussed.

1. INTRODUCTION

Saccharomyces cerevisiae or the budding yeast, better known as brewer's yeast or baker's yeast is a friendly eukaryotic microorganisms that has been used by mankind since ancient times. Mainly utilized in beer and wine making, as well as in bakery and pastry, it

[*] E-mail address: farcasanu.ileana@unibuc.ro (IC Farcasanu)
[†] E-mail address: mihaela.matache@gmail.com (M Matache)

seems like the universal microorganisms that has been known and utilized in all times, anywhere. In brewing, *S. cerevisiae* is used for bottom or top fermentation of malt to produce beers and ales. For wine making, the fermentation of glucose from grapes is initiated by naturally occurring yeasts present in the vineyards; nowadays many wineries use modern methods of strain maintenance and isolation. One yeast cell can ferment approximately its own weight of glucose per hour and under optimal conditions and is one of the few microorganisms that can survive in ethanol-containing environments (up to 20%, with 15 to 16 percent being the norm). The yeast's function in baking is to ferment sugars present in the flour or added to the dough. This fermentation gives off carbon dioxide and ethanol. The carbon dioxide that is produced by fermentation is trapped within dough bubbles resulting in the dough expanding, or rising.

Apart from the large applicability in food and beverage industries, *S. cerevisiae* is one of the most studied experimental microorganism. Towards the end of the 20th century, this simple eukaryote was no doubt in the frontline of genetics and molecular biology studies, possessing the first genome to be ever sequenced completely (Goffeau *et al.*, 1996; Dujon, 1996). The elegance of yeast genetics, the ease of manipulation of yeast, and finally the technical breakthrough of yeast transformation to be used in reverse genetics, have substantially contributed to the enormous attention that this organism acquired from the scientific community (Broach *et al.*, 1991; Guthrie and Fink, 1991, Broach *et al.*, 1991). Besides the huge advantage that it is easy to grow and it is not pathogenic, yeast has basic biological structures and processes that have been conserved throughout eukaryotic life, making it an attractive alternative to more complex or inapproachable systems. The intense research carried out on yeast to understand various biological processes at molecular level shifted the applicability of yeast cells to new fields. Thus, while still being in the prime light of fermentation technology, yeast is an important player in most biotechnology fields, including bioremediation (Feldmann, 2005).

Industrialization, along with its benefits, brought up pollution issues and environment awareness. Strict legislation and regulation imposed rules and norms, therefore the industrial wastes and waters cannot be discharged in the environment without a thorough removal of the possible contaminants. Cleaning polluted or contaminated sites is never an easy task, especially when classical physico-chemical approaches are expensive or inefficient, produce secondary pollution, or simply fail. As pollution-related issues become more stringent, cost-effective, sustainable and environmentally friendly methods for removal or degradation of hazardous substances are more and more needed. Bioremediation by microorganisms, algae or plants is often regarded as inexpensive, safe and efficient way of cleaning up wastes, sediments, and soils that have been contaminated. Hardly a natural born bioremediator, the potential use of *S. cerevisiae* in bioremediation is more and more considered. Apart from being totally non-pathogenic and easy to grow, the ease of yeast genetics makes it the ideal candidate to be engineered in order to obtain pollutant-tolerant or hyperaccumulating strains, as well as strains capable to biodegrade various chemicals due to metabolic activity or to endo/exogenous enzymatic hyperactivity. In this paper, the use of *S. cerevisiae* cells in various bioremediation or biodegradation processes is reviewed.

2. Bioremediation from a Yeasty Point of View

Bioremediation may be regarded as the disappearance of a hazardous substance from the environment due to an interaction between the substance and 1) one or more living species; 2) dead biomass; 3) a secreted biomolecule (usually enzyme) that mediates the degradation. The bioremediation actions can take place *in situ* or in bioreactors designed both for the hazardous substance and for the bioremediator. Bioremediation can be achieved through a plethora of mechanisms, but the main bioremediation actions (which can be also performed by yeast) are biosorption, bioaccumulation and biodegradation.

2.1. Biosorption

Biosorption implies binding of the pollutant to the surface of the cell, usually by non-covalent forces. Biosorption is regarded as a cost-effective biotechnology for the treatment of high-volume-and-low-concentration complex wastewater. Most times, biosorption refers to a property of certain types of inactive, dead, microbial biomass to bind and concentrate a substance from dilute aqueous solutions. Biomass exhibits this property, acting as an ion exchanger of biological origin. It is particularly the cell wall structure of certain algae, fungi and bacteria which was found responsible for this phenomenon. Among the biosorbents which have been considered during the past decades for pollutant removal, *S. cerevisiae* has received increasing attention due to the unique nature of its cell wall. The yeast has a rigid cell wall made of two major components, glucan and mannoproteins, which are present in roughly equal amounts. The glucan, which is made up of β-1,3- and β-1,6-linked glucose, is complexed with chitin to provide mechanical strength to the cell wall. The mannoproteins, which form the outer layer of the cell wall, are generally heavily glycosylated, and determine most of the surface properties of the cell wall (Zlotnick *et al.*, 1984; de Nobel *et al.*, 1990, 2000).

Many of the mannosyl residues are phosphorylated, providing the cell wall surface with ionizable groups, very useful for binding positively charged substances. Although many biosorption processes imply utilization of dead or inactivated biomass, biosorption occurs also in the case of living organisms. Biosorption capacities can be further improved by adding to the cell surface proteins or other ligands that specifically bind to the chemical to be removed. This can be done by genetic engineering, taking advantage of the possibility to anchor such proteins to the cell wall (Schreuder *et al.*, 1996).

2.2. Bioaccumulation

Bioaccumulation is another attractive possibility to remove pollutants from the environment. In contrast to biosorption, bioaccumulation can be achieved only by living cells, requiring metabolically driven continuous uptake that needs to occur with a rate greater than the excretion rate. The retention of a chemical within the living cells is a very important for a bioremediation action and obtaining resistant strains is vital for such a process. Bioaccumulation has bioremediation significance when correlated with increased (gained) tolerance to the pollutant and easy separation of the bioremediator from the site to be decontaminated.

2.3. Biodegradation

Biodegradation refers to the chemical transformation of a pollutant, preferably to a non-pollutant, non-toxic form. The biodegradable material is usually of organic structure that can be transformed by microorganisms during the xenobiotic metabolism; ideally the pollutant could be used by microorganisms as a carbon or nitrogen source. When such approach is not realistic, strains can be engineered through the addition of exogenous genes encoding enzymes that break down organic materials; this approach can be also used in combination with laboratory evolution approach. A series of engineered *S. cerevisiae* strains obtained for biodegradation or bioaccumulation purposes is summarized in Table 1. Multipurpose bioremediation yeasts can be created. Also, the process can be improved by the addition of genes for specific adhesion proteins that could immobilize the yeast on a solid support.

In the following part, studies concerning the possible use of *S. cerevisiae* cells for bioremediation processes are reviewed.

3. Bioremediation of Industrial Byproducts

3. 1. Byproducts from Ethanol Plants

Ethanol production has always been related with yeast. Ethanol produced by *S. cerevisiae* in the course of fermentation is considered an environmentally friendly compound, but paradoxically, the byproduct emissions from ethanol production plants are a public health concern. Many of these byproducts are classified as hazardous air pollutants and current treatment methods, mainly thermal-oxidation are costly, energy consuming, and may produce other undesirable byproducts. Ethanol plant byproducts include formaldehyde, acetaldehyde, ethanol, methanol, glycerol, acetic acid, and lactic acid. Apart from the unpleasant odors caused by aldehydes, chemicals such as formaldehyde and acetic acid are considered toxic. *S. cerevisiae* are able to utilize most the above byproducts as sole carbon and energy sources for growth, except methanol and formaldehyde; however, both contaminants can be utilized as secondary metabolites by cultures initially fed on either ethanol or acetic acid (Pirnie-Fisker and Woertz, 2007). Growth studies conducted in batch bottle studies (Veiga and Kennes, 2001; Woertz *et al.* 2001; Wunch *et al.* 1997) showed that it is possible to use two contaminants simultaneously as the presence of one contaminant did not inhibit the degradation of another. Growth and contaminant utilization was observed in cultures fed two contaminants simultaneously (Pirnie-Fisker and Woertz, 2007). *S. cerevisiae* is very attractive for ethanol plant byproduct remediation because it is a food grade yeast and is involved in the anaerobic fermentation of ethanol, so there would be no concern about it contaminating the ethanol production process. Moreover, fungi are able to withstand acidic environments. Therefore, using a fungal bioreactor may be a feasible treatment option for ethanol plant emission (Pirnie-Fisker and Woertz, 2007). Alternatively, genetically engineered *S. cerevisiae* cells overexpressing plant glutathione-dependent formaldehyde dehydrogenase (FALDH), could be used (Table 1). Such cells were shown to process formaldehyde to CO_2 and water, conferring formaldehyde tolerance to the yeast cells lacking endogenous FALDH (Achkor *et al.*, 2003).

3.2. Byproducts from Printing Press Air

The volatile organic compound (VOCs) composition of printing press air contains mostly ethanol, but also ethyl acetate, 1-propanol, 2-propanol, 1-methoxy-2-propanol and 3-ethoxy-1-propanol (Granström *et al.*, 2002). For VOCs removal Wu *et al.* (1998) constructed a biofilter for waste gases containing VOCs using the cultures of *Bacillus*, *Micrococcus*, *Acinetobacter* and yeast. On the basis of ethanol tolerance, Granström *et al.* (2002) studied the ability of *S. cervisiae* cells to remove the VOCs both individually and as a complex mixture in a microcultivation system and in continuous chemostat cultures with and without cell recycling, in a comparative study with *Candida guilliermondii*. The *Candida* yeast could use all the compounds as a carbon source while growth of *S. cerevisiae* was markedly slower on the methoxylated and ethoxylated propanols. The effect of recycling was studied to scale down the constructed pilot plant bioscrubber. The VOC composition of the gas stream from the printing press was used to design a minimal carbon limited medium for continuous chemostat cultures. Altogether six steady state chemostat cultivations were carried out, with or without cell recycling. The differences in the oxygen consumption between these two yeast strains suggested different metabolic pathways, when utilising VOCs. It is possible that *S. cerevisiae* utilises these compounds mainly through alcohol dehydrogenase enzymes (ADH) with acetate as an intermediate. Acetate is then metabolised in the TCA cycle at the expense of oxygen consumption and carbon dioxide production.

3.3. Byproducts from Olive Oil Plants

The byproducts from olive oil factories are quite voluminous and are rich in water less soluble compounds. Olive oil mill wastewaters (OMWWs) for instance make up to 50% by volume of milled olives and have high concentration of organic matter containing oil, aromatic compounds, sugars, and pectic substances (Handy *et al.*, 1992). Often the OMWW are poured into the soil or disposed of in sewage, causing soil and water pollution as well as changes in the microbial composition of the soil through their antibacterial activity (Borya *et al.* 1995; Tardioli *et al.* 1997) and phytopathogenic effects due to their high toxicity (Bressan *et al.* 2004). Laconi *et al.* (2007) used a mixture of four strains of the ligninolytic basidiomycete genus of *Pleurotus*, together with the yeast strains *S. cerevisiae* and *Kluyveromyces lactis*; the species of filamentous fungi *Oidodendron spp*. and *Penicillum spp*. were also employed as inocula in treated OMWW in an aerated open bioreactor. These strains were chosen because they were all edible fungi, grew fairly well in treated OMWW and produced a good quality biomass to be used as an animal feed integrator. *S. cerevisiae* was used not as lignolytic, but as a probiotic strain, supporting the good biomass formation in a three week fermentation process. In a separate study, a fermentative strain ressembling

Saccharomyces spp was isolated from a two-phase decanter system olive oil waste, but it is unclear whether it could be used both for fermentation of the waste and for biomass production (Giannoutsou *et al.*, 2004).

4. Biodegradation of Food Contaminants

4.1. Mycotoxins

Aflatoxins are naturally occurring mycotoxins that are produced by many species of *Aspergillus*, most notably *Aspergillus flavus* and *Aspergillus parasiticus*, which colonize a variety of grains, including maize, oilseeds, spices, groundnuts, and tree nuts. Additionally, when animals farmed for dairy production consume aflatoxin-contaminated feed, a metabolite, aflatoxin M_1, is excreted in the milk (Strosnider *et al.*, 2006). Aflatoxins can be acutely toxic, carcinogenic, mutagenic, teratogenic, immunosuppressive and exposure in food is a significant risk factor for human cell carcinoma (Wild and Gong, 2010). No animal species is immune to the acute toxic effects of aflatoxins, including humans. Naturally occurring mixes of aflatoxins have been classified as a Group 1 human carcinogen by the International Agency for Research on Cancer (IARC, 1993). Unfortunately, these contaminants can never be completely removed, and on that account, many studies have been carried out to explore an effective process of their detoxification to a threshold level. The biological transformation of aflatoxins by microbial systems to metabolites that are non-toxic or easily excreted was considered, although thought to be unrealistic (Smith, 1997). Some fungal strains, including *S. cerevisiae*, have the ability to resist to aflatoxins and even to biotransform or biodegrade them to less toxic compounds (El Shieck *et al.*, 2007). The *S. cerevisiae* cells were shown to grow in broth containing up to 500 ppb aflatoxin B1. Moreover, when co-cultivated with an aflatoxin-producing strain of *Aspergillus*, the yeast cells inhibited the production of aflatoxins B1, B2, G1 and G2 with 63.7, 51.0, 44.9 and 25.0%, respectively. The reason for reduction of aflatoxin levels can be attributed to physical competition for space and nutrition, competition for a nutrient that can be converted to aflatoxin, or degradation of the aflatoxin after its formation. In the case of *S. cerevisiae*, the mass specrometry analysis revealed degradation in both the furan and coumarin moieties, *i. e.* the core of the aflatoxin structures. Application of fungal biodegradation may be a way to detoxify aflatoxin-contaminated foods and this utility could be extended to aflatoxin bioremediation in infected plants. El Shieck *et al.* (2007) emphasized the new value of *S. saccharomyces* cells and its possible usage as a bioremedy for aflatoxin poisoning when other treatment approaches fail. Aflatoxins and other mycotoxins also bind to yeast cell walls, thus being removed from contaminated food products. In this line of usage, co-administration of yeast cell wall preparations and aflatoxin B1 or ochratoxin A decreased the extent, but not the rate of absorption of the toxins in rats (Firmin *et al.*, 2010), probably due to mycotoxin binding to the glucans in the cell wall (Yiannikouris, 2006).

S. cerevisiae cells were also reported to interact with other mycotoxins, like patulin (Yiannikouris, 2006). Patulin is a mycotoxin produced by a number of organisms involved in food spoilage, including *Penicillium expansum*, the cause of the blue rot storage disease of apples. Patulin has been isolated from infected apples, and the reported occurrences of patulin in commercial juice are believed to result from the use of apples with some degree of decay (Wilson and Nuovo, 1973). Disappearance of patulin during alcoholic fermentation of apple juice was reported when fermentation was mediated by *S. cerevisiae* strains (Stinson *et al.*, 1978; Moss and Long, 2002). Three commercial cider strains of *S. cerevisiae* can degrade patulin during active fermentative growth, but not when growing aerobically. The products of patulin degradation are more polar than patulin itself and remained in the clarified fermented cider, one of them having the same properties as the less toxic ascladiol (Moss and Long, 2002). In a separate study, zearalenone (ZEN), an estrogenic mycotoxin produced by several *Fusarium* species could be converted to a non-toxic compound by a transgenic yeasts through the expression of a synthetic lactonohydrolase gene (Takahashi-Ando *et al.*, 2005).

4.2. 3-Chloro-1,2-Propanediol

3-Monochloro-1,2-propanediol (3-MCPD) is one of the numerous chlorinated food contaminants known collectively as chloropropanols (Hamlet *et al.*, 2002). Using commercial yeast, Bel-Rhlid *et al.* (2004) reported a novel enzymatic dehalogenating activity of 3-chloro-1,2-propanediol under aerobic conditions, at 28°C. The biodegradation was possible at different pH values (6.2, 7.0, and 8.2), in the presence and absence of glucose, using racemic 3-MCPD at two different concentrations (7.3 μmol/L and 27 mmol/L). Optimal conversion (68%) of racemic (R,S)-3-MCPD at a concentration of 27 mmol/L was achieved after 48 h of reaction time, at pH 8.2, and in the presence of glucose. At a concentration of 7.3 μmol/L, 73% degradation was observed after 72 h, at pH 8.2 and in the absence of glucose. Under the same experimental conditions, the conversion of pure (*S*)-3-MCPD (85%) was higher than that of the (*R*)-enantiomer (60%). The pH value seems important enough, since more acidic pH favoured the formation of 3-MCPD in pilot-scale yeast fermented doughs, conditions under which the level of another food contaminant, acrylamide, decreased considerably (Sadd *et al.*, 2008).

5. Biodegradation of Aromatic Compounds

5.1. Aromatic Hydrocarbons

There are very few studies reporting the use of *S. cerevisiae* cells in biodegradation of aromatic hydrocarbons. Hofmann (1986) reported the yeast-mediated oxidation of naphthalene and anthracene, in a parallel study on *Candida utilis* and *S. cerevisiae*. Sutherland (1991) showed that *S. cerevisiae* have the ability to oxidize polyaromatic hydrocarbons (PHAs), a group of hazardous environmental pollutants, many of which are acutely toxic, mutagenic, or carcinogenic. Sutherland studied a diverse group of fungi, including *Aspergillus ochraceus*, *Cunninghamella elegans*, *Phanerochaete chrysosporium*, *S. cerevisiae*, and *Syncephalastrum racemosum*, showing that the PAHs anthracene, benz[a]anthracene, benzo[a]pyrene, fluoranthene, fluorene, naphthalene, phenanthrene, and pyrene, as well as several methyl-, nitro-, and fluoro-substituted PAHs, are metabolized by one or more of these fungi. Unsubstituted PAHs are oxidized initially to arene oxides, trans-dihydrodiols, phenols, quinones, and tetralones, while phenols and trans-dihydrodiols may be further metabolized, and thus detoxified, by conjugation with sulfate, glucuronic acid, glucose, or xylose, rendering them less toxic. High bioremediation potential could have the *S.cerevisiae* cells expressing laccases (Table 1). Laccases are multicopper oxidoreductases (benzenediol: oxygen oxidoreductases, E.C. 1.10.3.2) mediating a multitude of reactions, including ruptures of alkyl–aryl bonds, benzylic alcohol oxidation, and rupture of aromatic rings, etc. (Bourbonnais and Price, 1992). Fungal laccases are remarkable green catalysts with potential in the remediation of recalcitrant contaminants (PAHs, chlorophenols, polychlorinated biphenyls, dyes) (Alcalde *et al.*, 2002). Laccases with remarkable properties were expressed in *S.cerevisiae* cells exhibiting large enzymatic specificity, high activity and resisting a wide array of cosolvents at concentrations as high as 50%. (Bulter *et al.*, 2003; Necochea *et al.*, 2005; Zumárraga *et al.*, 2007).

5.2. Phenol

Phenol is a hazardous chemical, very corrosive and toxic. Much utilized in pharmaceutical, textile, paper, paint solvent, pesticide, dye, plastic, explosive industries, it poses a threat to the environment. Although reactive and easily oxidized, the complete removal from industries waste waters is not an easy task and chemical degradation very often results in other toxic derivatives. Being relatively simple, phenol is easily biodegraded to non-toxic CO_2 and H_2O by adapted microorganisms. Patel and Rajkumar (2008) reported the isolation of a phenol-degrading *S. cerevisiae* strain that was tolerant up to 800 mg/L phenol and had the ability to 8.57%- 100% degrade phenol in the range of concentration 800 mg/L-200 mg/L, during a 5-days incubation period.

5.3. 2,4- Dichlorophenol

Chlorinated phenols are widely used by the chemical industry as intermediate products in chemical synthesis (Balfanz and Rehm, 1991). They have high toxicity and are very difficult to degrade in natural conditions, posing serious ecological problems as environmental pollutants (Sharma *et al.*, 1997). The lignin-degrading white rot fungus *Phanerochaete chrysosporium* is capable of degrading a wide variety of organopollutants, including 2,4-dichlorophenol (2,4-DCP) by means of lignin peroxidase and manganese peroxidase (Hammel *et al.*, 1993; Valli and Gold, 1991). Ryu *et al.* (2008 a, b) increased the efficiency of fungus *P. chrysosporium* lignin peroxidase in the degradation of 2,4-DPC by expressing it as a secretive form in recombinant *S.cerevisiae* cells (Table 1). The resulting variants showed approximately 1.6-fold improved 2,4-dichlorophenol (2,4-DCP) degradation activity and stability against H_2O_2 compared with the parent strain.

Table 1. Heterologous proteins expressed in *S. cerevisiae* cells for bioremediation purposes

Gene/Protein	Source/Role	Localization of recombinant protein in S. cerevisiae cells	Use	Comments	Reference
CYP1A1, *CYP1A2*/Cytocrome P450 (A1, A2)	Rat/Key enzyme for the metabolism of PolyChlorinated Dibenzo-p-Dioxins (PCDD) in mammals	Microsomal fraction	Bioconversion of PCDD Prediction of drug metabolism in humans	Potential use for bioremediation of dioxin contaminated soils	Murakami et al., 1990 Sakaki et al., 2002
LiP /Lignin peroxidase (LiP)	Phanerochaete chrysosporium (white rot fungus)/Lignolysis at nonphenolic methoxy-substituted lignin subunits	Cell surface	Biodegradation and biotransformation of 2,4- dichlorophenol (DCP)	Functionally improved LiP for stability at excessH_2O_2 and increased DCP degradation	Ryu et al., 2008
ADH2 / glutathione-dependent formaldehyde dehydrogenase (FALDH)	Arabidopsis / formaldehyde metabolism in eukaryotes	Yeast homogenate	Detoxification of exogenous formaldehyde	Overexpression of Arabidopsis FALDH in yeast cells sfa1Δ (lacking endogenous FALDH gene) augmented the resistance to formaldehyde added exogenously	Achkor et al., 2003
LCC1 / Laccase α	Trametes versicolor UAMH 8272 / lignin degradation	Culture supernatant	Oxidation of 0.5 mM 2,6-dimethoxyphenol, 0.14 mM ABTS (2,20-azino-bis[3-ethylbenzthizoline-6-sulfonic acid]), or 1 mM tetramethylbenzidine.	Potential for bioremediation applications	Necochea et al., 2005

Table 1. (Continued)

LCC2 / Laccase β	*Trametes versicolor* / lignin degradation	Culture supernatant	Production of fuel ethanol from renewable raw materials	Strain resistant to phenolic fermentation inhibitors present in lignocellulose hydrolysates	Larsson *et al.*, 2005
Mt-LT2 / Laccase MtLT2	*Myceliophthora thermophila*	Culture supernatant	Lignocellulose treatment, construction of biosensors and biofuel cells, dyestuffs processing	Resisting a wide array of cosolvents at concentrations as high as 50% (v/v).	Bulter *et al.*, 2003 Zumárraga *et al.*, 2007
TAA / Taka-Amylase TAA-$(His)_6$-Cwp1	*Aspergillus oryzae* / starch hydrolysis	Cell wall surface	Enhanced biosorption of Cu^{2+}, Ni^{2+}, and Zn^{2+}	A fusion protein of hexa-histidine repeat (His) and (GPI)-anchor region of S. cerevisiae Cwp1	Kambe-Honjoh *et al.*, 2000
CUP1 / metallothionein (Cup1)	*S. cerevisiae* / cation binding	Cell wall surface	Enhanced biosorption of Ni^{2+}, and Cu^{2+}	A fusion protein between yeast invertase and Cup1 (Suc2-Cup1)	Nakajima *et al.*, 2001
AtPCS / Phytochelatin Synthase (PCS)	*Arabidopsis thaliana* / cation binding	Cytosol	As^{3+} removal	Increase in As^{3+} accumulation, even in the presence of excess Zn^{2+}	Singh *et al.*, 2007

5.4. Polychlorinated Dibenzo-*p*-Dioxins

Polychlorinated dibenzo-*p*-dioxins (PCDD) are dioxin derivatives with high toxicity, often associated with environmental pollution. Unlike many chlorinated compounds, PCDD are never produced as desired commercial products but occur as byproducts in the manufacture of organochlorides, in the incineration of chlorine-containing substances, in the bleaching of paper, and from natural sources such as volcanoes and forest fires (Beychok, 1987). Members of the PCDD family have been shown to bioaccumulate in humans and wildlife due to their lipophilic properties, and are known teratogens, mutagens, and suspected human carcinogens. Microbe degradation of PDCC is initiated by various oxidative mechanisms (Bumpus *et al.*, 1985; Beurskens *et al.*, 1995; Sato *et al.*, 1997; Armengaud *et al.*, 1998; Nojiri *et al.*, 1999) all essential for bioremediation. As the ability of natural microorganisms appears insufficient for the efficient degradation of dioxins, Sakaki *et al.* (2002) engineered *S. cerevisiae* cells to express the genes for mammalian cytocromes 450 (CYPs) which appear to be the be key enzymes for the metabolism of PCDDs in mammals (Hu and Bunce, 1999). Sakaki *et al.* (2002) obtained recombinant *S. cerevisiae* cells expressing either rat *CYP1A1* or *CYP1A2* (Table 1) that had the ability to act on mono-,di-,and tri-chloro dibenzo-*p*-dioxins through multiple reactions such as hydroxylation at an unsubstituted position, hydroxylation with migration of a chloride substituent, hydroxylation with elimination of a chloride substituent, as well as opening of dioxin ring. The remarkable biotransformation abilities indicate the possibility of application of the transgenic cells to bioremediation of contaminated soils with dioxins.

5.5. Dicarboxylic Acids and Derivatives

Dicarboxylic acid and their derivatives are widely used in chemical industries. As the *S. cerevisiae* cells can use dicarboxylic acid derivatives as carbon source (Begum *et al.*, 2003), attempts were made to use yeast cells in the biodegradation of such compounds. The ability of yeast to degrade extracellular dicarboxylic acid is based on the cell capacity to import dicarboxylic acids through simple diffusion, followed by decarboxylation during fermentation (Delcourt *et al.*, 1995; Ansanay *et al.*, 1996).

Phthalic acid esters (PAE), widespread in environmental samples (sediments, natural water, soils, plants, aquatic organisms) have deleterious effects on the living cells. The rate of photolysis and hydrolysis of PAE is very low, therefore the metabolic degradation by microorganisms is of utmost importance. Begum *et al.* (2003) were the first to use *S. cerevisiae* cells to degrade diethyl phthalate (DEP), di-*n*-butyl phthalate (DBP) and butyl benzyl phthalate (BBP) using lyophilized cells. Incubating yeast cells with 0.5 mg/L PAEs in diammonium hydrogen phosphate resulted in ester degradation within 72 (first-order kinetics) with the degradation rates DBP > BBP > DEP.

Dicarboxylic acid solid waste (DASW) containing phthalic acid, malic acid, quinone, saturated and unsaturated dicarboxylic esters are discharged during the crackdown of benzene over the catalyst vanadium at temperatures grater than 500ºC in dicarboxylic acid manufacturing. The dicarboxylic acid exhibit toxicological effects on the living organisms, therefore their release in the environment has to be carefully controlled. Arumugam *et al.* (2006) used ricehusk based mesoporous activated carbon (RHAC) as a carrier matrix for immobilization of yeast cells for the degradation of DASW in aqueous phase by two stage anaerobic processes, one consisting in yeast immobilized RHAC reactor and stage II consisting of the treatment in the upflow anaerobic hybrid reactor. In the first phase 0.5% (w/v) aqueous DASW was treated with in *S. cerevisiae* immobilized reactor and the yeast fermented samples were further treated in an upflow anaerobic reactor containing mixed culture of anaerobic bacteria immobilized in activated carbon. The activated carbon acts as a modulator by adsorbing the toxic compounds in the aqueous system, thereby regulating the concentration of free material and at the same time providing the yeast cells with the low soluble, hydrophobic compounds. The yeast strain used was tolerant to dicarboxylic acids and had been isolated from a distillery plant, a strain that fermented 0.5% DASW with substantial growth. The work highlights the use of *S. cerevisiae* immobilized reactor for the breakdown of the dicarboxylic acid wastes into volatile fatty acids, thus preventing the anaerobic bacterial populations from the toxic effects of VFA during upflow treatment. The first stage fermentation process seems to cause degradation of DASW with production of VFA (propionic and acetic acids as major constituents). Interestingly, mass spectrometry of fermented and anaerobically treated DASW revealed the possibility of breakdown of phthalic acids through the formation of benzoyl-CoA, while proton and ^{13}C NMR spectroscopy demonstrated the breakdown of DASW into aliphatic carboxylic compound suggesting that the combination of yeast fermentation and anaerobic-activated carbon hybrid reactors are efficient for the disposal of solid waste generating from dicarboxylic acid manufacturing industries (Arumugam *et al.*, 2006).

6. Biodegradation of Dyes

Dyes are widely used for various industrial purposes (textile, food, cosmetics, pharmaceutical industries) and large amounts are discharged in the form of waste water that may be further spilled into the environment. While the majority of dyes used nowadays are less toxic, a major problem with the dye wastes is the lingering color. The presence of color in water reduces light penetration, affecting the photosynthetic processes. Decolorization of the dye wastes through biotechnological processes can be achieved through either adsorption on the biomass or bioaccumulation. Using growing cultures in the bioremoval process eludes the problem of biomass production, which involves cultivation, harvesting, drying, processing, and storage prior to use. Thus, removal by living cells seems more appealing, although limitations apply in this case, too. There are practical limitations to biouptake, and the most significant is growth limitation when dye concentrations are too high. In spite of these limitations, biotechnological approaches to bioremediate dye wastes are receiving a great deal of attention. The use of *S. cerevisiae* cells to degrade and decolorize various dyes is summarized in Table 2.

Table 2. Use of *S. cerevisae* cells in the bioremediation of dyes

Dye commercial name	Dye type	Applicability	Removal system	Mechanism of action	Removal (%)	Comment	Reference
Remazol Blue, Remazol Black B, Remazol Red RB	Reactive Azo dyes	Waste water, batch, oxidation pond	Growing cells; carbon source: molasses	Bioaccumulation	> 62%	Slow	Aksu, 2002
Methyl Red	Azo dye, toxic	Anoxic conditions, batch	Biomass or cell extract	Enzymatic degradation by azo reductase	Complete; cells were potent 10 cycles	Fast	Jadhav *et al.*, 2007
Aniline Blue	Soluble Azo dye	Batch	Magnetically-modified cells	Biosorption	220 mg /g of dried magnetic adsorbent	Fast	Safaríková *et al.*, 2005
Malachite Green	Triphenyl methane	Anoxic conditions Batch	Magnetically modified baker's yeast, fresh biomass, growing cells	Biosorption Biosorption followed by biodegradation	85% in plain distilled water within 7 h, and about 95.5% in 5% glucose medium within 4 h	Fast	Safarík *et al.*, 2002 Jadhav and Govindwar, 2006
Crystal Violet, Methyl Violet Cotton Blue	Triphenyl methane	Anoxic conditions, batch	Fresh biomass, growing cells	Biosorption Changes in pH		Fast	Jadhav and Govindwar, 2006
Methylene Blue	Cationic dye	Batch	Poly(amic acid)-modified biomass	Biosorption	680.3 mg/g (13 times more efficient than the untreated biomass)	Fast	Yu *et al.*, 2010
Basic Magenta	Cationic dye	Batch	Poly(amic acid)-modified biomass	Biosorption	353.4 mg/g (7 times more efficient than the untreated biomass)	Fast	Yu *et al.*, 2010

6.1. Azo Dyes

Azo dyes are prominent class of colorants produced worldwide annually, used in many different industries, among which the most important are the textile, paper and leather, and to a lesser extent cosmetics, pharmaceuticals and food. The textile industry accounts for two-thirds of the total dyestuff market, consuming large amounts of azo dyes, due to the high demand of fabrics with brilliant colors. The azo dyes are remarkably durable, causing pollution once they are spilled into the environment. Moreover, some azo dyes are toxic or mutagenic (Holme, 1984). The decolorization of textile waste water is still a major problem because the azo dyes are difficult to remove by conventional wastewater treatment (chemical degradation, photo-degradation, adsorption employing activated carbon and various other sorbents, electrochemical oxidations, aerobic biodegradation and biosorption, etc.). Most of these processes are costly and slow and are based on oxidation, a process to which these dyes are supposed to be resistant to. Removal of azo dyes by biosorption, and even better by bioaccumulation was investigated, mainly in batch conditions. Bioaccumulation especially is of high interest and could be used for the removal of various kinds of textile dyes if the growing cells find sufficient carbon and nitrogen sources in the contaminated media.

6.1.1. Reactive Azo Dyes

Reactive dyes bind to cotton fibers by addition or substitution mechanisms under alkaline conditions and high temperatures. Due to the poor exhaustion properties, as much as 40% of the initial dyes remains unfixed and ultimately ends up in the dye effluent. Reactive azo dyes are electron deficient and therefore less susceptible to oxidative catabolism. Remazol Blue, Remazol Black B, Remazol Red RB are bioaccumulated by *S. cerevisiae* cells, at optimal rate at pH 3. Among the dyes examined, Remazol Black B bioaccumulation was motre significant, and higher bioaccumulation percentages were observed at lower concentrations of all dyes (Aksu, 2003).

6.1.2. Methyl Red

Cells *of S. cerevisiae* MTCC 463 have the capacity to degrade soluble azo dyes in plain distilled water. Thus, Jadhav *et al.* (2007) showed that methyl red (100 mg/L) is degraded completely within 16 min in plain distilled water under static anoxic condition at room temperature and the yeast cells retain the degrability capacity up to 10 cycles in plain distilled water.

6.2. Triphenylmethane Dyes

Triphenylmethane dyes are used extensively in textile industries for dying cotton, wool, silk, nylon, etc. (Culp and Beland, 1996). Four different dyes (malachite green, crystal violet, methyl violet and cotton blue) were tested for the study of decolourization pattern of triphenylmethane dyes using *S. cerevisiae.* The cells showed an ability to decolorize all dyes tested, with best activity on malachite green. Thus, the *S. cerevisiae* cells decolourize 85 ± 0.6% malachite green (100 mg/l) within 7 h by simultaneous biosorption (initially up to 3 h) and further biodegradation. Biodegradative activity is needed since autoclaved cells absorbed about 30% malachite green and did not show further decolourization activity (Jadhav and Govindwar, 2006). The *S. cerevisiae* MTCC 463 effectively decolourizes malachite green due to biotransformation, while methyl violet, crystal violet and cotton blue decolourize due to biosorption and pH changes. Removal of malachite green from synthetic effluents can be achieved by biosorption only, using magnetically modified baker's yeast cells (Safarík *et al.*, 2002).

6.3. Cationic Dyes

Cationic dyes can be removed from waters using negatively-charged materials. Yu *et al.* (2010) used baker's yeast to obtain a poly(amic acid)-modified biomass suitable for efficient adsorption of two cationic dyes: methylene blue and basic magenta. The modified biomass showed enhanced adsorption capacity (13 times for methylene blue and 7 times for basic magenta) compared to non-modified biomass. Remarkably, the adsorption capacity was not dependent on pH or ionic strength, making the modified biomass a good candidate for practical use.

7. Bioremediation of Heavy Metals

Perhaps the most numerous studies concerning the use of *S. cerevisiae* cells in bioremediation are related to heavy metals. This is not surprising, because the budding yeast has obvious affinity for cations. The outer layer of the cell wall is rich in mannproteins heavily glycosilated and many of the mannosyl residues are phosphorylated (Zlotnick *et al.*, 1984; de Nobel *et al.*, 1990, 2000), making the cell wall susceptible to cation binding through electrostatic forces. Metal ions that do not bind to cell surface can penetrate through the porous cell wall and reach the plasma membrane, where they can enter the cell by passive or active transport mechanisms. Many metal ions (Cu^{2+}, Co^{2+}, Mn^{2+}, Ni^{2+}, Zn^{2+}) are essential for normal metabolism, therefore cell has developed systems for active transport, accumulation and homeostasis, and many such mechanisms have been already described at molecular level. Other metals (Ag, As, Cd, Hg, Pb) do not have metabolic significance, but they can be recognized by various cell components.

Heavy metals are discharged in small quantities into environment through industrial activities, being a potential health hazard and a threat to water supplies. Removal of polluting metals from contaminated sites is often difficult and remediation through common physico-chemical techniques is expensive and unsuitable in case of voluminous effluents containing complexing organic matter and low metal contamination. Biotechnological approaches may be an alternative remediation choice, and the increase of environment pollution by heavy metals generated the interest to live organsims that tolerant to these metals, and even better, are able to remove the contaminant metals either by biosorption or by metabolic uptake. Some reports in which the *S. cerevisiae* cells were investigated for their potential use in heavy metal bioremediation are discussed bellow.

7.1. Heavy Metal Biosorption

The use of *S. cerevisiae* to remove metals from industrial waste streams is more and more considered as it is an inexpensive source of biomass, with an ability to accumulate a broad range of heavy metals under a wide range of external conditions (Blackwell *et al.*, 1995). Biosorption is relatively inexpensive, non-hazardous, and may permit recovery of the metals from the sorbing biomass. Biosorption of metal ions may take place by different passive metal-sequestering processes such as ion exchange, complexation, physical entrapment, and inorganic microprecipitation. Brewer biomass has been used in several studies for heavy metals removal (Avery and Tobin, 1992; Blackwell and Tobin, 1999; Fereiduni *et al.*, 2009; Ferraz *et al.*, 2004; Ferraz and Teixeira, 1999; Lin *et al.*, 2006; Marques *et al.*, 1999, 2000; Simon and Singleton, 1996; Volesky and May-Phillips, 1995; Zhang *et al.*, 2009). Some of these studies gave results that are summarized in Table 3.

Naeem *et al.* (2006) measured the proton, Cd, Pb, Sr, and Zn adsorption onto the *S. cerevisiae* in bulk adsorption experiments as a function of time, pH, surface:metal ratio, and ionic strength. Modelling the acid/base properties of the fungal cell, they found that the affinity of the fungal cells for the metal ions follows the trend: Pb > Zn > Cd > Sr. Biosorption is a rapid process, and when the cells are placed in contact with the metal containing effluent, the cations bind within minutes to the cell surface.

Once the metal ions are absorbed, there is the problem of cell separation after the contact with the effluent. Machado *et al.* (2008) used a flocculent strain to remove heavy metals from a synthetic effluent and they carried out the solid-liquid separation process using the rapid sedimentation ability of a heat inactivated flocculent strain. The flocculent strain was able to sediment in the presence of Cu^{2+}, Ni^{2+}, Zn^{2+}, Cd^{2+} and Cr^{3+}, achieving simultaneous removal of the heavy metals and cell separation. Although several cations could be efficiently removed, the use of flocculation strain seemed to be most suitable for Ni^{2+} removal (Machado 2008, 2009, 2010).

7.2. Engineering Cell Surface for Heavy Metal Higher Absorption

"Molecular display (Arming) technology" or "Cell surface engineering" make use of the power of molecular biology techniques to express heterologous proteins on the surface of microorganisms, proteins that are expected to offer novel functionality to cell wall (Georgiou *et al.*, 1997; Murai *et al.*, 1997; Shibasaki *et al.*, 2009). In *S. cerevisiae*, several secretory enzymes are known to localize mainly on the cell surface without being released into the medium (de Nobel and Barnett, 1991) and are mainly used for the directed expression of fusion proteins on the cell surface. One example is invertase (Suc2), a heavily glycosylated enzyme mainly retained in the periplasmic space as homo-octamer (Esmon *et al.*, 1987).

Table 3. Use of *S. cerevisiae* cells in heavy metal bioremediation

Metal	Strain	Removal mechanism	Removal capacity	Source of metal (concentration)	Comments	Reference
Cd	Biomass inactivated by heat and pretreated by ethanol	Biosorption	43.4% at 7.1 initial solution pH, 4.07 g/L biomass dosage, and 37 mg/L initial Cd^{2+}	Aqueous solution (11-42 mg/L)	Characteristics of the possible interactions between biosorbents and metal ions were evaluated by scanning electron microscope (SEM)	Fereidouni *et al.*, 2009
	Dried biomass modified by crosslinking cystine with glutaraldehyde	Biosorption	11.63 mg/g biomass	Aqueous solution	The loaded biosorbent was regenerated with HCl solution and could be used repeatedly six times.	Yu *et al.*, 2007
	pmr1Δ	Bioaccumulation	50% after the first batch, 94% after three consecutive batches	Synthetic effluent with nutrients (20-50 μM)	Mn^{2+} inhibited Cd^{2+} accumulation	Ruta *et al.*, 2010
Co	*pmr1Δ*	Bioaccumulation	63% after three consecutive batches	Synthetic effluent with nutrients (1-2 mM)		Ruta *et al.*, 2010
Cu	Ethanol and caustic soda-treated biomass	Biosorption		Aqueous solution	Increase of biosorption capacity due to exposure of carboxyl and amino groups on the surface of biomass sample	Zhang *et al.*, 2010
	Cells immobilized on the surface of chitosan-coated magnetic nanoparticles	Biosorption	96.8% Maximum adsorption capacity: 144.9 mg/g	Aqueous solution, 60 mg/L	Optimum pH adsorption: 4.5 Fast process, equilibrium reached within 1 hour.	Peng *et al.*, 2010
	EDTA-modified biomass	Biosorption	65 mg/ g (14 times higher than that of the unmodified biomass)	Removal in the presence of high concentration of K^+, Na^+, Ca^{2+} and Mg^{2+}	K^+, Na^+, Ca^{2+} and Mg2+ had little effects on the uptake. The adsorbed Cu^{2+} on the modified biomass could be effectively desorbed in an EDTA solution.	Yu *et al.*, 2008

Table 3. (Continued)

Hg	Dowex anion exchanger-immobilized-baker's yeast [Dae-yeast]	Biosorption	90.77-97.91+/-3.00-5.00% (pH 2) 90.00-93.40+/-4.00-5.00% (pH 4) 92.31-100.00+/-3.00-4.00% (pH 7)	Buffer solution (pH 4-9)	Surface modification of [Dae-yeast] was characterized by scanning electron microscopy (SEM) and infrared spectroscopy	Mahmoud *et al.*, 2009
Mn	Industrial waste biomass	Biosorption		Buffer solution (pH 4-9)	The contribution of the functional groups and lipids to Mn^{2+} biosorption was in the order: carboxylic acids > amines > lipids > phosphates. Blocking of sulfhydryl group did not have any significant effect on Mn^{2+} uptake.	Parvathi *et al.*, 2007
	pmr1Δ	Bioaccumulation	50% after the first batch 82% after three consecutive batches	Synthetic effluent with nutrients (2 mM)	In the presence of Cu^{2+} or Co^{2+}, Mn^{2+} was preferentially accumulated	Ruta *et al.*, 2010
Ni	Heat-inactivated flocculant strain	Biosorption	78% after the first batch, 96% after the second batch	Synthetic effluent	Cell separation does not require energy input	Machado *et al.*, 2008
	Heat-killed flocculant strain	Biosorption	89%	Two simulated effluents and validated one real electroplating effluent (303 mmoles/L)		Machado *et al.*, 2010
	Biomass inactivated by heat and pretreated by ethanol	Biosorption	65.5% at 7.1 initial solution pH, 4.07 g/L biomass dosage, 16 mg/L initial Ni^{2+}	Aqueous solution (11-42 mg/L)	Characteristics of the possible interactions between biosorbents and metal ions were evaluated by scanning electron microscope (SEM)	Fereidouni *et al.*, 2009

Table 3. (Continued)

	Deactivated protonated yeast	Biosorption	The biosorption capacity decreased from 9.8 to 9.3 mg/g at an initial Ni^{2+} concentration of 400mg/L, while at a lower initial concentration of 100mg/L, it decreased from 8.2 to 4.9 mg/g, as the temperature was increased from 27°C to 60°C	Aqueous solution 400mg/L, initial pH of 6.75.	Biosorption of Ni^{2+} onto baker's yeast was spontaneous and exothermic. Desorption studies revealed that the protonated yeast biomass can be regenerated using 0.1N HCl and reused.	Padmavathy, 2008
	W303-1A	Bioaccumulation	32 %	Aqueous solution with nutrients (1 mM)	Histidine-stimulated	Farcasanu *et al.*, 2005
Pb	Engineered strain	Biosorption	Specific, even in the presence of 3-fold molar excess of either Cd^{2+} or Zn^{2+}.	Aqueous solution, 75-300 µM	Short metal-binding NP peptides (harboring the CXXEE metal fixation motif of the bacterial Pb(2+)-transporting P1-type ATPases) displayed and covalently anchored to the cell wall.	Kotrba *et al.*, 2010
	EDTA-modified biomass	Biosorption	192.3 mg/ g (10 times higher than that of the unmodified biomass)	Removal in the presence of high concentration of K^+, Na^+, Ca^{2+} and Mg^{2+}	K^+, Na^+, Ca^{2+} and Mg2+ had little effects on the uptake. The adsorbed Pb^{2+} on the modified biomass could be effectively desorbed in an EDTA solution.	Yu *et al.*, 2008
	Dried biomass modified by crosslinking cystine with glutaraldehyde	Biosorption	45.86 mg/g biomass	Aqueous solution	The loaded biosorbent was regenerated with HCl solution and could be used repeatedly six times.	Yu *et al.*, 2007
U	Biomass	Biosorption	84-92% 2.4 mg/g biomass	Aqueous solution 0.1-0.5 mM	Temperature has no effect on the biosorption, while pH does have an influence, 4.5 being the best value.	Omar et al., 1996
Zn	Intact cells	Biosorption	74.8 to 654.8 µmol/g	batch, 0.08 and 0.8 mmol/L	Ion exchange was one of the mechanisms for Zn^{2+} biosorption	Chen and Wang, 2007

The *in vivo* expression of metal-binding peptides offers distinct advantages in obtaining potent metal-adsorbing microorganisms. Kambe-Honjoh *et al.* (2000) expressed a fusion protein of hexa-histidine repeat (His) and glycosylphosphatidylinositol (GPI)-anchor region of *Saccharomyces cerevisiae* Cwp1 with *Aspergillus oryzae* Taka-amylase A (TAA) on the yeast cell surface, obtaining recombinant cells that exhibited 1.6- to 2.8-fold higher adsorbing capacity for Cu^{2+}, Ni^{2+}, and Zn^{2+}. Nakajima *et al.* (2001) expressed *S. cerevisiae* metallothionein (Cup1) encoded by the *CUP1* gene (Winge *et al.*, 1985) fused to *S. cerevisiae* Suc2 as a carrier and anchor for the expression of Cup1 on the cell surface. Metallothioneins are highly conserved low-molecular- weight Cys-rich proteins that bind metal ions (e.g., Cu^{2+}, Ni^{2+}, Cd^{2+}, Hg^{2+}, and Ag^{+}) and sequester them in a biologically inactive form (Butt and Ecker, 1987; Hamer, 1986). Cells expressing invertase-metallothionein (Suc2-Cup1) fusion protein on the cell surface showed 1.6-fold and 3.5- fold increases in Ni^{2+} and Cu^{2+} ion adsorbing capacity when compared with the control.

To improve the biosorption capacity of the potential yeast biosorbent, Kotrba *et al.* (2010) displayed and covalently anchored to the cell wall of *S. cerevisiae* a short metal binding NP peptide, harboring the CXXEE metal fixation motif of the bacterial Pb^{2+}-transporting P1-type ATPases, thus increasing the Pb^{2+} biosorption capacity up to 5 fold. As the Pb^{2+} biosorption properties were not changed by the presence of 3-fold molar excess of either Cd^{2+} ori Zn^{2+}, the biosorption mechanism could be specifically upgraded by engineering of the biosorbent with an eligible metal-binding peptide.

7.3. Heavy Metal Bioaccumulation

Biosorption alone may not suffice for effective metal bioremediation, and application of active growing cells might be a better option due to their ability to use continuous metabolic uptake following physical adsorption. Moreover, the metals diffused into the cells during detoxification get bound to intracellular proteins or chelatins before being incorporated into vacuoles and other intracellular sites. These processes are often irreversible and ensure less risk of metal releasing back to the environment. Apart from this, using growing cultures in removal processes could avoid the need for a separate biomass production process, e.g. cultivation, harvesting, drying, processing and storage prior to the use. The advantages of using growing accumulating cells in heavy metal bioremediation were extensively reviewed by Malik (2004). Ideally, a living bioremediator would be tolerant to high concentrations of the pollutant and at the same time it would accumulate it in sufficient amount to contribute significantly to pollutant removal. *S. cerevisiae* cells are mediocre accumulators, therefore it is expected that only cells grown under special conditions or engineered cells would be suitable for bioremediation through metabolic uptake. Farcasanu *et al.* (2005) showed that exogenously L-histidine augmented both Ni^{2+} tolerance and bioaccumulation, increasing the capacity of the growing cells to remove Ni^{2+} from synthetic effluents; however, the applicability of this approach was limited by the high cost of histidine. Mutants tolerant to Co^{2+} or Ni^{2+} were reported, but with no remarkable accumulation, and hence removal capacities (Farcasanu *et al.*, 2008 a, b). Alternatively, sensitive, but accumulating strains can be utilized. For example, a null mutant *pmr1Δ* strain was investigated for the ability to remove Mn^{2+}, Cu^{2+}, Co^{2+}, or Cd^{2+} from synthetic effluents. Due to increased metal accumulation, the mutant strain was more efficient than the wild-type in removing Mn^{2+},

Cu^{2+}, or Co^{2+} from synthetic effluents containing 1-2 mM cations, with a selectivity Mn^{2+} > Co^{2+} > Cu^{2+} and also in removing Mn^{2+} and Cd^{2+} from synthetic effluents containing 20–50 μM cations, with a selectivity Mn^{2+} > Cd^{2+} (Ruta *et al.*, 2010).

Transgenic cells expressing proteins known to be involved in metal binding or trafficking may also be an option, and plant phytochelatins or metallothioneins are certainly interesting candidates. For example, engineered yeast cells expressing *Arabidopsis thaliana* phytochelatin synthase exhibited enhanced specific arsenic accumulation even in the presence of excess Zn^{2+} (Singh *et al.*, 2008). Although such approaches are remarkable, one important issue needs to be addressed: genetically modified cells would clearly restrict the applicability of the method in the field, unless strictly controlled conditions are to be implemented.

7.4. Heavy Metal Recovery

Once the metals are removed from contaminated sites, it is important to design processes for metal recovery. Park and Choi (2002) reported Pb, Cu and Cd recovery from calcium alginate beads containing immobilized *S. cerevisiae* cells. The cell-suspension immobilized beads retained the initial metal-uptake capacity after 20 repeated batches of adsorption and desorption, but the fraction of metal desorbed from the beads by 1 M HCl soln. was only 70% of the adsorbed metal. The beads, which had been contained for 14 successive days in the 0.5% (w/v) CaCl2 soln. at 4° just after 20 cycles of adsorption/desorption, retained the initial metal-uptake capacity after 30 repeated cycles, and > 90% of the copper and cadmium adsorbed on the beads was desorbed by the 1 M HCl soln. The recycling of rare metals from wastes is of utmost importance and strains designed to bind specifically to such metals are very interesting. For example, ModE protein derived from *Escherichia coli*, which is a molybdate-dependent transcriptional regulator with the ability to bind molybdate as a form of soluble molybdenum, was displayed on the yeast cell surface by alpha-agglutinin-based cell surface display system for the adsorption and recovery of molybdate (Nishitani *et al.*, 2010). Thus, a system was created that was suitable not only for adsorption of molybdate as a rare metal bioadsorbent but also for the easy recovery of molybdate located on the cell surface. A similar mechanism could be specifically upgraded by engineering of the cell surface with a suitable metal-binding system.

8. Concluding Remarks

As human activities get more diverse, non-invasive ways to protect and clean the environment are more and more needed. In this direction, *S. cerevisiae* is invaluable: completely non-pathogenic and safe, un-pretentious concerning nutriment requirement, easily removable once the job is finished. And thanks to the power of genetic engineering, adjustable to a variety of unprecedented conditions. And although in use for thousands of years, it's still on track!

REFERENCES

Achkor H, Díaz M, Fernández, MR, Biosca JA, Parés X, Martínez MC. (2003) *Enhanced formaldehyde detoxification by overexpression of glutathione-dependent formaldehyde dehydrogenase from Arabidopsis*. Plant Physiol 132: 2248-2255.

Aksu Z (2003) *Reactive dye bioaccumulation by Saccharomyces cerevisiae.* Process Biochem 38:1437-1444.

Alcalde M, Bulter T, Arnold FH (2002) *Colorimetric assays for biodegradation of polycyclic aromatic hydrocarbons by fungal laccases*. J Biomol Screen 7:547–553.

Ansanay V, Denquin S, Camarasa C, Schaeffer V, Grivet J, Blondin B, Salmon J, Barre P (1996) *Malolactic fermentation by engineered Saccharomyces cerevisiae is compared with engineered Saccharomyces pombe*. Yeast 12:215-225.

Armengaud J, Happe B, Timmis KN (1998) *Genetic analysis of dioxin dioxygenase of Sphingomonas sp. Strain RW1: catabolic genes dispersed on the genome.* J Bacteriol 180:3954-3966.

Arumugam GK, Somasundaram S, Kassey VB, Ganesan S (2006) *Immobilized cell reactors in mineralization of dicarboxylic acid solid waste.* Biodegradation 17:559-570.

Avery SV, Tobin JM (1992) *Mechanism of strontium uptake by laboratory and brewing strains of Saccharomyces cerevisiae.* Appl Environ Microbiol 58:3883-3889.

Balfanz J, Rehm HJ (1991) *Biodegradation of 4-chlorophenol by adsorptive immobilized Alcaligenes sp. A-72 in soil.* Appl Microbiol Biotechnol 35:662–668.

Barton, J. W., Vodraska, C. D., Flanary, S. A. and Davison, B. H. *"Solubility of toluene, benzene and TCE in high-microbial concentration systems.,"* Chemosphere (73:11), 2008, pp. 1737-1740.

Begum A, Katsumata H, Kaneco S, Suzuki T, Ohta K (2003) *Biodegradation of phthalic acid esters by bakery yeast Saccharomyces cerevisiae.* Bull Environ Contam Toxicol 70:255-261.

Bel-Rhlid R, Talmon JP, Fay LB, Juillerat MA (2004) *Biodegradation of 3-chloro-1,2-propanediol with Saccharomyces cerevisiae.* J Agric Food Chem 52:6165-6169.

Beurskens JEM, Toussaint, de Wolf J. Van der Steen JMD, Slot PC, Commandeur LCM, Parsons R (1995) *Dehalogenation of chlorinated dioxins by an anaerobic microbial consortium from sediment*. Environ Toxicol Chem 14:939-943.

Beychok MR (1987) *A data base for dioxin and furan emissions from refuse incinerators.* Atmospheric Environment 21:29-36.

Blackwell KJ, Singleton I, Tobin JM (1995) *Metal cation uptake by yeast: a review.* Appl Microbiol Biotechnol 43:579-584.

Blackwell KJ, Tobin, JM (1999) *Cadmium accumulation and its effects on intracellular ion pools in a brewing strain of Saccharomyces cerevisiae.* J Ind Microbiol Biotechnol 23:204-208.

Borya R, Alba J, Garrido SE, Martinez LM, Garcìa MP, Monteoliva M, Ramos Cormenzana A (1995) *Effect of aerobic pretreatment with Aspergillus terreus on the anaerobic digestion of olive-mill wastewater*. Biotechnol Appl Biochem 22:233-246.

Bourbonnais R, Price MG (1992) *Demethylation and delignification of kraft pulp by Trametes versicolor laccase in the presence of 2,20-azinobis-(3-ethylbenzthiazoline-6-sulphonate)*. Appl Microbiol Biotechnol 36:823–827.

Broach JR, Pringle JR and Jones EW. *The Molecular and Cellular Biology of the Yeast Saccharomyces.* Cold Spring Harbor Laboratory Press, Cold Spring Harbor, New York, 1991.

Bressan M, Liberatore L, d'Alessandro N, Tonucci L, Belli C, Ranalli M (2004) *Improved combined chemical and biological treatments of olive oil wastewaters.* J Agric Food Chem 52:1228-1233.

Bulter T, Alcalde M, Sieber V, Meinhold P, Schlachtbauer C, Arnold FH (2003) *Functional expression of a fungal laccase in Saccharomyces cerevisiae by directed evolution.* Appl Environ Microbiol 69:987-995.

Bumpus JA, Tien M, Wright D, Aust SD (1985) *Oxidation of persistent environmental pollutants by a white rot fungus.* Science 228:1434-1436.

Butt TR, Ecker DJ (1987) *Yeast metallothionein and applications in biotechnology*. Microbiol Rev 51:351-364.

Chen C, Wang JL (2007) *Characteristics of Zn2+ biosorption by Saccharomyces cerevisiae.* Bioremed Environ Sci 20:478-482.

Culp SJ, Beland FA (1996) *Malachite green: a toxicological review.* J Am Coll Toxicol 15: 219–238.

Delcourt F, Taillandier P, Vidal F, Strehaiano P (1995) *Influence of pH, malic acid and glucose concentrations on malic acid consumption by Saccharomyces cerevisiae.* Appl Microbiol Biotechnol 43:321-324.

Dujon, B. (1996) *The yeast genome project: what did we learn?* Trends Genet. 12: 263- 270.

El-Shiekh HH, Mahdy HM, El-Aaser MM (2007) *Bioremediation of aflatoxins by some reference fungal strains.* Pol J Microbiol 56:215-223.

Esmon PC, Esmon BE, Schauer IE, Taylor A, Schekman R (1987) *Structure, assembly, and secretion of octameric invertase.* J Biol Chem 262:4387-4394.

Farcasanu IC, Mizunuma M, Nishiyama F, Miyakawa T (2005) *Role of L-histidine in conferring tolerance to Ni2+ in Sacchromyces cerevisiae cells.* Biosci Biotechnol Biochem 69:2343-2348.

Farcasanu IC, Oprea E, Paraschivescu C, Ruta L, Avramescu S (2008a) *Characterization of Saccharomyces cerevisiae mutants resistant to high concentrations of Co2+: a primary step to bioremediation by removal and recovery of Co2+ from waste waters*. Revista de Chimie 59:1041-1044.

Farcasanu IC, Paraschivescu C, Ruta L, Oprea E, Avramescu S. (2008b) *Manipulation of Ni2+ tolerance of Sacchromyces cerevisiae cells: a primary step to bioremediation by removal and recovery of Ni2+ from contaminated waters*, Revue Roumaine de Chimie 53:647-651.

Feldmann H (2005) *Yeast transport.* In: Yeast Molecular Biology, A short compendium on basic features and novel aspects. Adolf-Butenandt-Institute, University of Munich, Available at http://biochemie.web.med.uni-muenchen.de/Yeast_Biol/

Fereidouni M, Daneshi A, Younesi H (2009) *Biosorption equilibria of binary Cd(II) and Ni(II) systems onto Saccharomyces cerevisiae and Ralstonia eutropha cells: application of response surface methodology*. J Hazard Mater 168:1437-48.

Ferraz AI, Teixeira JA (1999) *The use of flocculating brewer's yeast for Cr(III) and Pb(II) removal from residual wastewaters*. Bioprocess Eng 21:431-437.

Ferraz AI, Tavares T, Teixeira JA (2004) *Cr(III) removal and recovery from Saccharomyces cerevisiae.* Chem Eng J 105:11-20.

Firmin S, Gandia P, Morgavi DP, Houin G, Jouany, JP, Bertin, G, Boudra H (2010) *Modification of aflatoxin B1 and ochratoxin A toxicokinetics in rats administered a yeast cell wall preparation*. Food Addit Contam Part A Chem Anal Control Expo Risk Assess 27:1153-1160.

Georgiou G, Stathopoulos C, Daugherty PS, Nayak AR, Iverson BL, Curtiss R3rd (1997) *Display of heterologous proteins on the surface of microorganisms: From the screening of combinatorial libraries to live recombinant vaccines*. Nature Biotechnol 15:29-34.

Giannoutsou EP, Meintanis C, Karagouni AD (2004). *Identification of yeast strains isolated from a two-phase decanter system olive oil waste and investigation of their ability for its fermentation*, Bioresour Technol 93:301-306.

Goffeau, A, Barrell BG, Bussey H, Davis RW, Dujon B, Feldmann H, Galibert F, Hoheisel JD, Jacq C, Johnston M, Louis EJ, Mewes HW, Murakami Y, Philippsen P, Tettelin H, Oliver SG. (1996) *Life with 6000 genes*. Science, 274:546-567.

Granström T, Lindberg P, Nummela J, Jokela J and Leisola M. (2002) *Biodegradation of VOCs from printing press air by an on-site pilot plant bioscrubber and laboratory scale continuous yeast cultures*. Biodegradation 13:155-162.

Guthrie C and Fink GR *Guide to Yeast Genetics and Molecular Biology, Methods in Enzymology, Vol. 169,* Academic Press, San Diego, 1991.

Hamer, D. H. (1986) *Metallothionein*. Annu Rev Biochem, 55:913-951.

Hamlet CG, Sadd PA, Crews C, Velísek J, Baxter DE (2002) *Occurrence of 3-chloro-propane-1,2-diol (3-MCPD) and related compounds in foods: a review*. Food Addit Contam 19:619-631.

Hammel KE, Jensen Jr. KA, Mozuch MD, Landucci LL, Tien M, Pease EA (1993) *Lignolysis by a purified lignin peroxidase*. J Biol Chem 268:12274–12281.

Handy M, Festino C, Aubart C (1992) *Anaerobic digestion of olive oil mill wastewater in fully mixed reactors and in fixed film reactors*. Process Biochem 27:37–42.

Hofmann KH (1986) *Oxidation of naphthalene by Saccharomyces cerevisiae and Candida utilis*. J Basic Microbiol 26:109-111.

Holme JA (1985) *Xenobiotic metabolism and toxicity in primary monolayer cultures of hepatocytes*. NIPH Ann 8:49-63.

Hu K, Bunce NJJ (1999) *Metabolism of polychlorinated dibenzo-p-dioxins by rat liver microsomes*. Biochem Mol Toxicol 13:307-315.

IARC, *International Agency for Research on Cancer Monographs on the Evaluation of Carcinogenic Risks to Humans*. Aflatoxins: Occurring Aflatoxins (Group 1), Aflatoxin M_1 (Group 2B) Vol. 56; Lyon, France: 1993, p. 445.

Jadhav JP, Govindwar SP (2006) *Biotransformation of malachite green by Saccharomyces cerevisiae MTCC 463*. Yeast 23:315-323.

Jadhav JP, Parshetti GK, Kalme SD, Govindwar SP (2007) *Decolourization of azo dye methyl red by Saccharomyces cerevisiae MTCC 463*. Chemosphere 68:394-400.

Kambe-Honjoh H, Ohsumi K, Shimoi H, Nakajima H, Kitamoto K (2000) *Molecular breeding of yeast with higher metal-adsorption capacity by expression of histidine-repeat insertion in the protein anchored to the cell wall*. J Gen Appl Microbiol 46:113-117.

Kotrba P, Ruml T (2010) *Surface display of metal fixation motifs of bacterial P1-type ATPases specifically promotes biosorption of Pb(2+) by Saccharomyces cerevisiae*. Appl Environ Microbiol 76:2615-2622.

Laconi S, Molle G, Cabiddu A, Pompei R. (2007) *Bioremediation of olive oil mill wastewater and production of microbial biomass.* Biodegradation 18:559-566.

Lin YH, Hwang SCJ, Shih WC, Chen KC (2006). *Development of a novel microorganism immobilization method using anionic polyurethane.* J Appl Polym Sci 99:738-743.

Machado MD, Janssens S, Soares HMVM, Soares EV (2009) *Removal of heavy metals using a brewer's yeast strain of Saccharomyces cerevisiae: advantages of using dead biomass.* J Appl Microbiol 106:1792-1804.

Machado MD, Santos MSF, Gouveia C, Soares HMVM, Soares EV (2008) *Removal of heavy metals using a brewer's yeast strain of Saccharomyces cerevisiae: the flocculation as a separation process.* Bioresour Technol 99:2107-2115.

Machado MD, Soares EV, Soares HM (2010) *Removal of heavy metals using a brewer's yeast strain of Saccharomyces cerevisiae: chemical speciation as a tool in the prediction and improving of treatment efficiency of real electroplating effluents.* J Hazard Mater 180:347-353.

Mahmoud ME, Yakout AA, Osman MM (2009) *Dowex anion exchanger-loaded-baker's yeast as bi-functionalized biosorbents for selective extraction of anionic and cationic mercury(II) species.* J Hazard Mater 164:1036-1044.

Malik A (2004) *Metal bioremediation through growing cells.* Environ Int 30:261-278.

Marques PA, Pinheiro HM, Teixeira JA, Rosa MF (1999) *Removal efficiency of Cu^{2+}, Cd^{2+} and Pb^{2+} by waste brewery biomass: pH and cation association effects.* Desalination 124:137-144.

Marques PA, Rosa MF, Pinheiro HM (2000) *pH effects on the remove of Cu2+, Cd2+ and Pb2+ from aqueous solution by waste brewery biomass.* Bioprocess Eng 23:135-141.

Moss MO, Long MT (2002) *Fate of patulin in the presence of the yeast Saccharomyces cerevisiae.* Food Addit Contam 19:387-399.

Murai T, Ueda M, Yamamura M, Atomi H, Shibasaki Y, Kamasawa N, Osumi M, Amachi T, Tanaka A (1997) *Construction of a starch-utilizing yeast by cell surface engineering.* Appl Environ Microbiol 63:1362-1366.

Murakami H,Yabusaki Y, Sakaki T, Shibata M, *Ohkawa H (1990)* J Biochem 108:859-865.

Naeem A, Woertz JR Fein JB (2006) *Experimental measurement of proton, Cd, Pb, Sr, and Zn adsorption onto the fungal species Saccharomyces cerevisiae.* Environ Sci Technol 40:5724-5729.

Nakajima H, Iwasaki T, Kitamoto K (2001) *Metalloadsorption by Saccharomyces cerevisiae cells expressing invertase-metallothionein (Suc2-Cup1) fusion protein localized to the cell surface.* J Gen Appl Microbiol 47:47-51.

Necochea R, Valderrama B, Díaz-Sandoval S, Folch-Mallol JL, Vázquez-Duhalt R, Iturriaga, G (2005) *Phylogenetic and biochemical characterisation of a recombinant laccase from Trametes versicolor.* FEMS Microbiol Lett 244:235-241.

Nishitani T, Shimada M, Kuroda K, Ueda M (2010) *Molecular design of yeast cell surface for adsorption and recovery of molybdenum, one of rare metals.* Appl Biochem Biotechnol 86:641-648.

de Nobel JG, Klis FM, Munnik T, Priem J, van den Ende H. (1990) *An assay of relative cell wall porosity in Saccharomyces cerevisiae, Kluyveromyces lactis and Schizosaccharomyces pombe.* Yeast 6:483-90.

de Nobel JG and Barnett JA (1991) *Passage of molecules through yeast cell walls: A brief essay—review.* Yeast 7:313-323.

de Nobel H, van Den Ende H, Klis FM. (2000) *Cell wall maintenance in fungi.* Trends Microbiol 8:344-345.

Nojiri H, Nam JW, Kosaka M, Morii KI, Takemura T, Furihata K, Yamane H, Omori T (1999) *Diverse oxygenations catalyzed by carbazole 1,9a-dioxygenase from Pseudomonas sp. Strain CA10.* J Bacteriol 181:3105-3113.

Omar NB, Merroun ML, González-Muñoz MT, Arias JM (1996) *Brewery yeast as a biosorbent for uranium.* J Appl Bateriol 81:283-287.

Padmavathy V (2008) *Biosorption of nickel(II) ions by baker's yeast: kinetic, thermodynamic and desorption studies.* Bioresour Technol 99:3100-3109.

Park JK, Choi SB (2002) *Metal recovery using immobilized cell suspension from a brewery.* Kor J Chem Eng 19:68-74.

Parvathi K, Nareshkumar R, Nagendran R (2007) *Manganese biosorption sites of Saccharomyces cerevisiae.* Environ Technol 28:779-784.

Patel R, Rajkumar S (2009) *Isolation and characterization of phenol degrading yeast.* J Basic Microbiol 49:216-219.

Peng Q, Liu Y, Zeng G, Xu W, Yang C, Zhang J (2010) *Biosorption of copper(II) by immobilizing Saccharomyces cerevisiae on the surface of chitosan-coated magnetic nanoparticles from aqueous solution.* J Hazard Mater 177:676-682.

Pirnie-Fisker EF and Woertz JR (2007) *Degradation of ethanol plant by-products by Exophiala lecanii-corni and Saccharomyces cerevisiae in batch studies.* Appl Microbiol Biotechnol 74:902--910.

Ruta L, Paraschivescu C, Matache M, Avramescu S, Farcasanu IC (2010) *Removing heavy metals from synthetic effluents using "kamikaze" Saccharomyces cerevisiae cells.* Appl Microbiol Biotechnol 85:763-771.

Ryu K, Hwang SY, Kim KH, Kang JH, Lee, EK (2008a) *Functionality improvement of fungal lignin peroxidase by DNA shuffling for 2,4-dichlorophenol degradability and H2O2 stability.* J Biotechnol 133:110-115.

Ryu K, Kang JH, Wang L, Lee EK (2008b) *Expression in yeast of secreted lignin peroxidase with improved 2,4-dichlorophenol degradability by DNA shuffling.* J Biotechnol 135:241-246.

Sadd PA, Hamlet CG, Liang L (2008) *Effectiveness of methods for reducing acrylamide in bakery products.* J Agric Food Chem 56:6154-6161.

Safarík I, Ptáčkova L, Safaríková M (2002) *Adsorption of dyes on magnetically labeled baker's yeast cells.* Eur Cells Mater 3:52-55.

Safaríková M, Ptácková L, Kibriková I, Safarík I (2005) *Biosorption of water-soluble dyes on magnetically modified Saccharomyces cerevisiae subsp. uvarum cells.* Chemosphere 59:831-835.

Sakaki T, Shinkyo R, Takita T, Ohta M, Inouye K (2002) *Biodegradation of polychlorinated dibenzo-p-dioxins by recombinant yeast expressing rat CYP1A subfamily.* Arch Biochem Biophys 401:91-98.

Sato SI, Nam JW, Kasuga K, Nojiri H, Yamane H, Omori T (1997) *Identification and characterization of genes encoding carbazole 1,9a-dioxygenase in Pseudomonas sp. strain CA10.* J Bacteriol 179:4850-4858.

Schreuder MP, Mooren AT, Toschka HY, Verrips CT, Klis FM. (1996) *Immobilizing proteins on the surface of yeast cells.* Trends Biotechnol 14:115-120.

Sharma HA, Barber JT, Ensley HE, Polito MA (1997) *A comparison of the toxicity and metabolism of phenol and chlorinated phenols by Lemna gibba, with special reference to 2,4,5-trichlorophenol.* Environ Toxicol Chem 16:346-350.

Shibasaki S, Maeda H, Ueda M (2009) *Molecular display technology using yeast–arming technology.* Anal Sci 25:41-49.

Simon P, Singleton I (1996) *A method to increase silver biosorption by an industrial strain of Saccharomyces cerevisiae.* Appl Microbiol Biotechnol 45:278-285.

Singh S, Lee W, Dasilva NA, Mulchandani A, Chen W (2008) *Enhanced arsenic accumulation by engineered yeast cells expressing Arabidopsis thaliana phytochelatin synthase.* Biotechnol Bioeng 99:333-340.

Smith JE, Lewis CW, Anderson JG, Solomns GL (1994) *Micotoxins in human nutrition and health, EUR Report No 16048,* Luxemburg, European Comission.

Stinson EE, Osman SF, Huhtanen CN, Bills DD (1978) *Disappearance of patulin during alcoholic fermentation of apple juice.* Appl Environ Microbiol 36:620-622.

Strosnider H, Azziz-Baumgartner E, Banziger M, Bhat RV, Breiman R, Brune M, DeCock K, Dilley A, Groopman J, Hell K, Henry SH, Jeffers D, Jolly C, Jolly P, Kibata GN, Lewis L, Liu X, Luber G, McCoy L, Mensah P, Miraglia M, Misore A, Njapau H, Ong CN, Onsongo MT, Page SW, Park D, Patel M, Phillips T, Pineiro M, Pronczuk J, Rogers HS, Rubin C, Sabino M, Schaafsma A, Shephard G, Stroka J, Wild C, Williams JT, Wilson D. (2006) *Workgroup report: public health strategies for reducing aflatoxin exposure in developing countries.* Environ Health Perspect 114:1898-1903.

Sutherland JB (1992) *Detoxification of polycyclic aromatic hydrocarbons by fungi.* J Ind Microbiol 9:53-61.

Takahashi-Ando N, Tokai T, Hamamoto H, Yamaguchi I, Kimura M (2005) *Efficient decontamination of zearalenone, the mycotoxin of cereal pathogen, by transgenic yeasts through the expression of a synthetic lactonohydrolase gene.* Appl Microbiol Biotechnol 67:838-844.

Tardioli S, Bannè E, Santori F (1997) *Species-specific selection on soil fungal population after olive mill waste-water treatment.* Chemosphere 34:2329-2336.

Valli K, Gold MH (1991) *Degradation of 2,4-dichlorophenol by the lignin-degrading fungus Phanerochaete chrysosporium.* J Bacteriol 173:345-352.

Veiga MC, Kennes C (2001) *Parameters affecting performance and modeling of biofilters treating alkylbenzene-polluted air.* Appl Microbiol Biotechnol 55: 254-258

Volesky B, May-Phillips HA (1995) *Biosorption of heavy metals by Saccharomyces cerevisiae.* Appl Microbiol Biotechnol 42:797-806.

Wild CP, Gong YY (2010) *Mycotoxins and human disease: a largely ignored global health issue.* Carcinogenesis 31:71-82.

Wilson DM, Nuovo GJ (1973) *Patulin production in apples decayed by Penicillium expansum.* Appl Microbiol 26:124-125.

Winge DR, Nielson KB, Gray W, Hamer D (1985) *Yeast metallothionein: Sequence and metal-binding properties.* J Biol Chem 260:14464-14470.

Woertz, J. R., Kinney, K. A., Szaniszlo P. J. (2001) *A fungal vapor-phase bioreactor for the removal of nitric oxide from waste gas streams.* J Air Waste Manage Assoc 51:895-902

Wu G, Chabot JC, Caron JJ, Heitz M (1998). *Biological elimination of volatile organic compounds from waste gases in a biofilter.* Can Water Air Soil Pollut 101:69-78.

Wunch KG, Feibelman, T, Bennett JW (1997) *Screening for fungi capable of removing benzo[a]pyrene in culture.* Appl Microbiol Biotechnol 47:620-624

Yiannikouris A, André G, Poughon L, François J, Dussap CG, Jeminet G, Bertin G, Jouany JP (2006) *Chemical and conformational study of the interactions involved in mycotoxin complexation with beta-D-glucans.* Biomacromolecules 7:1147-1155.

Yu JX, Li BH, Sun XM, Yuan J, Chi RA (2010) *Poly (amic acid)-modified biomass of baker's yeast for enhancement adsorption of methylene blue and basic magenta.* Appl Biochem Biotechnol 160:1394-1406.

Yu J, Tong M, Sun X, Li B (2007) *Cystine-modified biomass for Cd(II) and Pb(II) biosorption.* J Hazard Mater 143:277-284.

Yu J, Tong M, Sun X, Li B (2008) *Enhanced and selective adsorption of Pb2+ and Cu2+ by EDTAD-modified biomass of baker's yeast.* Bioresour Technol 99:2588-2593.

Zhang Y, Fan C, Meng Q, Diao Z, Dong L, Peng X, Ma S, Zhou Q (2009) *Biosorption of Pb(2+) by Saccharomyces cerevisiae in static and dynamic adsorption tests.* Bull Environ Contam Toxicol 83:708-712.

Zhang Y, Liu W, Xu M, Zheng F, Zhao M (2010) *Study of the mechanisms of Cu2+ biosorption by ethanol/caustic-pretreated baker's yeast biomass.* J Hazard Mater 178:1085-1093.

Zlotnik H, Fernandez MP, Bowers B, Cabib E. (1984) *Saccharomyces cerevisiae mannoproteins form an external cell wall layer that determines wall porosity.* Bacteriol. 159:1018-1026.

Zumárraga M, Bulter T, Shleev S, Polaina J, Martínez-Arias A, Plou FJ, Ballesteros A, Alcalde, M (2007) *In vitro evolution of a fungal laccase in high concentrations of organic cosolvents.* Chem Biol 14:1052-1064.

In: Bioremediation
Editor: Alexander C. Mason

ISBN: 978-1-61122-730-7

Chapter 12

MICROBIAL ASPECTS OF ANAEROBIC DIGESTION FOR BIOGAS PRODUCTION

A. B. Chaudhari, P. C. Suryawanshi[1], *and R. M. Kothari*[2]
[1]School of Life Sciences, North Maharashtra University, Jalgaon 425 001, India
Jain R&D, Jain Irrigation Systems Ltd., Jalgaon 425 001 India
[2]Rajiv Gandhi Institute of Information Technology and Biotechnology, Bharati Vidyapeeth University, Pune 411 046 India

ABSTRACT

Biogas production through anaerobic digestion of biodegradable solid and liquid waste appears to be a sustainable approach. It involves hydrolysis of waste by a consortium of aerobes, acidogenesis by facultative microbes, followed by acetogenesis by both facultative or obligate anaerobes and finally methanogenesis through acetotrophic/hydrogenotrophic/ methylotrophic reactions. For this purpose, habitats of facultative and obligate anaerobes have been reviewed, classification of methanogens undertaken, their substrate-specificity examined and metabolic pathways for methane production from a variety of substrates discussed through major and minor metabolic pathways, including the role of syntrophs. Together, these microbes provide efficiency of digestion and if their integrated processes are not attended in time, it leads to deterioration in methane production and at times digester failure. Finally, eco-friendly applications for biogas before and after up-gradation are considered, concluding intricacies of the process characterized by ecological and economical gains to control pollution.

1.0 INTRODUCTION

Fuel costs have been rising steadily for a number of years, reaching to a peak value of US $ 137/barrel with concomitant increase in taxation burden, leading to a double financial load to be borne by the users. On this background, bio-degradable wastes are being viewed as renewable energy sources through biogas route. Its production technologies appear attractive by treating biodegradable liquid and solid waste, since energy required for anaerobic treatment is minimal compared to energy required for aerobic treatment. Secondly, even a basic construction using mostly used materials, produces biogas if a few simple design rules and operational procedures are followed [Gerardi, 2003] Biogas production is achievable in small scale, at moderate cost, in minimal gestation period, obviating the need to supply energy to outlying areas such as farms, farm houses, farm-based agro-industry, picnic spots, vacation resorts etc. To achieve this objective, intricacies of biochemical reactions catalyzed by 4 consortia of microbial population need to be understood. Therefore, microbial aspects of biogas production constitute the theme of this article.

2.0 Outline of Biogas Production

Anaerobic digestion is broadly a four-step, but extremely intricate process, involving a consortium of aerobic, facultative and strictly anaerobic microbes, often inter-linked by their individual substrate and product specificities. Thus, anaerobic digestion can be divided into the following microbial/enzymatic processes.

(i) Hydrolysis: Photo-synthetically fixed bio-polymers (carbohydrates, proteins and lipids) are broken down to respective monomers (glucose, amino acids, fatty acids, glycerol etc.) by extracellular hydrolases (α-amylase, amyloglucosidase, proteases and lipases) produced by aerobic microbial consortia [Shah et al. 1986; Shah et al. 1989; Upadhyay et al. 1989; Shah et al. 1990 ; Ramamurthy et al. 1991 ; Kaur et al. 1993]

(ii) Acidogenesis: Mixture of these monomers is converted into lower molecular weight volatile fatty acids (VFAs) such as acetic acid, propionic acid, butyric acid etc. by another set of extracellular enzymes from acidogenic bacteria (acidogens), mostly facultative, besides producing ammonia (NH_3), carbon dioxide (CO_2) and hydrogen sulfide (H_2S) as byproducts.

(iii) Acetogenesis: The C_3-C_6 VFAs are converted into acetic acid (51 %), H_2 and CO_2 (19 %) through biochemical processes (acetogenesis) catalyzed by a number of acetogenic microorganisms (acetogens, obligate H-producing acetogens), which are either facultative or obligate anaerobes.

(iv) Methanogenesis: Finally, methanogenic bacteria (methane-forming archaebacteria) produce methane from acetic acid (aceticlastic/acetotrophic reaction), H_2 and CO_2 (hydrogenotrophic reaction) and methylotrophic reaction by a series of biochemical processes, clubbed under methanogenesis.

The emerging end product, biogas/digester gas/*gobar* gas is a gaseous mixture comprised chiefly of 60-65 % CH_4 (the combustible component), 30-35 % CO_2, making up more than 95 % of the total, the balance being traces of N_2, H_2, CO, NH_3, H_2S, methyl-mercaptans etc. Thus, anaerobic digestion is a complex set of several biochemical processes in which different types of bacteria synergistically digest an organic waste in aerobic and oxygen-free environment, resulting in the production of biogas. Besides biogas, the process gives treated water to serve as a liquid manure and solid soil conditioner (organic manure).

At this stage, it was worthwhile to look into habitats of these microbes, especially the facultative and strictly anaerobes, since hydrolytic bacteria, are ubiquitous, being aerobic.

3.0 Occurrence of Facultative and Obligate Anaerobes

It is now established that sites where oxidation- reduction potential (ORP) is between -200 and -400 millivolts (mV), facultative and obligate methanogens are found. While facultative anaerobes grow between +200 and -200 mV ORP, methanogens grow best in an environment with an ORP of less than -300 mV. While generation time for the facultative bacteria (acetogens) is 15-30 min, methane-forming bacteria are 3 days at 35 °C or 50 days at 10°C. While organotrophs obtain their nourishment from organic substrates, chemoautotrophs obtain the same through inorganic substrates. Such natural ecosystems (habitats) are given below.

3.1 Habitats

Obligate anaerobes like methanogens usually occur in mixed cultures, where non-methanogens hydrolyze a wide variety of organic matter. They are abundant in habitats where electron acceptors such as O_2, NO_3^-, SO_4^{2-} and Fe^{3+} are limiting.

(i) Sludges from anaerobic digesters are rich in number and diversity of bacteria (more than 10^{16} bacteria/ml, comprising of saccharolytic bacteria (~ 10^8 cells ml^{-1}), proteolytic bacteria (~ 10^6 cells ml^{-1}), lipolytic bacteria (~ 10^6 cells ml^{-1}) and methane-forming bacteria (~ 10^8 cells ml^{-1})
(ii) Cattle dung and septic tanks (human excreta) contain at least 300 species of facultative and obligate anaerobes.
(iii) Anoxic sediments in deep water reservoirs, lakes, river bottoms, deep sea, black mud lakes, swamps, marshy sediments etc.
(iv) Gastro-intestinal tracts/ rumen of herbivores animals (cow, goat, sheep, pigs and termites)
(v) Industrial landfills/ municipal waste disposal sites.
(vi) Geo-thermal vents, volcanic vents and hot water/sulfur springs.
(vii) Soils irrigated for long periods as in paddy, sugarcane and banana farms/ flooded soils.

3.2 Factors Conditioning the Habitats

The following factors promote habitats for rich anaerobic micro-flora.

3.2.1 Anoxic Conditions

Methanogens are generally absent from the water column of un-stratified lakes and rivers because convection currents rapidly aerate the deep waters. However, diffusion of O_2 between the layers of stratified lakes is often too slow to maintain oxic conditions in the lower layers. Since physical structure of sediments limits oxygen dispersive mechanism(s), deeper sediments are often anoxic and harbor methanogens. Oxygen diffusion through water being slower than through the soil, water-saturated zone is often anoxic. As air-saturated water at 20 °C contains less than 0.3 mM O_2, even small amount of organic pollution causes groundwater to become anoxic and suitable for habitat of methanogenic populations.

3.2.2 Anoxic Micro-Environment

In axenic culture, methanogenic bacteria are extremely sensitive to even small amounts of O_2. However, in natural habitats, activities of other organisms protect methanogens in an apparently oxic environment. For instance, methanogens have been isolated from large dental caries and sub-gingival plaques in human mouth [Brusa et al. 1987; Belay et al. 1988] Methanogens may survive in such environments, because O_2-uptake by aerobic and euryoxic bacteria creates anoxic micro-environment.

3.2.3 Anoxic Pockets

Soil environment may also be divided into vadose zone (also termed as unsaturated zone, a portion of earth between the land surface and zone of saturation), which is not saturated with water and saturated zone below the water table. In the vadose zone, O_2 diffusion is rapid. Therefore, anoxic vadose zones are extremely rare, although anoxic micro-environment may occur. Thus, the activity of methanogens in the vadose zone is limited to anoxic pockets only.

3.2.4 Sulphate Limiting Environment

If sulfate becomes limiting, methane replaces H_2S as the reduced product and organic carbon is disproportionated to CO_2 and methane. The fate of the methyl group of acetic acid is an indicator of whether methanogenesis is a dominant catabolic pathway in an eco-system. If acetic acid is catabolized by sulphate reducing bacteria (SRBs), CO_2 is formed from the methyl group of acetic acid. In a methanogenic system, acetic acid is catabolized by the aceticlastic reaction, which forms methane from the methyl group. However, this test is not applicable in some thermophilic eco-systems, where acetic acid is oxidized to CO_2 prior to formation of methane. Thus, methanogenic degradation of [2-^{14}C]-acetic acid leads to $^{14}CH_4$ formation, whereas sulfidogenic degradation leads to $^{14}CO_2$ and the ratio of these labeled gaseous products indicates if methanogenesis predominates or otherwise [Winfrey and Zeikus, 1979] Beneath the areas of higher productivity, such as kelp forests, sulfate may be limiting in the deep organic sediments. Under these conditions, aceticlastic and H_2 or formic acid using methanogens develop.

3.2.5 Sulphate-sufficient environment

In anoxic environments, presence of NO_3^-, SO_4^{2-} and Fe^{3+} inhibits methanogenesis by allowing other organisms to out-compete methanogens for reduced substrates. For instance, in the presence of sulfate, SRBs utilize H_2 at concentrations lower than the minimum concentration, which can be utilized by methanogens [Kristjansson et al. 1982; Lovley, 1985]. Presumably, the ability of SRBs to out-compete methanogens is a direct consequence of more positive reduction potential of SO_4^{2-} compared to that of CO_2. In environments with sufficient quantities of sulfate, H_2S is predominantly the reduced product and major fate of biodegradable organic carbon is oxidation to CO_2. Sulfidogenesis normally dominates in estuarine, marine and hyper-saline sediments, where sulfate diffuses from overlying water.

3.2.6 Sulphate-Surplus Environment

In surface layers where sulfate is in excess, some methanogenesis may occur from methylated compounds such as tri-methylamine and di-methyl sulfide. Tri-methylamine is not easily utilized by sulfidogenic bacteria, while it is rapidly utilized by methanogens to generate methane, CO_2 and ammonia. Tri-methylamine is often found in marine sediments, where it is formed from betaine, glycine or other related osmo-protectants, which are produced by marine plants and bacteria to balance osmolarity of their cytoplasm with that of the seawater. Thus, tri-methylamine has been termed as a non-competitive substrate for methanogens. Tri-methylamine-degrading methanogens from marine environments constitute the family Methanosarcinaceae. As a corollary, methanogens isolated from hyper-saline environments use tri-methylamine as a catabolic substrate.

3.2.7 Nitrate-Rich Environment

When nitrate ions are used to degrade organic compounds in the wastes, microbial de-nitrification occurs and N_2 is produced.

$$6NO_3^- + 5CH_3OH \longrightarrow 3N_2 + 5CO_2 + 7H_2O + 6OH^-$$

As nitrogen is an essential element for living organisms, non-availability of a suitable nitrogen source often limits bacterial growth. Many archaea are able to reduce nitrate (nitrate reducing bacteria, NRBs) by assimilatory or respiratory pathways, although dissimilatory nitrate reduction being much more frequent than nitrate assimilation [Cabello et al. 2004].

In addition to ammonium assimilation, members of archaea can drive all the reductive pathways of the N-cycle, through either dissimilatory reactions, such as nitrate respiration and denitrification, or assimilatory pathways like N_2 fixation and nitrate assimilation. Because archaea are in extreme environments, they sustain the N-cycle in these ecosystems. In spite of this, to date, there is little information on inorganic nitrogen metabolism in archaea, especially on nitrate reduction. Significant differences between the bacterial and archaeal nitrate-reducing systems are emerging. Three different nitrate-reducing systems (Nas, Nar and Nap) have been described in bacteria [Berks et al. 1995; Moreno et al. 1999; Richardson et al. 2001].

Assimilatory nitrate reductases (Nas) are usually cytoplasmic enzymes which are repressed by ammonium ions and use either NADH or ferredoxin as physiological electron donor. The NADH-Nas enzymes of *Klebsiella* and *Rhodobacter capsulatus* are dimers composed of a small FAD-containing diaphorase and a large catalytic subunit with one Fe-S centre and a molybdo-bismolybdopterin guanine dinucleotide (Mo-bisMGD) cofactor, whereas cyanobacterial ferredoxin-dependent enzymes are monomers [Moreno et al. 1999; Richardson et al. 2001].

Membrane-bound nitrate reductases (Nar) are involved in anaerobic nitrate respiration and denitrification, being negatively regulated by O_2 and unaffected by ammonium ions.

The periplasmic Nap systems of many Gram-negative bacteria participate in different processes, like redox balancing, scavenging nitrate in nitrate-limited environments, and aerobic or anaerobic denitrification [Potter et al. 2001; Gavira et al. 2002; Ellington et al. 2003]. The following are examples of notable NRBs (Table 1).

Table 1. Profile of NRBs

NRB	Reference
Aeropyrum pernix	
Archaeoglobus fulgidus	Springer et al. 1998
Azoarcus anaerobius	
Garciella nitratireducens	Tello et al. 2003
Paracoccus solventivorans	Siller et al. 1996
Nitratiruptor tergarcus	Nakagawa et al. 2005
Nitratifactor salsuginis	
Petrobacter succinatimandens	Salinas et al. 2004
Pyrobaculum aerophilum	
Pyrococcus furiosus	
Pyrococcus horikoshii	
Rhodobacter capsulatus	Hougardy et al. 2000
Rhodobacter sphaeroides	
Rhodoferax fermentans	
Rhodopseudomonas rhenobacensis	
Sulfolobus solfataricus	
Sulfolobus tokadaii	Haridon et al. 2006
Thermoplasma acidophilum	
Vulcanibacillus modesticaldus	

Thus, from the above information it seems that habitats are intimately interlinked with the preference of a substrate by the methanogens. In view of a variety of habitats, wide variety of substrates and syntrophic dependence on acetogens, classification of methanogens facilitates their understanding to throw more light on their utility in evolving real life applications.

3.3 Classification of Methanogens

This is a complex subject, complexity being a function of their (i) shape and size, (ii) substrate-specificity, (iii) optimal pH, (iv) optimal temperature, (v) activator requirements etc. However, an effort is made to classify them in as many details as was possible. For the ease of understanding, four classes of methanogens (*Methanobacteria, Methanococci, Methanopyri* and *Methanomicrobia*) are further classified under Order, Family, Genus and Species (Table 2).

From Table 2, it is clear that a variety of methanogens exist, differing in their habitat, morphology, substrate-specificity, set of enzymes and dependence on syntrophs for optimal generation of biogas. Therefore, their metabolic pathways also differ, an account of which follows.

Table 2. Classification of methanogens

Kingdom: Archea; Phylum: Euryarchaeota

Order	Family	Genus	Species
Class: Methanobacteria			
Methanobacteriales	*Methanobacteriaceae*	*Methanobacterium*	*aarhusense*
			alcaliphilum
			beijingense
			bryantii
			congolense
			espanolae
			formicicum
			ivanovii
			oryzae
			palustre
			subterraneum
			thermoaggregans
			uliginosum
		Methanobrevibacter	*acididurans*
			arboriphilicus
			curvatus
			cuticularis
			filiformis
			gottschalkii
			millerae
			olleyae
			oralis
			ruminantium
			smithii
			thaueri
			woesei
			wolinii
		Methanosphaera	*cuniculi*
			stadtmanae
		Methanothermobacter	*defluvii*
			marburgensis
			thermoautotrophicus
			thermoflexus
			thermophilus
			wolfeii
	Methanothermaceae	*Methanothermus*	*fervidus*
			sociabilis

Table 2. (Continued)

Class: Methanococci			
Methanococcales	*Methanococcaceae*	*Methanococcus*	*aeolicus*
			deltae
			maripaludis
			vannielii
			voltae
		Methanothermococcus	*okinawensis*
			thermolithotrophicus
	Methanocaldococcaceae	*Methanocaldococcus*	*fervens*
			indicus
			infernus
			jannaschii
			vulcanius
		Methanotorris	*formicicus*
			igneus
Class: Methanopyri			
Methanopyrales	*Methanopyraceae*	*Methanopyrus*	*kandleri*
Class: Methanomicrobia			
Methanosarcinales	*Methanosarcinaceae*	*Methanomicrococcus*	*blatticola*
		Methanococcides	*alaskense*
			burtonii
			methylutens
		Methanohalobium	*evestigatum*
		Methanohalophilus	*halophilus*
			mahii
			portucalensis
		Methanolobus	*bombayensis*
			oregonensis
			taylorii
			tindarius
			vulcani
		Methanomethylovorans	*hollandica*
			thermophila

Table 2. (Continued)

		Methanosalsum	*zhilinae*
		Methanosarcina	*acetivorans*
			baltica
			barkeri
			frisia
			lacustris
			mazei
			semesiae
			siciliae
			thermophila
			vacuolata
	Methanosaetaceae	*Methanosaeta*	*concilii*
			harundinacea
			thermoacetophila
		Methanothrix	*thermophila*
			soehugenii
Methanomicrobiales	Unclassified Methanomicrobiales	*Methanocalculus*	*halotolerans*
			taiwanense
			pumilus
			chunghsingensis
		Methanolinea	*tarda*
	Methanocorpusculaceae	*Methanocorpusculum*	*aggregans*
			bavaricum
			labreanum
			parvum
			sinense
	Methanomicrobiaceae	*Methanoculleus*	*bourgensis*
			chikugoensis
			marisnigri
			palmaeoli
			submarinus
			thermophilicus
		Methanofollis	*aquaemaris*
			formosanus
			liminatans
			tationis
		Methanogenium	*cariaci*
			frigidum
			marinum
			organophilum
		Methanolacinia	*paynteri*
		Methanomicrobium	*mobile*
		Methanoplanus	*endosymbiosus*
			limicola
			petrolearius

3.4 Metabolic Pathways for Methane Production

The methanogenic decomposition of organic matter into methane requires coordinated catabolic activities of different kinds of microbial consortia, which form a microbial food chain or web. Briefly, it includes mainly (i) hydrolytic, (ii) fermentative (acidogenic), (iii) syntrophic/ acetogenic and (iv) methanogenic bacteria. Of these categories, methane producing microbes are strictly anaerobic and constitute the final group in the consortia. Since methanogens are substrate-specific, it was desirable to deal with substrates prior to their pathways of methanogenesis.

Table 3. Profiles of methane production as a function of substrates

Sr. No.	Substrate	Microbe-catalyzing the reaction	Product	Reference
1	$4CO + H_2O$	*Methanobacterium thermoautotrophicus*	$CH_4 + 3CO_2$	Zeikus and Wolfe, 1972
2	$4H_2 + CO_2$	*Methanobacterium oryzae*	$CH_4 + 2H_2O$	Julian et al. 2000
3	4HCOOH	*Methanococcus maripaludis*	$CH_4 + 3CO_2 + 2H_2O$	Jones et al. 1983
4	$4CH_3COOH$	*Methanosaeta harundinacea*	$4CH_4 + 4CO_2$	Ma et al. 2006
5	$4CH_3CH_2COOH + 2H_2O$	*Syntrophobacter wolinii and Methanogenium organophilum*	$3CH_4 + 4CH_3COOH + CO_2$	Boone and Bryant, 1980; Widdel et al. 1988
6	$CH_3CH_2CH_2COOH + 2H_2O$	*Syntrophobacter wolfei*	$2CH_3COOH + 2H_2$	McInerney et al. 1981
7	$2CH_3(CH_2)_3COOH + CO_2 + 2H_2O$	*Methanobacterium suboxydans*	$CH_4 + 2CH_3COOH + 2CH_3CH_2COOH$	Stadtman and Barker, 1951
8	$CH_3(CH_2)_4COOH + 4H_2O$	*Syntrophobacter wolfei*	$3CH_3COOH + 4H_2$	McInerney, 1979
9	$4CH_3OH$	*Methanomethylovorans thermophila*	$3CH_4 + CO_2 + 2H_2O$	Jiang et al. 2005
10	$2CH_3CH_2OH + CO_2$	*Pelobacter venetianus*	$CH_4 + 2CH_3COOH$	Schink and Stieb, 1983
11	$4\ CH_3NH_2 + 2H_2O$	*Methanococcoides methylutens*	$3CH_4 + CO_2 + 4NH_3$	Sowers and Ferry, 1983
12	$2\ CH_3CH_3NH + 2H_2O$	*Methanomicrococcus blatticola*	$3CH_4 + CO_2 + 2NH_3$	Sprenger et al. 2000
13	$4(CH_3)_3\text{-}N + 6H_2O$	*Methanomicrococcus blatticola*	$9\ CH_4 + 3CO_2 + 4NH_3$	Sprenger et al. 2000
14	$CH_3CH(NH_3^-)\ COO- + 2H_2O$	*Methanospirillum hungetei JF1 and Eubaterium acidaminophilum*	$CH_3COO- + NH_4^+ + CO_2 + 2H_2$	Schink, 1997

3.4.1 Substrates for Methanogens

Substrates used by methanogenic bacteria as carbon and energy include CO, H_2/CO_2, formic acid, acetic acid, high M.W. fatty acids, methanol and methylamines (Table 3). While most methanogens can grow on H_2/CO_2, certain species fail to metabolize H_2/CO_2. For example, *Methanococcides methylutens* grows only on methylamines [Sowers and Ferry, 1983] or *Methanomethylovorans thermophila* grows only on methanol [Jiang et al. 2005], while *Methanosarcina* TM-1 [Zinder and Mah, 1979], *Methanothrix soehugenii* and *Methanolobus tindarius* [Konig and Stetter, 1982] can grow on methanol as well as methylamines.

3.4.2 Major Pathways of Methanogenesis

Metabolic pathway from acetic acid contributes around 65 % methane production by methanogens. The biochemical reactions involve (i) activation of acetic acid to acetyl-CoA by acetic acid kinase, (ii) cleavage of acetyl-CoA to yield methyl and carbonyl groups, of which the later is oxidized to CO_2 by acetyl-CoA synthase in the presence of tetra-hydro-sarcinopterin, (iii) transfer of the cleaved methyl group to 2-mercapto-ethane-sulfonic acid by activated methyl reductase or methyl cobalamin:HS-CoM methyl-tranferase, (iv) oxidation of the methyl group to CO_2 by formyl-methanofuran dehydrogenase/5,10-methylene-tetrahydro-methanopterin dehydrogenase (*M. barkeri*) or formyl- methnao-furan: tetrahydro-methanopterin-formyl-transferase [5,10-methenyl tetrahydro-methanopterin cyclohydrolase and F_{420}-dependent 5,10-methylene tetrahydro-methanopterin dehydrogenase (*M. thermophila*)] and (v) reduction of CO_2 through oxidation of hydrogen by hydrogenase.

Studies on anaerobic digestion showed that several eco-systems contain acetic acid as a major methanogenic precursor. In most eco-systems, of the total methane generated, 70-74 % methane is derived from the methyl group of acetic acid and 30-33 % methane from CO_2 reduction through the oxidation of H_2 or formic acid generated during the conversion of starting organic substrates to acetic acid. This is achieved through the activities of (i) hydrogenotrophic, (ii) acetotrophic and (iii) methylotrophic methanogens. In this process, methanogens cleave acetic acid and methyl group is reduced to methane with electrons derived from oxidation of carbonyl group to CO_2 [Ferry, 1992] This pathway is supported by (i) Boone (1991), who described microbial ecology in many biomass digesters with abundant presence of *Methanosarcina* and (ii) Mladenovska and Ahring (2000) who reported the presence of the same species from commercial reactors, treating different kinds of organic wastes. These pathways are depicted in Figure 1.

3.4.2.1 Hydrogenotrophic Methanogens

The hydrogenotrophic methanogens use hydrogen to convert carbon dioxide to methane. By converting carbon dioxide to methane, these organisms help to maintain a low partial hydrogen pressure in an anaerobic digester that is required for acetogenic bacteria. Some hydrogenotrophic methanogens use carbon monoxide to produce methane.

$$CO_2 + 4H_2 \longrightarrow CH_4 + 2H_2O$$

$$4CO + 2H_2O \longrightarrow CH_4 + 3CO_2$$

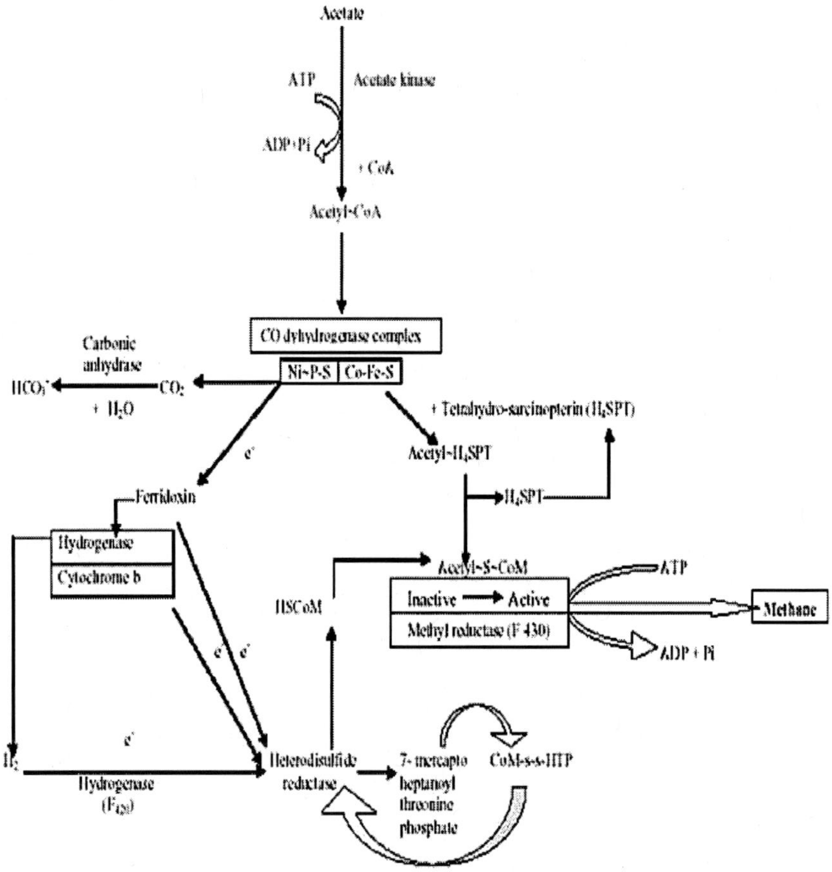

Figure 1. Metabolic pathway for methane production from acetate in methanogenic bacteria.

3.4.2.2 Acetotrophic Methanogens

The acetotrophic methanogens "split" acetate into methane and carbon dioxide. The carbon dioxide produced from acetate may be converted by hydrogenotrophic methanogens to methane.

$$4CH_3COOH \longrightarrow 4CO_2 + 2H_2$$

The acetotrophic methanogens reproduce more slowly than the hydrogenotrophic methanogens and are adversely affected by the accumulation of hydrogen. Therefore, the maintenance of a low partial hydrogen pressure in an anaerobic digester is favorable for the activity of not only acetate-forming bacteria, but also acetotrophic methanogens. Under a relatively high hydrogen partial pressure, acetate and methane production are reduced.

3.4.2.3 Methylotrophic Methanogens

Methylotrophic methanogens (methylotrophs) are widespread in the environment. They grow on substrates that contain the methyl group ($-CH_3$) such as methanol (CH_3OH) or methylamines [$(CH_3)_3-N$]. These methanogens produce methane directly from methyl groups and not from CO_2.

$$3CH_3OH + 6H \longrightarrow 3CH_4 + 3H_2O$$

$$4(CH_3)_3 - N + 6H_2O \longrightarrow 9CH_4 + 3CO_2 + 4NH_3$$

In particular, *Methylobacterium* sp. are common colonizers of plant leaves. Other prominent methylotrophic bacteria are *Methylobacillus flagellatum* KT, *Methylobacterium* sp. and *Methylobacterium extorquens*.

Metabolism of the C_1 found in methylotrophs could be viewed as a special pathway for the production, inter-conversion and transfer of C_1 units in a number of biosynthetic reactions. Consequently, formaldehyde consumption pathways are unique components of the methylotrophic metabolism. Most methylotrophs possess a previously unknown C_1 pathway, dependent on the cofactor tetrahydromethanopterin (a tetrahydrofolate analog) for the oxidation of formaldehyde to CO_2 [Chistoserdova *et al.* 1998; Vorholt, 2002] Alternatively, methylotrophs also possess formaldehyde activating enzyme, dependent on glutathione [Goenrich *et al.* 2002] Thus, pathways dependent on tetrahydromethanopterin as well as glutathione may throw some light on our understanding of the mode of functioning of appropriate C_1-unit partitioning into assimilatory and dissimilatory metabolism as well as growth under mixotrophic conditions through systems approach. Structures of these two enzymes have given insights into different catalytic mechanism of formaldehyde activation [Acharya *et al.* 2005; Neculai *et al.* 2005], indicating that C_1 metabolism is tightly interwoven with central carbon metabolism. Uniquely, many proteins induced during methylotrophic growth are also present for a multi-carbon source, thereby sustaining coherent contribution of methylotrophic pathways to overall metabolism of the bacterium in odd conditions [Laukel *et al.* 2004]

3.4.3 Minor Pathways of Methanogenesis

About half genera of methanogens can metabolize formic acid, first by oxidizing it to H_2/CO_2 through formic acid dehydrogenase and then by reducing CO_2 to methane [Daniels et al. 1984] *Methanosarcina barkeri* is metabolically most versatile species, being able to grow on H_2/CO_2, acetic acid, methanol and methylamines by producing methane [Mah and Smith, 1981]

3.4.3.1 Syntrophic Relationship

The metabolic activities of methanogens actually rely on a balanced/symbiotic relationship with proton reducing acetogens for the consistent supply of acetic acid, H_2 and formic acid, while acetogens in turn depend on methanogens for the removal of inhibitory/toxic level of H_2 and formic acid [Lastella et al. 2002] This relationship has been termed as syntrophic, meaning ‘feeding together’. These syntrophic bacteria (syntrophs) convert the metabolic products of the first group (ethanol, propionic acid, butyric acid, iso-butyric acid, valeric acid, iso-valeric acid, benzoic acid etc.) into immediate methanogenic precursors (acetic acid and H_2/CO_2 or formic acid) so as to prevent their accumulation, only in association with a hydrogen-utilizing methanogen. Any inhibition of activities between the syntrophs may initially lead to inefficient methanogenesis and further accumulation of VFAs, causing the digester failure. Hence, low levels of hydrogen (10^{-3}-10^{-5} atm) and formic acid have to be maintained via methanogenesis or sulfate reduction during acetogenesis in order to allow the reaction to proceed in a forward direction.

Thus, methanogenic reactions are crucial for steady performance of the digester. In turn, the process of methanogenesis critically depends on syntrophic association between acetogenic and methanogenic bacterial populations for the conversion of VFAs to methane [Hori et al. 2006] The syntrophic acetogens oxidize fatty acids, ethanol and benzoic acid to produce acetic acid together with generation of hydrogen or formic acid, by utilizing methanogen partner for inter-species electron transfer [Lowe et al. 1993]. Table 4 summarizes various syntrophs.

Table 4. Syntrophic acetogens catabolizing fatty acids

Sr. No.	Microbe	Fatty acid (substrate)	Symbiotic methanogen	End product	Reference
1	*Syntrophobacter wolinii*	Propionic acid	*Desulfovibrio* sp.	Acetic acid + H2	Boone and Bryant, 1980
2	*Smithella propionica*	Propionic acid	*Methanospirillum hungatei* JF1	Acetic acid + butyric acid	Liu et al. 1999
3	*Syntrophomonas wolfeii*	Butyric acid	*Methanospirillum hungatei, Methanobacterium* sp. *formicicum*	Acetic acid + H2	McInerney et al. 1981
4	*Syntrophomonas* sp.	Unsaturated/saturated long chain fatty acids	*Desulfovibrio* sp.	Acetic acid	Sousa et al. 2007

3.4.3.2 β-Oxidation of Fatty Acids

The syntrophic associations are a consortium of species involved in the (i) β-oxidation of fatty acids of even carbon number to acetic acid, propionic acid and H_2, (ii) β-oxidation of fatty acids of odd carbon number to acetic acid, propionic acid and H_2 and (iii) decarboxylation of propionic acid to acetic acid, CO_2 and H_2 [Boone and Bryant, 1980) Syntrophic propionic acid oxidation mainly occurs via either randomizing methyl-malonyl-coenzyme A (CoA) pathway or non-randomizing pathway through butyric acid [Liu et al. 1999) The pathway of propionic acid conversion in a co-culture of 2 syntrophs *Smithella propionica* and *Methanospirillum hungatei* JF1 has revealed various steps [Liu et al. 1999]; viz. (i) condensation of C_2 of propionic acid to the carboxyl group of another propionic acid, (ii) rearrangement of methyl group and transfer of oxygen to the C_3 of the intermediate, (iii) cleavage of 3-lectohexanoic acid, yielding butyric acid and acetic acid and (iv) syntrophic β-oxidation of butyric acid to acetic acid by butyryl-CoA dehydrogenase and crotonase.

3.4.3.3 Isomerization and β-Oxidation

Wu *et al.* (1994) examined syntrophic degradation of normal and branched chain fatty acids with syntrophic tri-culture of (i) fatty acid degrading Gram positive rod shaped strain IB, (ii) *Methanobacterium formicicum* T1N and (iii) *Methanosarcina mazei.* It showed (i) conversion of valeric acid and 2-methyl butyric acid to propionic acid and methane, (ii) degradation of caproic acid and 4-methyl valeric acid, as also heptanoic acid and caprylic acid to propionic acid and methane, (iii) reversible isomerization between butyric acid and isobutyric acid and (iv) formation of isobutyric acid and butyric acid during valeric acid conversion. This study strongly indicated isomerization and β-oxidation pathway during syntrophic degradation of branched fatty acids to propionic acid and methane. On this background, the representative reactions starting from different substrates, but leading to methane formation are summarized in Table 3. Having familiarized with various pathways using various substrates metabolized by various methanogens, it is appropriate to consider overall efficiency of anaerobic digestion, irrespective of the above variables.

3.5 Efficiency of Anaerobic Digestion and its Products

Nearly 30 % of the waste is digested by primary clarifiers through the consortium of hydrolytic aerobes rich in amylolytic, proteolytic and lipolytic activities and about 50 % is transformed and then transferred to anaerobic digesters. By and large, (i) 80-90 % BOD removal, (ii) 1.5 x BOD of COD removal (mg/L), (iii) 0.5 m^3 biogas production/kg of COD removal (65 % of it being methane) and (iv) sludge production of 0.05-0.1 kg kg^{-1} of COD removal is achievable. Typical gas production rate ranges from 24-32 m^3 kg^{-1} VS degraded and its heat value is about 1000 BTU/ft^3. Calorific value of the biogas is about 6 kWh/m^3, which corresponds to about half a liter of petro-diesel. The net calorific value further depends on the efficiency of burners or appliances. Methane is the valuable component from the view-point of using biogas as a fuel.

The production of biogas is not a steady stream; it is very little in the early stage, highest during the middle stage and reduced towards the end stage, since only recalcitrant ingredients remain to be digested. Total volume of biogas produced increases as loading increases. More gasification also occurs upon longer retention time (RT) The increased yield of biogas may offset the cost of an additional digester or enhancing the capacity of the existing digester required to provide longer RT. However, the composition of biogas does not change significantly by increasing either loading of VS or RT.

At the end of anaerobic digestion, three principal products emerge. They are biogas, digestate (solid fraction from the effluent) and wastewater. Many digesters have ancillary processes to treat and manage these by-products. While biogas could be used *per se* either for heating or enrichment to serve as an automotive fuel, solid fraction could be composted to achieve C:N ratio between 16-20 for soil amendment. The wastewater could be used in fertigation as a liquid manure as practiced in several EU countries.

3.6 Deterioration of Methanogenesis

In most cases of deterioration of methanogenesis, fatty acids such as propionic acid, butyric acid, isobutyric acid, valeric acid etc. were found to be accumulated and inhibitory to several essential biochemical reactions in the anaerobic digestion [Hanaki et al. 1981] The net result was lowering the rate of production of methane gas. If the digester remained un-attended, these fatty acids gradually accumulate. Their higher concentration causes (i) microbial stress/population reduction, (ii) further inhibition of the catabolism of VFAs, (iii) decrease in pH, (iv) unpleasant rancid odor and (v) eventually long lag period for revival [Pind et al. 2003; Duran et al. 2006; Hori et al. 2006] It further induces a dynamic transition of hydrogenotrophic methanogens, particularly *Methanothermobacter* and *Methanoculleus* population [Hori et al. 2006] Therefore, for an effective digester performance, continuous conversion of these short or long chain fatty acids to acetic acid and finally to methane is highly desirable.

3.7 Eco-Friendly Applications

In the developing countries, biogas from simple home/farm-based anaerobic digesters offers low cost, clean energy. In the developed countries, an increasing environmental pressure on solid waste disposal has prompted anaerobic digestion not only for reducing the waste volume, but for simultaneously also availing biogas for on-site applications, viz. biogas can be converted into either mechanical or electrical energy to operate agitators, aeration blowers, pumps etc., besides home/office heating, lift irrigation, hot water/steam production and dehydration of agri-produce.

In Germany, biogas is burnt to produce electricity, usually with a reciprocating engine or micro-turbine. Surplus electricity after farm operations is sold to the local grid as green energy to avail subsidies. Since biogas does not contribute to atmospheric CO_2, as such coming from an organic source with a short carbon cycle, it is considered as green energy. It burns more cleanly than coal and emits less CO_2 per unit of energy. Finally, disease-causing pathogens are destroyed and the volume of disposed waste products is reduced significantly. An intangible contribution of anaerobic digestion is by not letting methane content of biogas produced loose; it has averted the major cause of the greenhouse effect. Therefore, in totality, methane is environmental-friendly source of energy.

3.8 Conclusion

Knowledge of the fundamental processes involved in methane production is necessary to be able to plan, build and operate biogas plants. First, anaerobic fermentation involves the activities of four different bacterial communities. Next, the process of biogas-production depends on various parameters. For example, a change in ambient temperature (to lower or higher side) has a negative effect on bacterial activity. As with any pure gas, characteristic properties of biogas are pressure-and temperature-dependent, as also affected by moisture content. Here, the factors of main interest are: (i) change in volume as a function of temperature and pressure, (ii) change in water-vapor content as a function of temperature and pressure and (iii) change in calorific value as a function of temperature and pressure.

Reviewed By

Dr. Karmaraj Yadav, Deputy Director, Biotechnology Division, Mahatma Gandhi Institute of Rural Industrialization (MGIRI), Wardha, Maharashtra 442 001, India.

REFERENCES

Acharya, P., Goenrich, M., Hagemeier, C. H., Demmer, U., Vorholt, J. A., Thauer, R. K. & Ermler, U. (2005). *How an enzyme binds the C1 carrier tetrahydro-methanopterin: Structure of the tetrahydro-methanopterin-dependent formaldehyde-activating enzyme (Fae) from Methylobacterium extorquens* AM1. J Biol Chem, 280, 13712-13719.

Belay, N., Johnson, R., Rajagopal, B. S., Macario, E. C. & Daniels, L. (1988). *Methanogenic bacteria from human dental plaque*. Appl Environ Microbiol, 54, 600–603.

Berks, B. C., Ferguson, S. J., Moir, J. W. B. & Richardson, D. J. (1995). *Enzymes and associated electron transport systems that catalyze the respiratory reduction of nitrogen oxides and oxy-anions.* Biochim Biophys Acta, 1232, 97–173.

Boone, DR. *Ecology of methanogenesis.* In: Rogers, JE; William, WB editors. Microbial production and consumption of greenhouse gases: Methane, nitrogen oxides and halo-methanes. USA: American Society Microbiology; 1991; 57-70.

Boone, D. R. & Bryant, M. P. (1980). *Propionate-degrading bacterium, Syntrophobacter wolinii* sp. nov. gen. nov., *from methanogenic ecosystems*. Appl Environ Microbiol, 40, 626-632.

Brusa, T., Conca, R., Ferrara, A., Ferrari, A. & Pecchioni, A. (1987). Presence of Methanobacteria in human sub-gingival plaque. J Clin Periodontol, 14, 470–471.

Cabello, P., Roldan, M. D. & Vivian, C. M. (2004). *Nitrate reduction and the nitrogen cycle in archaea.* Microbiology, 150, 3527–3546.

Chistoserdova, L., Vorholt, J. A., Tauter, R. K. & Lidstrom, M. E. (1998). *C1 transfer enzymes and coenzymes linking methylotrophic bacteria and ethanogenic archaea.* Science, 281, 99-102.

Daniels, L., Sparling, R. & Sprott, G. D. (1984). *Bioenergetics of methanogenesis*. Biochim Biophys Acta, 768, 113-163.

Duran, M., Tepe, N., Yurtsever, D., Punzi, V. L., Bruno, C. & Mehta, R. J. (2006). *Bioaugmenting anaerobic digestion of bio-solids with selected strains of Bacillus, Pseudomonas and Actinomycetes species for increased methanogenesis and odor control.* Appl Microbiol Biotechnol, 73, 960-966.

Ellington, M. J. K., Richardson, D. J. & Ferguson, S. J. (2003). *Rhodobacter capsulatus gains a competitive advantage from respiratory nitrate reduction during light–dark transitions.* Microbiology, 149, 941–948.

Ferry, J. G. (1992). *Methane from acetate.* J Bacteriol, 174, 5489-5495.

Gavira, M., Roldan, M. D., Castillo, F. & Moreno, V. C. (2002). *Regulation of nap gene expression and periplasmic nitrate reductase activity in the phototrophic bacterium Rhodobacter sphaeroides DSM158.* J Bacteriol, 184, 1693–1702.

Gerardi, MH. *The microbiology of anaerobic digesters. Wastewater microbiology series.* New York: Wiley-Interscience; 2003.

Goenrich, M., Bartoschek, S., Hagemeier, C. H., Griesinger, C. & Vorholt, J. A. (2002). *A glutathione-dependent formaldehyde-activating enzyme (Gfa) from Paracoccus denitrificans detected and purified via two-dimensional proton exchange NMR spectroscopy*. J Biol Chem, 277, 3069-3072.

Hanaki, K., Matsuo, T. & Nagase, M. (1981). *Mechanism of inhibition caused by long chain fatty acids in anaerobic digestion process.* Biotechnol Bioengg, 23, 1591-1610.

Haridon, S. L., Miroshnichenko, M. L., Kostrikina, N. A., Tindall, B. J., Spring, S., Schumann, P., Stackebrandt, E., Osmolovskaya, B. E. A. & Jeanthon, C. (2006). *Vulcanibacillus modesticaldus gen. nov., sp. nov., a strictly anaerobic, nitrate-reducing bacterium from deep-sea hydrothermal vents*. Int J Syst Evol Microbiol, 56, 1047–1053.

Hori, T., Haruta, S., Ueno, Y., Ishii, M. & Igarashi, Y. (2006). *Dynamic transition of a methanogenic population in response to the concentration of volatile fatty acids in a thermophilic anaerobic digester*. Appl Environ Microbiol, 72, 1623-1630.

Hougardy, A., Tindall, B. J. & Klemme, J. H. (2000). *Rhodopseudomonas rhenobacensis sp. nov., a new nitrate-reducing purple non-sulfur bacterium*. Int J Syst Evol Microbiol, 50, 985–992.

Jiang, B., Parshina, S. N., van Doesburg, W., Lomans, B. P. & Stams, A. J. M. (2005). *Methanomethylovorans thermophila sp. nov., a thermophilic, methylotrophic methanogen from an anaerobic reactor fed with methanol*. Int J Syst Evol Microbiol, 55, 2465-2470.

Jones, W. J., Paynter, M. J. B. & Gupta, R. (1983). *Characterization of Methanococcus maripaludis sp. nov., a new methanogen isolated from salt marsh sediment*. Arch Microbiol, 135, 91-97.

Julian, C., Patel, B. K. C., Ollivier, B., Garcia, J. L. & Roger, P. A. (2000). *Methanobacterium oryzae sp. nov., a novel methanogenic rod isolated from a Philippines rice-field*. Int J Syst Bacteriol, 50, 525-528.

Kaur, J., Ramamurthy, V. & Kothari, R. M. (1993). *Characterization of oat lipase for lipolysis of rice bran oil*. Biotechnol Letts, 15, 257-262.

Konig, H. & Stetter, K. O. (1982). *Isolation and characterization of Methanolobus tindarius, sp. nov., a coccoid methanogen growing only on methanol and methylamines*. Zbl. Bakt. Mikrobiol Hyg Callg, 3, 478-490.

Kristjansson, J. K., Schönheit, P. & Thauer, R. K. (1982). *Different Km values for hydrogen of methanogenic bacteria and sulfate reducing bacteria*. Arch Microbiol, 131, 278–282.

Lastella, G., Testa, C., Cornacchia, G., Notornicola, M., Voltasio, F. & Sharma, V. K. (2002). *Anaerobic digestion of semi-solid organic waste: biogas production and its purification*. Energy Conserv Manag, 43, 63-75.

Laukel, M., Rossignol, M., Borderies, G., Völker, U. & Vorholt, J. A. (2004). *Comparison of the proteome of Methylobacterium extorquens AM1 grown under methylotrophic and non-methylotrophic conditions*. Proteomics, 4, 1247-1264.

Liu, Y., Balkwill, D. L., Aldrich, H. C., Drake, G. R. & Boone, D. R. (1999). *Characterization of the anaerobic propionate degrading syntrophs Smithella propionica gen. nov., sp. nov. and Syntrophobacter wolinii*. Int J Syst Bacteriol, 49, 545–556.

Lovley, D. R. (1985). *Minimum threshold for hydrogen metabolism in methanogenic bacteria*. Appl Environ Microbiol, 49, 1530–1531.

Lowe, S. E., Jain, M. K. & Zeikus J. G. (1993). *Biology, ecology and biotechnological applications of anaerobic bacteria adapted to environmental stresses in temperature, pH, salinity or substrates*. Microbiol Rev, 57, 451-509.

Ma, K., Liu, X. & Dong, X. (2006). *Methanosaeta harundinacea sp. nov., a novel acetate-scavenging methanogen isolated from a UASB reactor*. Int J Syst Microbiol, 56, 127-131.

Mah, RA; Smith, MR. The methanogenic bacteria. In: Starr, MP, Stolp, H, Truper, HG, Balows, A, Schlegel, HG editors. *The prokaryotes: A handbook on habitats, isolation and identification of bacteria*. New York: Springer Verlag; 1981; 260- 296.

McInerney, M. J., Bryant, M. P. & Pfennig, N. (1979). *Anaerobic bacterium that degrades fatty acids in syntrophic association with methanogens*. Arch Microbiol, 122, 129-135.

McInerney, M. J., Bryant, M. P., Hespell, R. B. & Costerton, J. W. (1981). *Syntrophomonas wolfei gen. nov., an anaerobic, syntrophic, fatty acid-oxidizing bacterium.* Appl Environ Microbiol, 41, 1029-1039.

Mladenovska, Z. & Ahring, B. K. (2000). *Growth kinetics of thermophilic Methanosarcina sp., isolated from full-scale biogas plants treating animal manures.* FEMS Microbial Ecology, 31, 225-229.

Moreno, V. C., Cabello, P., Martınez, L. M., Blasco, R. & Castillo, F. (1999). *Prokaryotic nitrate reduction: Molecular properties and functional distinction among bacterial nitrate reductases.* J Bacteriol, 181, 6573–6584.

Nakagawa, S., Takai, K., Inagaki, F., Horikoshi, K. & Sako, Y. (2005). *Nitratiruptor tergarcus gen. nov., sp. nov. and Nitratifractor salsuginis gen. nov., sp. nov., nitrate-reducing chemolithoautotrophs of the ε-Proteobacteria isolated from a deep-sea hydrothermal system in the Mid-Okinawa trough.* Int J Syst Evol Microbiol, 55, 925–933.

Neculai, A. M., Neculai, D., Griesinger, C., Vorholt, J. A. & Becker, S. (2005). *A dynamic zinc redox switch.* J Biol Chem, 280, 2826-2830.

Pind, P. F., Angelidaki, I. & Ahring, B. K. (2003). *Dynamics of the anaerobic process: Effects of volatile fatty acids.* Biotechnol Bioengg, 82, 791-801.

Potter, L., Angove, H., Richardson, D. J. & Cole, J. A. (2001). *Nitrate reduction in the periplasm of Gram-negative bacteria.* Adv Microb Physiol, 45, 51–112.

Ramamurthy, V., Upadhyay, C. M. & Kothari, R. M. (1991). *An optimized protocol for the preparation and application of acid protease.* J Biotechnol, 21, 187-196.

Richardson, D. J., Berks, B. C., Russell, D. A., Spiro, S. & Taylor, C. J. (2001). *Functional, biochemical and genetic diversity of prokaryotic nitrate reductases.* Cell Mol Life Sci, 58, 165–178.

Salinas, M. B., Fardeau, M. L., Cayol, J. L., Casalot, L., Patel, B. K. C., Thomas, P., Garcia, J. L. & Ollivier, B. (2004). *Petrobacter succinatimandens* gen. nov., sp. nov., *a moderately thermophilic, nitrate-reducing bacterium isolated from an Australian oil well.* Int J Syst Evol Microbiol, 54, 645–649.

Schink, B. (1997). *Energetics of syntrophic cooperation in methanogenic degradation.* Microbiol. Molbiol Rev, 61, 262–280.

Schink, B. & Stieb, M. (1983). *Fermentative degradation of polyethylene glycol by a new strictly anaerobic Gram-negative non-spore forming bacterium, Pelobacter venetianus sp. nov.* Appl Environ Microbiol, 45, 1905–1913.

Shah, D. N., Shah, V. D., Nehete, P. N. & Kothari, R. M. (1986). *Isolation of Bacillus licheniformis mutant for stable production profiles of alkaline protease.* Biotechnol Letts, 8, 103–106.

Shah, N. K., Shah, D. N., Upadhyay, C. M., Nehete, P. N., Kothari, R. M. & Hegde, M. V. (1989). *An economical, upgraded, stabilized and efficient preparation of amyloglucosidase.* J Biotechnol, 10, 267-276.

Shah, N. K., Upadhyay, C. M., Nehete, P. N., Kothari, R. M. & Hegde, M. V. (1990). *An economical, upgraded, stabilized and efficient preparation of α-amylase.* J Biotechnol, 16, 97-108.

Siller, H., Rainey, F. A., Stackebrandt, R. & Winter, J. (1996). *Isolation and characterization of a new Gram-negative, acetone-degrading, nitrate-reducing bacterium from soil, Paracoccus solventivorans sp. nov.* Int J Syst Bacteriol, 46, 1125-1130.

Sousa, D. Z., Smidt, H., Alves, M. M. & Stams, A. J. M. (2007). *Syntrophomonas zehnderi* sp. nov., *an anaerobe that degrades long-chain fatty acids in co-culture with Methanobacterium formicicum*. Int J Syst Evol Microbiol, 57, 609-615.

Sowers, K. R. & Ferry, J. G. (1983). *Isolation and characterization of a methylotrophic marine methanogen, Methanococcoides methylutens* gen. nov., sp. nov. Appl Environ Microbiol, 45, 684-690.

Sprenger, W. W., van Belzen, M. C., Rosenberg, J., Hackstein, J. H. P. & Keltjens, J. T. (2000). *Methanomicrococcus blatticola* gen. nov., sp. nov., *a methanol- and methylamine-reducing methanogen from the hindgut of the cockroach Periplaneta americana.* Int J Syst Bacteriol, 50, 1989-1999.

Springer, N., Ludwig, W., Philipp, B. & Schink, B. (1998). *Azoarcus anaerobius* sp. nov., *a resorcinol degrading, strictly anaerobic, denitrifying bacterium.* Int J Syst Evol Microbiol, 48, 953–956.

Stadtman, T. C. & Barker, H. A. (1951). *Studies on the methane fermentation. VIII. Tracer experiments on fatty acid oxidation by methane bacteria.* J Bacteriol, 61, 67-80.

Tello, E. M., Fardeau, M. L., Lveda, J. S., Fernande, L. Z., Cayol, J. L., Thomas, P. & Ollivier, B. (2003). *Garciella nitratireducens* gen. nov., sp. nov., *an anaerobic, thermophilic, nitrate- and thiosulfate-reducing bacterium isolated from an oil-field separator in the Gulf of Mexico.* Int J Syst Evol Microbiol, 53, 1509-1514.

Upadhyay, C. M., Nehete, P. N. & Kothari, R. M. (1989). *A lipase preparation with dual pH optima, wide temperature optima and broad substrate specificity for multiple applications.* Biotechnol Letts, 11, 793-796.

Vorholt, J. A. (2002). *Cofactor-dependent pathways of formaldehyde oxidation in methylotrophic bacteria.* Arch Microbiol, 178, 239-249.

Widdel, F., Rouviere, P. E. & Wolfe, R. S. (1988). *Classification of secondary alcohol-utilizing methanogens including a new thermophilic isolate.* Arch Microbiol, 150, 477-481.

Winfrey, M. R. & Zeikus, J. G. (1979). *Microbial methanogenesis and acetate metabolism in a meromictic lake.* Appl Environ Microbiol, 37, 213–221.

Wu, W. M., Jain, M. K., Hickey, R. F. & Zeikus, J. G. (1994). *Anaerobic degradation of normal- and branched-chain fatty acids with four or more carbons to methane by a syntrophic methanogenic triculture.* Appl Environ Microbiol, 57, 2220-2226.

Zeikus, J. G. & Wolfe, R. S. (1972). *Methanobacterium thermoautotrophicus* sp. nov., *an anaerobic, autotrophic, extreme thermophile.* J Bacteriol, 109, 707-713.

Zinder, S. H. & Mah, R. A. (1979). *Isolation and characterization of a thermophilic strain of Methanosarcina unable to use* H_2-CO_2. Appl Environ Microbiol, 38, 996-1008.

In: Bioremediation
Editor: Alexander C. Mason
ISBN: 978-1-61122-730-7

Chapter 13

KNOWING MORE ABOUT METALS, MICROBES AND ENVIRONMENT INTERACTIONS: HOW TO IMPROVE WASTEWATER BIOTREATMENTS

Matías Barrionuevo, María Alejandra Daniel, Luciana Garavaglia, Natividad López, Nerina Méndez, Griselda Sosa, Roberto Candal, Silvia Cerdeira, Helena Ceretti, Silvana Ramírez, Eduardo Reciulschi, Anita Zalts and Diana L. Vullo

[1]Área Química, Instituto de Ciencias,
Universidad Nacional de General Sarmiento, Buenos Aires, Argentina
[2]Área Microbiología, Depto. de Química Biológica, Facultad de Ciencias Exactas y Naturales (UBA), Ciudad Universitaria, Buenos Aires, Argentina
[3]INQUIMAE-CONICET, Facultad de Ciencias Exactas y Naturales (UBA), Ciudad Universitaria, Buenos Aires, Argentina
[4]ECyT-UNSAM, Campus Miguelete, Buenos Aires, Argentina

ABSTRACT

Through evolution, microorganisms have developed effective mechanisms that help them regulate their cellular function in response to changes in its microenvironment. One of these survival strategies is metal resistance, which becomes a useful tool for biotreatment of metal loaded wastewaters. For the development of a successful biotreatment, studies on metal-microbes, microbe-surfaces and metal-environment interactions are required. The correct design of laboratory scale bioreactors for electroplating wastes, containing Cd(II), Zn(II), Ni(II) or Cr(VI), is the first step of our research project. In that way, the use of metal resistant indigenous bacteria either in suspension or in packed bed reactors is the chosen methodology to fulfil our general aim. Several aspects should be studied in order to implement high-efficient processes: a) bacterial interactions with Cd(II) and Zn(II) in terms of biosorption mechanisms; b) biotransformation of Cr(VI) to Cr(III); c) Ni(II)-microbe interactions; d) metal influence on bacterial motility; e) bacterial capacity of biofilm development and microbial community behaviour and f) complexing capacity of environment components with metals.

1. INTRODUCTION

The Industrial Revolution, which began in the late 1700s, continued through the 20th century with the pace of technologic discovery and advancement increasing continually. The search of a better life style for an increasing global population has led to dramatic changes in uses and consumption of natural resources. As world population increases and the world economy expands, so does the demand for natural resources. The intensive use of some of these resources, like oil and coal, which are consumed when used, sooner or later will lead to their depletion. Other resources, like metals, have the potential for almost infinite recovery and reuse.

The demand for metals is based on the services that people receive from metal and metal-containing products, e.g., construction, transportation, machinery, consumer goods, communications, health services, and electrical power. There are evidences of the recovery of metals from their sources in the earth in the Near East about 6000 years ago. The amounts of most of the metals mined and smelted before 1900 were tiny in comparison with the amounts extracted during the 20th century. Also, the period from 1900 to 2000 was marked by significant changes in the various industries that consume metals and by major historic events and technological advances that affected the way metals are produced and used. Not only there is an increase in the amounts of metals in use, but also there is a growing variety of materials required for industries. Johnson et al. (2007) mentioned the case of electronics: Intel circuit board is an example of this increase in metal variety. In the 1980s the board contained 11 elements, 4 more were added in the 1990s, and in the 2000s the number of elements in use was over 60. Experimentation and innovation in metallurgy, as well as metals production technology, resulted in the ability to produce metals of high purity, some alloys of high strength, others with corrosion resistance, and some with resistance to abrasion and erosion in chemically aggressive and high temperature regimes. The technologic advancements that have been incorporated into consumer applications to make everyday living easier also have assisted industry to become more efficient, to produce more with fewer resources. Almost all industrial activities are related in one or another way to metals.

Metals as copper, nickel, chromium and many others, once extracted, remain in use and can be recycled. For some metals, their mining was so intensive, that a shift in their sources can be expected in a near future: mines with the higher metal content will be the discarded materials and not the original minerals. Although, some metals, e.g. zinc, have an inherently dissipative use (in galvanizing of steel, iron, and brass in brake linings (Gordon et al., 2006), for example, where it is impractical, or even impossible in some instances, to collect and recycle the metal), leading to metal losses in waste repositories or by dissipation to the environment.

Zinc in low concentrations is generally considered to be nontoxic. Consequently, recycling of zinc is driven primarily by the production cost of primary metal and the price of zinc. This metal is necessary to modern living, and, in tonnage produced, stands fourth among all metals in world production - being exceeded only by iron, aluminium, and copper. About 75% of zinc used is consumed as metal, mainly as a coating to protect iron and steel from corrosion (galvanized metal), as alloying metal to make bronze and brass, as zinc-based die casting alloy, and as rolled zinc. The remaining one-fourth is consumed as zinc compounds mainly by the rubber, chemical, paint, and agricultural industries (http://minerals.usgs.gov/minerals/pubs/commodity/zinc/).

Being similar to zinc in chemistry, cadmium is associated in nature with zinc (and, less closely, with lead and copper) and is extracted mainly as a by-product of the mining and processing of zinc. World production of cadmium in 2000 was 19,700 metric tons. Cadmium's pattern of consumption has changed considerably during the past several decades. Almost 75% of the cadmium consumed goes into nickel-cadmium (NiCd) batteries. Most of the remainder is used in colorants and pigments, coatings, and stabilizers for organic polymers, but small quantities are used for more than 30 other uses. Although the NiCd battery is a growing application, some of the other uses are being phased out, or at least constrained, by concern about cadmium's toxicity and, consequently, its potential for pollution of the environment. Considering that many of cadmium's uses are essentially dissipative, this concern is understandable. As awareness of the health dangers of heavy metals in general, and cadmium in particular, has grown, Governments have imposed restrictions on the use of cadmium for some purposes, and more are expected in the future. Nonetheless, overall world cadmium consumption actually grew, if very slowly, during the past three decades (Butterman et al., 2000).

Total world mine production of nickel increased steadily from 7,500 metric tons in 1900 to 2 million tons by 2000. Nickel is primarily sold for first use as refined metal (cathode, powder, briquet, etc.) or ferronickel. Most metallic nickel (about 65%) is used to make stainless steel. Another 12% goes into superalloys (e.g., Inconel 600) or nonferrous alloys (e.g., cupronickel), with high corrosion and temperature resistance. The aerospace industry is a leading consumer of nickel-base superalloys: turbine blades, discs and other critical parts of jet engines, as well as electrodes, magnets, electronics, surgical implants, are fabricated from superalloys. Nickel-base superalloys are also used in land-based combustion turbines, such those found at electric power generation stations. The remaining 23% of consumption is divided between alloy steels, rechargeable batteries, catalysts and other chemicals, coinage, foundry products, and plating. Consumption of nickel in batteries has risen dramatically since 1995, but still accounts for less than 5% of world nickel use (Eisler, 1998). Nickel enters natural waterways from waste water because it is poorly removed by treatment processes. Human activities that contribute to nickel loadings in aquatic and terrestrial ecosystems include mining, smelting, refining, alloy processing, scrap metal reprocessing, fossil fuel combustion, and waste incineration (http://minerals.usgs.gov/minerals/pubs/commodity/nickel/).

Chromium has a wide range of uses in chemicals, metals, and refractory materials. Its use in iron, nonferrous alloys, and steel is for enhancing hardness or resistance to corrosion and oxidation. Production of stainless steel and nonferrous alloys are two of its more critical applications. Other applications are in alloy steel, catalysts, leather processing, pigments, plating of metals, refractory materials, and surface treatments.

Chromium is an essential trace element for human health. Some chromium compounds, however, are acutely toxic, chronically toxic, and/or carcinogenic. Chromium concentrations in non-biological materials are elevated in the vicinity of industrial operations and municipal waste treatment facilities where chromium is a significant component of wastes discharged into the environment (Eisler, 1986).

Over the course of the past century, the attitude that the Earth was nothing more than resources to be used changed slowly. The increased use of metals produces environmental concerns. Most industrial effluents contain trace amounts of a variety of metals, many of which are classified as priority pollutants (EC Proposal, 2000) because of their persistence in the environment and their ecotoxicity. As they are indestructible, and cannot be chemically or biologically degraded to innocuous end products in the soil and water environments, it is important that wastewaters are properly treated before released. Strict environmental regulations compel industries to shift to cleaner production methods, demanding the development of environmental friendly, low cost and efficient treatment techniques for metal rich effluents. As technological changes take place in the industrial processes, the waste water composition will change accordingly, so treatment also need to be flexible to adapt to current requirements with minimal investment.

Several processes have been used and developed over the years to remove dissolved metals from wastewaters (Vullo et al., 2007). The most conventional techniques used for waste water treatment are based on chemical precipitation, usually using lime, adsorption, coagulation/flotation, sedimentation, filtration, membrane processes, electrochemical techniques, ion-exchange, biological processes and chemical reactions. Although these methods are simple and relatively cheap, most of them present some disadvantages such as a difficult separation of the solid, the need of an adequate disposal of the resulting toxic sludge and the ineffectiveness for low metal concentration. In many cases, these methods are combined to get better results or to adapt to special requirements.

Life on Earth has evolved in the presence of metals. Metals are important in many biochemical processes in any kind of living organism, and no less than 25 of the 95 naturally occurring elements of the periodic table have an essential biological function (Hermoso et al., 2009). Nonetheless, many metals exhibit toxic effects. Through evolution, microorganisms have developed effective mechanisms that help them regulate their cellular functions in response to changes in its microenvironment. In between the presence of metals in the bulk liquid and the actual biological effects, complex processes as biouptake and biosynthesis play an important role. One of these survival strategies is metal resistance, which becomes a useful tool for biotreatment of metal loaded wastewaters. For the development of a successful biotreatment, knowledge on metal-microbes, microbe-surfaces and metal-environment interactions is required. Furthermore, these interactions are not independent. Metal bioavailability, thus toxicity, will be modulated by speciation and metal-ligand lability in the environment; different metals can compete for the same uptake sites, affecting the conditional affinity; the organisms can actively decrease or increase the number of transporters in response to its environment, affecting the maximum uptake rates.

The proper design of laboratory scale bioreactors for electroplating wastes containing Cd(II), Zn(II), Ni(II) or Cr(VI) was the first step of our research project. Several aspects of the previously mentioned interactions were taken into account to achieve high metal removal efficiencies. The use of metal resistant indigenous bacteria, either in suspension or in packed bed reactors, was the chosen methodology to fulfil our main goal. In order to implement highly-efficient biotreatment processes, we describe different aspects of our research work, such as: a) bacterial interactions with Cd(II) and Zn(II) in terms of biosorption mechanisms; b) biotransformation of Cr(VI) to Cr(III); c) Ni(II)-microbe interactions; d) metal influence on bacterial motility; e) bacterial capacity of biofilm development and microbial community behaviour, and f) metal complexing capacity of environmental components.

2. BACTERIAL INTERACTIONS WITH CD(II) AND ZN(II): BIOSORPTION AND ITS ENVIRONMENTAL RELEVANCE

Biosorption is a survival strategy of microorganisms based on metal-binding capacity of functional groups present in external cellular structures like exopolysaccharides (EPS), cell walls or other biopolymers (Guibaud et al., 2008, 2009, Vijayaraghavan, 2008). These microbial interactions decrease the amount of bioavailable metal in the microenvironment, reducing toxic effects on cells. Biosorption can be used as an immobilization technique for metal removal from wastewaters using either living or non-living biomass (Chen et al., 2005; Davis et al., 2003; Volesky, 2001; Wang et al., 2009). Biomass capabilities to immobilize metals rely on the type of biomass, the solution's chemical composition and the physicochemical conditions. After loading, the metals can be stripped from the biological matrix and the system may be regenerated for further sorption-desorption cycles thus opening the possibility for metal recovery.

Pseudomonas veronii 2E is an indigenous bacterium isolated from a polluted ecosystem which was selected as study case by its highest biosorptive capacity (Vullo et al., 2007, 2008). Cd(II) and Zn(II) retention by bacterial suspensions at regulated pH=7.5 (10 mM HEPES buffer) could be evaluated by adsorption isotherms construction and by changes in electrical surface properties in terms of electrophoretic mobility measurements (Méndez et al., 2011).

The behaviour of this strain towards Cd(II)-Zn(II) mixtures is particularly interesting since both metals can be found together in electroplating wastewaters. Adsorption isotherms were constructed under the same experimental conditions as previous experiments (Vullo et al., 2008, Méndez et al., 2011) but with Cd(II)-Zn(II) mixtures and were compared to the corresponding isotherm performed with each metal alone. Figure 1 shows the obtained results testing different Cd(II) and Zn(II) proportions to give two different total metal initial concentrations: 0.1 mM and 0.5 mM. Regarding the plots of q (mmol metal (Cd or Zn)/g biomass dry weight) vs. metal (Cd or Zn) equilibrium concentration (mM), it can be extracted that there is a bacterial interaction with each metal, under the experimental conditions. It seems that Cd(II) or Zn(II) in mixtures are biosorbed independently since an overlapping is observed in the obtained couples of adsorption isotherms.

When total q (Cd+Zn mmol/ g biomass dry weight) is plotted vs. total metal (Cd+Zn, mM) in equilibrium (Figure 2), a trend is clearly visible either with 0.5 mM or 0.1 mM total initial concentrations: q values decrease with Zn concentration increase. Adsortive capacity of *Pseudomonas veronii* 2E is higher for Cd(II) than for Zn(II), indicating a greater number of adsorption sites available for Cd(II) than for Zn(II).

Electrophoretic mobility (μ) measurements were carried out with bacterial suspensions in contact with Cd(II)-Zn(II) mixtures (combinations to give 0.1 or 0.5 mM total concentration) with 10 mM HEPES, pH=7.5 and 10 mM KCl (Méndez et al., 2011). Figure 3.B shows that when total initial concentration is 0.5 mM, electrophoretic mobility values increase following the increasing Zn(II) concentration and decrease following the Cd(II) concentration increase. The obtained values of μ are consistent with the μ values registered when each metal was studied in separate solutions. When total initial concentration is 0.1 mM (Figure 3.A), μ represents a mean value between those obtained with Zn and Cd at each individual concentration.

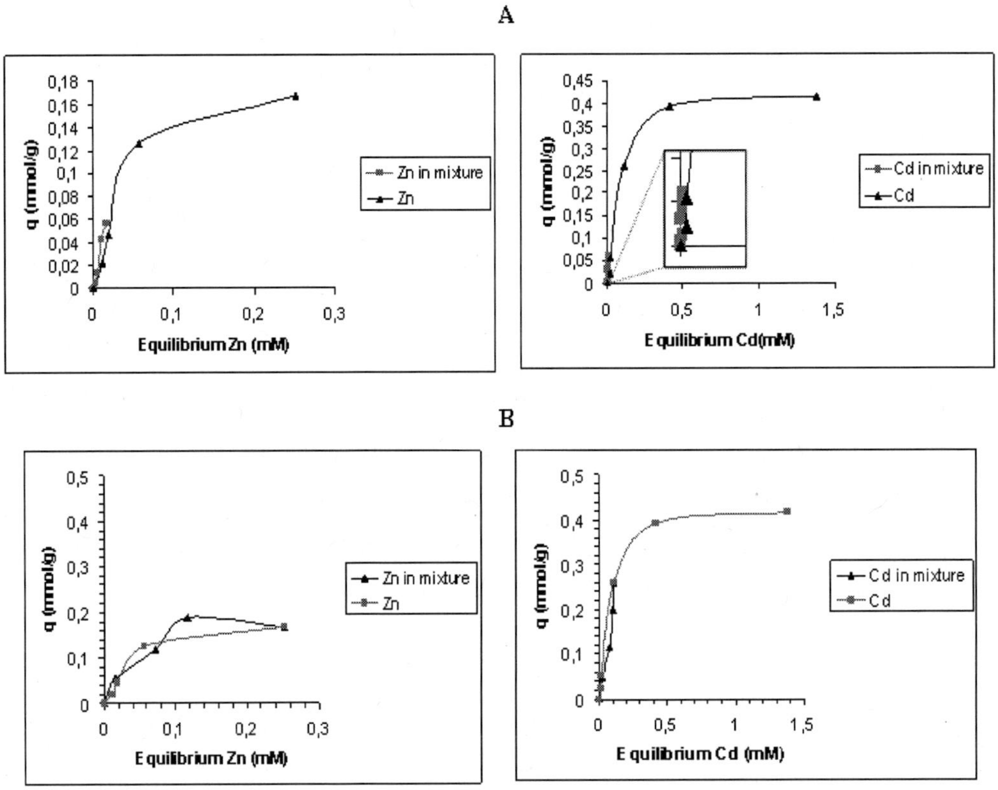

Figure 1. Adsoption isotherms of *Pseudomonas veronii* 2E in presence of Cd(II), Zn(II) or Cd(II)-Zn(II) mixtures with 0.1 mM (A) or 0.5 mM (B) total metal initial concentrations.

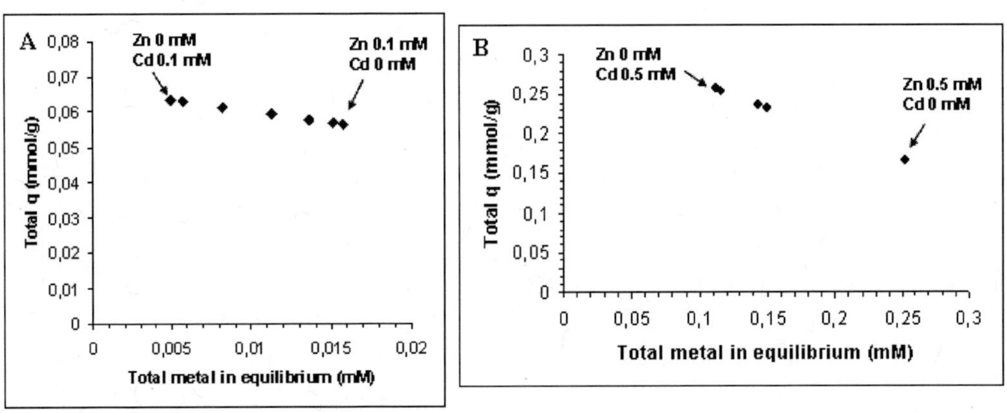

Figure 2. Total q variation as function of Cd(II)+Zn(II) equilibrium concentration in 0.1 mM (A) or 0.5 mM (B) total initial concentrations.

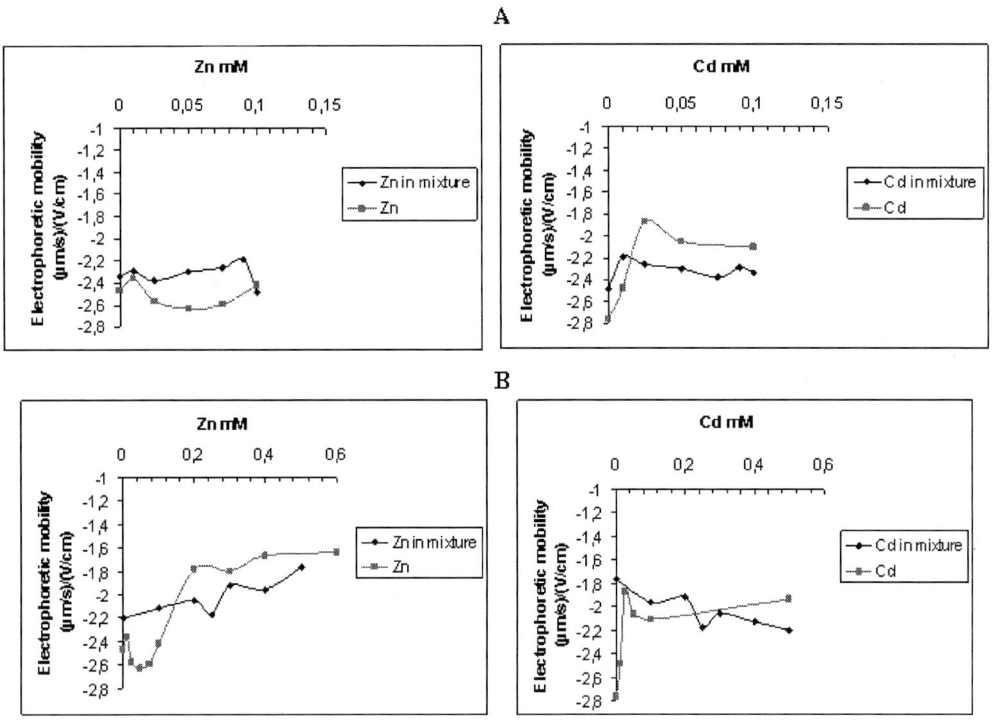

Figure 3. Electrophoretic mobility of *Pseudomonas veronii* 2E in presence of Cd(II), Zn(II) or Cd(II)+Zn(II) with 0.1mM (A) or 0.5 mM total metal concentrations (B).

These results suggest the existance of at least some different binding sites for Zn(II) and Cd(II) and reinforce the idea that all the adsorption sites are available for Cd(II) but only a fraction is available for Zn(II).

3. Cr(VI) Biotransformation and Cr(III) Immobilization

Microbiological detoxification of waste containing Cr(VI) has gained importance lately, due to the efficiency of bacterial reduction of the highly toxic Cr(VI) into the comparatively less toxic Cr(III). The details of the reaction mechanism are still unclear. Bioreduction of Cr(VI) can occur directly as a result of microbial metabolism (enzymatic) or indirectly through a bacterial metabolite. In aerobic conditions, most of the chromate reductases reported at the moment are cytosolic and reduce Cr(VI) to Cr(III) inside and outside the citoplasmatic membrane (Ackerley et al., 2004; Opperman and van Heerden, 2007; Thacker et al., 2007). Some studies showed that the Cr(VI) reducing activity of intact cells of chromium resistant bacteria, isolated from soils contaminated with dichromate, may be useful for Cr(VI) detoxification in polluted environments. In wastewaters, Cr(III) must be immobilized by precipitation as $Cr(OH)_3$. Cr(VI) bioreduction was studied in five indigenous strains either in axenic or mixed cultures. Maximal biotransformation yields were obtained at stationary phase of batch cultures in PYG medium (PYG in g/L: casein peptone, 2.5; yeast extract, 1.25; glucose, 0.5) and are resumed in Table 1. It can be noticed that a 74.32% Cr(VI) was removed by *Klebsiella oxytoca* P2, but total chromium determinations revealed that this metal remained in solution.

Table 1. Maximal Cr(VI) removal in axenic or mixed batch cultures

Strain	Cr(VI) removal (%)
Klebsiella oxytoca P2	74.32
Klebsiella ornithinolytica 1P	65.01
Ralstonia taiwanensis M2	48.31
Pseudomonas veronii 2E	44.47
Delftia acidovorans AR	38.83
Klebsiella ornithinolytica 1P + *Pseudomonas veronii* 2E	72.88
Pseudomonas veronii 2E + *Klebsiella oxytoca* P2	59.82
Klebsiella ornithinolytica 1P + *Klebsiella oxytoca* P2	65.65
Klebsiella ornithinolytica 1P + *Pseudomonas veronii* 2E + *Klebsiella oxytoca* P2	54.67

A precipitation step at pH>8 was needed to make sure chromium separation as $Cr(OH)_3$. Different precipitant agents were assayed, but only with 0.1 M Na_2CO_3 a quantitative Cr recovery was achieved (Garavaglia et al., 2010).

Chromate reductase activity was found in this strain. Reduction of Cr(VI) was detected at 32ºC using whole cells at different growth stages, resuspended in 0.05 M phosphate buffer (pH= 7.25) or 10 mM HEPES buffer (pH=7.0). Table 2 shows the results obtained for chromate reductase activity measurements. The maximum reduction was obtained using cell cultures at exponential growth stage (t=6hs) and the presence of NADH increases significantly the effect. No significant reduction was detected in the absence of the cells (abiotic reduction) in spite of the presence of the electron donor. Reduction experiments using HEPES buffer show smaller values for abiotic and biotic biotransformation than when using phosphate buffer in absence of NADH. NADH was slightly stronger reductor in HEPES buffer than in phosphate buffer. Reduction experiments using buffer TRIS-HCl (100mM pH= 7.0) were also performed to compare reduction yields, but non-reproducible results were obtained in this case. Interestingly, this is the first report of chromate reductase activity registered in *Klebsiella* genus.

Table 2. Chromate reductase activity in whole cell suspensions of *Klebsiella oxytoca* P2

Sample	HEPES Buffer 10 mM		PHOSPHATE Buffer 50 mM	
	mg/ml Cr(VI)	mmol/ml min	mg/ml Cr(VI)	mmol/ml min
Buffer + 5 mM Cr(VI)	47.16	-	46.87	-
Buffer + 5 mM Cr(VI) + 9.2 mM NADH	43.44	$4.81\ 10^{-4}$	42.44	$7.10\ 10^{-4}$
9.2 mM NADH + resuspended cells in buffer	-	-	-	-
5 mM Cr(VI) + resuspended cells in buffer	46.44	$1.15\ 10^{-4}$	42.44	$7.10\ 10^{-4}$
5 mM Cr(VI) + 9.2 mM NADH + resuspended cells in buffer	37.30	$1.47\ 10^{-3}$	38.44	$1.35\ 10^{-3}$

4. Ni(II)-Microbe Interactions

Bacterial retention by either biosorption or bioaccumulation mechanisms is an innovative and eco-friendly technology for Ni(II) removal from industrial effluents. *Pseudomonas veronii* 2E, *Delftia acidovorans* AR, *Ralstonia taiwanensis* M2, *Klebsiella oxytoca* P2 and *Klebsiella ornithinolytica* 1P were tested for Ni(II) tolerance and Ni(II) removal in batch cultures. Minimal Inhibitory Concentration (MIC) experiments were performed in PYG and M9 broths (PYG in g/L: casein peptone, 2.5, yeast extract, 1.25, glucose, 0.5; M9 in g/L: K_2HPO_4, 7.3, KH_2PO_4, 3, NH_4Cl, 1, NaCl, 0.5, $MgSO_4$, 0.120, $CaCl_2.2H_2O$, 0.013, glucose, 5) and, in all cases, strain adaptation to Ni(II) presence was needed to make possible bacterial growth. This "domestication" step was not necessary in previous studies with the other metals. Table 3 shows MIC results for each strain.

Table 3. Ni(II) MIC for the isolates either in PYG or M9 culture media

Microorganisms	MIC Ni(II)	
	PYG	M9
Klebsiella ornithinolytica 1P	0.80mM	--------
Pseudomonas veronii 2E	0.80mM	0.80mM
Delftia acidovorans AR	1.55mM	0.02mM
Ralstonia taiwanensis M2	7.74mM	0.80mM
Klebsiella oxytoca P2	65.04mM	0.20mM

Bacterial growth was tested in presence of different Ni(II) concentrations in PYG broth, according to MIC results. Figure 4 shows a decreasing specific growth rate with Ni(II) initial concentration increase, coincident also with a biomass yield decrease, when 0.8 mM or 1.5 mM Ni(II) is present, except with M2 strain. Similar results were observed when M9 broth was used.

Total Ni(II) concentrations in culture supernatants were determined by Adsorptive Stripping Voltammetry (AdSV) after total organic matter elimination by photooxidation in acidic conditions, using a complexing agent to form an electroactive complex. AdSV is used when the metal to be tested has a low solubility in mercury, as is the case of Ni(II). Ni(II) quantification is performed at pH=9.2 (0.1M borate buffer) in the presence of dimethylglyoxime ($DMGH_2$). The complex $Ni(DMGH)_2$ is adsorbed on mercury at -0.6V (vs Ag/AgCl) and reduced during a cathodic potential scan; peak current or charge can be used as analytical signals.

Results are shown in Figure 5. No significant changes were observed in Ni(II) concentrations during batch culture evolution. Some explanations can be inferred from these results:

1. Ni(II) is complexed by medium components and complexes were not eliminated with the photooxidation process; or
2. Ni(II) is complexed by a bacterial metabolite not necessarily released by the presence of the metal in culture and photooxidation process was not efficient enough to destroy the complex; or
3. Efflux mechanisms are present in bacteria, so as soon as Ni(II) penetrates inside the cells, it is immediately transported outside, either as free or complexed species.

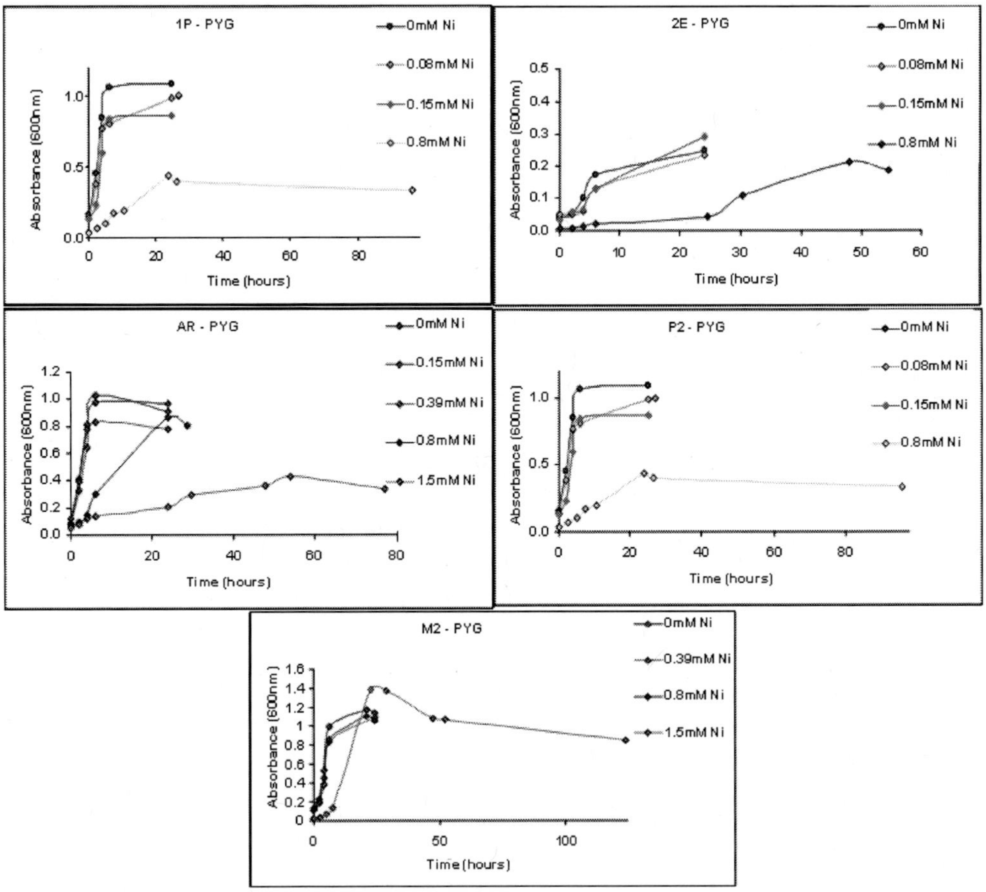

Figure 4. Bacterial growth evolution in presence of Ni(II) at PYG batch cultures of: *Klebsiella ornithinolytica* 1P, *Pseudomonas veronii* 2E, *Delftia acidovorans* AR, *Klebsiella oxytoca* P2 and *Ralstonia taiwanensis* M2.

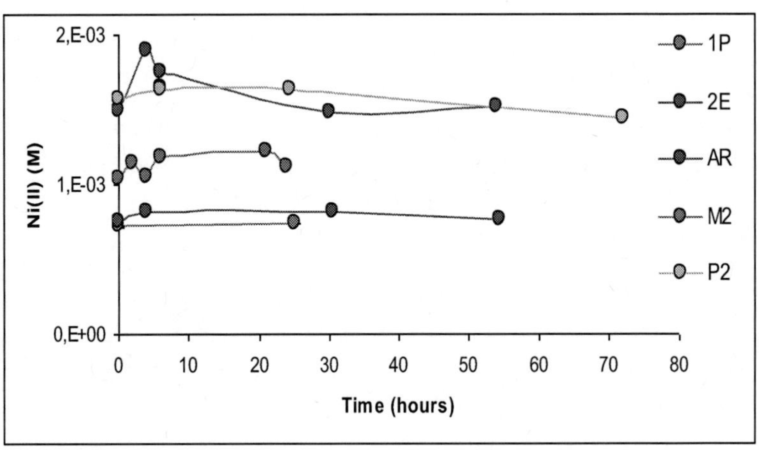

Figure 5. Total Ni(II) concentration in supernatants of batch cultures of : *Klebsiella ornithinolytica* 1P, *Pseudomonas veronii* 2E, *Delftia acidovorans* AR, *Klebsiella oxytoca* P2 and *Ralstonia taiwanensis* M2.

5. Bacterial Motility in Presence of Metals

The development of easy-to implement and rentable remediation alternatives, such as the use of indigenous bacteria to biosorb Cu(II), Cd(II) and Zn(II) or biotransform Cr(VI) in wastewaters, is essential to create options for the industry to diminish polluted discharges. Fixed-bed reactors are usually designed for this purpose, where the biofilm establishment and maintenance play the most important roles. Biofilm development strictly relies on different types of bacterial motility: swimming, swarming, twitching and chemotaxis, which can be altered by the microenvironmental conditions (Jefferson, 2004; Kato et al., 2008). An integrated study on the different effects of Cu(II), Cd(II), Zn(II) and Cr(VI) on swimming, swarming and chemotaxis of *Pseudomonas veronii* 2E, *Delftia acidovorans* AR and *Ralstonia taiwanensis* M2 was performed to improve biofilm development for metal loaded wastewater biotreatment in fixed-bed bioreactors. Swimming and swarming experiments were carried out in semisolid media (PYG plus 0.3% agar and 0.45% agar, respectively) with different metal concentrations as well as chemotactic response evaluation, using a modified "chemical in plug" assay with low temperature jellying point agarose, based on the characteristics of each strain. Clearly, the effects were independent, since a delay in swimming did not mean an inhibition of swarming or viceversa, nor a change in chemotaxis showed a correlation to swimming and swarming changes. A summary of the obtained results are shown in Table 4. From Table 4, it can be concluded that *Pseudomonas veronii* 2E is a good candidate for biotreatment processes in fixed-bed bioreactors because no chemotaxis towards metals was observed. *Delftia acidovorans* AR could be suitable for wastewater biotreatments since no biofilm detachment is expected as a result of the presence of Cd(II), Zn(II) or Cr(VI). *Ralstonia taiwanensis* M2 exhibit a complex response to metals presence: it seems that *R. taiwanensis* M2 would not be as useful as expected for biofilm reactors. Belonging to *Ralstonia* genus, this strain resulted particularly interesting for future studies on resistance mechanisms (Mergeay et al., 2003).

Table 4. Metal influence on microbial motility

	Swimming	Swarming	Chemotaxis towards PYG	Chemotaxis towards metal
P. veronii 2E				
Cu(II)	Normal	Normal	Positive	Negative
Cd(II)	Normal	Inhibited	Enhanced	Negative
Zn(II)	Normal	Altered	Enhanced	Negative
Cr(VI)	Altered	Inhibited	Enhanced	Absent
D. acidovorans AR				
Cu(II)	Altered	Normal	Altered	Positive
Cd(II)	Altered	Inhibited	Normal	Absent
Zn(II)	Altered	Inhibited	Normal	Absent
Cr(VI)	Normal	Normal	Normal	Absent
R. taiwanensis M2				
Cu(II)	Normal	Enhanced	Normal	Positive
Cd(II)	Altered	Altered	Negative	Absent
Zn(II)	Altered	Altered	Normal	Absent
Cr(VI)	Normal	Enhanced	Negative	Absent

6. Biofilm Development for Wastewater Treatments in Packed Bed Reactors

Biofilm-mediated wastewater treatments represent a proficient and safer alternative, especially when metal retention is involved. Extracellular materials, usually present in bacterial biofilm structures, enhance microbial metal immobilization mechanisms such as biosorption or bioaccumulation. Bacterial attachment or detachment to matrices may be strictly related to environmental conditions, well described for *Pseudomonas aeruginosa* (Caiazza et al., 2007; Shrout, et al., 2006; Teitzel et al., 2003). *Pseudomonas veronii* 2E is an indigenous strain, isolated from polluted sediments of Reconquista River, which represents an interesting tool for metal removal in wastewater biotreatments (Vullo et al., 2008). Biofilm development of *P. veronii* 2E over glass under different experimental conditions requires a special study for a successful design of a high efficiency packed bed bioreactor. Cellular attachment was studied with PYG, PYG (1:2 dilution) and PYG (1:4 dilution) and with M9-0.5%glucose, M9-0.5%glutamate and M9-0.5%citrate at 28ºC or 32ºC. Samples were incubated during 0 hs (t_0), 7 hs. (t_1), 31 hs. (t_2) and 55 hs. (t_3), with medium renewal at 24 hs. intervals. In all cases, influence of 0.3µM Fe(II) presence was also tested. Adequate biofilm visualization was achieved following different techniques after crystal violet staining: 1. ethanol extraction and Absorbance (590nm) measurement; 2. direct microscopic observation and 3. sample scanning and image analysis by MATLAB program.

Images captured by scanning and processed by MATLAB program are shown in Figure 6. Violet intensity indicates biofilm thickness, so extracting the dye with ethanol and measuring the Absorbance (590 nm) of the extracts, helps to confirm the biofilm presence in the scanned images. These last results are shown in Figure 7. The analysis of both groups of results converged to the same conclusions: a) best biofilm development was obtained at 32ºC with PYG broth or M9-0.5%glucose-Fe(II)(3µM), b) in all cases, except in presence of PYG-Fe(II), cell density increased with incubation time, being necessary at least 55 hs. to achieve best cell density, c) biofilm development with M9-0.5%glutamate was higher than with the other carbon sources and d) biofilm detachment was observed after 31 hs. when PYG-Fe(II) was used.

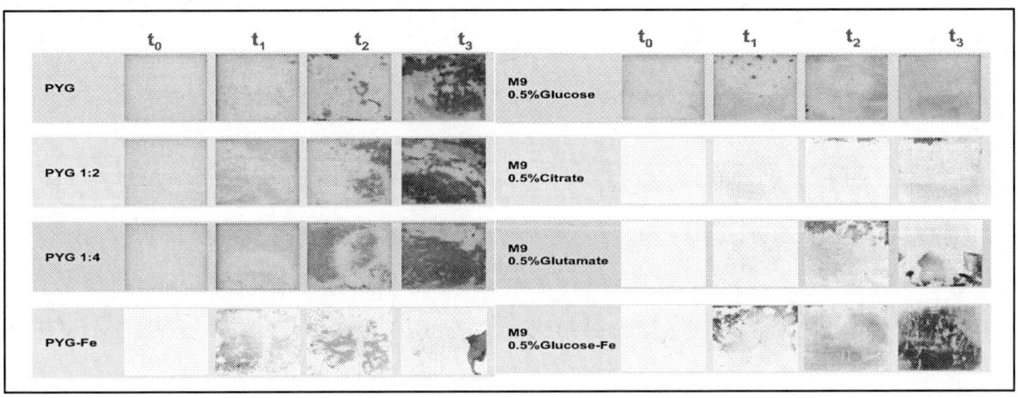

Figure 6. Images of biofilm development over coverslips captured by scanning and analyzed with MATLAB program. Incubation times: 0 hs (t_0), 7 hs. (t_1), 31 hs. (t_2) and 55 hs. (t_3).

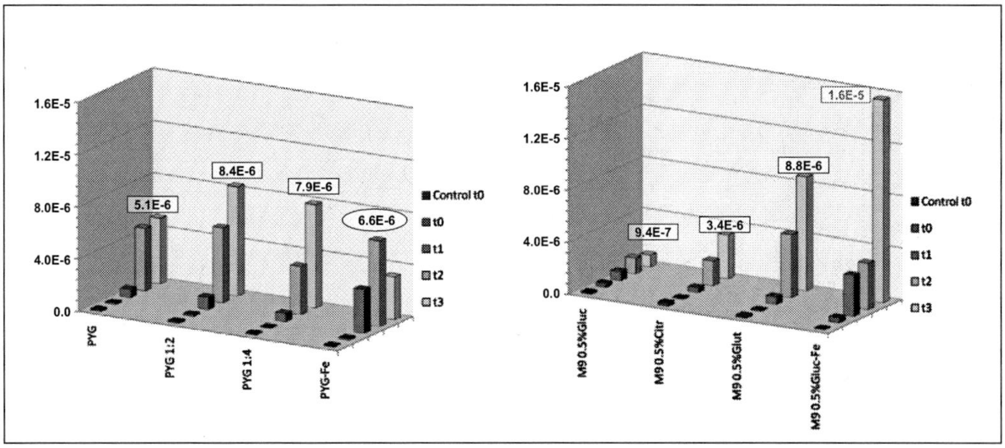

Figure 7. Analysis of biofilm development over coverslips by crystal violet staining and ethanol extraction. Incubation times: 0 hs (t_0), 7 hs. (t_1), 31 hs. (t_2) and 55 hs. (t_3).

7. Complexing Capacity of the Environment and its Influence on Biotreatment Process

The discharge of heavy metals into aquatic reservoirs has become a matter of global concern over the last few decades. Lead, chromium, mercury, zinc, cadmium, copper and nickel are usually found and are originated and disposed of in liquid industrial effluents by activities such as metal finishing, manufacture of electrical equipment, paints, alloys, batteries and pesticides.

Metals are of particular interest not only due to their toxicity and ubiquitousness but also to the fact that they can not be destroyed; instead they distribute in a variety of species in reservoirs like water and soil. In the case of water environments metals may exist in different chemical forms: aquo complexes, complexes with inorganic and organic ligands (either natural or anthropogenic), adsorbed on to suspended colloid particles, etc. This species distribution will be determined by the complexing properties of the ligands present in the particular system towards a specific metal ion. Thus determination of the concentration of ligands and the stability constant of the complexes is an important task since they define the complexing capacity (CC) of the system and govern metal toxicity and bioavalability (Fernandez-Pinas et al, 1991; Bridge et al, 1999).

During biotreatment of an industrial effluent, the liquid is contacted with a biofilm in the reactor to allow for metal retention. This can both occur on bacterial cells and/or extracellular materials (Guibaud et al., 2009; Prado Acosta et al., 2005). Considering the fact that metals in an industrial effluent are involved in a variety of species, a competition for the metal between dissolved ligands and “fixed” (surface) groups is established. A successful metal retention will occur when the interaction between the “fixed” groups and the specific metal is strong enough to overcome complexation with the dissolved ligand (which points to adsorption constants on the surface side and stability constants on the solution side) and also when the number of surface sites (i.e. surface concentration) is enough to allow for metal immobilisation. It follows that any successful treatment or biotreatment should consider both the complexing capacity of the solution to be treated and the retention capability of the immobilising material. Characterisation of both parts of the problem becomes a must.

Focussing on the interaction between a specific metal and a dissolved potential ligand, the characterisation involves determination of total ligand concentration, L_t, and the conditional stability constant, K'_f , which can be achieved by a variety of techniques (Pesavento et al., 2009). In any case, the technique should allow determination of the labile fraction of the metal in the presence of the ligand. Electroanalytical techniques are adequate since they allow distinction between different metallic species, i.e. fractions with different lability; knowledge of the thermodynamics of the complexation reaction (Scarponi et al., 1996) as well as an approach to the kinetic lability of the complex can be achieved (Van Leeuwen et al., 1989).

Amongst electroanalytical techniques, stripping methods are particularly interesting because the use of a preconcentration step allows achieving low detection limits (in the nanomolar range and lower) and adequate sensitivity. In the case of anodic stripping voltammetry (ASV) the metal signal is originated in the reaction of the labile metal fraction either diffusing to the electrode surface or forming in the diffusional layer through dissociation of a metal complex. The use of ASV to determine L_t and K'_f has been described (Pesavento et al., 2009; Scarponi et al., 1996) and can be applied to metal forming mercury amalgam (copper, cadmium, zinc, lead, etc). For metals that do not form mercury amalgams an alternative but not so direct method has been developed. It is based on the use of a competing ligand (X) combined with adsorptive stripping voltammmetry (Donnat and van den Berg, 1992; Pesavento et al., 2009). In this case a synthetic ligand X added to the system under study competes with a natural existing ligand L, to form an electroactive complex with the specific metal, MeX, that also must adsorb on mercury. The analytical signal is originated in the reaction of the adsorbed MeX.

Just to give a few examples, stripping techniques have been applied to CC determination in samples of sea, estuarine and freshwater (Donnat and van den Berg, 1992; Wisniewski Jakubaa et al., 2008; Meylan et al., 2004; Cobelo-García et al., 2003; Santos –Echeandía et al., 2008), industrial and domestic effluents (Van Veen et al., 2002); also model ligand solutions have been studied (Bruland, 1992; Xue and Sigg, 1998; Wisniewski Jakubaa et al., 2008; Wang and Chakrabarti, 2008; Laglera et al., 2007; Town and Filella, 2002; Domingos et al., 2004). Metal adsorption onto suspended particles has been evaluated in estuarine and surface water (Sigg et al., 1987; Minaberry and Gordillo, 2007). Also very interesting though less frequent is the study of more complex media like biological systems (Gonzalez Gil et al., 2003; Ceretti et al., 2006; Ceretti et al., 2010).

Biosorption is an interesting tool for metal removal from solutions. Most studies have been carried out using microbial systems using either whole cells, cell components or cell products and taking advantage of physicochemical processes such as adsorption, ion exchange and surface complexation, amongst others (Yin et al., 1999). Most of the work points to lab studies (Guibaud et al., 2009; Guibaud et al., 2008; Pal and Paul, 2008; Jefferson, 2004; Prado Acosta et al., 2005; Vijayaraghavan and Yun, 2008) and it feels like there is a long way to walk before transferring biosorption based technology to the industry (Gadd, 2009).

The measurement method and conditions greatly influence the data obtained (equilibrium conditions, sorption kinetics, ionic strength, solid/solution ration, still or stirred solution, etc).

Though a wide variety of biosorption systems have been used, comparison between results is difficult because of the different conditions.

The retention of any substrate (or release) on a solid is described by an isotherm, a curve describing the retention of a substance at various concentrations. A variety of models have been used to describe biosorption; in the case of a single component system Langmuir and Freundlich models are the most used; models derived from Langmuir and Freundlich models are used in multi components systems.

The Langmuir model (Langmuir, 1918) assumes that the solid has a limited sorption capacity, q_{max}. This model also considers that every site is occupied by only one molecule, that all sites are energetically identical and independent. The equation that describes the Langmuir isoterm is:

$$q = q_{max}\, LC/(1+LC)$$

which can be linearized to

$$q/C = q_{max}\, L - L.\, q$$

where q represents the adsorbate concentration in the solid, C is adsorbate concentration in the solution (once equilibrium has been reached). L is a constant that quantifies the affinity of the adsorbate for the solid.

There is quite an important number of publications where data fit to one of the previous models is performed. Although nice curve fits can be observed for example in Langmuirian type isoterms suggesting that all binding sites have the same affinity, it is quite simplistic in the case of cell walls or biomass components where a variety binding groups (carboxyl, amine, phosphate, etc) with different acid-base characteristics, coexist.

The Freundlich model (Freundlich, 1909) is empirical and is based in the equation

$$q = F\, C^{n}$$

which can be linearized to

$$\log q = \log F + n \log C$$

q and C have been previously defined. F and n are both constants; n is dimensionless and F units are $L.Kg^{-1}$. As can be observed, this isoterm does not reach a plateau.

Once the sorption system has been characterised, how can we decide if it will be effective in retaining a specific adsorbate present in the solution, considering the conditions and the complexing capacity of the solution? Both the complexing and the sorption characteristics should be compared, but that is not possible under the present situation. Conditional stability constant and affinity constant (Langmuir model) or F constant (Freundlich model) can not be compared. However, a recent work shows that characterization of surface ligands in colloidal particles suspended in surface waters can be performed in terms of total surface ligand concentration, Lt, and stability constant, K´f, just like in the case of dissolved ligands (Minaberry and Gordillo, 2007). In this way soluble and surface ligands are characterized by similar parameters and decision on whether some sorbent material will succeed to retain a certain substrate can be made just by knowing two stability constants (one corresponding to the system to be treated and the second one to the adsorbent material) and the total ligand concentrations in each phase.

An interesting system currently studied in our lab is the so called "second rinse" water (SRW) in the local galvanoplastic industry. This effluent is the product of rinsing with water the plated objects emerging from the electroplating bath; this process occurs in big rinsing pools and great volumes of liquid effluents are produced, ca. 20m^3 per day per pool (SAyDS, 2010).

As a result of this process metal ions (for example nickel, cadmium, zinc, etc) and organic substances (polyelectrolytes, surfactants, etc) derived from the galvanic process are dissolved and it is quite obvious that both metal and organic content are expected to increase with use along the day. This effluent should be treated for disposal or directly disposed after use in case it meets Argentinean Regulations (Decree PEN 999/92). It is important to consider that these organic substances behave as possible metal ligands and that also there is a varying chemical environment depending on the use of the rinsing pools.

Keeping in mind the relevance of complexing capacity, we are dealing with the study of the SRW for further treatment. Samples of this effluent were taken every hour along a working day and characterized: pH was measured; also lead and zinc were detected in the samples (UV digestion, metal measurements by square wave anodic stripping voltammetry). Results are resumed in the following figures (Figure 8 and Figure 9).

Both metal content and pH values increase with use. Metal concentrations are low so that metal recovery is not interesting from an economic point of view; however concentrations are not low enough to meet effluent disposal criteria. Also pH should be adjusted (specifically, the solution should be acidified).

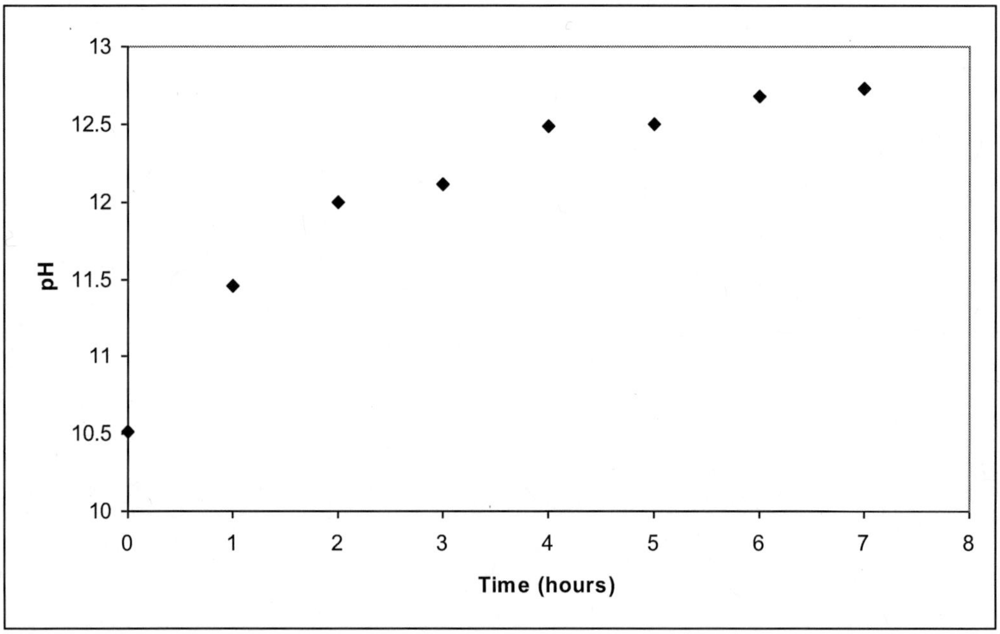

Figure 8. pH changes accompanying the use of the rinse bath.

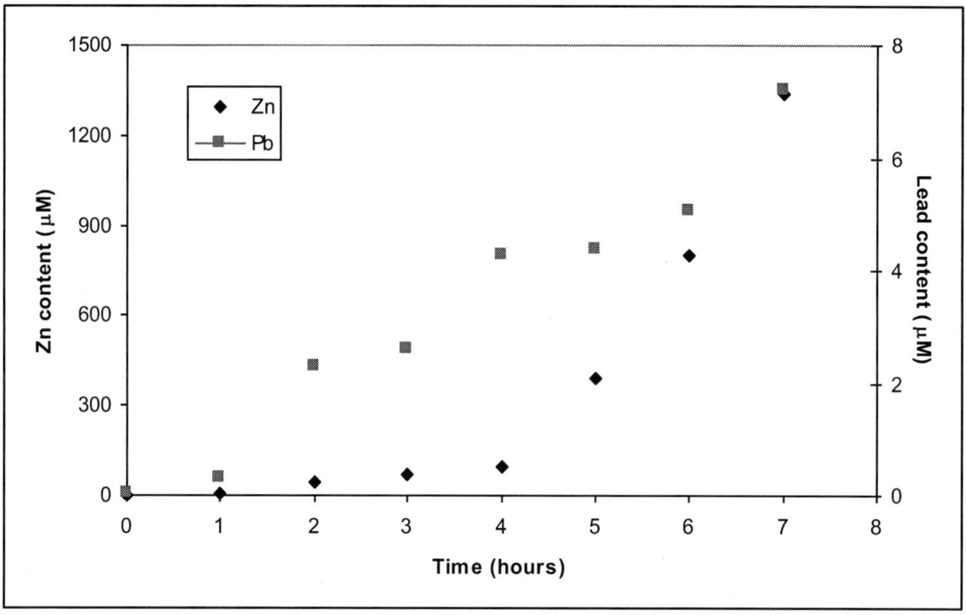

Figure 9. Zn(II) and Pb(II) contents accompanying the use of the rinse bath.

After these initial results, complexing capacity studies towards zinc were started. These were performed at constant pH = 7.5 (buffer HEPES) which is an intermediate value in the pH range recommended for disposal. Thus, the aim is to get information on labile zinc (bioavailable) fraction existing at this pH. This was performed by an ASV monitored titration of the specific fraction which was previously exposed to Chelex 100 to release the potential ligands from their zinc and lead complexes.

Results obtained so far in three of eight fractions show that i) ligand concentration increases on time, as expected: 4.95μM (0hs), 96.7μM (1hs), 377μM (2hs) and 2300μM (7hs); ii) these fractions show only one class of ligands (in the analytical window) with $K'_f = 1.1 .10^6$ iii) according to this stability constant, ligands interacting with zinc may be classified as weak.

Further steps involve characterization of all fractions and the solid that will be used in the treatment process.

8. CONCLUSIONS

In this chapter, work in progress on bioreactors design using indigenous metal resistant bacteria, as well as related parameters, is described. The final goal of these studies is to achieve the transference of the biosorption technology to the galvanoplastic industry, a good example of metal containing waste water producer.

In this sense, two parallel approaches were applied. On one hand, as exhaustive knowledge of the metal resistant microbiota is required, their metal removal efficiencies, bacteria – metal and bacteria-bacteria interactions, electrophoretic mobility, metal retention and immobilization, chemotactic behaviour and biofilm formation were studied. The second approach was focused on the characterization of the galvanoplastic industries effluents, in order to optimize the whole technological process.

Acknowledgments

This work was supported by the Universidad Nacional de General Sarmiento and Agencia Nacional de Promoción Científica y Tecnológica (ANPCyT), PICTO N°36782. We are specially grateful to Comisión de Investigaciones Científicas y Técnicas de la Provincia de Buenos Aires (CIC) and Universidad Nacional de General Sarmiento for the student's financial support.

References

Ackerley, D. F., Gonzalez, C. F., Keyhan, M., Blake II, R. Matin, A. (2004). Mechanism of chromate reduction by the *Escherichia coli* protein, NfsA, and the role of different chromate reductases in minimizing oxidative stress during chromate reduction. *Environmental Microbiology*, 6(8), 851–860.

Bridge, T. A.M., White, C., Gadd, G. M. (1999) Extracellular metal-binding activity of the sulphate-reducing bacterium *Desulfococcus multivorans. Microbiology*, 145, 2987-2995.

Bruland, K. (1992). Complexation of cadmium by natural organic ligands in the central North Pacific. *Limnology and Oceanography*, 37 (5), 1008-1017.

Butterman, W.C., Reston, J.P., (2000). Mineral Commodity Profiles – Cadmium, U.S. Geological Survey, Open-File Report 02-238 (*http://pubs.usgs.gov/of/2002/of02-238/of02-238.pdf.).*

Caiazza, N., Merrit, J., Brothers, K., O´Toole, G. (2007). Inverse regulation of biofilm formation and swarming motility by *Pseudomonas aeruginosa* PA14. *Journal of Bacteriology,* 189(9), 3603-3612.

Ceretti, H.M., Vullo, D.L., Zalts, A., Ramírez, S.A. (2006). Cadmium Complexation in culture media. *Electroanalysis*, 18(5), 493-498

Ceretti, H.M., Vullo, D.L., Zalts, A., Ramírez, S.A. (2010). Effect of bacterial growth in the complexing capacity of a culture medium supplemented with cadmium(II). World *Journal of Microbiology and Biotechnology*, 26, 847–853.

Chen, X.C., Wang, Y.P., Lin, Q., Shi, J.Y., Wu, W.X., Chen, Y.X. (2005). Biosorption of copper(II) and zinc(II) from aqueous solution by *Pseudomonas putida* CZ1, *Colloids and Surfaces B: Biointerfaces,* 46, 101-107.

Cobelo-García, A., Prego, R., Nieto, O. (2003). Chemical speciation of dissolved lead in polluted environments. A case study: the Pontevedra Ria (NW Spain). *Ciencias Marinas*, 29, 377-388.

Davis, T.A., Volesky, B., Mucci, A. (2003). A review of the biochemistry heavy metal biosorption by brown algae. *Water Research*, 37, 4311-4330.

Domingos, R.F., Benedetti, M.F., Croué, J.P., Pinheiro, J.P. (2004). Electrochemical methodology to study labile trace metal/natural organic matter complexation at low concentration levels in natural waters. *Analytica Chimica Acta*, 521, 77–86.

Donnat, J.R., van den Berg C.M.G. (1992). A new cathodic stripping voltammetric method for determining organic copper complexation in seawater. *Marine Chemistry*, 38, 69-90.

EC Proposal for a European Parliament and Council Decision Establishing the List of Priority Substances in the Field of Water Policy. Brussels, Belgium, European Commission; COM (2000) 47 Final.

Eisler, R. (1986). Chromium hazards to fish, wildlife, and invertebrates: a synoptic review; Contaminant Hazard Reviews (Report Nº 6), *Biological Report* 85(1.6).

Eisler, R. (1998). Nickel hazards to fish, wildlife, and invertebrates: a synoptic review; Contaminant Hazard Reviews (Report Nº 34) *Biological Science Report USGS/BRD/BSR-1998-0001.*

Fernandez-Pinas, F., Mateo, P., Bonilla, I. (1991). Binding of cadmium by cyanobacterial growth media: Free ion concentration as a toxicity index to the cyanobacterium *Nostoc* UAM 208. *Archives of Environmental Contamination and Toxicology*, 21, 425-431.

Freundlich, H. (1909). *Kapillarchemie. Akademische Verlagsgessellschaft*, Leipzig, Germany.

Gadd, G.M. (2009). Biosorption: critical review of scientific rationale, environmental importance and significance for pollution treatment. *Journal of Chemical Technology & Biotechnology,* 84, 13–28.

Garavaglia, L., Cerdeira, S.B., Vullo, D.L (2010). Chromium (VI) biotransformation by β- and γ-Proteobacteria from natural polluted environments: a combined biological and chemical treatment for industrial wastes. *Journal of Hazardous Materials*, 175, 104–110.

Gonzalez Gil, G., Jansen, S., Zandvoort, M.H., van Leewen, H.P. (2003). Effect of yeast extra ton speciation and bioavailability of nickel and cobalto in anaerobic reactors. *Biotechnology and Bioengineering*, 82, 134-142.

Gordon, R.B., Bertram, M.,. Graedel, T.E, (2006). *Metal Stocks and Sustainability, PNAS*, 103, 1209-1214.

Guibaud, G, Bordas, F., Saaid, A., D'abzac, P., Van Hullebusch, E. (2008). Effect of pH on cadmium and lead binding by extracellular polymeric substances (EPS) extracted from environmental bacterial strains. *Colloids and Surfaces B: Biointerfaces*, 63, 48–54.

Guibaud, G., van Hullebusch, Bordas, F., d´Abzac, P., Joussein, E. (2009). Sorption of Cd(II) and Pb(II) by exopolymeric substances (EPS) extracted from activated sludges and pure bacterial strains: Modeling of the metal/ligand ratio effect an role of the mineral fraction. *Bioresource Technology*, 100, 2959-2968.

Hermoso, F.G., Bartacek, J., Janseri, S., Lens P.N.L. (2009). Metal supplementation to UASB bioreactors: from cell-metal interactions to full-scale application. *Science of Total Environment*, 407, 3652-3667.

Jefferson, K. (2004). What drives bacteria to produce a biofilm? *FEMS Microbiology Letters*, 236, 163-173.

Johnson, J, Harper, E.M., Lifset, R., Graedel, T.E. (2007). Dining at the Periodic Table: Metals Concentrations as They Relate to Recycling. *Environmental Science and Technology,* 41, 1759-1765.

Kato, J., Kim, H.E., Takiguchi, N., Kuroda, A., Ohtake, H. (2008). *Pseudomonas aeruginosa* as a model microorganism for investigation of chemotactic behaviours in ecosystem. *Journal of Bioscience and Bioengineering*, 106, 1-7.

Laglera, L.M., Battaglia, G., van den Berg, C.M.G. (2007). Determination of humic substances in natural waters by cathodic stripping voltammetry of their complexes with iron. *Analytica Chimica Acta*, 599, 58–66.

Langmuir, I. (1918). The adsorption of gases on plane surfaces of glass, mica and platinum. *Journal of American Chemical Society*, 40, 1361-1403.

Méndez, M., Ramírez, S.A.M., Ceretti, H.M., Zalts, A, Candal, R, Vullo, D.L. (2011). *Pseudomonas veronii* 2E surface interactions with Zn(II) and Cd(II). *Global Journal of Environmental Science and Technology* (in press).

Mergeay, M., Monchy, S., Vallaeys, T., Auquier, V., Benotmane, A., Bertin, P., Taghavi, S., Dunn , J., van der Lelie, D., Wattiez, R.(2003). *Ralstonia metallidurans*, a bacterium specifically adapted to toxic metals: towards a catalogue of metal-responsive genes, *FEMS Microbiology Reviews*, 27, 385-410.

Meylan, S., Odzak, N., Behra, R., Sigg, L. (2004). Speciation of copper and zinc in natural freshwater: comparison of voltammetric measurements, diffusive gradients in thin films (DGT) and chemical equilibrium models. *Analytica Chimica Acta* , 510, 91-100.

Minaberry, Y.S., Gordillo, G.J. (2007). Complexing capacity of natural waters carrying a great amount of suspended matter. *Chemosphere*, 69 (9), 1465-1473.

Opperman, D.J., van Heerden, E. (2007). Aerobic Cr(VI) reduction by *Thermus scotoductus* strain SA-01. *Journal of Applied Microbiology*, 103, 1907-1913.

Pal, A., Paul, A.K. (2008). Microbial extracellular polymeric substances: central elements in heavy metal bioremediation. *Indian Journal of Microbiology*, 48, 49-64.

Pesavento, M., Alberti, G., Biesuz R. (2009). Analytical methods for determination of free metal ion concentration, labile species fraction and metal complexation capacity of environmental waters: A review. *Analytica Chimica Acta*, 631, 129–141.

Prado Acosta, M., Valdman, E., Leite, S.G.F., Battaglini, F., Ruzal, S.M. (2005). Biosorption of copper by *Paenibacillus polimyxa* cells and their exopolysaccharide, *World Journal of Microbiology and Biotechnology,* 21, 1157-1163.

Santos –Echeandía, J., Laglera, L.M., Prego, R., van den Berg, C.M.G. (2008). Copper speciation in estuarine waters by forward and reverse titrations. *Marine Chemistry*, 108, 148-158.

SAyDS: Secretaría de Ambiente y Desarrollo Sustentable. Programa de vigilancia y control. Contaminación hídrica industrial. Estudio de caso: Establecimientos Potencialmente Generadores de Metales Pesados Área Metropolitana- http://*www.caru.org.uy/13.pdf*.

Scarponi, G., Capodaglio, G., Barbante, C., Cescon, P. Elemental Speciation. In S. Caroli (Ed.) *Bioinorganic Chemistry* Wiley, Chichester, 1996, 363.

Shrout, J., Chopp, D., Just, C., Hentzer, M., Givskov, M., Parsek, M. (2006). The impact of quorum sensing and swarming motility on *Pseudomonas aeruginosa* biofilm formation is nutritionally conditional. *Molecular Microbiology*, 62(5), 1264-1277.

Sigg, L., Sturm, M., Kistler, D. (1987). Vertical transport of heavy metals by settling particles in Lake Zurich. *Limnology and Oceanography*, 32, 112-130.

Teitzel, G.M., Parsek, M.R. (2003). Heavy metal resistance of biofilm and planktonic *Pseudomonas aeruginosa. Applied and Environmental Microbiology*, 69(4), 2313-2320.

Thacker, U., Parikh, R., Shouche, Y., Madamwar, D. (2007). Reduction of chromate by cell – free extract of *Brucella sp.* Isolated from Cr (VI) contaminated sites. *Bioresource Technology*, 98, 1541 -1547.

Town, R.M., Filella, M. (2002). Crucial role of the detection window in metal ion speciation analysis in aquatic systems: the interplay of thermodynamic and kinetic factors as exemplified by nickel and cobalt. *Analytica Chimica Acta*, 466, 285–293.

Van Leeuwen, H., Cleven, R., Buffle J. (1989). Voltammetric techniques for complexation measurements in natural aquatic media. *Pure & Applied Chemistry*, 61, 255-274.

Van Veen, E., Burton, E., N., Comber, S., Gardner, M. (2002). Speciation of Copper in sewage effluents and its toxicity to *Daphnia Magna*. *Environmental Toxicology and Chemistry*, 21 (2), 275-280.

Vijayaraghavan, K., Yun Y. (2008). Bacterial biosorbents and biosorption. *Biotechnology Advances*, 26, 266-291.

Volesky, B. (2001). Detoxification of metal-bearing effluents: biosorption for the next century. *Hydrometallurgy*, 59, 203-216.

Vullo, D.L., Ceretti, H.M., Daniel, M. A., Ramírez, S.A.M., Zalts, A. (2008). Cadmium, zinc and copper biosorption mediated by *Pseudomonas veronii* 2E. *Bioresource Technology*, 99, 5574-5581.

Vullo, D.L., Ceretti, H.M., Ramírez, S.A.M., Zalts, A. Heavy metals and microorganisms in the environment: taking advantage of reciprocal interactions for the development of a wastewater treatment. In Myung-Bo Kim Eds. *Progress in Environmental Microbiology*, Nova Science Publishers, Inc., New York (2007), 111-149.

Wang, J., Chen, C. (2009). Biosorbents for heavy metals removal and their future. *Biotechnolology Advances*, 27, 195-226.

Wang, R., Chakrabarti, C.L. (2008). Copper speciation by competing ligand exchange method using differential pulse anodic stripping voltammetry with ethylenediaminetetraacetic acid (EDTA) as competing ligand. *Analytica Chimica Acta*, 614, 153-160.

Wisniewski Jakubaa, R., Moffett, J.W., Saito, M.A. (2008). Use of a modified, high-sensitivity, anodic stripping voltammetry method for determination of zinc speciation in the North Atlantic Ocean. *Analytica Chimica Acta*, 614, 143–152.

Xue, H., Sigg, L. (1998). Cadmium speciation and complexation by natural organic ligands in fresh water. *Analytica Chimica Acta*, 363, 249-259.

Yin, P., Yu, Q., Jin, B., Ling Z. (1999). Biosorption removal of cadmium from aqueous solution by using pretreated fungal biomass cultured from starch wastewater. *Water Research*, 33, 1960–1963.

In: Bioremediation
Editor: Alexander C. Mason

ISBN: 978-1-61122-730-7

Chapter 14

SALT MARSHES' MEANING ON NITROGEN REMEDIATION

Ana I. Sousa[1,2*], Ana I. Lillebø[2], Nils Risgaard-Petersen[3], Miguel A. Pardal[4] and Isabel Caçador[1]

[1]CO – Centre of Oceanography, Faculty of Sciences, University of Lisbon, Campo Grande, 1749-016 Lisbon, Portugal

[2]CESAM – Centre for Environment and Marine Studies, Department of Chemistry, University of Aveiro, Campus de Santiago, 3810-193 Aveiro, Portugal

[3]Institute of Biological Science, Center for Geomicrobiology, Aarhus University, Ny Munkegade, Building 1540, DK-8000 Aarhus C, Denmark

[4]CEF – Centre for Functional Ecology, Department of Life Sciences, University of Coimbra, Apartado 3046, 3001-401 Coimbra, Portugal

ABSTRACT

Nitrogen is ubiquitous in the environment, being present in the atmosphere, lithosphere, hydrosphere and biota. The fast increase in human population and human activities over the course of the 20th century has lead to an increase in the loading of land-derived nitrogen, from diffuse sources with anthropogenic origin, to coastal areas, namely transitional waters. Thus, estuaries are subdue to large discharges of this nutrient, which has different consequences, namely eutrophication. Salt marshes have important ecological functions as bio-stabilizers, i.e. decreasing tidal currents and wave action and enhancing settling of suspended matter. Halophytes sustain an elevate productivity and provide habitats. Moreover, nitrogen incorporation in salt marsh plants' biomass (N fixation/primary production) gives them the ability to auto-remediate estuarine systems. The availability of the excess of nitrogen, due to eutrophication, to different biological processes is determined by its conversion in different chemical forms through biological and physico-chemical processes. These can be incorporation of inorganic forms (NO_3^-, NH_4^+) and transformation into organic forms (e.g. aminoacids, proteins), biomass mineralization, nitrification and denitrification (NO_3^- conversion to N_2 gas), which return

[*]Corresponding author: A.I.Sousa (aisousa@fc.ul.pt)

the reactive nitrogen to its main reservoir, the atmosphere. Thus, denitrification is another process that contributes to remove reactive nitrogen from the estuarine environment, contributing to the reduction of eutrophication. In this book chapter, nitrogen sequestration, nitrogen uptake and denitrification will be discussed as natural remediation processes provided by salt marshes, contributing to maintain the ecosystem health.

INTRODUCTION

Nitrogen can be found in sedimentary and crystalline rocks but the major fraction of nitrogen at Earth's surface is in the atmosphere (e.g. Tolstikhin & Marty, 1998). In addition, nitrogen is an essential element of all biological life, since it is associated to proteins pool (organic nitrogen), being often the limiting nutrient for net primary production (NPP) in estuarine and marine systems (Matamala & Drake, 1999). The nitrogen cycle between the environmental compartments and the biota is quite complex and involves several processes mediated by microorganisms, i.e., nitrogen (N_2 gas) fixation, nitrogen incorporation of inorganic forms (NO_3^-, NH_4^+) and transformation into organic forms (e.g. aminoacids, proteins), nitrogen mineralization, nitrification and denitrification (NO_3^- conversion to N_2 gas). So, in non-human-impacted environments, reactive nitrogen results from biological nitrogen fixation. In addition, between nitrogen fixation and denitrification, reactive nitrogen is utilized by ecosystems and distributed among earth's reservoirs (i.e. atmosphere, ocean, terrestrial and marine biota and soil/sediment organic matter) (Galloway, 1998). According to Galloway (1998) review, and considering non-human-impacted environments, most of the land-derived nitrogen loads to coastal environments could be denitrified in estuarine and shelf regions. Therefore, under these circumstances, the cycle should be approximately in balance with little accumulation of reactive nitrogen in coastal areas. However, the supply of reactive nitrogen to global terrestrial ecosystems has increased as a consequence of world population and human activity (e.g. Galloway, 1998; Galloway et al., 2008). Altering the nitrogen cycle as a result of the production and industrial use of artificial nitrogen fertilizers worldwide has induced environmental problems, such as eutrophication of terrestrial and aquatic systems (e.g. Hauxwell & Valiela, 2004; Gruber & Galloway, 2008).

Concerning the management of coastal systems, namely transitional waters, there is an increasing need to remedy long-lasting adverse effects of human interventions and much effort has been put into reducing eutrophication. Towards this goal, conservation and restoration of coastal areas, has become a priority during the last decades (e.g. Valiela et al., 2000; McLusky & Elliott, 2004; Lillebø et al., 2007). Once in the aquatic system land-derived nitrogen loads may be trapped in biota or in the sediment, or passed to its main compartment, the atmosphere, reducing the availability of reactive nitrogen. Although there are still uncertainties concerning the fate of all land-derived nitrogen (Galloway et al., 2004), denitrification seems to be an important reactive nitrogen sink, meaning that even in heavily altered regions, rivers, although important sources of nitrogen to coastal systems, represent small sources of reactive nitrogen to the open ocean (Galloway et al., 2008). Thus, primary production, organic nitrogen burial and denitrification constitute processes that may trap reactive nitrogen in estuaries (Nedwell et al., 1999; Valiela, 2000; Sousa et al., 2008, 2010). Hence, considering the role of primary producers in this process, salt marshes may have a crucial part in nitrogen balance. Thus, the central question of this chapter is to address the salt marshes' meaning on nitrogen remediation, by intercepting the land-derived nitrogen and buffering the loading of reactive nitrogen to the open ocean.

Nitrogen Compounds as Environmental Pollutants: Eutrophication and Greenhouse Gases

As stated by Gruber & Galloway (2008) "There is compelling evidence that human alteration of the nitrogen cycle is negatively affecting human and ecosystem health".

As a result of energy and food production, the nitrogen oxides and ammonia emitted are efficiently spread in the atmosphere and deposited on the terrestrial and aquatic systems in a form that is readily available to primary producers, thereby stimulating productivity (Philipott et al., 2009). As an outcome of increasing global population and increasing human activities coastal systems, namely transitional waters, became subjected to increasing of land-derived nitrogen loads, often leading to eutrophication (e.g. Hauxwell & Valiela, 2004; Lillebø et al., 2007; Howarth, 2008). In fact, eutrophication has been classified as a worldwide agent of change for coastal ecosystems (NRC, 2000). Therefore, nutrient cycling in coastal ecosystems is a crucial function performed by salt marshes (Nixon, 1981), acting as transformers of nutrients. Salt marsh plants have similar nutrient requirements to non-saline-tolerant species and like these species they need a well-developed root- system for anchoring (Amos et al., 2004) and an efficient uptake of nutrients (Boorman, 2003).

In non-human-impacted environments the nitrogen cycle also involves a greenhouse gas out coming from biochemical pathways mediated by microorganisms such as denitrifying bacteria. A small percentage of the reactive nitrogen passing through the denitrification process is released as nitrogen dioxide (N_2O) gas, which is a powerful greenhouse gas in Earth's atmosphere (Schlesinger, 2009). In human-impacted environments the increase in N_2O gas results from the high nitrogen availability (Nielsen et al., 2009), low oxygen availability (de Bie et al., 2002; Stief et al., 2002) and variation of other abiotic factors/environmental variables such as temperature, pH, organics, NO_2, salinity, sulphur availability (Sørensen et al., 1980; Nielsen et al., 2009). N_2O increase in the atmosphere has clear impacts on the environment such as the destruction of stratospheric ozone (Dickinson & Cicerone, 1986; Schlesinger, 2009) and the warming potential of the atmosphere (Seitzinger, 1984), which is 200 to 300 times greater than CO_2 (Kramlich & Linak, 1994, in Nielsen et al., 2009). Moreover, several studies have reported adverse human health effects, particularly among susceptible populations, due to long-term exposure to nitrogen dioxide below the current exposure limits (Latza et al., 2009).

Salt Marshes: Functions and Services

Salt marshes occur in low-energy environments, and are usually restricted to relatively sheltered areas, characterised mainly in five physiographic situations: in estuaries, in saline lagoons, behind barrier islands, at the heads of sea lochs, and on beach plains; allowing in all cases the accumulation of fine sediments (Best et al., 2007). The development of salt marsh vegetation is dependent on the presence of intertidal mudflats and other supplies of sediment, being naturally dynamic systems (Boorman, 2003). Salt marsh halophytes' communities are spatially distributed according to the marsh's topography, which determines the frequencies and durations of tidal submersion, the physical and chemical characteristics of the sediment and the interspecific competition conditions (Lefeuvre et al., 2003). Salt marsh ecosystem shows a clear zonation according to the frequency of inundation, being the halophyte species adapted to regular immersion by the tides. These halophytes may act as sediment traps, playing an important role in the settling of suspended matter (e.g. Boorman, 2003). In non-human-impacted intertidal salt marshes, plant diversity tends to decrease from low marsh towards high marsh, while primary production follows an opposite pattern (Lefeuvre et al., 2003). Thus, salt marshes are commonly characterized by a relatively small number of highly productive marsh species (Lefeuvre et al., 2003).

Salt marsh plants possess a well-developed aerenchyma system, which confers on them the ability to transport oxygen to the belowground parts (Maricle & Lee, 2002), where it is used for root respiration and oxidation of the rhizosphere. Therefore, these plants may interfere with benthic nutrient cycles, as well as modifying redox potentials and pH within the sediment, which lastly eventually influences processes such as nutrient adsorption to particles and several steps of the nitrogen cycle (i.e. ammonification, nitrification, denitrification and N_2 fixation) (e.g. Pedersen et al., 2004; Sousa et al., 2008; Philipott et al., 2009). In addition, during high tide salt marshes are important nursery sites for juvenile fish and crustaceans (e.g. Jin et al., 2007; Green et al., 2009) with economical value. In this manner, salt marshes are among the most productive ecosystems in the world (e.g. McLusky & Elliott, 2004), and the multiple services provided by these ecosystems have already been estimated in terms of economical value (Constanza et al., 1997). Ecosystem services represent *"the benefits human populations derive, directly or indirectly, from ecosystem functions"* (Constanza et al., 1997), or as defined by the Ecological Society of America, "*the processes by which the environment produces resources that we often take for granted*" (http://www.esa.org/), or even the "*components of nature, directly enjoyed, consumed, or used to yield human wellbeing*" (Boyd & Banzhaf, 2007). Accordingly, salt marsh plants' biomass and productivity, as well as denitrification are important ecosystem services provided by salt marshes, by reducing the impact of land-derived nitrogen loads and, consequently, eutrophication (Valiela et al., 2000; Sousa et al., 2008; Philipott et al., 2009). The productivity of salt marshes in non-human-impacted environments is affected by several inter-related parameters, such as the tidal pattern (flooding frequency and duration), soil salinity, temperature, nutrient availability, oxygen levels, sediment type, marsh maturity (Ibañez et al., 2000; Valiela et al., 2000; Boorman, 2003; Sousa et al., 2008). However, anthropogenic disturbance (e.g. habitat disruption and fragmentation, pollution, climate change, namely storm events, and coastal development) has lead to the reduction of salt marsh areas worldwide. This is now an issue of major concern, and several studies on the ecology of estuaries have emphasized the negative consequences of its disappearance (e.g. Valiela et al., 2000; Boorman, 2003; Lefeuvre et al., 2003; Best et al., 2007; Jin et al., 2007; Green et al., 2009; Wieski et al., 2010).

NITROGEN INCORPORATION AND SEQUESTRATION IN SALT MARSHES

In warm-temperate estuaries, salt marshes are often colonized by the species *Spartina maritima*, *Scirpus maritimus*, *Halimione portulacoides*, *Sarcocornia fruticosa*, and *Sarcocornia perennis*. These species are adapted to different physicochemical characteristics, colonizing upper, middle, or lower marshes, but they all have an important role, contributing to the ecosystem productivity. In addition, despite differences in their physiology and annual biological cycle, these species may function as nitrogen buffers (Caçador et al., 2007; Sousa et al., 2010). Salt marshes' functioning, as sinks and/or sources of nitrogen, seems to depend on several factors such as stage of succession, tidal energy, salinity, nutrient supply, namely nitrogen uptake and fixation, oxygen release, and nutrient production and losses due to mineralization and nitrification-denitrification (Valiela et al., 2000; Eyre & Ferguson, 2002; Boorman, 2003; McGlathery et al., 2004). The role of salt marshes acting as a sink or source of nutrients attending to marshes maturity has been discussed elsewhere (Valiela et al., 2000; Boorman, 2003; Sousa et al., 2008; Lillebø et al., in press). Nevertheless, in the case of *Spartina maritima* salt marsh (Sousa et al., 2008, Lillebø et al., in press), results concerning nitrogen were in agreement with the revision carried out by Boorman (2003). This author concludes that mature marshes have higher nutrients content both in the plants and the rhizosediment, due to the accumulation of litter on the sediment surface and to the net belowground primary production (NBPP), which enhance the steady build-up in the surface level of the marsh.

When comparing the nitrogen cycling in salt marshes colonized by five different species, one can see from figure 1 that net primary production (NPP) is species specific.

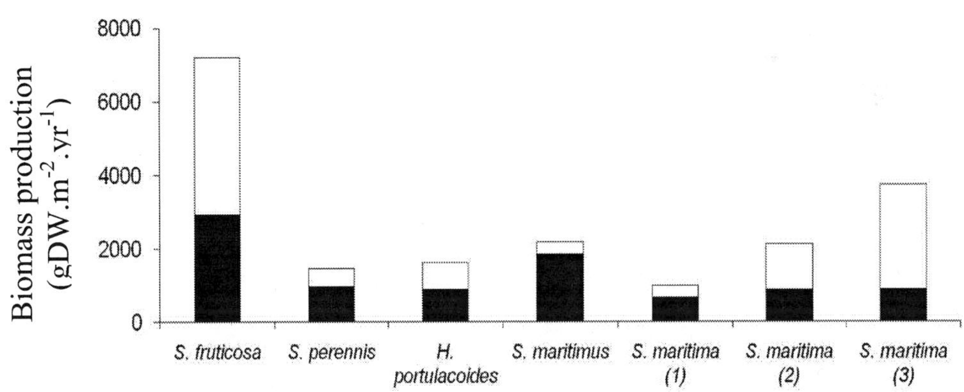

Figure 1. Biomass production for aboveground (white column) and belowground (black column) organs of five different salt marsh species (data from Sousa et al., 2008, 2010). Biomass production, dry weight (gDW.m^{-2}.yr^{-1}).

Moreover, the contribution of the NAPP (net aboveground primary production) and NBPP also seem to be specific for each species (Sousa et al., 2010). Nitrogen content in plant tissue is directly related to the nitrogen availability. So, above and belowground biomass and nitrogen standing stocks reflect the nitrogen content and availability to each studied plant (Sousa et al., 2010). Results from figure 2 show that most of the sediments colonized by different species are enriched by land-derived nitrogen loads (i.e. external sources), increasing the amount of nitrogen retained in the sediment. Comparing plants, *S. fruticosa, S. perennis* and *H. portulacoides* contributed most to nitrogen sequestration (Sousa et al., 2010). Thus, although, salt marsh species share potentially the same ecosystem services, their contribution seems to be species specific. Moreover, when comparing the three marshes colonised by *Spartina maritima* one can see that the provided services (i.e nitrogen incorporation in biomass and organic nitrogen burial) do not have the same value (Figure 2). This means that local physicochemical and environmental factors also have to be considered. To sum up, this information is of crucial importance to evaluate the impact of anthropogenic disturbances on the services provided by salt marshes, but also, for the management and restoration actions in salt marshes, namely to enhance salt marshes auto-remediation capacity.

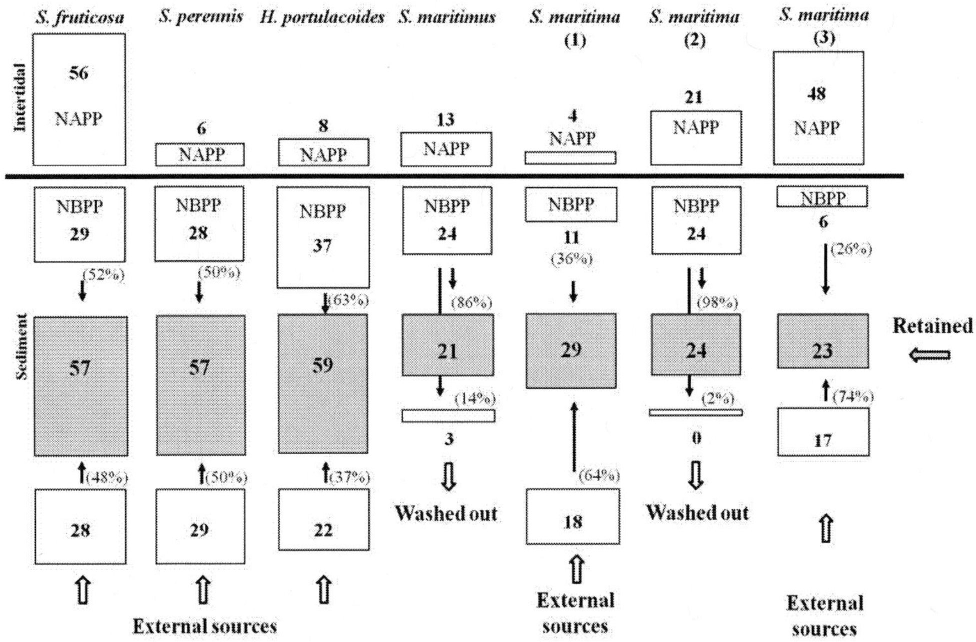

Figure 2. Nitrogen cycling in salt marshes colonized by five different salt marsh species. NAPP - net aboveground primary production/standing stock in the aboveground, NBPP - net belowground primary production/standing stock in the belowground and the nitrogen sequestration in the sediment (due to the belowground plant production and some external sources) are shown in the scheme. The numbers in bold refer to g DW N m^{-2} y^{-1} (data from Sousa et al., 2008, 2010).

Denitrification in Salt Marshes

Denitrification, the stepwise reduction of nitrate (NO_3^-) and nitrite (NO_2^-) to the gaseous nitric oxide (NO), nitrous oxide (N_2O) and, mainly, dinitrogen (N_2), is a process mediated by anaerobic bacteria. Under suboxic conditions, these bacteria can use NO_3 (or NO_2) as a terminal electron acceptor in respiration. Denitrification process occurs in the sediment suboxic surface and includes denitrification of the NO_x (NO_3+NO_2) of bottom water and of

NO_x produced via sedimentary nitrification (this step is called coupled nitrification-denitrification).

Nitrification (the microbial aerobic oxidation of NH_4^+ and NO_2^- to NO_3^-) is an important step of the nitrogen cycle that occurs in the oxic surface sediment. The product of this process (NO_3^-) may be later on denitrified (coupled nitrification-denitrification). The potential nitrification is affected by environmental factors like temperature, oxygen and NH_4^+ availability, but also by the microbiological community, i.e, number and activity of nitrifying bacteria (Henrikson et al., 1981), and it is an essential step for the coupled nitrification-denitrification (Koop-Jacobsen & Giblin, 2009).

Salt marshes have been recognized as important players on the nitrogen cycle by enhancing the removal of the excess of reactive nitrogen through denitrification (e.g. Teal & Howes, 2000; Valiela & Cole, 2002; Galloway et al., 2008). In addition, by counteracting eutrophication in coastal areas (Seitzinger, 1988), this process constitutes an ecosystem service provided by salt marshes (Constanza, 1997). Moreover, denitrification is considered the only significant biochemical pathway, mediated by denitrifying bacteria, that regenerates N_2 (Schlesinger, 1997; Jaffe, 2000). This process enables the reduction of the loading of reactive nitrogen to the open ocean, up to 40-50% of the inorganic nitrogen (Seitzinger, 1988), meaning that rivers become small sources of reactive nitrogen to the open ocean (Galloway et al., 2008).

Several studies have been carried out in order to quantify denitrification in freshwater wetlands (e.g. Merrill & Cornwell, 2000; Risgaard-Petersen, 2003; Trimmer et al., 2003; Sundbäck et al., 2006) but much less studies have addressed salt marshes (e.g. Valiela & Teal 1979; Koch et al., 1992; White & Howes, 1994; Erickson et al., 2003; Poulin et al., 2007). Table 1 summarizes denitrification values in salt marshes, using several quantification methods. Within these studies, different techniques, namely the ^{15}N-isotope pairing (Nielsen, 1992), acetylene inhibition (Sørensen, 1978) and N_2/Ar ratio change (via Membrane Inlet Mass Spectrometry) (Kana et al., 1994) have been applied, which may restrict the comparison between systems. Moreover, from table 1 one can see that denitrification rates can be highly variable. This variability results from the natural variability within the system, given the number of abiotic and biotic variables (from microorganisms to macrofauna, and from microphytobenthos to macrophytes) that may have a cross effect in these biogeochemical processes (e.g. Lillebø et al., 1999; Gilbert et al., 2003; Erickson et al., 2003; Risgaard-Petersen, 2003; Hou et al., 2007). Nitrate and oxygen availability, easily degradable organic carbon, temperature, light, and water retention time, are some of the abiotic factors influencing denitrification rates in aquatic ecosystems (Seitzinger, 1988; Cornwell et al., 1999; Piña-Ochoa & Álvares-Cobelas, 2006; Silvennoinen et al., 2008). Plants can influence denitrification rates (e.g. Erickson et al., 2003; Koop-Jacobsen & Giblin, 2009) due to O_2 diffusion through the aerenchyma and creation of oxic micro-zones surrounding the roots and rhizomes at a certain depth in the sediment, which enhances coupled nitrification-denitrification (Piña-Ochoa & Álvares-Cobelas, 2006; Koop-Jacobsen & Giblin, 2009). Moreover, as most of the processes occurring in salt marshes seem to be species specific (e.g. biomass production and nitrogen sequestration capacity), more comparable results are needed to evaluate the capacity of salt marshes auto-remediation through denitrification, regarding the excess of land-derived nitrogen loads.

Table 1. Seasonal or annual denitrification in salt marshes: vegetated sediment ability to remove N in salt marshes through denitrification. These studies were performed using different techniques (single values correspond to the average, while ranges correspond to minimum and maximum values recorded)

<table>
<tr><th rowspan="2">Salt marsh location</th><th colspan="4">Denitrification (µmol N.m⁻².h⁻¹)</th><th rowspan="2">Plant species</th><th rowspan="2">Technique</th><th rowspan="2">Reference</th></tr>
<tr><th>Autumn</th><th>Winter</th><th>Spring</th><th>Summer</th></tr>
<tr><td>Atlantic coast of USA</td><td colspan="4"></td><td>Spartina alterniflora</td><td>^{15}N</td><td>Valiela & Teal, 1979</td></tr>
<tr><td>Colne Point, England</td><td colspan="4">8.3</td><td>Puccinelia sp. Halimione sp.</td><td>$^{15}N_2$ generated from $^{15}NO_3$</td><td>Abd Aziz & Nedwell, 1986[a]</td></tr>
<tr><td>Great Sippewissett marsh, MS, USA</td><td colspan="4">75</td><td>Spartina alterniflora</td><td>$^{15}NH_4$ retention mass balance</td><td>White & Howes, 1994</td></tr>
<tr><td>Great Sippewissett Marsh, MA, USA</td><td colspan="4">17 - 496</td><td>Spartina alterniflora</td><td>$^{15}NH_4$ retention mass balance</td><td>Hamersley & Howes, 2005</td></tr>
<tr><td rowspan="4">Venice lagoon, Italy</td><td>125-250</td><td>-</td><td>21 -125</td><td>7-21</td><td>Limonium serotinum</td><td rowspan="4">^{15}N-IPT</td><td rowspan="4">Erickson et al., 2003</td></tr>
<tr><td>-</td><td>-</td><td>25</td><td>-</td><td>Limonium serotinum</td></tr>
<tr><td>-</td><td>-</td><td>32</td><td>-</td><td>Juncus maritimus</td></tr>
<tr><td>-</td><td>-</td><td>28</td><td>-</td><td>Halimione portulacoides</td></tr>
<tr><td>Pointe-au-Père, Canada</td><td>11 - 25</td><td>6 - 15</td><td>-</td><td>18 - 42</td><td>Spartina alterniflora</td><td>^{15}N-IPT</td><td>Poulin et al., 2007</td></tr>
<tr><td>Louisiana, USA</td><td colspan="4">4.2</td><td>Spartina alterniflora</td><td>Acetylene inhibition</td><td>Smith et al. 1983; De Laune et al., 1989[a]</td></tr>
<tr><td>Torridge river estuary, England</td><td colspan="4">62</td><td>Halimione portulacoides</td><td>Acetylene inhibition</td><td>Koch et al., 1992</td></tr>
<tr><td>Newport river estuary, NC, USA</td><td colspan="4">8</td><td>Spartina alterniflora</td><td>Acetylene inhibition</td><td>Thompson et al., 1995</td></tr>
<tr><td>Georgia, USA</td><td colspan="4">98</td><td>Spartina alterniflora</td><td>N_2 sediment profiles</td><td>Haines et al., 1977[a]</td></tr>
<tr><td>Great Sippewissett marsh, MS, USA</td><td colspan="4">36 – 92</td><td>Spartina alterniflora</td><td>Bell-jarr/ N_2 production</td><td>Kaplan et al., 1979[a]</td></tr>
</table>

[a] in White & Howes, 1994.

THE CASE STUDY OF *SPARTINA MARITIMA* SALT MARSH

Spartina maritima is distributed along the coasts of western, southern, and southeastern Europe, and western Africa. It is an herbaceous perennial plant, with a C_4 type of photosynthetic mechanism (Nieva et al., 1999; Gray & Mogg, 2001) and colonizes low marshes. However, in some marshes, and through an interspecific hybridization process with the introduced *Spartina alterniflora,* has resulted the fertile allopolyploid *Spartina anglica* (Ayres & Strong, 2001), altering European marshes species composition. In other systems, this species has been replaced by the invasive *Spartina densiflora* Brongn. (Poaceae) (e.g. Cambrollé et al., 2008). Moreover, as many marshes processes seem to be species-specific (e.g. Lillebø et al., 2006; Sousa et al., 2010) it is important to address the services provided by this European native species. Thus, in the present chapter/section we focus on the Mondego estuary *Spartina maritima* salt marsh meaning on nitrogen remediation (named as *Spartina* (3) in Figure 2).

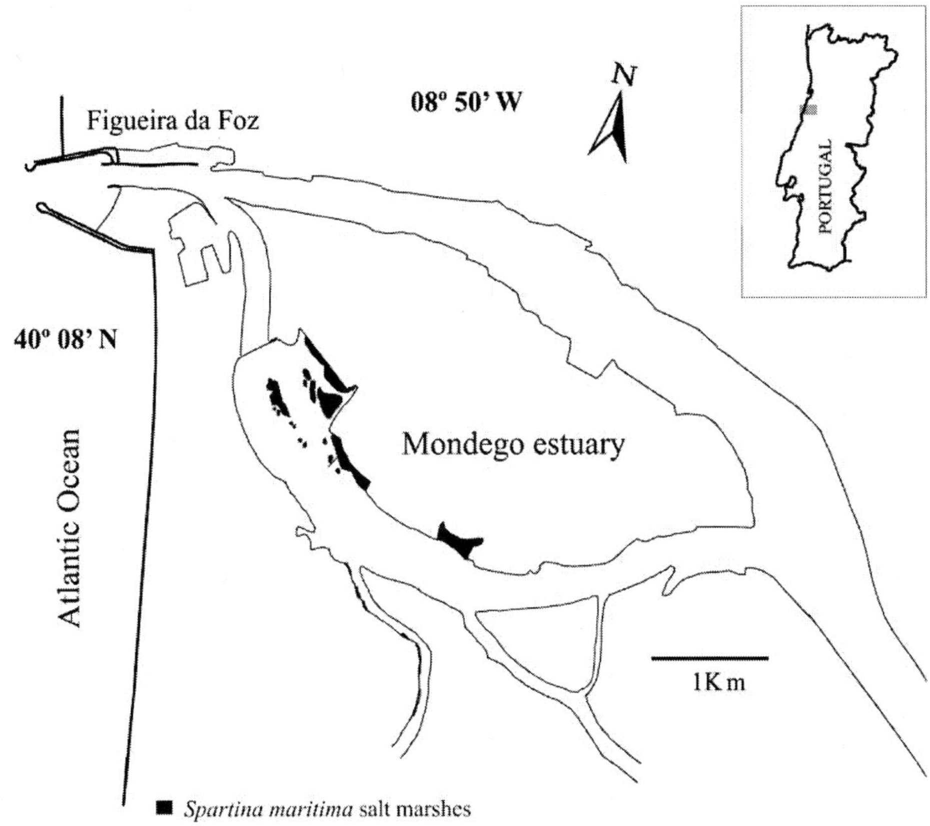

Figure 3. Mondego estuary with the location of *Spartina maritima* salt marshes.

The Mondego estuary (40°08N, 08°50W) is located in the southern European Atlantic margin (Portugal). *S. maritima* was sampled from homogeneous stands every two months for one year, and the rhizosediment characterized for organic matter content as % LOI (loss on ignition) and granulometry (see Sousa et al., 2008, 2010). Aboveground and belowground net biomass productions (respectively, NAPP and NBPP) were calculated following De la Cruz & Hackney (1977). The methods applied for the calculation of the nitrogen pools, nitrogen standing stocks and nitrogen driven from external sources are described in Sousa et al. (2008, 2010).

For denitrification quantification, five sediment cores were collected seasonally (each one containing one or two shoots – plant biomass was as similar as possible). According to a previous study (Sousa et al., unpublished data), maximum denitrification in a *S. maritima* marsh located two hundred kilometres south (Tagus estuary, 38°49N, 08°56'W) occurs in winter. This corresponds to the period when potential nitrification and nitrate concentration are higher and the competition between primary producers and nitrifiers is lower (Sousa et al., unpublished data). Thus, only this season will be considered. Denitrification measurements were performed according to the isotope pairing technique described in Nielsen (1992). The method assumptions (*^{15}N- IPT assumptions*) (Nielsen, 1992) were verified, and the incubation procedures for O_2 consumption and denitrification rates' quantification followed the description by Cabrita & Brotas (2000) and Dalsgaard et al. (2000).

Sediment characteristics (% LOI and granulometry), annual biomass production and turnover rate for aboveground and belowground of *Spartina maritima* are summarized in table 2.

Table 2. Sediment organic matter content (% LOI, loss on ignition) and granulometry in *Spartina* rooted sediment, and annual biomass production ($gN.m^{-2}.y^{-1}$) and turnover rate for aboveground and belowground biomass (data from Sousa et al., 2008, 2010)

Sediment characteristics		
Organic Matter (% LOI)	Jun	7.3
	Aug	10.4
	Oct	8.7
	Dec	8.1
	Feb	6.0
	Apr	6.5
	Annual mean	7.8±1.6
Granulometry	Silt	6%
	Clay	
	Fine sand 63µm-125µm	56%
Spartina maritima annual biomass production and turnover rate		
Aboveground (NAPP)	2833 (76%)	
Belowground (NBPP)	895 (24%)	
Total	3728	
Aboveground	0.50	
Belowground	0.51	

Bold numbers within brackets represent the percentage of the above- or belowground biomass productions, in relation to the total biomass production. Turnover rate is shown in terms of biomass

The sediment is mainly constituted by fine sand (56 % of grain size between 63 µm-125 µm) and the annual mean of % LOI was 7.8 ± 1.6. In this *Spartina maritima* marsh the NAPP was comparatively higher than NBPP, representing 76 % of NPP. However, the above- and belowground biomass turnover rates were similar, respectively 0.50 and 0.51 (Table 2). Figure 4 shows a schematic representation of *Spartina maritima* primary production, and organic nitrogen burial and daily winter denitrification within the marsh. Denitrification was estimated as 60 ± 18 μmol N m^{-2} h^{-1} during daylight and 120±17 μmol N m^{-2} h^{-1} during night. Considering, for the winter period, that days only have 10 hours of light conditions, daily denitrification will be approximately/about 2280 μmol N m^{-2} day^{-1}.

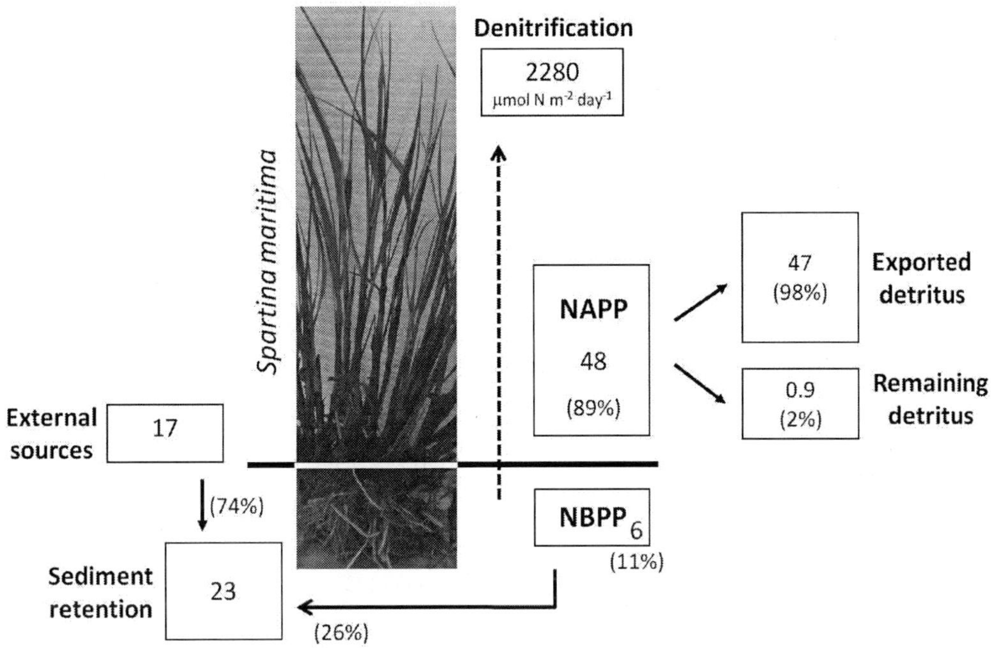

Figure 4. Schematic representation of *Spartina maritima* primary production, organic nitrogen burial (data from Sousa et al., 2010) and denitrification within the marsh in a winter day.

Analysing the nitrogen cycling in *S. maritima* salt marshes it can be seen that it may follow preferentially different pathways (e.g. primary production, organic N burial, Figure 2). The three analyzed *S. maritima* salt marshes (Figure 2) show a similar capacity for nitrogen retention in the sediment. However, in (1) and (3) plants show a buffer capacity for nitrogen from external sources, whilst in (2) 2% of the nitrogen in the sediment is washed out. Nevertheless, in this marsh the NBPP is comparatively higher. Even though we are comparing N cycling of one plant species and its rhizosediment, this cycling is also determined by abiotic factors such as physico-chemical parameters (e.g. oxygen availability, salinity, tidal inundation, nutrient availability) that strongly depend on the salt marsh topography and hydrology (Sousa et al., 2008), rather than on the plant species.

The meaning of *S. maritima* salt marshes on nitrogen remediation, by intercepting the land-derived nitrogen, is reflected on *S. maritima* primary production, organic N burial and denitrification process (Figure 4). Comparing denitrification in *S. maritima* Mondego salt marsh, with the rates on table 1 (considering only the ones determined with ^{15}N addition), one can see that denitrification rates in *Spartina maritima* were comparatively higher than in *Spartina alterniflora* (Valiela & Teal 1979; White & Howes 1994; Poulin et al., 2007) and within the range of *Limonium serotinum* seasonal variation (Erickson et al., 2003). Denitrification in *S. maritima* was also higher than in *Juncus maritimus,* in *Halimione portulacoides* (Erickson et al., 2003) or in *Puccinelia/Halimione* (Aziz & Nedwell 1986, in

White & Howes, 1994). Two main reasons may explain these differences: i) species specific interactions with other biota (e.g. competition with nitrifiers for nitrate); ii) geographical and seasonal environmental characteristics (i.e. abiotic factors).

This study shows that through nitrogen incorporation in biomass, organic nitrogen burial and denitrification *Spartina maritima* contributes to nitrogen remediation, promoting the ecosystem health. The comparatively higher denitrification rates (at least in winter) and the nitrogen sequestration capacity show that in the Mondego estuary, this European native species offers a valuable ecological service. These results also emphasize the negative consequences of an eventual disappearance of *Spartina maritima* salt marshes.

CONCLUSION

The relationship between marsh functions and environmental factors is complex Boorman (2003). However, the described ecosystem processes represent salt marsh services, which should gain greater importance considering main anthropogenic threats to these coastal areas, i.e. the reduction or fragmentation of salt marsh areas (e.g. Best et al., 2007) and the increasing of the land-derived nitrogen loading (e.g. Gruber & Galloway, 2008; Philipott et al., 2009). The present chapter highlights that biomass production and nitrogen sequestration by salt marsh plants is a species-specific process, meaning that, although salt marsh species potentially share the same ecosystem services, their contribution for bioremediation seems to be species specific. Although within the same system there may be some spatial variation, the native European salt marsh species *Spartina maritima* contributes to nitrogen remediation by intercepting the land-derived nitrogen and buffering the loading of reactive nitrogen to the open ocean through nitrogen incorporation in biomass, organic nitrogen burial and denitrification.

ACKNOWLEDGMENTS

Authors acknowledge the funding from the Portuguese FCT (Foundation for Science and Technology) to CESAM, to the project PTDC/MAR/67752/2006; FCOMP-01-0124-FEDER-007378, and the grant given to A.I. Sousa (Ref. SFRH/BD/23634/2005).

REFERENCES

Amos, C., Cappucci, S., Bergamasco, A., Umgiesser G., Bonardi, M., Cloutier, D., Flindt, M. R., De Nat, L. & Cristante, S. (2004). *The stability of tidal flats in Venice Lagoon – the results of in situ measurements using two benthic annular clumes*. Journal of Marine Systems, 51(1-4), 211-241.

Best, M., Massey, A. & Prior A. (2007). *Developing a saltmarsh classification tool for the European water framework directive*. Marine Pollution Bulletin, 55, 205–214.

Boorman, L.A. (2003). *Saltmarsh Review. An overview of coastal saltmarshes, their dynamic and sensitivity characteristics for conservation and management*. JNCC, Peterborough.

On-line version at http://www.jncc.gov.uk/pdf/jncc334.pdf assessed at 14th October 2009.

Boyd, J. & Banzhaf, S. (2007). *What are ecosystem services?* Ecological Economics, 63(2–3), 616–626.

Cabrita, M. T. & Brotas, V. (2000). *Seasonal variation in denitrification and dissolved nitrogen fluxes in intertidal sediments of the Tagus estuary.* Marine Ecology Progress Series, 202, 51–65.

Caçador I., Costa, A. L. & Vale, C. (2007). *Nitrogen sequestration capacity of two salt marshes from the Tagus estuary.* Hydrobiologia, 587, 137-145.

Cambrollé J., Redondo-Gómez S., Mateos-Naranjo E. & Figueroa, M. E. (2008). *Comparison of the role of two Spartina species in terms of phytostabilization and bioaccumulation of metals in the estuarine sediment.* Marine Pollution Bulletin, 56, 2037-2042.

Cornwell, J. C., Kemp, W. M. & Kana, T. M. (1999). *Denitrification in coastal ecosystems: methods, environmental controls, and ecosystem level controls, a review.* Aquatic Biology, 33, 41–54.

Costanza, R., d'Arge, R., de Groot, R., Farber, S., Grasso, M., Hannon, B., Limburg K., Naeem, S., O' Neill, R. V., Paruelo, J., Raskin, R. G., Sutton P. & van den Belt, M. (1997). *The value of the world's ecosystem services and natural capital.* Nature, 387, 353-360.

Dalsgaard, T., Nielsen, L. P., Brotas, V., Viaroli, P., Underwood, G., Nedwell, D.B., Sundbäck, K., Rysgaard, S., Miles, A., Bartoli, M., Dong, L., Thornton, D. C. O., Ottosen, L. D. M., Castaldelli, G. & Risgaard-Petersen, N. (2000) *Protocol handbook for NICE - Nitrogen Cycling in Estuaries: A project under the EU research programme: Marine Science and Technology (MAST III).* National Environmental Research Institute, Silkeborg, Denmark.

de Bie, M. J. M., Middelburg, J. J., Starink, M. & Laanbroek, H. (2002) *Factors controlling nitrous oxide at the microbial community and estuarine scale.* Marine Ecology Progress Series, 240, 1–9.

Dickinson, R. E. & Cicerone, R.J. (1986). *Future Global Warming from Atmospheric Trace Gases. Nature,* 319, 109-15.

Eriksson, P. G., Svensson, J. M. & Carrer, G.M. (2003). *Temporal changes and spatial variation of soil oxygen consumption, nitrification and denitrification rates in a tidal salt marsh of the Lagoon of Venice, Italy.* Estuarine, Coastal and Shelf Science, 58, 861-871.

Eyre, B. D. & Ferguson, A. J. P. (2002). *Comparison of carbon production and decomposition, benthic nutrient fluxes and denitrification in seagrass, phytoplankton, benthic microalgae- and macroalgaedominated warm-temperate Australian lagoons.* Marine Ecology Progress Series, 229, 43-59.

Galloway, J. N. (1998) *The global nitrogen cycle: changes and consequences.* Environmental Pollution, 102, 15-24.

Galloway, J. N., Dentener, F. J., Capone, D. G., Boyer, E. W., Howarth R. W., Seitzinger, S. P. & Asner, S. P. (2004). *Nitrogen cycles: past, present, future.* Biogeochemistry, 70, 153–226.

Galloway, J. N., Townsend A. R., Erisman, J. W., Bekunda, M., Cai, Z., Freney, J. R., Martinelli, L. A., Seitzinger, S. P., Sutton, M. A. (2008). *Transformation of the nitrogen cycle: Recent trends, questions, and potential solutions.* Science, 320, 889–892.

Gilbert, F., Aller, R. C. & Hulth, S. (2003). *The influence of macrofaunal burrow spacing and diffusive scaling on sedimentary nitrification and denitrification: an experimental simulation and model approach.* Journal Marine Research, 61, 101-125.

Gray, A. J. & Mogg, R. J. (2001). *Climate impacts on pioneer salt marsh plants.* Climate Research, 18, 105 –112.

Green, B. C., Smith, D. J., Earley, S. E., Hepburn, L. J. & Underwood, G. J. C. (2009). *Seasonal changes in community composition and trophic structure of fish populations of five salt marshes along the Essex coastline, United Kingdom.* Estuarine, Coastal and Shelf Science, 85, 247–256.

Gruber, N. & Galloway, J. N. (2008). *An Earth-system perspective of the global nitrogen cycle.* Nature, 451, 293-296.

Hamersley, M. R. & Howes, B. L. (2005). *Coupled nitrification–denitrification measured in situ in a Spartina alterniflora marsh with a 15NH4+ tracer.* Marine Ecology Progress Series, 299, 123-135.

Hauxwell, J. & Valiela, I. (2004). *Effects of nutrient loading on shallow seagrass-dominated coastal systems: patterns and processes.* In S. Nielsen, G. Banta, M. Pedersen (Eds.), Estuarine nutrient cycling: The influence of primary producers (pp. 59-92). The Netherlands: Kluwer Academic Publishers.

Henriksen, K., Hansen, J.I. & Blackburn, T.H. (1981). *Rates of nitrification, distribution of nitrifying bacteria, and nitrate fluxes in different types of sediment from Danish waters.* Marine Biology, 61, 299-304.

Hou, L. J., Liu, M., Xu, S. Y., Ou, D. N., Yu, J., Cheng, S. B., Lin, X. & Yang, Y, (2007). *The effects of semi-lunar spring and neap tidal change on nitrification, denitrification and N2O vertical distribution in the intertidal sediments of the Yangtze estuary*, China. Estuarine, Coastal and Shelf Science, 73, 607-616.

Howarth, R., 2008. *Coastal nitrogen pollution: A review of sources and trends globally and regionally.* Harmfull Algae 8, 14-20.

Ibañez, C., Curcó , A., Day Jr., J. W. & Prat, N. (2000). *Structure and productivity of microtidal mediterranean coastal marshes.* In M. P. Weinstein & D. A. Kreeger (Eds.), Concepts and Controversies in Tidal Marsh Ecology (pp.107-136). Dordrecht, The Netherlands: Kluwer Academic Publishing.

Jaffe, D. A. (2000) The nitrogen cycle. In M. C. Jacobson, R. J. Charlson, H. Rodhe, & G. H. Orians (Eds.), *Earth system science (pp. 322–342).* San Diego, California, USA: Academic Press.

Jin, B., Fu, C., Zhong, J., Li, B., Chen, J. & Wu, J. (2007). *Fish utilization of a salt marsh intertidal creek in the Yangtze River estuary, China.* Estuarine, Coastal and Shelf Science, 73, 844-852.

Kana, T. M., Darkangelo, C. Hunt, M. D., Oldham, J. B., Bennett, G. E. & Cornwell, J. C., (1994). *A membrane inlet mass spectrometer for rapid high precision determination of N2, 02, and Ar in environmental water samples.* Analytical Chemistry, 66, 4166-4170.

Koch, M. S., Malby, E., Oliver, G. A., Bakker, S. A. (1992). *Factor controlling denitrification rates of tidal mudflats and fringing salt marshes in south-west England.* Estuarine, Coastal and Shelf Science, 34, 471–485.

Koop-Jakobsen, K. & Giblin, A. E. (2009). *Anammox in Tidal Marsh Sediments: The Role of Salinity, Nitrogen Loading, and Marsh Vegetation.* Estuaries and Coasts 32, 238–245.

Latza, U., Gerdes S., Baur X. (2009). *Effects of nitrogen dioxide on human health: Systematic review of experimental and epidemiological studies conducted between 2002 and 2006.* International Journal of Hygiene and Environmental Health, 212, 271–287.

Lefeuvre, J-C., Laffaille, P., Feunteun, E., Bouchard, V. & Radureau, A. (2003). *Biodiversity in salt marshes: from patrimonial value to ecosystem functioning.* The case study of the Mont-Saint-Michel bay. Comptes Rendus Biologies 326, s125-s131.

Lillebø, A. I., Flindt, M. R., Pardal, M.A. & Marques, J. C. (1999). *Population structure, dynamics and production of Hydrobia ulvae (Pennant) (Mollusca: Prosobranchia) along an eutrophication gradient in the Mondego estuary (Portugal).* Acta Oecologica, 20(4), 289-304.

Lillebø, A. I., Flindt, M. R., Pardal, M.A. & Marques, J. C. (2006). *The effect of Zostera noltii, Spartina maritima and Scirpus maritimus on sediment pore-water profiles, in a temperate intertidal estuary.* Hydrobiologia, 555, 175-183.

Lillebø, A. I., Sousa, A. I., Flindt, M. R., Pereira, E., Duarte, A. C., Pardal, M. A. & Caçador, I. (in press). *Nutrient cycling in salt marshes: an ecosystem service to reduce eutrophication.* In C. D. Webber (Ed.), Eutrophication: ecological effects, sources, prevention and reversal. New York, USA: Nova Science Publishers, Inc. ISBN 978-1-61728-911-8

Lillebø, A. I., Teixeira, H., Pardal, M. A. & Marques, J. C. (2007). *Applying water quality status criteria to a temperate estuary (Mondego, Portugal) before and after the mitigation measures to reduce eutrophication symptoms.* Estuarine, Coastal and Shelf Science, 72, 177-187.

Maricle, B. R., Lee, R. W. (2002). *Aerenchyma development and oxygen transport in the estuarine cordgrasses Spartina alterniflora and S. anglica.* Aquatic Botany, 74,109–120.

Matamala, R. & Drake, B.G. (1999). *The influence of atmospheric CO2 enrichment on plant-soil nitrogen interactions in a wetland plant community on the Chesapeake Bay.* Plant and Soil, 210, 93-101.

McGlathery, K.J., Sundbäck, K., Anderson, I.C., 2004. *The importance of primary producers for benthic nitrogen and phosphorus cycling.* In S. Nielsen, G. Banta, M. Pedersen (Eds.), Estuarine nutrient cycling: The influence of primary producers (pp. 1-15). The Netherlands: Kluwer Academic Publishers.

McLuscky, D.S. & Elliot, M. (2004). *The Estuarine Ecosystem.* Ecology, Threats, and Management (third edition). UK: Oxford University Press, UK, 214 pp.

Merrill, J. Z. & Cornwell, J. C. (2000). *The role of oligohaline marshes in estuarine nutrient cycling.* In M. P. Weinstein & D. A. Kreeger (Eds.), Concepts and Controversies in Tidal Marsh Ecology (pp.425-444). Dordrecht, The Netherlands: Kluwer Academic Publishing.

National Research Council (NRC) (2000). *Clean Coastal Waters: Understanding and Reducing the Effects of Nutrient Pollution (pp. 165–176).* Washington, DC, USA.

Nedwell, D. B., Jickells, T. D., Timmer, M. & Sanders, R. (1999). *Nutrients in estuaries. In D. B. Nedwell & D. G. Raffaelli (Eds), Estuaries.* Advances in Ecological Research 29 (pp 43-92).

Nielsen, L. P. (1992) *Denitrification in sediment determined from nitrogen isotope pairing.* FEMS Microbiology Ecology, 86, 357-362.

Nielsen, M., Gieseke, A., de Beer, D. & Revsbech, N. P., (2009). *Nitrate, nitrite, and nitrous oxide transformations in sediments along a salinity gradient in the Weser Estuary.* Aquatic Microbial Ecology, 55, 39-52.

Nieva, F. J. J., Castellanos, E. M., Figueroa, M. E. & Gil F. (1999). *Gas exchange and chlorophyll fluorescence of C3 and C4 salt marsh species*. Photosynthetica, 36(3), 397-406.

Nixon, S. W. (1981) *Remineralization and nutrient cycling in coastal marine ecosystems*. In B. J. Neilson & L. E. Cronin (Eds), Estuaries and nutrients (pp. 111-138). Clifton. NJ: Humana Press.

Pedersen, M.F., Nielsen, S.L. & Banta, G.T., 2004. *Interactions between vegetation and nutrient dynamics in coastal marine ecosystems: an introduction*. In S. Nielsen, G. Banta, M. Pedersen (Eds.), Estuarine nutrient cycling: The influence of primary producers (pp. 1-15). The Netherlands: Kluwer Academic Publishers.

Philippot, L., Hallin, S., Börjesson, G. & Baggset, E. M. (2009). *Biochemical cycling in the rhizosphere having an impact on global change*. Plant and Soil, 321(1-2), 61-81.

Piña-Ochoa, E. & Álvarez-Cobelas, M. (2006) *Denitrification in aquatic environments: a cross-system analysis*. Biogeochemistry, 81, 111-130.

Poulin, P., Pelletier, E. & Saint-Louis, R. (2007) *Seasonal variability of denitrification efficiency in northern salt marshes: An example from the St. Lawrence Estuary*. Marine Environmental Research, 63, 490–505.

Risgaard-Petersen, N. (2003). *Coupled nitrification-denitrification in autotrophic and heterotrophic estuarine sediment: on the influence of benthic microalgae*. Limnology and Oceanography, 48, 93–105.

Schlesinger, W.H. (1997) *Biogeochemistry*, 2nd ed. San Diego, California, USA: Academic Press

Schlesinger, W.H. (2009). *On the fate of anthropogenic nitrogen*. Proc. Natl. Acad. Sci. USA, 2009. 106: 203-208. 30th August 2010. www.pnas.org/cgi/content/full/0810193105/DCSupplemental.

Seitzinger, S. P. (1988) *Denitrification in fresh and coastal marine ecosystems: ecological and geochemical significance*. Limnology and Oceanography, 33, 702–724.

Seitzinger, S. P., Nixon, S. W. & Pilson, M. E., 1984. *Denitrification and nitrous oxide production in coastal marine ecosystems*. Limnology and Oceanography, 29, 73-83.

Silvennoinen, H., Liikanen, A., Torssonen, J., Stange, C. F. & Martikainen, P. J. (2008). *Denitrification and N2O effluxes in the Bothnian Bay (northern Baltic Sea) river sediments as affected by temperature under different oxygen concentrations*. Biogeochemistry, 88, 63-72.

Sørensen, J. (1978). *Denitrification rates in a marine sediment as measured by the acetylene inhibition technique*. Applied Environmental Microbiology, 36, 139–143.

Sørensen, J., Tiedje, J. M. & Firestone, R. B. (1980). *Inhibition by sulfide of nitric and nitrous oxide reduction by denitrifying Pseudomonas fluorescens*. Applied Environmental Microbiology, 39, 105–108.

Sousa, A. I., Lillebø, A. I., Caçador, I. & Pardal, M. (2008). *Contribution of Spartina maritima to the reduction of eutrophication in estuarine systems*. Environmental Pollution, 156, 628–635.

Sousa, A. I., Lillebø, A. I., Pardal, M. & Caçador, I. (2008). *Productivity and nutrient cycling in salt marshes: Contribution to ecosystem health*. Estuarine, Coastal and Shelf Science, 87, 640-646.

Stief, P., De Beer, D. & Neumann, D. (2002). *Small-scale distribution of interstitial nitrite in freshwater sediment microcosms: the role of nitrate and oxygen availability, and sediment permeability.* Microbial Ecology, 43, 367–378.

Sundbäck, K., Miles, A. & Linares, F. (2006). *Nitrogen dynamics in nontidal littoral sediments: Role of microphytobenthos and denitrification.* Estuaries and Coasts, 29(6), 1196-1211.

Teal, J. M. & Howes, B. L. (2000). *Salt marsh values: retrospection from the end of the century.* In M. P. Weinstein & D. A. Kreeger (Eds.), Concepts and Controversies in Tidal Marsh Ecology (pp. 9-22). Dordrecht, The Netherlands: Kluwer Academic Publishing.

Tolstikhin, I. N. & Marty, B. (1998). *The evolution of terrestrial volatiles: a view from helium, neon, argon and nitrogen isotope modelling.* Chemical Geology, 147, 27-52.

Trimmer, M., Gowen, R. J. & Stewart, B. M. (2003). *Changes in sediment processes across the western Irish Sea front.* Estuarine, Coastal and Shelf Science, 56, 1011-1019.

Valiela, I. & Cole, M.L. (2002). *Comparative evidence that salt marshes and mangroves may protect seagrass meadows from land-derived nitrogen loads.* Ecosystems, 5, 92–102.

Valiela, I. & Teal, J. M. (1979) *The nitrogen budget of a salt marsh ecosystem.* Nature, 280, 652-656.

Valiela, I., Cole, M. L., Mcclelland, J., Hauxwell, J., Cebrian, J. & Joye, S. B., (2000). *Role of salt marshes as part of coastal landscapes.* In M. P. Weinstein & D. A. Kreeger (Eds.), Concepts and Controversies in Tidal Marsh Ecology (pp. 23-38). Dordrecht, The Netherlands: Kluwer Academic Publishing.

White, D.S. & Howes, B. L. (1994). *Long-term 15N-nitrogen retention in the vegetated sediments of a New England salt marsh.* Limnology and Oceanography, 39(8), 1878-1892.

Wieski, K., Guo, H., Craft, C. B. & Pennings S.C. (2010). *Ecosystem Functions of Tidal Fresh, Brackish, and Salt Marshes on the Georgia Coast.* Estuaries and Coasts, 33, 161-169.

In: Bioremediation
Editor: Alexander C. Mason

ISBN: 978-1-61122-730-7

Chapter 15

BIOREMEDIATION OF CHROMIUM CONTAMINATED SOILS BY ACTINOMYCETES: MECHANISMS INVOLVED

Marta A. Polti[1,2]***, María J. Amoroso***[1,3] ***and Carlos M. Abate***[1,2,3]

[1]Planta Piloto de Procesos Industriales y Microbiológicos (PROIMI-CONICET), Argentina

[2]Facultad de Ciencias Naturales e IML. UNT, Argentina

[3]Facultad de Bioquímica, Química y Farmacia. UNT. Avenida Belgrano y Pasaje Caseros, T4001MVB, Tucumán, Argentina

In natural water and subsurface soils chromium occurs in two major oxidation states: III and VI. Cr(VI) is approximately 1,000-fold more cytotoxic and mutagenic than Cr(III). As the application of Cr is extensive in several industries, chromium-associated pollution is an increasing problem. Development of suitable methods for cleaning up such contaminated soils is an important topic of environmental recovery and protection, particularly regarding methods for remediation of Cr(VI)-contaminated soils. At present there exist physicochemical technologies to remove Cr(VI) from industrial waste, but these methods are costly and require much energy and specific equipment. Therefore the employment of lower-cost biotechnology could be an important alternative.

The three processes, by which the microorganisms interact with Cr(VI), are biosorption, bioaccumulation and enzymatic. Biosorption is a metabolism-independent process and thus can be performed by both living and dead microorganisms. Bioaccumulation involves metabolism dependent active uptake of metals. Biological transformation of Cr(VI) to Cr(III) by enzymatic reduction is a means of chromium decontamination. This biological reduction may provide a less costly and environmentally friendly approach to remediation. Actinomycetes are the dominant population in soil. Metal-resistant actinobacteria, and their potential use for bioremediation strategies, have been described.

Recent progress has been made studying metal resistance in streptomycetes isolated from polluted areas. The general strategy for control of Cr (and other heavy metal) pollution has relied upon dissimilatory metal reduction, that is the uptake of toxic and permeable Cr(VI) molecules by microorganisms and plants and their subsequent bioremediation and conversion into less toxic trivalent. In particular, the mechanism of Cr(VI) resistance observed in *Streptomyces* sp. MC1, a strain able to remove bioavalible Cr(VI) from soil, includes adsorption coupled with reduction to Cr(III), and finally, Cr(III) bioaccumulation. This mechanism have special relevance to remediation of Cr(VI) contaminated environments.

INTRODUCTION

Soils are the target of thousands of contaminants that vary in composition and in concentration. These contaminants enter the system as a result of a wide range of actions such as intentional applications, inadequate residue disposal, accidental wastes and inappropriate use [1].

Soil may become polluted with high concentrations of metals by either a natural phenomenon such as proximity to an ore body, or, principally, as a consequence of industrial activities that eliminate residues in the soil that in long terms, promote their accumulation [2,3].

Contaminants from diffuse sources are commonly heavy metals [4]. After being deposited on the soil surface, soil use becomes an important factor that determines the vertical and/or horizontal distribution of pollutants. In agricultural soils, the dilution with uncontaminated subsoil by plowing decreases the concentration of contaminants. As a consequence of diffuse contamination, soils may lose fundamental functions with a decrease in their overall environmental quality [5]. The Thematic Strategy for Soil Protection of the European Commission (2006) recognizes diffuse pollution as a threat to the soil quality [2].

Therefore, anthropogenic inputs associated with agricultural practices, mineral exploration, industrial processes and solid waste management are important contributors to heavy metal contamination of natural ecosystems [6-8].

CHROMIUM

Chromium is the 24th element on the periodic chart. It is situated between vanadium and manganese and has an average atomic weith of 52. Chromium is a naturally occurring element and is the 21st most abundant element in the Earth's crust [9]. Chromium (Cr) is found in rocks, soil, plants, animals, volcanic dust and gases. Cr exist in oxidation states ranging from -2 to +6. Hexavalent chromium, Cr(VI), and trivalent chromium, Cr(III), are ecologically important because they are the most stable oxidation states in natural environments [10,11].

Cr(III) is an essential micronutrient for proper glucose metabolism, because it stimulates the enzyme system and stabilizes nucleic acids [12].

Cr(VI) is more mobile and soluble in water than Cr(III), which is relatively inert, chemically more stable and less bioavailable due to its negligible permeability to biomembranes [11]. Cr(VI) is toxic and mutagenic to most organisms [13], it induces acute and chronic toxicity, neurotoxicity, dermatotoxicity, genotoxicity, carcinogenicity, immunotoxicity, and general environmental toxicity due to its strong oxidizing potential [14]. Cr(VI) compounds are approximately 1,000 fold more cytotoxic and mutagenic than Cr(III)

[15]. Cr(VI) to Cr(III) reduction, therefore, represents a significant immobilization mechanism [16].

Cr(VI) is taken up through the membrane sulfate transport channels in cells of sulfate-utilizing organisms. Under normal physiological conditions, Cr(VI) reacts spontaneously with the intracellular reductants to generate the short-lived intermediates Cr(V) and/or Cr(IV), free radicals and the end-product Cr(III). Cr(V) undergoes a one-electron redox cycle to regenerate Cr(VI) and transferring the electron oxygen. The process produces a reactive oxygen species (ROS) that easily combines with DNA–protein complexes. Cr(IV) would bind to cellular materials and deteriorate their normal physiological functions. In humans, several trauma are associated with Cr(VI) exposure, including nasal irritation and ulceration, skin irritation, eardrum perforation and lung carcinoma. Furthermore, Cr(VI) can accumulate in the placenta, impairing fetal development in mammals [17].

Dana Devi et al. [18] confirmed that the function of DNA as a template in both replication and transcription is altered by Cr(VI)-induced DNA damage. Bagchi et al. [16] demonstrate that hexavalent chromium results in enhanced formation of ROS, including superoxide anion, hydroxyl radicals and nitric oxide, decreased cell viability, increased cellular and genomic hepatic DNA fragmentation, enhanced intracellular oxidized states, membrane damage with leakage of lactate dehydrogenase, activation of protein kinase C, apoptotic and necrotic cell death.

In view of the seriousness of Cr(VI) and its alarming effects on human health, it has been listed as a priority pollutant and classified as a class A human carcinogen by the US Environmental Protection Agency (USEPA) [19].

Chromium is one of the most widely used metals in a variety of industrial processes, such as steel production, wood preservation, leather tanning, metal corrosion inhibition, paints and pigments, and it is mainly used as chromate or dichromate [20]. For nearly all of these uses, the chemical properties of chromium are integral to its effectiveness. Industrial effluents containing chromium compounds are released directly or indirectly into natural water resources, mostly without proper effluent treatment, resulting in anthropogenic contamination of pristine environments [10, 12, 17, 21, 22].

Chromium has been released inadvertently and intentionally into the environment and has become one of the most frequently detected ground water and soil contaminants at hazardous waste sites. Cr(VI) as chromate and dichromate is considered one of the principal pollutants in the United States by the Environment Protection Agency [23].

Transformations of Cr in complex and dynamic soil systems are governed by several physical, chemical, and biological factors. The influence of soil physical factors in relation to Cr transformations is not documented in detail, except for a few reports. The adsorption and desorption processes are governed largely by the bulk and particle density of the soils and the type and amount of clay and organic matter content. The chemical factors dominate and play a major role in Cr transformations in soil. Soil pH plays a significant role in controlling the dynamics of Cr redox reactions [24, 25]. Soil pH along with redox potential determine the nature of Cr prevalent in the soil [26]. Low pH favors the formation of stable cationic Cr(III) species whereas at higher pH, especially in alkaline soils, anionic Cr(VI) formation is favored. At low pH, Cr(III) precipitates or tightly binds to a variety of ligands such as hydroxyls, humates, and phosphates present in soil. Humic substances and organic matter play a major role in the reduction of Cr(VI) to Cr(III). Organic matter during its oxidation reduces Cr(VI). Citric and fulvic acids and water-soluble extracts of air-dried soils form soluble complexes with Cr(III) [27].

Once it has been reduced to Cr(III), nearly all of the Cr should remain stable in this form indefinitely, since there are relatively few mechanisms for its reoxidation. The only redox couple in natural systems that can oxidize Cr(III) to Cr(VI) is Mn(IV)/Mn(II) [28-31]. Oxidation of chromium occurs in soils that are high in Mn(IV) and oxides and low in organic matter content, conditions fairly specific and only a few cases of Cr(III) oxidation are reported in literature. [32].

These factors are of great importance in assessing potential environmental hazards and remediation strategies for ecosystems with high levels of natural or anthropogenic chromium.

Remediation Technologies- Bioremediation

Cr(VI) has persistence in soils because of sorption to soil particles, which increases the potential risk of human exposure [33]. Because of the high accumulation of metals in roots and leaves, plants growing on metal-contaminated sites are not suitable for consumption by animals or humans [34].

Development of suitable methods for cleaning up such contaminated soils is an important topic of environmental recovery and protection, particularly regarding methods for remediation of Cr(VI)-contaminated soils. At present there exist physicochemical technologies to remove Cr(VI) from industrial waste, but these methods are costly and require much energy and specific equipment. Therefore the employment of lower-cost biotechnology could be an important alternative.

Remediation methods that involve the excavation of chromium contaminated soils are known to generate airborne dust that creates an additional exposure hazard to field workers and to the general public. To address these environmental concerns and health risks, various groups have proposed to facilitate in situ reduction of chromium. Chemical reduction of chromium by these methods is known to decrease the toxicity and bioavailability of this metal drastically. Recently, methods have been reported for remediation of Cr in contaminated soils usingelectro-kinetic remediation and ion exchange using polymer membranes. All these methods have certain merits and demerits, are limited to small scale facilities, and require special monitoring techniques like ion chromatography, Cr speciation, atomic absorption [38].

Technical applicability, plant simplicity and cost-effectiveness are the key factors that play major roles in the selection of the most suitable treatment system for inorganic effluent. All the factors mentioned above should be taken into consideration in selecting the most effective and inexpensive treatment in order to protect the environment [39].

Reduction alters the valence and may consequently alter the toxicity and environmental mobility of a metal. Therefore conversion of Cr(VI) to Cr(III) could be an effective method of combating Cr(VI) pollution. The existing treatment processes for chromium detoxification generally involve aqueous reduction of Cr(VI) by a reductant and subsequent pH adjustment to neutral ranges so as to precipitate the less soluble Cr(III). This process requires large amounts of chemicals and energy and hence is not economically feasible, and besides it can be a potential source of metal pollution through the resultant metal-containing chemical sludge [40].

In the environment, Cr(VI) contamination alters the structure of soil microbial communities [17, 41, 42].

The presence of Cr(VI) in the environment plays a selective pressure on microflora. Most microorganisms are sensitive to Cr(VI) toxicity, but some groups possess resistance mechanisms to Cr(VI) and can tolerate high levels. A relationship was found between total chromium content of soil and the presence of metal-tolerant/resistant bacteria. The bacterial chromate resistance is generally combined to plasmids, but it can also be coupled to chromosomal DNA [12].

The three processes, by which the microorganisms interact with toxic metals, are biosorption, bioaccumulation and enzymatic reduction [43]. Biosorption is a metabolism-independent process and thus can be performed by both living and dead microorganisms. On the other hand, microbial heavy metal accumulation generally comprehends two phases, an initial rapid phase of biosorption followed by slower, metabolism dependent active uptake of metals. During the bioaccumulation, many features of a living cell, like intracellular sequestration followed by localization within cell components, metallothionein binding, metal accumulation, extracellular precipitation and complex formation can occur [44, 45].

Bioremediation is appropriate for large scale application on heterogeneous environments, such as ground water, soil sludge and industrial waste [46-48]. Microbial remediation in the environment may be limited by metal toxicity [49].

Biotransformation of Cr(VI) to Cr(III) using bacteria is the most pragmatic approach with a well-established feasibility in bioremediation. Reduction of Cr(VI) has been demonstrated in various bacterial species including *Bacillus* sp. [50, 51, *Pseudomonas* sp. [13, 52], *Escherichia coli* [53], *Desulfovibrio* sp. [54], *Microbacterium* sp. [55] *Shewanella* sp. [56,57] and *Arthrobacter* sp. [58, 11]. Bacterial enzymes such as hydrogenases [59], nitroreductases [60] and quinone reductases [61] have been reported to exhibit chromate reductase activity. These enzymes commonly show a NADH:flavin oxidoreductase activity and can use Cr(VI) as electron acceptor [61].Ishibashi et al. [62] suggested that the ability to reduce chromate may be a secondary function for Cr(VI) reductases, which have a different primary role other than Cr(VI) reduction. These secondary functions may be related to the bacterial enzymatic adaptation as a result of the relatively recent increase of Cr(VI) content in the environment due to anthropogenic activities [63].

ACTINOMYCETES AS BIOREMEDIATION TOOLS

Actinomycetes represent an important constituent of the microbial population in most soils. Their metabolic diversity and specific growth characteristics, like mycelium formation and relatively rapid colonization of selective substrates, make them well-suited agents for bioremediation [64]. Among the soil filamentous microorganisms, streptomycetes represent up to 20 % of soil bacteria [65]. They are predominantly found in soil as spores, which are resistant to desiccation and starvation and can germinate and grow into a mycelial state for brief periods of time when nutrients became available. Streptomycetes are known for their ability to produce a wide variety of secondary metabolites, including many pigments and antibiotics. Due to this active secondary metabolism, streptomycetes also may be a good source for the identification of heavy metal binding components with possible future biotechnological application [66]. Metal-resistant actinobacteria, and their potential use for bioremediation strategies, have been described before [64-71].

Recent progress has been made studying Cr(VI) resistance in actinomycete isolated from polluted areas [46, 47, 70-75].

Cr(VI) reduction and accumulation by actinomycetes was study principally on complex culture medium. The first report on Cr(VI) reduction by *Streptomyces* was from Das and Chandra [76], later, Cr(VI) reduction and/or accumulation was determined for several actinomyces, as *Streptomyces griseus* [46], *Arthobacter* [50, 58, 74]) and *Corynebacterium sp*. [12].

On minimal culture medium, Cr(VI) removal takes more time than in complex medium. Amoroso et al. [68] reported that metal resistance and biosorption capability may be widespread among actinomycetes growing in contaminated environments. Cr(VI) bioaccumulation by *Streptomyces* strains was also revealed [39, 77]. Polti et al. [68] carried out a systematic screening on Cr(VI) resistance by actinomycetes on minimal medium. These authors found 11 strains resistant towards 10 mM Cr(VI) and, three strains could grow in the presence of up to 17 mM Cr(VI). They determined Cr(VI) removal by nine *Streptomyces* strains and one *Amycolatopsis* strain in minimal syntetic medium.

Long-term incubation on chromium-amended media is also an important factor to improve biopsortion and bioreduction. *Streptomyces* sp. MC1, a sugar cane isolated, increased significantly it Cr(VI) reduction ability after 70 d of incubation, moreover, no Cr(III) re-oxidation could be observed [71].

Visualization of metal deposits by means of optic or electron microscopy together with histochemical methods has been widely used in higher eukaryotes including human cells. Recently, the same methods have been applied to detect metal accumulation in microorganisms and in some cases, X-ray microanalysis has been proposed as another approach for visualization of heavy metal deposits in bacteria [66, 75].

The low cost staining method was applied for the first time to bacteria to detect chromium accumulation with high sensitivity and specificity [75]. Polti et al. [75] confirmed Cr(III) accumulation inside cells of *Streptomyces* sp. MC1. Also, EDXA allowed to detect chromium the *Streptomyces* sp. MC1 surface [75]. These authors suggest that *Streptomyces* sp. MC1 reduced Cr(VI) to Cr(III), and after that, chromium is sequestered from the culture medium and accumulated within the cell.

Poopal and Laxman [78] charaterized the chromate reductase activity of Streptomyces griseus NCIM 2020, this activity was heat labile and was lost on dialysis which was restored on addition of NADH. Addition of NADH/NADPH enhanced chromate reduction by 2–3-fold and the enzyme had no preference for NADH or NADPH. The enzyme was optimally active at 28 ◦C and pH 7.

On the other hand. Polti et al. [75] carried out a partial chracterization of a chromate reductase from *Streptomyces* sp. MC1. The enzyme is intracellular or linked to the membrane and showed activity within an extensive temperature (20 to 40 ºC) and pH (5.0 to 9.0) range, with optimum activity between 28 and 30 ºC and pH 7. Chromate reductase needed NAD(P)H as electron donors, while NADH was the most efficient cofactor.

Studies of soil bioremediation generally are carried out in soil supplemented with a organic substrate, because this improves the process [35, 36, 37]. Other strategy includes the slurry preparation to increase the water bioavailability and the aeration of soil Desjardin et al. [39]. However, these soil modifications could increase the general cost of the bioremediation process.

The strain *Streptomyces* sp. MC1 could remove Cr(VI) from soil samples without adding any substrate and at normal humidity level (20%). It is important also to notice that this strain could slightly less remove Cr(VI) even in non sterile soil, indicating that bioremediation activity by *Streptomyces* sp. MC1 was not inhibited by the natural soil microbial flora [71]. There is not other studies about Cr(VI) removal under theses conditions using actinomycetes.

CONCLUSION

The general strategy for bioremediation of Cr pollution has relied upon dissimilatory metal reduction, that is the uptake of toxic and permeable Cr(VI) molecules by microorganisms and plants and their conversion into less toxic trivalent. The actinomycetes *Streptomyces* sp. MC1 is a strain able to decrease bioavalible Cr(VI) from soil. The mechanism involves adsorption coupled with reduction to Cr(III), and finally, Cr(III) bioaccumulation.

The chromate reductase enzyme of *Streptomyces* sp. MC1 is active in the cell free extract; and it could be utilized in bioremediation processes at no compatible conditions with the cell reproduction.

Cr(VI) reduction and accumulation by *Streptomyces* sp. MC1 represents a useful mechanism for bioremediation in soil samples. Biotechnological processes could be improved using this actinomycete and it could not only be applied on solid systems such as soil but also on liquid or semi-liquid systems such as waste water or slurry. A particularly important factor is that the system is environment friendly.

REFERENCES

[1] Kavamura, V. N. & Esposito, E. (2010). *Biotechnological strategies applied to the decontamination of soils polluted with heavy metals.* Biotechnology Advances, 28, 61-69.

[2] Glick, B. R. (2003). *Phytoremediation: synergistic use of plants and bacteria to clean up the environment.* Biotechnology Advances, 21, 383-393.

[3] Zhang, G. L.; Yang, F. G.; Zhao, Y. G.; Zhao, W. J.; Yang, J. L.; Gong, Z. T. (2005). *Historical change of heavy metals in urban soils of Nanjing, China during the past 20 centuries.* Environment International, 31, 913-919.

[4] Tremolada, P.; Villa, S.; Bazzarin, P.; Bizzotto, E,.; Comolli, R.; Vighi, M. (2008). *Pops in mountain soils from the Alps and Andes: suggestions for a 'precipitation effect' on altitudinal gradients.* Water Air and Soil Pollution, 188, 93-109.

[5] Fabietti, G.; Biasioli, M.; Barberis, R.; Ajmone-Marsan, F. (2010). *Soil contamination by organic and inorganic pollutants at the regional scale: the case of Piedmont, Italy.* Journal of Soils and Sediments, 10, 290-300.

[6] Bilos, C.; Colombo, J. C.; Skorupka, C. N.; Rodrigues, P. M. J. (2001). *Sources, distribution and variability of airbone trace metal in La Plata City area, Argentina.* Environmental Pollution, 111, 149-158.

[7] Keane, B.; Collier, M. H.; Shann, J. R.; Rogstad, S. H. (2001). *Metal content of dandelion (Taraxacum officinale) leaves in relation to soil contamination and airborne particulate matter.* The Science of the Total Environment, 281, 63-78.

[8] Alumaa, P.; Kirso, U.; Petersel, V.; Steinnes, E. (2002). *Sorption of toxic heavy metals to soil.* International Journal of Hygiene and Environmental Health, 204, 375-376.

[9] Barnhart, J. (1997). *Chromium chemistry and implications for environmental fate and toxicity.* Journal of Soil Contamination, 6, 561-568.

[10] Cefalu, W. T. & Hu, F. B. (2004). *Role of chromium in human health and in diabetes.* Diabetes Care, 27, 2741-2751.

[11] Megharaj, M.; Avudainayagam, S.; Naidu, R. (2003). *Toxicity of hexavalent chromium and its reduction by bacteria isolated from soil contaminated with tannery waste.* Current Microbiology, 47, 51-54.

[12] Viti, C.; Pace, A.; Giovannetti, L. (2003). *Characterization of Cr (VI)-resistant bacteria isolated from chromium-contaminated soil by tannery activity.* Current Microbiology, 46, 1-5.

[13] Ganguli, A. & Tripathi, A. K. (2002) *Bioremediation of toxic chromium from electroplating effluents by chromate-reducing Pseudomonas aeruginosa A2Chr in two bioreactors.* Applied Microbiology and Biotechnology, 58, 416-420.

[14] Bagchi, D.; Stohs, S. J.; Downs, B. W.; Bagchi, M.; Preuss, H. G. (2002). *Cytotoxicity and oxidative mechanisms of different forms of chromium.* Toxicology, 180, 5-22.

[15] Biedermann, K. A & Landolph, J. R: (1990). *Role of valence state and solubility of chromium compounds on induction of cytotoxicity, mutagenesis, and anchorage independence in diploid human fibroblasts.* Cancer Research, 50, 7835-7842.

[16] Bagchi, D., Bagchi, M., Stohs, S. J. (2001). *Chromium (VI)-induced oxidative stress, apoptotic cell death and modulation of p53 tumor suppressor gene.* Molecular and Cellular Biochemistry, 222, 149-158.

[17] Cheung, K. H. & Gu, J. D. (2007). *Mechanism of hexavalent chromium detoxification by microorganisms and bioremediation application potential: A review.* International Biodeterioration and Biodegradation, 59, 8-15.

[18] Dana Devi, K.; Rozati, R.; Saleha Banu, B.; Jamil, K.; Grover, P. (2001). *In vivo genotoxic effect of potassium dichromate in mice leukocytes using comet assay.* Food and Chemical Toxicology, 39, 859-865.

[19] Costa, M. & Klein, C. B. (2006) *Toxicity and carcinogenicity of chromium compounds in humans.* Critical Reviews in Toxicology, 36, 155-163.

[20] Baldi, F.; Vaughan, A. M.; Olson, G. J. (1990) *Chromium (VI) resistant yeast isolated from a sewage treatment plant receiving tannery wastes.* Applied and Environmental Microbiology, 56, 913-918.

[21] Shakoori, A. R.; Makhdoom, M.; Haq, R. U. (2000). *Hexavalent chromium reduction by a dichromate-resistant gram-positive bacterium isolated from effluents of tanneries.* Applied Microbiology and Biotechnology, 53, 348-351.

[22] Ryan, M. P.; Williams, D. E.; Chater, R. J.; Hutton, B. M.; McPhail, D. S. (2002) *Why stainless steel corrodes.* Nature, 415, 770-774.

[23] Deng, B.; Lan, L.; Houston, K.; Brady, P. V. (2003). *Effects of clay Minerals on Cr(VI) reduction by organic compounds*. Environmental Monitoring and Assessment, 84, 5-18.

[24] Bartlett, R. J. & James, B. R. (1988). *Mobility and bioavailability of chromium in soils.* Advances in Environmental Science and Technology, 20, 267-304.

[25] Losi, M. E.; Amrhein, C., Frankenberger, W. T. (1994). *Environmental biochemistry of chromium.* Reviews of Environmental Contamination & Toxicology, 136, 92-121.

[26] Rai, D.; Eary, L. E.; Zachara, J. M. (1989). *Environmental chemistry of chromium.* Science of the Total Environment, 86, 15-23.

[27] Kamaludeen, S. P. B.; Megharaj, M.; Juhasz, A. L.; Sethunathan, N.; Naidu, R. (2003). *Chromium–Microorganism Interactions in Soils: Remediation Implications*. Reviews of Environmental Contamination and Toxicology, 178, 93-164.

[28] Eary, L. E. & Rai, D. (1987). *Kinetics of chromium(III) oxidation to chromium(VI) by reaction with manganese-dioxide.* Environmental Science & Technology, 21, 1187-1193.

[29] Palmer, C. D. & Puls, R.W. (1994). *Natural attenuation of hexavalent chromium in groundwater and soil. USEPA Rep. 540/S-94/505.* U.S. Gov. Print. Office, Washington, DC.

[30] James, B.R. (1996). *The challenge of remediating chromium-contaminated soil.* Environmental Science & Technology, 30, A248-A251.

[31] Oliver, D. S.; Brockman, F. J.; Bowman, R. S.; Kieft, T. L. (2003). *Microbial reduction of hexavalent chromium under vadose zone conditions.* Journal of Environmental Quality, 32, 317-324.

[32] Kozuh, N.; Stupar, J.; Gorenc, B. (2000). *Reduction and oxidation processes of chromium in soils*. Environmental Science and Technology, 34, 112-119.

[33] Stewart, M. A.;. Jardine, P. M.; Brandt, C. C.; Barnett, M. O.; Fendorf, S. E.; McKay, L. D.; Mehlhorn, T. L.; Paul, K. (2003). *Effects of contaminant concentration, aging, and soil properties on the bioaccessibility of Cr(III) and Cr(VI) in soil.* Soil and Sediment Contamination, 12, 1-21.

[34] Sinha, S.; Gupta, A. K. (2005). *Translocation of metals from fly ash amended soil in the plant of Sesbania cannabina L. Ritz: Effect on antioxidants*. Chemosphere, 61, 1204-1214.

[38] Gondal, M. A.; Hussain, T.; Yamani, Z. H.; Baig, M. A. (2009). *On-line monitoring of remediation process of chromium polluted soil using LIBS.* Journal of Hazardous Materials, 163. 1265-1271.

[39] Kurniawan, T. A.; Chan, G. Y. S.; Lo, W. H. ; Babel, S. (2006). *Physicochemical treatment techniques for waste waterladen with heavy metals.* Chemical Engineering Journal, 118, 83-98.

[40] A.R. Shakoori, M. Makhdoom, R.U. *Haq, Hexavalent chromium reduction by a dichromate-resistant gram positive bacterium isolated from effluents of tanneries*, Appl. Microbiol. Biotechnol. 53 (2000) 348–351

[41] Zhou, J.; Xia, B.; Treves, D. S.; Wu, L. Y.; Marsh, T. L. O'Neill, R. V., Palumbo, A. V.; Tiedje, J. M. (2002). *Spatial and resource factors influencing high microbial diversity in soil.* Applied and Environmental Microbiology, 68, 326-334.

[42] Turpeinen, R.; Kairesalo, T.; Häggblom, M. M. (2004). *Microbial community structure and activity in arsenic-, chromium- and copper contaminated soils.* FEMS Microbiology Ecology, 47, 39-50.

[43] Srinath, T.; Verma, T.; Ramteke P. W.; Garg, S. K. (2002). *Chromium (VI) biosorption and bioaccumulation by chromate resistant bacteria.* Chemosphere, 48, 427-435.

[44] Gadd, G. M. (2004). *Microbial influence on metal mobility and application for bioremediation.* Geoderma, 122, 109,119.

[45] Malik A. (2004). *Metal bioremediation through growing cells.* Environment International, 30, 261-278.

[46] Laxman, R. S. & More, S. (2002). *Reduction of hexavalent chromium by Streptomyces griseus.* Minerals Engineering, 15, 831-837.

[47] Polti, M. A.; Amoroso, M. J.; Abate, C. M. (2010). *Chromate reductase activity in Streptomyces sp. MC1.* The Journal of General and Applied Microbiology, 56, 11-18.

[48] Ksheminskaa, H.; Fedorovycha, D.; Babyaka, L.; Yanovycha, D.; Kaszyckib, P.; Koloczekb, H. (2005). *Chromium (III) and (VI) tolerance and bioaccumulation in yeast: a survey of cellular chromium content in selected strains of representative genera.* Process Biochemistry, 40, 1565-1572.

[49] Tseng, J. K. & Bielefeldt, A. R. (2002). *Low-temperature chromium(VI) biotransformation in soil with varying electron acceptors.* Journal of Environmental Quality, 31, 1831-1841.

[50] Camargo, F. A. O.; Bento, F. M.; Okeke, B. C.; Frankenberger, W. T. (2004). *Hexavalent chromium reduction by an actinomycete, Arthrobacter crystallopoietes ES 32.* Biological Trace Element Research, 97, 183-194.

[51] Liu, Y. G.; Xu, W. H.; Zeng, G. M.; Li, X.; Gao, H. (2006). *Cr(VI) reduction by Bacillus sp. Isolated from chromium landfill.* Process Biochemistry, 41, 1981-1986.

[52] Park, C. H.; Keyhan, M.; Wielinga, B.; Fendorf, S.; Matin, A. (2000). *Purification to homogeneity and characterization of a novel Pseudomonas putida chromate reductase.* Applied and Environmental Microbiology, 66, 1788-1795.

[53] Bae, W. C.; Lee, H. K.; Choe, Y. C.; Jahng, D. J.; Lee, S. H.; Kim, S. J.; Lee, J. H.; Jeong, B. C. (2005). *Purification and Characterization of NADPH-Dependent Cr(VI) Reductase from Escherichia coli ATCC 33456.* The Journal of Microbiology, 43, 21-27.

[54] Mabbett, A. N. & Macaskie, L. E. (2001). *A novel isolate of Desulfovibrio sp. with enhanced ability to reduce Cr(VI).* Biotechnology Letters, 23, 683-687.

[55] Pattanapipitpaisal, P.; Brown, N. L.; Macaskie, L. E. (2001). *Chromate reduction by Microbacterium liquefaciens immobilised in polyvinyl alcohol.* Biotechnology Letters, 23, 61-65.

[56] Myers, C. R.; Carstens, B. P.; Antholine, W. E.; Myers, J. M. (2000). *Chromium(VI) reductase activity is associated with the cytoplasmic membrane of anaerobically grown Shewanella putrefaciens MR-1.* Journal of Applied Microbiology, 88, 98-106.

[57] Vaimajala, S.; Peyton, B. M.; Apel, W. A.; Peterson, J. N. (2002). *Chromate reduction in Shewanella oneidensis MR-1 is an inducible process associated with anaerobic growth.* Biotechnology Progress, 18, 290-296.

[58] Asatiani, N. V.; Abuladze, M. K.; Kartvelishvili, T. M.; Bakradze, N. G.; Sapojnikova, N. A.; Tsibakhashvili, N. Y.; Tabatadze, L. V.; Lejava, L. V. *et al.* (2004). *Effect of chromium(VI) action on Arthrobacter oxydans.* Current Microbiology, 49, 321-326.

[59] Chardin, B.; Giudici-Orticoni, M. T.; DeLuca, G.; Guigliarelli, B.; Bruschi, M. (2003). *Hydrogenases in sulfate-reducing bacteria function as chromium reductase.* Applied Microbiology and Biotechnology, 63, 315-321.

[60] Kwak, Y. H.; Lee, D. S.; Kim, H. B. (2003). *Vibrio harveyi nitroreductase is also a chromate reductase.* Applied and Environmental Microbiology, 69, 4390-4395.

[61] González, C. F.; Ackerley, D. F.; Lynch, S. V.; Matin, A. (2005). *ChrR, a soluble quinone reductase of Pseudomonas putida that defends against H2O2.* The Journal of Biological Chemistry, 280, 22590-22595

[62] Ishibashi, Y.; Cervantes, C.; Silver, S. (1990). *Chromium reduction in Pseudomonas putida.* Applied and Environmental Microbiology, 56, 2268-2270.

[63] Silver, S. & Phung, L. T. (1996). *Bacterial heavy metal resistance: new surprises.* Annual Review of Microbiology, 50, 753-790.

[64] Albarracín, V. H.; Amoroso, M. J.; Abate, C. M. (2005). *Isolation and characterization of indigenous copper resistant actinomycete strains.* Chemie der Erde/Geochemistry, 65, 145-156.

[65] Kothe, E.; Bergmann, H.; Buchel, G. (2005). *Molecular mechanisms in bio-geointeractions: from a case study to general mechanisms.* Chemie der Erde/Geochemistry, 65, 7-27.

[66] Albarracín, V. H.; Winik, B.; Kothe, E.; Amoroso, M. J.; Abate, C. M. (2008). *Copper bioaccumulation by the actinobacterium Amycolatopsis sp. AB0.* Journal of Basic Microbiology, 48, 323-330.

[67] Albarracín, V. H.; Avila, A. L.; Amoroso, M. J.; Abate, C. M. (2008). *Copper removal ability by Streptomyces strains with dissimilar growth patterns and endowed with cupric reductase activity.* FEMS Microbiology Letters, 288, 141-148.

[68] Amoroso, M. J.; Castro, G. R.; Carlino, F. J.; Romero, N. C.; Hill, R. T.; Oliver, G. (1998). *Screening of heavy metaltolerant actinomycetes isolated from the Salí River.* The Journal of General and Applied Microbiology, 44, 129-132.

[69] Schmidt, A.; Haferburg, G.; Siñeriz, M.; Merten, D.; Buchel, G.; Kothe, E. (2005). *Heavy metal resistance mechanisms in actinobacteria for survival in AMD contaminated soils.* Chemie der Erde/Geochemistry, 65, 131-144.

[70] Polti, M. A.; Amoroso, M. J.; Abate, C. M. (2007). *Chromium (VI) resistance and removal by actinomycete strains isolated from sediments.* Chemosphere, 67, 660-667.

[71] Polti, M. A.; García, R. O.; Amoroso, M. J.; Abate, C. M. (2009). *Bioremediation of chromium(VI) contaminated soil by Streptomyces sp. MC1.* Journal of Basic Microbiology, 49, 285-292.

[72] Francisco, R.; Alpoim, M. C.; Morais, P. V. (2002). *Diversity of chromium-resistant and -reducing bacteria in a chromium-contaminated activated sludge.* Journal of Applied Microbiology, 92, 837-843.

[73] Richards, J. W.; Krumholz, G. D.; Chval, M. S.; Tisa, L. S. (2002). *Heavy metal resistance patterns of Frankia strains.* Applied and Environmental Microbiology, 68, 923-927.

[74] Camargo, F. A. O.; Bento, F. M.; Okeke, B. C.; Frankenberger, W. T. (2003). *Chromate reduction by chromium-resistant bacteria isolated from soils contaminated with dichromate.* Journal of Environmental Quality, 32, 1228-1233.

[75] Polti, M. A.; Amoroso, M. J.; Abate, C. M. (2010) *Intracellular chromium accumulation by Streptomyces sp. MC1.* Water Air Soil Pollut DOI 10.1007/s11270-010-0401-5

[76] Das, S. & Chandra, A. L. (1990). *Chromate reduction in Streptomyces.* Experientia, 46, 731-733.

[77] Amoroso, M. J.; Castro, G. R.; Durán, A.; Peraud, O.; Oliver, G.; Hill, R. T. (2001). *Chromium accumulation by two Streptomyces spp. isolated from riverine sediments*. Journal of Industrial Microbiology and Biotechnology, 26, 210-215.
[78] Poopal, A. C. & Laxman, R. S. (2009). *Studies on biological reduction of chromate by Streptomyces griseus.* Journal of Hazardous Materials, 169, 539-545.

In: Bioremediation
Editor: Alexander C. Mason

ISBN: 978-1-61122-730-7

Chapter 16

CANDIDA BIOSURFACTANT-ENHANCED REMOVAL HYDROPHOBIC ORGANIC POLLUTANTS

Leonie A. Sarubbo*[*] *and Galba M. Campos-Takaki
Nucleous of Research in Environmental Sciences, Center of Sciences and Technology,
Catholic University of Pernambuco, Pernambuco, Brazil

ABSTRACT

The present interest in hydrocarbon biodegradation is motivated in large part by their presence as environmental pollutants. The biodegradation of hydrocarbons has a high ecological significance, as it constitutes the major process for remediation of contaminated areas. Micro-organisms with the ability to degrade many hydrocarbons have been described, and their mechanisms of action have been studied. The bioremediation of soils contaminated with hydrophobic pollutants is limited by the poor availability of these contaminants to micro-organisms. Surfactants can help, by solubilization or emulsification, to release hydrocarbons sorbed to soil organic matter and increasing the aqueous concentrations of hydrophobic compounds, resulting in higher mass transfer rates. Microbially-derived surfactants offer several advantages and interest in these biosurfactants continues to grow. Microorganisms, having diverse synthetic capabilities, offer potential in expanding the range of surfactants with novel compounds, thereby providing new possible applications. The main classes of biosurfactant are glycolipids, lipoamino acids and lipopeptides, polymers (lipoproteins, lipopoly-saccharides), phospholipids, monoglycerides and diglycerides, and fatty acids. The most important advantage of biosurfactants over chemical surfactants is probably their ecological acceptability. Biosurfactants are biodegradable and thus problems of toxicity and accumulation in natural ecosystems are avoided. In the environmental sector, bio-surfactants have potential applications in bioremediation and waste treatment because of their inherent degradability. Few works have investigated the optimization of yeast biosurfactant production, mainly within the last decade, revealing that the potential of such microorganisms is not well explored in the industrial field. This chapter focuses on the successfully application of *Candida* biosurfactants produced in inexpensive media in the removal of hydrophobic contaminants.

[*] E-mail: leonie@unicap.br

INTRODUCTION

Oil pollution accidents have become numerous and have caused ecological and social catastrophes (Burger, 1993; Shaw, 1992; Burns *et al.,* 1993). The ability of biosurfactants to emulsify hydrocarbon-water mixtures has been widely reported (Broderick & Cooney, 1982; Harvey *et al.,* 1990; Oberbremer *et al.,* 1990; Wasko & Bratt, 1990; Francy *et al.,* 1991; Zhang & Miller, 1992). These emulsification properties have also been demonstrated to enhance hydrocarbons degradation in the environment, hence making them potentially useful tools for oil spill pollution-control (Atlas & Bartha, 1992; Atlas, 1993; Bertrand *et al.,* 1994).

Microbial surfactants are complex molecules synthesized by bacteria, yeasts, and fungi, as amphipatic molecules, possess both lipophilic tail (commonly a fatty acid chain), and hydrophilic head (carboxylate groups, amino groups, phosphate groups, or carbohydrates). They are the active ingredients found in soaps and detergents with ability to concentrate at the air–water interface and are commonly used to separate oily materials from a particular media (Mukherjee *et al.*, 2006; Deleu & Paquot, 2004). Surfactants are classified as ionic and non-ionic with varying chemical structures according to their hydrophilic group. For industrial applications, the ionic surfactants are classified based on the charge they carry when dissociated in water at their neutral pH. These classifications are namely anionic, cationic and zwitterionic or amphoteric (Desai & Banat, 1997; Gallert & Winter, 2002). Surfactants increase the aqueous solubility of nonaqueous phase liquids (NAPLs) by reducing their surface/interfacial tension at air–water and water–oil interfaces. As the interfacial tension is reduced and the aqueous surfactant concentration increased, the monomers aggregate to form micelles (Gautam & Tyagi, 2006. The concentration at which micelles first begin to form is known as the critical micelle concentration (CMC). This concentration corresponds to the point where the surfactant first shows the lowest surface tension. Many physical properties used to characterize surfactants depend on the CMC such as; emulsion formation, oil solubilization, foaming and detergency, interfacial and surface tensions. These properties may be used to assess surfactant suitability for environmental remediation such as soil washing (Cortis & Ghezzehei, 2007).

Surfactants produced from chemically based materials are known as synthetic surfactants and those from biologically based materials are biosurfactants. Biosurfactants can be synthesized by many different microorganisms and are grouped into six major classes based on the producing microorganism. These classes are glycolipids, phospholipids, polysaccharide–lipid complexes, lipoproteins–lipopetides, hydroxylated and cross-linked fatty acids, and the complete cell surface (Kosaric *et al.*, 1987; Maneerat, 2005). Bognolo (1999) concludes that almost all biosurfactants are either non-ionic or anionic, as no studies report biosurfactants with cationic structures. Biosurfactant production, commercial applications, potential in biodegradation of hydrocarbons and metals removal from soil has been studied by Banat (1995; 2006), Banat *et al.* (2000), Maier & Soberon-Chavez (2000), Makkar & Cameotra (2002) and Mulligan *et al.* (2001).

According to Mata-Sandoval *et al.* (2000), important characteristics of biosurfactant are related with properties for use in the remediation of contaminated soils as:

- Small average minimum surface tension,
- High average emulsifying activity,

- Low CMC to enhance the mobilization of great amounts of pollutant through micelle transport, and
- High affinity for hydrophobic compounds but still be hydrophilic enough to minimize partition into soil.

So, chemically-synthesized surfactants have been used in the oil industry to aid the clean up of oil spills, as well as to enhance oil recovery from oil reservoirs. These compounds are not biodegradable and can be toxic to the environment. Biosurfactants, however, have been shown in many cases to have equivalent emulsification properties and are biodegradable. Thus, there is an increasing interest in the possible use of biosurfactants in mobilizing heavy crude oil, transporting petroleum in pipelines, managing oil spills, oil-pollution control, cleaning oil sludge from oil storage facilities, soil/sand bioremediation and microbial enhanced oil recovery (MEOR) (Aksu, 2000; Mulligan, 2005).

So, the largest market for environmental use of biosurfactant is the petroleum industry, considering the use in microbial enhanced oil recovery (MEOR), and in special in the remediation of sites contaminated by oil spills (Mata-Sandoval, 2000; Ron & Rosenberg, 2002; Rahman & Gakpe, 2008).

Despite the advantages of the biosurfactants over their synthetic counterparts, microbial surfactants are not yet competitive with those produced by the chemical industry, but efforts should be made on the different production aspects to find suitable and economic substrates and to develop new strategies to increase the volumetric productivity. Fermentation medium can represent approximately 30% of the cost for a microbial fermentation (Deleu & Paquot, 2004). Nevertheless, much effort in process optimization at the engineering and biological levels has been done, and, for some applications, biosurfactants can be produced from several inexpensive waste substrates, thereby decreasing their production cost (Lang & Wullbrandt, 1999). A great variety of alternative raw materials is currently available as nutrients for industrial fermentations, namely various agricultural and industrial byproducts and waste materials.

Biosurfactants have been produced specially by bacteria and yeasts. Though biosurfactants produced by *Pseudomonas* and *Bacillus* strains have been extensively used in MEOR experiments (Banat, 2005; Wu *et al.*, 2008; Silva et al., 2010), some *Candida* surfactants have also been found very effective (Sarubbo *et al.*, 2007; Sobrinho *et al.*, 2008; Coimbra *et al.*, 2009; Batista *et al.*, 2010). Moreover, few studies describe the production of biosurfactants from yeast species. We have investigated in our laboratories for over a decade, several species of yeast of the genus *Candida* that produce anionic biosurfactants from low-cost substrates with attractive properties which make them suitable for application in the petroleum industry, as it will be described below.

REMOVAL OF CONTAMINANTS POLLUTANTS BY CANDIDA SURFACTANTS

The combined behavior of two immiscible liquids such as oil and water that result in the formation of the emulsions is important in the application of surfactants in oil-contaminated soil treatment. When one liquid is dispersed into the other, the small droplets provide a large amount of interfacial surface area and hence greater interfacial free energy in the system. With two pure liquids, the droplets may rapidly coalesce and two separate phases will form which minimize the interfacial area. With the surfactants, emulsions formed may stabilize by reducing the interfacial tension and decreasing the rate of coalescence (Sing *et al.*, 1990). The behavior of the emulsion is related to the equilibrium phase behavior of the oil/water/surfactant system from which it is made. The complex fluid dispersions generally formed and stabilized by adsorbing surfactants onto air-liquid interface are known as foams. These dispersed fluids contain small bubbles with large surface areas, which can be stabilized using surfactants (Sarubbo *et al.*, 1999; 2001).

Experimental studies have been performed in our laboratories to assess the potential of biosurfactants for their applications as MEOR agents. In order to show the efficacy of some of the biosurfactant producing strains, the results of laboratory scale simulated experiments on emulsification index and percentage of oil recovered from sand samples are presented in Tables 1 and 2. The emulsification index (E24) was determined by adding 6ml crude of hydrophobic substrates to 4 ml culture broth in a graduated tube, followed by vortexing at high speed for 2 min. The emulsion stability was determined after 24 h. The emulsification index is thus defined as the height of the emulsion layer, divided by the total height, multiplied by 100 (Cooper & Goldenberg, 1987).

Laboratory studies typically utilize samples containing the desired substrate. These substrates have been utilized to demonstrate the usefulness of the yeasts biosurfactants in oil recovery from sand. The Biosurfactants suitability for enhanced oil recovery was carried out using impregnated sand with 10% petroleum or motor oil. Fractions of 20.0 g of contaminated sand were submitted to the following treatments: addition of 40.0 ml of distilled water (control), addition of 40.0 ml of the metabolic broth free of cells or addition of the isolated biosurfactant at or above its CMC. Samples were incubated on a rotary shaker (150 rpm) for 24 h at 27 °C and then centrifuged at 5000 g for 20 min to separate the laundering solution and the sand. The amount of oil residing in the sand after the impact of biosurfactant was gravimetrically determined as the amount of material extracted from the sand by an organic solvent (Luna *et al.*, 2009b).

The biosurfactant produced by *C. glabrata* strain UCP 1002 grown in cotton seed oil and glucose was the first tested for motor oil removal The results demonstrated that the biosurfactant solution at 2.5% concentration (at the CMC) was capable to remove 84% of the oil adsorbed in the sand, while the distilled water (control) removed 56% of the contaminated oil (Luna *et al.*, 2009b). In another study, the biosurfactant from *C. sphaerica* UCP0995 cultivated in two agroindustrial by-products at 0.1% concentration (above the CMC) was capable to remove 65% of the oil adsorbed in the sand. The biosurfactant solution at the CMC (0.08%) removed 55% of the oil, while the solution under the CMC (0.05%) removed around 30%. The distilled water (control) removed only 14.5% of the contaminated oil (Sobrinho *et al.*, 2008).

Table 1. Effectiveness of various *Candida* species surfactant solutions (cell-free broth) in emulsifying hydrophobic substrates

Candida specie	Cultivation medium	Hydrophobic substrate	Emulsification index (%)	Reference
C. glabrata UCP1002	7.5% cotton seed oil plus 5.0% glucose in mineral salt medium	Coton seed oil	100.0	Sarubbo et al., (2006)
C. glabrata UCP1002	7.5% cotton seed oil plus 5.0% glucose in mineral salt medium	Hexadecane	65.0	Sarubbo et al., (2006)
C. lipolytica UCP0988	10% canola oil plus 10% Glucose in mineral salt medium	Hexadecane	18.0	Sarubbo et al., (2007)
C. lipolytica UCP0988	10% canola oil plus 10% Glucose in mineral salt medium	Corn oil	30.0	Sarubbo et al., (2007)
C. lipolytica UCP0988	2.5% ground-nut oil refinery in mineral salt medium	Motor oil	100.0	Rufino et al., (2007)
C. lipolytica UCP0988	2.5% ground-nut oil refinery in mineral salt medium	Hexadecane	50.0	Rufino et al., (2007)
C. lipolytica UCP0988	2.5% ground-nut oil refinery in mineral salt medium	kerosene	45.0	Rufino et al., (2007)
C. sphaerica UCP0995	5.0% ground-nut oil refinery residue plus 2.5% corn steep liquor in distilled water	Motor oil	54.0	Sobrinho et al., (2008)
C. lipolytica UCP0988	4% ground-nut oil refinery residue plus 1% glutamic acid in mineral salt medium	Motor oil	79.0	Rufino et al., (2008)

Table 1. (Continued)

C. sphaerica UCP0995	4% ground-nut oil refinery residue plus 4% corn steep liquor in yeast salt water	Motor oil	85.0	Luna et al., (2009a)
C. sphaerica UCP0995	4% ground-nut oil refinery residue plus 4% corn steep liquor in yeast salt water	Corn oil	80.0	Luna et al., (2009a)
C. sphaerica UCP0995	4% ground-nut oil refinery residue plus 4% corn steep liquor in yeast salt water	Cotton seed oil	70.0	Luna et al., (2009a)
C. glabrata UCP1002	7.5% cotton seed oil plus 5% glucose in mineral salt medium	Hexadecane	66.0	Luna et al., (2009b)
C. lipolytica UCP0988	2.5% refinery residue plus 2.5% corn steep liquor in distilled water	Motor oil	92.0	Coimbra et al., (2009)
C. lipolytica UCP0988	2.5% refinery residue plus 2.5% corn steep liquor in distilled water	Corn oil	45.0	Coimbra et al., (2009)
C. guilliermondii UCP0992	2.5% refinery residue plus 2.5% corn steep liquor in distilled water	Motor oil	82.0	Coimbra et al., (2009)
C. guilliermondii UCP0992	2.5% refinery residue plus 2.5% corn steep liquor in distilled water	Corn oil	30.0	Coimbra et al., (2009)
C. tropicalis UCP0996	2% waste frying oil in mineral salt medium	Lubrication oil	93.0	Batista et al., (2010)
C. tropicalis UCP0996	2% waste frying oil in mineral salt medium	Motor oil	87.0	Batista et al., (2010)

Table 2. Effectiveness of various *Candida* surfactant solutions in releasing hydrophobic contaminants from sand

Candida specie	Cultivation medium	Biosurfactant solution	Surface tension (mN/m)	Hydrophobic contaminant	Removal of the contaminant (%)	Reference
C. glabrata UCP1002	7.5% cotton seed oil plus 5% glucose in mineral salt medium	Isolated biosurfactant at 1.5%	31.0	Motor oil	63.6	Luna et al., (2009b)
C. glabrata UCP1002	7.5% cotton seed oil 5% glucose in mineral salt medium	Isolated biosurfactant at 2.5%	31.0	Motor oil	84.0	Luna et al., (2009b)
C. lipolytica UCP0988	2.5% refinery residue plus 2.5% corn steep liquor in distilled water	Cell-free broth	35.0	Petroleum	97.4	Coimbra et al., (2009)
C. lipolytica UCP0988	2.5% refinery residue plus 2.5% corn steep liquor in distilled water	Cell-free broth	32.0	Motor oil	86.2	Coimbra et al., (2009)
C. guilliermondii UCP0992	2.5% refinery residue plus 2.5% corn steep liquor in distilled water	Cell-free broth	30.0	Petroleum	97.9	Coimbra et al., (2009)
C. guilliermondii UCP0992	2.5% refinery residue plus 2.5% corn steep liquor in distilled water	Cell-free broth		Motor oil	89.8	Coimbra et al., (2009)

Table 2. (Continued)

C. sphaerica UCP0995	5.0% ground-nut oil refinery residue plus 2.5% corn steep liquor in distilled water	Isolated biosurfactant at 0.1%	26.0	Motor oil	65.0	Sobrinho et al., (2008)
C. sphaerica UCP0995	5.0% ground-nut oil refinery residue plus 2.5% corn steep liquor in distilled water	Isolated biosurfactant at 0.08%	26.0	Motor oil	55.0	Sobrinho et al., (2008)
C. sphaerica UCP0995	5.0% ground-nut oil refinery residue plus 2.5% corn steep liquor in distilled water	Isolated biosurfactant at 0.05%	26.0	Motor oil	30.0	Sobrinho et al., (2008)
C. lipolytica UCP0988	Cassava flour wastewater (10.0%), Ammonium sulphate (0.2%), and Urea (0.1%)	Cell-free broth	26.3	Motor oil	81.7	Andrade et al., (2009)
C. tropicalis UCP0996	2% waste frying oil in mineral salt medium	Cell-free broth	34.0	Motor oil	95.1	Batista et al., (2010)
C. tropicalis UCP0996	2% waste frying oil in mineral salt medium	Cell-free broth	34.0	Petroleum	97.2	Batista et al., (2010)
C. glabrata UCP1002	5% vegetable fat waste in mineral salt medium	Cell-free broth	24.0	Motor oil	92.6	Gusmão et al., (2010)

Apart from the industrial applications of biosurfactants envisaged, their application in the oil industry is one of the potential uses which require lower purity specifications so that whole culture broth could be used (Van Hamme *et al.* 2006). Then, two reasons motivated us to examine the behavior of the biosurfactants containing cell-free broths in a soil-water-hydrophobic contaminant system. The first is that the use of crude surfactants in oil washing would be more cost-effective than using purified biosurfactants and the second is the importance of studying forms of application next to real environmental conditions, as the adsorption of surfactants onto soil surfaces is influenced by many factors that alter the CMC values obtained with clear water, such as the ionic strength of pore waters in these systems. The studies have been giving very interesting results, as described as follows.

Results obtained for the cell-free culture medium containing the biosurfactants produced by the yeasts *C. guilliermondii* UCP0992 and *C. lipolytica* UCP0988 cultivated in industrial residues showed high ability of these molecules in removing motor oil and petroleum adsorbed in the samples of sand tested. These two yeasts biosurfactants removed more than 93% of the petroleum adsorbed in sand samples (Coimbra *et al.*, 2009). Results obtained for the removal of petroleum and motor oil adsorbed in the samples of sand tested by the cell-free culture medium from *C. tropicalis* obtained after 144 hours showed removal values around 80% (Batista *et al.*, (2010). In another study the cell-free broth from *C. glabrata* UCP1002 grown in vegetable fat waste removed 92.65% of the motor oil adsorbed in sand.

Presently there are no cost-effective techniques to mass produce biosurfactants for use in bioremediation, detergents, etc. Current bioremediation experiments primarily use synthetic surfactants to remove organic contaminants adsorbed to solid surfaces to facilitate biodegradation (Muthusamy *et al.* 2008). Thus, the application costs of the cell-free broth from *C. glabrata* UCP1002 grown in vegetable fat waste were evaluated in the system containing sand-water-hydrophobic contaminant. It could be observed that the biosurfactant solution at 1% had shown a removal of 71.36% (US$ 495.60 ton of contaminated sand), showing that a new and effective biosurfactant had been produced by *C. glabrata* strain UCP1002, as described by Gusmão *et al.* (2010).

It is important to note that most of these biosurfactants are produced from industrial residues as substrates. Thus, industrial wastes offers potential substitutes of non renewable resources with renewable resources, and they can be used to recover value added bioproduct, i.e. biosurfactants, which in the future might replace synthetic surfactants in a number of different applications, especially in the petroleum industry (Shah *et al.*, 2007).

CONCLUSIONS

The usefulness of yeasts biosurfactants in the emulsification of aqueous hydrocarbon mixtures has been clearly demonstrated. Yeasts biosurfactants in many cases have proved to be more effective than chemical surfactants and have the added benefit of being biodegradable. Laboratory studies have shown that the addition of biosurfactant mixtures alone may be useful for recovering these contaminants from the environment. Therefore, in ecological terms, the use of yeasts biosurfactants is obvious for closed systems but remains speculative in the open environment. Preliminary findings from the few investigations carried out to date, however, seem promising. The precise mechanism of enhanced oil recovery is unclear due to the lack of controls in some cases, with the unforeseen difficulties usually encountered and insufficient analyses in other cases. Future efforts in strain improvement and development with the aid of genetic engineering are expected to help progress in this emerging technology.

REFERENCES

Aksu, Z. (2000) *Application of biosorption for the removal of organic pollutants: a review.* Process Biochem., 40, 328-350.

Andrade, R. F. S., Luna, J. M. ; Rufino, R. D., Albuquerque,C. D. C., Sarubbo, L. A., Campos-Takaki, G. M. (2009) *Surface active agent produced by Candida lipolytica using cassava flour wastewater as substrate. In: In: Antonio Mendez-Vilas. (Org.).* Current Research Topics in Applied Microbiology and Microbial Biotechnology. EUA: WordScientific Publishing Co. Pte. Ltd., p. 701-705.

Atlas, R. M., Bartha, R. (1992). *Hydrocarbon biodegradation and oil-spill bioremediation.* Adv. Microb. Eco., 12, 287-338.

Atlas, R. M. (1993) Bacteria & bioremediation of marine oil-spills. Oceanus, 36, 71-81.

Banat I. M. (1995) *Biosurfactant production and possible uses in microbial enhanced oil recovery and oil pollution remediation.* Biores Technol. 51, 1-12.

Banat, I. M. (2006) *Bioremediation of petroleum sludge using bacterial consortium with biosurfactant.* In: Environmental Bioremediation Technologies. Eds. SINGH, S.N.; TRIPATHI, R. D., Springer Publication, pp. 391-408.

Banat I. M., Makkar R., Cameotra, S. (2000) *Potential commercial applications of microbial surfactants.* Appl Microbiol Biotechnol 53, 495–508.

Batista, R. M., Rufino, R. D., Luna, J. M., Souza, J. E. G., Sarubbo, L. A. (2010) *Effect of Medium Components on the Production of a Biosurfactant from Candida tropicalis Applied to the Removal of Hydrophobic Contaminants in Soil.* Water Env. Res., 82, 418-425.

Bertrand, J. C., Bonin, P., Goutx, M., Gauthier, M.; Mille, G. (1994). *The potential application of biosurfactants in combatting hydrocarbon pollution in marine environment.Res.* Microbiol. 145, 53-56.

Bognolo, G. (1999) *Biossurfactants as emulsifying agents for hydrocarbons.* Coll. Surf. A 152, 41-52.

Broderick, L. S., Cooney, J. J. (1982) *Emulsification of hydrocarbons by bacteria from freshwater ecosystems.* Dev. Ind. Microbiol. 23, 425-434.

Burger, A. E. (1993) *Estimating the mortality of seabirds following oil-spills -- effects of spill volume.* Mar. Poll. Bull. 26, 140-143.

Burns, K. A., Garrity, S. D., Levings, S. C. (1993) *How many years until mangrove ecosystems recover from catastrophic oil-spills.* Mar. Poll. Bull. 26, 239-248.

Coimbra, C. D., Rufino, R. D., Luna, J. M., Sarubbo, L. A. (2009) *Studies of the cell surface properties of Candida species and relation with the production of biosurfactants for environmental applications.* Curr. Microbiol. 58, 245-251.

Cooper, D. G., Goldenberg, B. G. (1987) *Surface active agents from two Bacillus species.* Appl. Environ. Microbiol. 53, 224-229.

Cortis, A., Ghezzehei, T. A. (2007) *On the transport of emulsions in porous media.* J Col Interface Sci. 313, 1-4.

Deleu, M., Paquot, M. (2004) *From renewable vegetables resources to microorganisms: new trends in surfactants.* Comptes Rendus Chimie, 7, 641-646.

Desai, J. D., Banat, I. M. (1997) *Microbial production of surfactants and their commercial potential. Microbiol.* Mol. Biol. Rev. 61, 47-64.

Francy, D. S., Thomas, J. M., Raymond, R. L., Ward, C. H. (1991) *Emulsification of hydrocarbons by subsurface bacteria.* J. Ind. Microbiol. 8, 237-46.

Gallert, C., Winter, J. (2002) *Solid and liquid residues as raw materials for biotechnology.* Review Article. 89, 483-496.

Gautam, K. K., Tyagi, V. K. (2006) *Microbial Surfactants: a review.* J Oleo Sci. 55, 155-166.

Gusmão, C. A. B., Rufino, R. D., Sarubbo, L. A. (2010) *Laboratory production and characterization of a new biosurfactant from Candida glabrata UCP1002 cultivated in vegetable fat waste applied to the removal of hydrophobic contaminant.* World J. Microbiol. Biotechnol., 26, 1683-1692.

Harvey, S., Elashvili, I., Valdes, J. J., Kamely, D. & Chakrabarty, A. M. (1990). *Enhanced removal of Exxon Valdez spilled oil from Alaskan gravel by a microbial surfactant.* Bio/Tech. 8, 228-30.

Kosaric, N., Cairns, W.L., Gray, N. C.C. (1987) *Biosurfactants and Biotechnology.* New York: M. Dekker.

Lang, S., Wullbrandt, D. (1999) *Rhaminose Lipids—Biosynthesis, Microbial Production and Application Potential.* Appl. Microbiol. Biotechnol. 51, 22–32.

Luna, J. M., Rufino, R. D., Sarubbo, L. A., Campos-Takaki, G. M. (2009a) *Produção de biossurfactante em meio de baixo custo formulado com água do mar.* Exacta 6, 209 - 215.

Luna J. M., Sarubbo, L. A., Campos-Takaki, G. M. (2009b) *A new biosurfactant produced by Candida glabrata UCP1002: characteristics of stability and application in oil recovery.* Brazilian Arch. Biol. Technol., 52, 785-793.

Maier, R. M., Soberon-Chavez, G. (2000) *Pseudomonas aeruginosa rhamnolipids: byosynthesis and potential applications.* Appl. Micobiol. Biotechnol. 54, 625-633.

Makkar, R. S., Cameotra, S. S. (2002) *An update on the use of unconventional substrates for biosurfactant production and their new applications.* Appl. Microbiol. Biotechnol. 58, 428-434.

Maneerat, S. (2005) *Production of biosurfactants using substrates from renewable-resources.* Songklanakarin J. Sci. Technol. 27, 675-683.

Mata-Sandoval, J. C., Karns, J., Torrents, A. (2000) *The influence of surfactants and biosurfactants on the bioavaibility of hydrophobic organic pollutants in subsurface environments.* Rev. Int. Contam. Ambient. 16, 193-203.

Mukherjee, S., Das, P., Sen, R. (2006) *Towards commercial production of microbial surfactants.* Trends Biotechnol., 24, 509-515.

Mulligan, C. N. (2005) *Evironmental applications for biosurfactants.* Environ. Pollution, 133, 183-190.

Mulligan, C. N., Yong, R. N., Gibbs, B. F. (2001) *Surfactant-enhanced remediation or contaminated soil: a review.* Eng Geol. 60, 371-380.

Muthusamy, K., Gopalakrishnan, S., Ravi, T. K., Sivachidambaram, P. (2008) *Biosurfactants: properties, commercial production and application.* Curr. Sci, 94, 736-747.

Oberbremer, A., Muller-Hurtiq, R., Wagner, E. (1990) *Effect of the addition of microbial surfactants on hydrocarbon degradation in a soil population in a stirred reactor.* Appl. Microbiol. Biotech. 32, 485-489.

Rahman, P. K. S. M., Gakpe, E. (2008) *Production, characterization and applications of biosurfactants-review.* Biotechnol. 7, 360-370.

Ron, E. Z., Rosenberg, E. (2002) *Biosurfactants and oil bioremediation.* Curr. Opinion Biotechnol., 13, 249-252.

Rufino, R. D., Sarubbo, L.A., Benicio, B.N., Campos-Takaki, G.M. (2008) *Experimental design for the production of tensio-active agent by Candida lipolytica.* J. Ind. Microbiol. Biotechnol., 35, 907-914.

Rufino, R. D., Sarubbo, L. A., Campos-Takaki, G. M. (2007) *Enhancement of stability of biosurfactant produced by Candida lipolytica using industrial residue as substrate.* World J. Microbiol. Biotechnol., 23, 729-734.

Sarubbo, L. A., Luna, J. M., Campos-Takaki, G. M. (2006) *Production and stability studies of the bioemulsifier obtained from a new strain of Candida glabrata UCP 1002.* Electronic J Biotechnol. 9, 400-406.

Sarubbo, L. A., Farias, C. B. B., Campos-Takaki, G. M. (2007) *Co-utilization of canola oil and glucose on the production of a surfactant by Candida lipolytica.* Curr. Microbiol., 54, 68-73.

Sarubbo, L. A., Marçal, M. C. R., Neves, M. L. C., Silva, M. P. C., Porto, A. L. F., *Campos-Takaki, G. M. (2001) Bioemulsifier production in batch culture using glucose as carbon source by Candida lipolytica.* Appl. Biochem. Biotechnol., 95, 59-67.

Sarubbo, L. A., Porto, A. L. F., Campos-Takaki, G. M. (1999) *The use of babassu oil as substrate to produce bioemulsifiers by Candida lipolytica.* Canadian J. Microbiol., 45, 423-426.

Shah, V., Jurjervic, M., Badia, D. (2007) *Utilization of restaurant waste oil as a precursor for sophorolipid production.* Biotechnol. Progress, 23, 512-515.

Shaw, D. G. (1992). *The Exxon-valdez oil-spill – ecological and social consequences.* Environ. Conserv., 19, 253-258.

Silva, S. N. R. L., Farias, C. B. B., Rufino, R. D., Luna, J. M., Sarubbo, L. A. (2010) *Glycerol as substrate for the production of biosurfactant by Pseudomonas aeruginosa Ucp0992.* Coll Sur. B., 79, 174-183.

Sing, M., Saini, V.S., Adhikari, D.K., Desai, J.D., Sista, V.R. (1990) *Production of bioemulsifier by a SCP-producing straing of Candida tropicalis during hydrocarbon fermentation.* Biotechnol. Lett., 12, 743-746.

Sobrinho, H. B. S., Rufino, R. D.; Luna, J. M., Salgueiro, A. A., Campos-Takaki, G. M., Leite, L. F. C., Sarubbo, L. A. (2008) *Utilization of two agroindustrial by-products for the production of a surfactant by Candida sphaerica UCP0995.* Process Biochem., 43, 912-917.

Van Hamme, J. D., Singh, A., Ward, O. W. (2006) *Physiological aspects. Part 1 in a series of papers devoted to surfactants in microbiology and biotechnology.* Biotechnol. Adv., 24, 604-620.

Wasko, M. P., Bratt, R. P. (1990) *Properties of a biosurfactant produced by the fuel contaminant Ochrobactrum anthropii.* Inter. Biodeter. 27, 265-73.

Wu, J-Y., Yeh, K-L., Lu, W-B., Lin, C-L, Chang, J-S. (2008) *Rhamnolipid production with indigenous Pseudomonas aeruginosa EM1 isolated from oil- contaminated site.* Biores. Technol., 99, 1157-1164.

Zhang, Y., Miller, R. M. (1992). *Enhanced octadecane dispersionmand biodegradation by a Pseudomonas rhamnolipid msurfactant (biosurfactant).* Appl. Environ. Microbiol.,m58, 3276-82.

In: Bioremediation
Editor: Alexander C. Mason

ISBN: 978-1-61122-730-7

Chapter 17

BIOREMEDIATION WITH MARINE ALGAE: A CASE STUDY OF THE SODA ASH INDUSTRY

R. N. Jadeja[1], Wenjie Ren[2] and Qixing Zhou[2]
[1]Department of Chemistry, Faculty of Science, The Maharaja Sayajirao University of Baroda, Vadodara, India
[2]Key Laboratory of Terrestrial Ecological Process, Institute of Applied Ecology, Chinese Academy of Sciences, Shenyang 110016, China

Marine algae also known as seaweeds are available in three different colors; Rodophyta (red), Pheophyta (brown) and Chlorophyta (green). They can withstand the extreme polluted conditions and are potential source of phyco-colloids. Agar-agar, alginic acid and carragenan to name a few are the important phyco-colloids extracted from the cellular wall of these algae. Due to economic importance and abundant availability, the seaweeds are important candidate for the *insitu* bioremediation as they can be artificially cultivated in the polluted site. The present chapter aims at the exploration of the different Indian Marine Alga for the potential bioremediation of the soda ash industry effluent both in laboratory condition and in the field. Some of the results from nutrient absorption capacity experiments conducted in the author's laboratory will be discussed. In addition, the interactions of the alga with the effluent and its impact on phyco-colloid availability and diversity will also be presented.

1.1 MARINE ALGAE OR SEAWEEDS

Seaweeds are a widely available source of biomass as over two million tonnes are either harvested from the oceans or cultured annually for food or phycocolloid production, especially in the Asia-Pacific region (Zemke-White and Ohno, 1999). Seaweeds (Fig.1) are found throughout the world's oceans and come in three basic colours: brown (Phaeophyta), red (Rhodophyta), and green (Chlorophyta). Brown and red seaweeds are almost exclusively marine but the vast majority of green-coloured seaweeds are freshwater and terrestrial. The brown colour of the Phaeophyta results from the dominance of the xanthophyll pigment fucoxanthin which masks the other pigments while the red colour of the Rhodophyta is due to the presence of the pigment phycoerythrin which reflects red light and absorbs blue light. In

Asian countries, several species of seaweed are used as human food, to provide nutrition and a peculiar taste. Fresh dried seaweed is extensively consumed, especially by people living in coastal areas. They are of nutritional interest as they are low calorie foods but rich in vitamins, minerals and dietary fibre (Noda, 1993; Oohusa, 1993). Currently, human consumption of green algae (5%), brown algae (66.5%) and red algae (33%) is high in Asia, mainly Japan, China and Korea (Oohusa, 1993). However, demand for seaweed as food has, now also increased in North America, South America and Europe. The different species consumed have great nutritional value as they are source of proteins, carbohydrates, minerals and vitamins. Seaweeds are also called sea vegetables because of their high nutritive value. Seaweeds are traditionally used for human and animal nutrition.

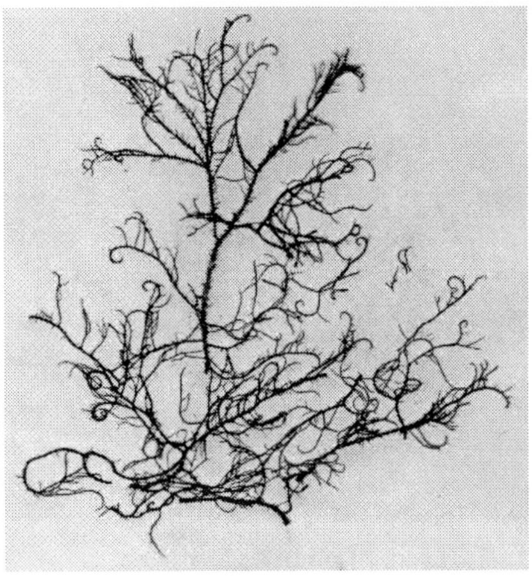

Hypnea musciformis

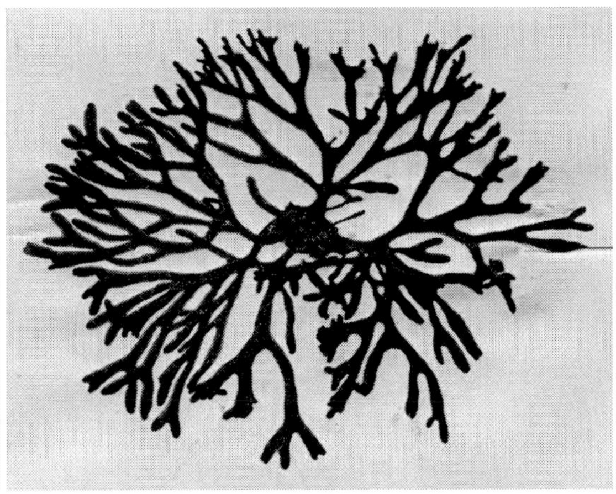

Codium dwarkese

Figure 1. Some of the seaweeds used in the present study.

Historically, seaweeds were harvested from natural populations. The increasing food demand has led to a dramatic growth of the seaweed industry since the 20th century. The seaweeds are the source of phyco-colloids like agar, alginic acid and carrageenan which are extracted from the cellular wall of Rodophyceae (red algae) and Pheophyceae (brown algae). These products are difficult to synthesize chemically because of the formidable chemical barriers and therefore for these commercially important products we have to depend on seaweed resources.

Gujarat has the longest available seacoast in India and the substratum is rocky in many parts of Gujarat, which provides the suitable environment for algal growth. Out of more than 300 species found along the coast.

Bioaccumulation is an important process through which chemicals can affect living organisms. An increase in the concentration of a chemical in a biological organism overtime may occur compared to the chemical's concentration in the environment. Bioaccumulation occurs when an organism absorbs a toxic substance at a rate greater than that at which the substance is lost. Several process including uptake, storage and elimination are involved during bioaccumulation. Bioaccumulation results from a dynamic equilibrium between exposure from the outside environment and uptake, excretion, storage, and degradation within an organism. Understanding of the dynamic process of bioaccumulation is a critical consideration in the regulation of chemicals such as aquatic metals.

Algae may not only be significant for biomonitoring studies and could also be a useful phytoremediation technology to restore water quality due to high bioaccumulation abilities. The algae species and amounts can directly reflect the water quality. Until now, however, most of the bioremediation studies have focused on the green seaweeds Ulva, the red seaweed Gracilaria (Anderson et al., 1999; Hernandez et al., 2006; Naldi and Viaroli, 2002; Suzuki et al., 2005; Troell et al., 1997) and the kelp, Laminaria (Chopin et al., 2003) for nutrient bioremediation.

1.2 Soda Ash Industry

Soda ash is the common name for the technical grade anhydrous sodium carbonate (Na_2CO_3). In the eighteenth century soda ash was produced by LeBlanc process based on roasting salt cake with carbon and limestone. The synthetic procedures available for the soda ash are Solvay process, Akzo Dry Lime Process, Dual Process and New Ashai Process. The most accepted technology for producing soda ash is the ammonia-soda or Solvay process developed by Ernest Solvay in 1861. The Solvay process is widely accepted because it requires low investment and has low maintenance costs compared to other manufacturing processes for soda ash. Of the six soda ash manufacturing units in India, four units use the Solvay process. This process produces 91.2% of the soda ash produced in India. The standard ammonia soda process is based on the following chemical reaction:

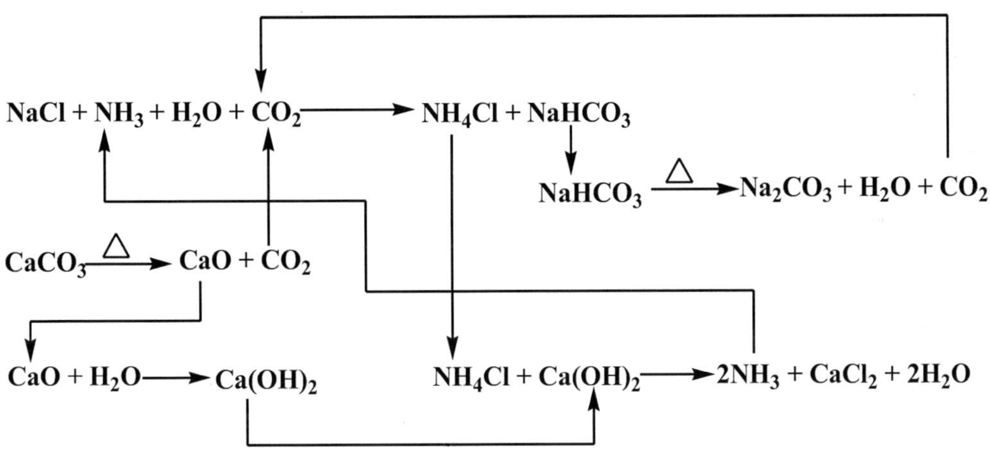

As the Solvay process is commercially viable, most of the chemical industries, including some in the Gujarat, India, produce the soda ash by this process. The main raw material as can been seen from the chemical reactions are a brine solution (source of salt), lime stone (source of carbon dioxide) and ammonia. Most of the ammonia is used in the process; however, some of the ammonia still remains in the effluent. To observe the pollution standard (Table 1) many industries dilute the generated effluent with the seawater and then discharges in the sea, sometime via a submarine outfall.

Table 1. Norms for effluent disposal – Solvay process (marine recipient body)[a]

Parameter	
pH	6.5-9
Temperature	45°C or less
Oil and grease	2 mg/ L
Total suspended solids	500 mg/ L
Ammoniacal nitrogen	5 mg /L
Bio-assay	96 hrs 90% survival

[a] Source: Central pollution control board.

1.3 Phycoremediation Experiments

Phycoremediation may be defined in a broad sense as the use of macroalgae or microalgae for the removal or biotransformation of pollutants, including nutrients and xenobiotics from wastewater and CO_2 from waste air. This field has evolved from the early work done by Oswald (Oswald et al., 1963) for the use of microalgae for tertiary treatment of municipal wastewater to many other applications in which microalgae and macroalgae are cultivated and utilized for specific bioremediation needs. Thus, phycoremediation comprises several applications: (a) nutrient removal from municipal wastewater and effluents rich in organic matter; (b) nutrient and xenobiotic compounds removal with the aid of algae-based biosorbents; (c) treatment of acidic and metal wastewaters; (d) CO_2 sequestration; (e) transformation and degradation of xenobiotics; and (f) detection of toxic compounds with the aid of algae-based biosensors.

As we have discussed in the last section, the main pollutant from the effluent of the soda ash industry is the ammonia. The ammonia will not only affect the pH of the seawater but at the same time affect many of the aqueous organisms like fish etc. The author has seen the deadly effects of ammonia on the fish during his visit to the site polluted from such effluent. Similar time of ammonical effluent releases caused massive deaths of shellfish, which were washed up on the beaches of Lusi, resulting in economic losses of at least US $630,000. Therefore, it is very much essential to carryout the remediation of the discharged ammonia. Nutrient removal with the aid of algae compares very favorable to other conventional technologies (Delanoue et al., 1992). With this aim, series of experiments were carried out in the laboratory to find out the bioaccumulation abilities of the seaweed growing in the polluted area. The seaweed was collected from the natural unpolluted site and was brought to the laboratory. The plants were thoroughly cleaned to remove the epiphytes and other contaminations. They were finally put in the sterilized seawater for few days before using for the bioaccumulation experiments. The calculated quantity of the each seaweed species was then kept in the different treatments having different concentration of the ammonia generated either chemically or directly putting the effluent collected from the soda ash industry. The ammonia in the seawater will be converted to nitrate and nitrite by the seaweeds as shown in following chemical reactions:

$$2NH_4^+ + 3O_2(g) \rightarrow 2NO_2^- + 4H^+ + 2H_2O$$

$$2NO_2^- + O_2(g) \rightarrow 2\,NO_3^-$$

The concentration of the nitrate and ammonia along with other parameters such as dissolved oxygen and pH were measured by the known methods (Jadeja and Tewari, 2007), every 24 hours after introducing the seaweeds in the ammoniated environment. The results obtained are shown in Fig.2. Generally, for seaweeds that have been studied, ammonium uptake rates exceed those for nitrate at ecologically relevant concentrations. At very high concentrations (N30–50 mM), ammonium may be toxic to some seaweeds(Waite and Mitchell, 1972).

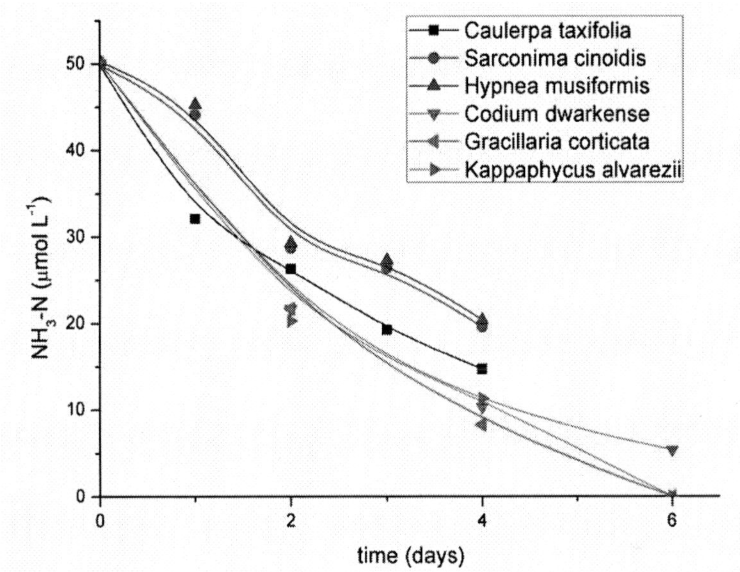

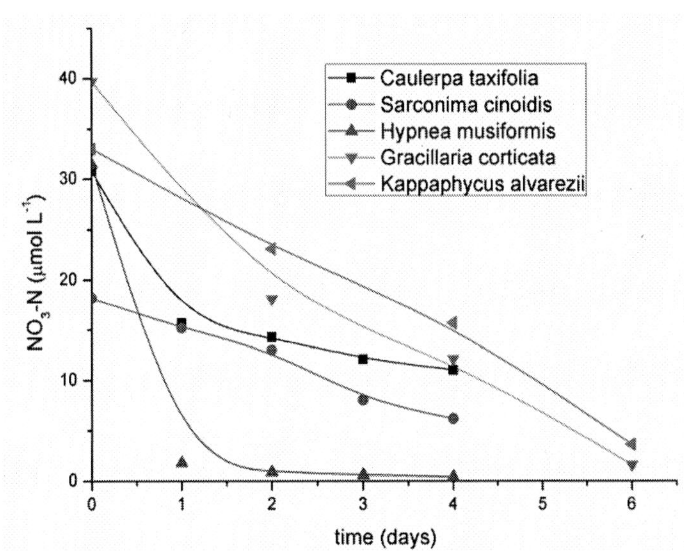

Figure 2. Bioaccumulation of ammonia and its oxidized products by the various species of seaweeds at different concentration.

Nutrient concentrations and the form of the limiting nutrient (NH_4-N vs. NO_3-N) may influence uptake rates and growth of seaweed (Harrison and Hurd, 2001). In the ocean-based farm, greater reduction of NH_4-N (50–94%) than NO_3-N (21–38%) by *P. yezoensis* was observed. An annual yield of 802.95 kg dry wt ha^{-1} was obtained when NH_4-N was the principal nutrient in the water. In laboratory experiments, Wu et al. (1984) also reported that their strain of *P. yezoensis* had a preference for NH_4-N rather than NO_3-N. Higher uptake rates of NH4-N were also observed in *Gracilaria tenuistipitata* (Haglund and Pedersen, 1993) and in *U. lactuca* (Neori et al., 1996). Other seaweeds can take up NH_4-N preferentially over NO_3-N and therefore NH_4-N inhibits the uptake of NO_3-N by up to 50% (DeBoer, 1981). However, Carmona et al. (2006) and Hanisak (1990) showed that Porphyra species and *Gracilaria tikvahiae* had similar growth rates irrespective of the nitrogen sources.

The percentage reduction of ammonia in red seaweeds like *Gracilaria corticata* and *Kappaphycus alvarezii* were 100% both in ammonia in seawater as well as ammonia from effluent while Ingeneral in green seaweeds like *Codium dwarkese* the reduction was higher in effluent affected seawater than the same concentrations of ammonia in seawater. Red seaweed- *Kappaphycus alvarezii* having branched finger like thick thallus could comparatively oxidized, absorb and accumulate higher concentration of ammonia and its oxidized products than the green alga- *Codium dwarkese*, having similar morphology (Fig. 1), at the different concentrations of initial ammonia. Therefore, the thicker cylindrical fronds of algae irrespective of green, brown or red seaweeds were better for this purpose than thin filamentous fronds/ algae having higher plant type structure at studied concentrations of ammonia. (Table 2).

Table 2. Oxidation of ammonia to its oxidized products and accumulation of nitrogen by the different form of thallus of seaweeds at 4.29gL^{-1} initial density and 25 and 50μmolL^{-1} initial concentration of ammonia in seawater on 4th day of experiment

Seaweed species	Codium dwarkense		Hypnea musciformis		Kappaphycus alvarezii	
Type of thallus	Branched finger like thick fronds		Thin branched filamentous		Branched finger like thick fronds	
Parameter/ Concentration (μmolL^{-1})	25	50	25	50	25	50
NH_4-N(%)	91.06	89.22	47.43	59.44	100	100
NO_2-N(%)	+49.28	14.71	+79.1	+28.6	100	100
NO_3-N(%)	+41.52	+38.58	97.85	98.73	92.48	89.11
TN(%)	61.97	59.75	32.93	56.06	36.92	92.31
Accumlation of N	37.09	21.71	61.88	72.15	66.85	62.61

Red algae are considered efficient in taking up nutrients rapidly due to their mechanisms for storing large reserves of nutrients. This behavior, however, seems species specific as reflected by the wide range of nutrient reduction efficiency of these seaweeds (2–90%) as summarized by Troell et al. (2003).

The genus *Gracilaria* (Rhodophyta) has been demonstrated to be the most attractive candidate for intensive culture because of its ability to achieve high yields and produce commercially valuable products. *Gracilaria* species, being an efficient nutrient pump, offer both high bioremediation efficiency and commercial value in established markets, such as agar-agar, human consumption, and fodder for other high-valued aquaculture organisms, such as abalone (Chopin et al., 2001; Fei, 2004; Neori et al., 2004). The cultivation of the agarophytic red alga *Gracilaria* has become economically important in some regions of the world, such as south Asia, South America and southern Africa (Santelices and Doty, 1989). Rapid development in seaweed cultivation was related to many desirable biological features, such as easy anchoring of their thalli in substrata, high regeneration capacity, high growth rates and high agar yield (Buschmann et al., 1994).

Nutrient uptake efficiencies by seaweeds are relatively low in open-water systems due to the 3-D hydrographic nature of the water flow (Troell et al., 1997; 2003). Further studies on the open-water bioremediation potential of seaweeds in dynamic open-water environments are still needed (Neori et al., 2004; Troell et al., 1997).

1.4 Effect of Soda Ash Effluent on Phyco Colloids

The impact of the effluent released by the soda ash industry on important red and brown macro algal species *Gelidiella acerosa*, *Gracilaria corticata*, *Soleria robusta*, *Sargassum tenerrimum*, *Padina tetrastromatic* was studied (Jadeja and Tewari, 2009). The study examined the effect of effluent discharge on availability of biomass and percentage of phyco-colloids extraction such as agar, alginic acid and carrageen of these commercial seaweeds. There is a considerable effect on extractable quantity of phyco-colloid of the seaweeds due to uptake of polluted water. The per gram yield of phyco-colloids decreased due to effect of soda ash industry effluent but overall yield will be more because of greater biomass production in the effluent affected region. All the studied species exhibit tolerance against the effluent affected environment especially *Gracilaria corticata* and thus can be cultivated on a large scale in effluent affected area to get a double benefit of cleaning the environment while producing a phyco-colloids as a by product.

1.5 Effect of Soda Ash Effluent on Protein Content of the Seaweeds

Seaweeds are also called as sea vegetables because of their high nutritive values. Seaweeds are traditionally used in human and animal nutrition. Their protein contents differ according to the species and seasonal conditions. In some green seaweed such as the species belonging to the genus *Ulva*, the protein content can represent between 10 and 26%(dry weight) of the plant. For instance, the species *Ulva pertusa*, which is frequently consumed under the name of "ao-nori" by the Japanese people, has a high protein level between 20 and 26% (dry weight) (Fujiwara-Arasaki et al., 1984).

Another independent study (Jadeja and Tewari, 2008) was therefore carried out in the author's laboratory to check the effect of soda ash industry effluent on the protein content of the seaweed *Ulva faciata* and *Chaetomorpha antennina*. Study shows that the effluent has positive effect on the protein content of the alga and thus these species can be used to reduce the effect of soda ash industry pollution because the rise of upto 35% protein level is found in these species of alga due to uptake of polluted water.

1.6 Impact of Soda Ash Effluent on Biodiversity of Seaweeds

The impact of soda ash industry effluent on macroalgal distribution, abundance, species composition, diversity and standing stock was studied, *in situ*, on tropical seaweeds of Arabian Sea, Gujarat was the interest of authors (Zemke-White and Ohno, 1999). *Caulerpa sertulariodes* and *Champia somalensis* were most resistant and had a preference for soda ash industry effluent as they were recorded only near the effluent discharge point. *Codium dwarkense* and *Dictyota bartayresiana* also had similar characteristics as they were recorded only 1 km away from the effluent discharge point. *Caulerpa racemosa, Chamaedoris auriculata, Valoniopsis pachynem* and *Stoechosperumum marginatum* were most sensitive as they were recorded only from the control site. *Ulva lactuca, Caulerpa taxifolia, Gracilaria corticata* and *Laurencia cruciata* were recorded from all sampling sites indicating their adoptability to varying degrees of pollution.

References

Anderson, R.J., Smit, A.J., Levitt, G.J. *Hydrobiologia.* 1999, 399, 455-462.

Buschmann, A.H., Mora, O.A., Gomez, P., Bottger, M., Buitano, S., Retamales, C., Vergara, P.A., Gutierrez, A. *Aquacul. Eng.* 1994, 13, 283-300.

Carmona, R., Kraemer, G.P., Yarish, C. *Aquaculture.* 2006, 252, 54-65.

Chopin, T., Bastarache, S., Belyea, E., Haya, K., Sephton, D., Martin, J.L., Eddy, S., Stewart, I. *J. Phycol.* 2003, 39, 10-10.

Chopin, T., Buschmann, A.H., Halling, C., Troell, M., Kautsky, N., Neori, A., Kraemer, G.P., Zertuche-Gonzalez, J.A., Yarish, C., Neefus, C. *J. Phycol.* 2001, 37, 975-986.

Delanoue, J., Laliberte, G., Proulx, D. *J. Appl. Phycol.* 1992, 4, 247-254.

Fei, X.G. *Hydrobiologia.* 2004, 512, 145-151.

Fujiwara-Arasaki, T., Mino, N., Kuroda, M. *Hydrobiologia.* 1984, 116-117, 513-516.

Haglund, K., Pedersen, M. *J. Appl. Phycol.* 1993, 5, 271-284.

Hanisak, M.D. *Hydrobiologia.* 1990, 204, 79-87.

Harrison, P.J., Hurd, C.L. *Cah. Biol. Mar.* 2001, 42, 71-82.

Hernandez, I., Perez-Pastor, A., Vergara, J.J., Martinez-Aragon, J.F., Fernandez-Engo, M.A., Perez-Llorens, J.L. *Aquaculture.* 2006, 252, 43-53.

Jadeja, R.N., Tewari, A. J. Hazard. *Mater.* 2007, 147, 148-154.

Jadeja, R.N., Tewari, A. J. Hazard. *Mate*r. 2008, 151, 559-561.

Jadeja, R.N., Tewari, A. J. Hazard. *Mater.* 2009, 162, 498-502.

Jadeja R.N., Tewari, A. Indian *J. Mar. Sci.* (In Press).

Naldi, M., Viaroli, P. *J. Exp. Mar. Biol. Ecol.* 2002, 269, 65-83.

Neori, A., Chopin, T., Troell, M., Buschmann, A.H., Kraemer, G.P., Halling, C., Shpigel, M., Yarish, C. *Aquaculture.* 2004, 231, 361-391.

Neori, A., Krom, M.D., Ellner, S.P., Boyd, C.E., Popper, D., Rabinovitch, R., Davison, P.J., Dvir, O., Zuber, D., Ucko, M., Angel, D., Gordin, H. *Aquaculture.* 1996, 141, 183-199.

Noda, H. *J. Appl. Phycol.* 1993, 5, 255-258.

Oohusa, T. *J. Appl. Phycol.* 1993, 5, 155-159.

Oswald, W.J., Golueke, C.G., Cooper, R.C., Gee, H.K., Bronson, J.C. 1963. *Air and Water Pollution,* 7, 627-&.

Santelices, B., Doty, M.S. *Aquaculture.* 1989, 78, 95-133.

Suzuki, Y., Kametani, T., Maruyama, T. *Water Res*. 2005, 39, 1803-1808.

Troell, M., Halling, C., Nilsson, A., Buschmann, A.H., Kautsky, N., Kautsky, L. *Aquaculture.* 1997, 156, 45-61.

Troell, M., Halling, C., Neori, A., Chopin, T., Buschmann, A.H., Kautsky, N., Yarish, C. *Aquaculture,* 2003, 226, 69-90.

Waite, T., Mitchell, R. *Bot. Mar*. 1972, 15, 151-&.

Wu, C.Y., Zhang, Y.X., Li, R.Z., Penc, Z.S., Zhang, Y.F., Liu, Q.C., Zhang, J.P., Fan, X. *Hydrobiologia*, 1984, 116, 475-477.

Zemke-White, W.L., Ohno, M. *J. Appl. Phycol*. 1999, 11, 369-376.

In: Bioremediation
Editor: Alexander C. Mason

ISBN: 978-1-61122-730-7

Chapter 18

BIOSORPTION OF TRIVALENT ARSENIC BY FACULTATIVE MARINE *ASPERGILLUS NIGER*

Anjana K. Vala*[*] *and Rajesh Patel
Department of Physics, Bhavnagar University, Bhavnagar-364 022, India

ABSTRACT

Metal (oid) contaminants present in the industrial effluents have toxic effects on the ecosystem. Arsenic is one such toxic contaminant increasingly distributed in the environment. While traditional metal removal processes have one or the other limitation, biosorption could be a potentially cost-effective alternative for treatment of wastewaters. Various biomaterials have been examined for metal removal efficiency. However, facultative marine fungi are comparatively less explored. In this study, As (III) biosorption ability of facultative marine *Aspergillus niger* fungal biomass was examined as a function of initial concentration of arsenic. Heat killed biomass of *A. niger* was used as biosorbent. Biosorption increased with increasing arsenic concentration. The fungus could remove more than 90% As at all test concentrations. The maximum biosorption value recorded was 108.083 mg/g at As concentration 600 mg/L. The data suggested that facultative marine *A. niger* is an efficient biosorbent and can be harnessed for arsenic remediation.

Keywords: arsenic, biosorption, bioremediation, facultative marine fungi, *Aspergillus niger*

INTRODUCTION

Arsenic is a toxic contaminant increasingly distributed in the environment especially due to anthropogenic activities including metal smeltering, coal combustion, mine tailings, hide tanning waste, waste incineration, cement works, dyes, chemical weapons and arsenic pesticides, etc. Arsenic exposure can induce a variety of adverse physiological responses including carcinogenesis, reproductive and developmental defects, nephropathies, and

[*] anjana_vala@yahoo.co.in

neuropathies (Goyer, 1996). Arsenate [As(V)] and arsenite [As(III)] are the most common inorganic forms of arsenic in the environment, where As(III) is more toxic. Based on USEPA's evaluation (USEPA, 2001) inorganic arsenic is regarded as a major environmental pollutant, and more stringent standards have been introduced for arsenic in many countries (US EPA, 2001). Hence, there has been a growing interest in development of arsenic decontamination techniques. A number of techniques have been employed conventionally, however, most of them are less effective, expensive and generate sludge. Microorganisms offer an attractive option for removal of such a toxicant. Fungi have been reported as efficient metal removal entities (Coulibaly et al., 2003; Kapoor and Viraraghavan, 1995; Pokhrel and Viraraghavan, 2006), facultative marine fungi are very less explored for this purpose (Vala and Upadhyay, 2008). Especially, very few reports are available on harnessing them for arsenic removal (Vala, 2010).

MATERIALS AND METHODS

Preparation of the Fungal Biosorbent

The test fungus *Aspergillus niger* was isolated from waters of Bhavnagar coast (Lat. 21° 45' N and Long. 72°. 14' E), Gulf of Cambay, West Coast of India. It was maintained by subculturing on potato-dextrose-agar medium, prepared in aged seawater and distilled water at a ratio of 3:1 (Schlieper, 1972). Fungal biomass was cultivated in liquid medium for 7 days at 28±2 °C. The medium was prepared in aged seawater and distilled water at a ratio of 3:1. After 7 days incubation, the fungal biomass was harvested from the medium by filtration and washed several times with deionized water. After washing, the biomass was dried at 90 °C in an oven for 24 h. The dried biomass was powdered using a mortar and pestle (Karunasagar et al., 2003) and stored in a desiccator until use.

Effect of Initial As (III) Concentration

To evaluate the effect of initial As (III) concentration (Ceq) on adsorption behavior of As by dried *A. niger* mycelial biomass, aliquots (50 mL) of 5, 10, 15, 20,25, 50, 100, 200, 300, 400, 500 and 600mg/L concentrations of sodium arsenite solutions were added to 100 mL Erlenmeyer flasks with a fixed biomass weight of 4g/L. The flasks were agitated on a rotary shaker (180 rpm) for 2h. The samples were filtered through 0.45 μm millipore filters. All the experiments are carried out in triplicates.

Arsenic Analysis

Removal of As from the aqueous phase was estimated using Hydride Generation Atomic Absorption Spectrophotometry (Shimadzu 6300). As content removed was calculated using the difference between the initial As concentration and that in the sample after respective incubation period. The amount of As sorbed onto the biomass, which represents the metal uptake was expressed as mg/g biomass (dry weight).

RESULTS AND DISCUSSION

In this experiment As (III) biosorption characteristics of *Aspergillus niger* fungal biomass were examined as a function of initial metal ion concentration. The heat killed biomass was used in the determination of optimum conditions. Figure 1 shows the effect of initial concentration on the adsorption capacity of As (III) onto the biosorbent.

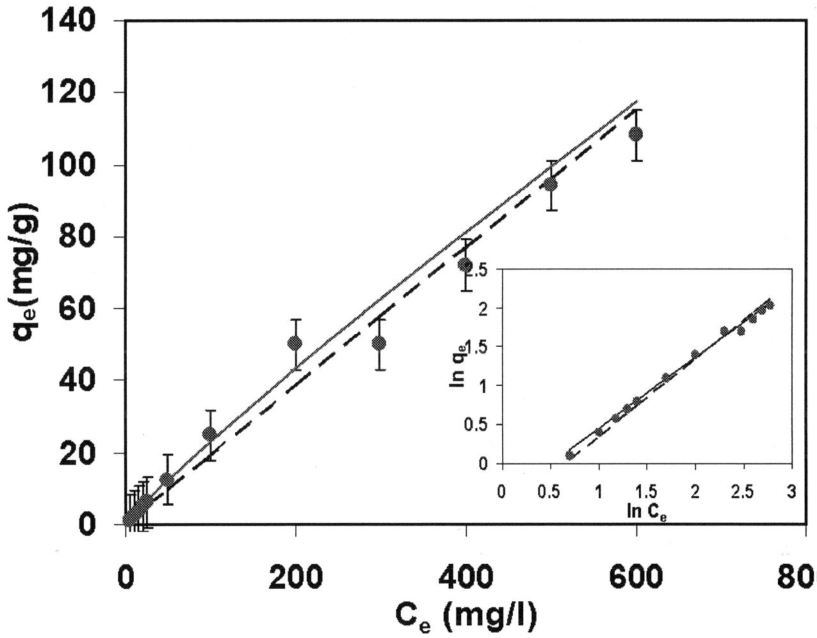

Figure 1. Effect of metal ion concentration on the biosorption of As (III) is shown. The solid line shows Freundlich model and dashed line shows the Langmuir model. The inset shows the linearity of the fit in log scale. The parameters obtained from both the models are $K_F = 0.35$, $n = 1.1$, $q_m = 4\times10^2$ mg/g and $b = 5.5\times10^{-6}$ l/mg.

The process of biosorption involves biomass as a solid phase and a liquid phase containing metal ions. Affinity of biomass for metals determines the ion distribution between solid and liquid phases. To obtain maximum adsorption capacities for the biosorbents, the initial test concentrations of As(III) were increased up to 600 mg/L in an aqueous solution. The experimental maximum adsorption capacities of As(III) onto the fungal biosorbent was 108.083 mg/g biosorbent. Even at the highest supplied concentration no saturation was observed this is a noteworthy observation. The test fungus *A. niger* when examined for its tolerance towards As (III) it could even withstand 1000mg/L of As (III) concentration and exhibited remarkable As removal potential (>20mg/g) when exposed to 100mg/L As. Hence, the fungus has indigenous As removal efficiency and supporting the hypothesis that facultative marine fungi could be a promising metal bioremediation agent. A strain of *A. niger* could biosorb 100% Cr(VI) at 200 mg/l concentration within about 24,000 min (Park et al., 2005). Say et al., 2003 observed that adsorption capacities of *Penicillium purpurogenum* for As (III) was 35.6 mg/g.

Equilibrium studies that give capacity of the adsorbent and the equilibrium relationship between adsorbent and adsorbate are described by biosorption isotherms, which are usually the ratio between the quantity adsorbed and the remaining in solution at fixed temperature at equilibrium. These experimental results are analyzed using following two models.

The Freundlich Isotherm: The Freundlich isotherm model was chosen to estimate the adsorption intensity of the sorbent towards the adsorbent. The empirical equation used is,

$$q_e = K_F C_e^{1/n} \tag{1}$$

where, q_e is the biosorption capacity of metal (mg/g), C_e is the equilibrium concentration of metal ion in an aqueous solution (mg/L), K_F and n are Freundlich biosorption isotherm constants. The Freundlich isotherm equation frequently gives an adequate description of biosorption over a restricted range of metal ion concentration. The values of 1/n, less than unity is an indication that significant adsorption takes place at low concentration. The higher the K_F value the greater the adsorption intensity. The values of K_F = 0.35 and n = 1.1 is obtained from the fit. The regression coefficient obtained is R^2 = 0.9863. These parameters are obtained from the fit to the experimental results shown in Figure 1. The values of K_F and n show ease of separation of heavy metal ions from metal contaminated water with high adsorption capacity as reported by Goyal et al., 2008 also.

The Langmuir Isotherm: The Langmuir isotherm was chosen for the estimation of maximum bisorption capacity corresponding to metal ion concentration. This equation is given as,

$$q_e = \frac{q_{\max} b C_e}{1 + bC_e} \tag{2}$$

where, q_{max} is the maximum biosorption capacity (mg/g), b is the binding constant (l/mg). From the fit the parameters obtained are q_{max} is 4×10^2 and b is 5.5×10^{-6}. These parameters are obtained from the fit to the experimental results shown in Figure 1. Holan and Volesky (1995) had observed 351mg/g lead accumulation by fungus *Absidia orchidis*. In the present study *A. niger* has exhibited promising As removal potential. Hence, despite earlier pessimistic reports on *Aspergillus* biosorption, the test fungus seems to sorb arsenic well, as observed for Au, Co, Th, and Zn also (Volesky 1994; Kuyucak and Volesky, 1988; Gadd et al., 1988). The data suggested that facultative marine *A. niger* is an efficient biosorbent and can be harnessed for arsenic bioremediation.

CONCLUSION

For efficient bioremediation applications, laboratory experiments are one of the important prerequisites. In this study, the biosorption of arsenic from aqueous solutions onto the heat-killed biomass of facultative marine fungus *A. niger* was studied in the concentration range of 5 – 600 mg/L. Mathematical models for the experimental results suggested that *A. niger* has potentials for arsenic removal from contaminated waters.

ACKNOWLEDGMENT

Thanks are due to the Department of Science and Technology, Government of India, for financial support sanctioned under SERC Fast Track Proposals for Young Scientists Scheme (No. SR/FT/L-99/2006) to AKV.

REFERENCES

Coulibaly, L; Gourene, G & Agathos, N.S. (2003). *Utilization of fungi for biotreatment of raw wastewaters.* Afri *J Biotechnol, 2(12)*: *620*-630.

Gadd, G.M.; White C. & De Rome L. 1988. *Heavy metal and radionuclide uptake by fungi and yeasts.* In : P. R. Norris and D. P. Kelly, eds., Biohydrometallurgy A. Rome, Chipperham, Wilts, U.K. p. 104.

Goyal P.; Sharma P.; Srivastava S. & Srivastava M. M. (2008). *Saraca indica leaf powder for decontamination of Pb: removal, recovery, adsorbent characterization and equilibrium modeling.* Int. J. Environ. Sci. Tech., 5 (1), 27-34.

Goyer, R.A. 1996. *Toxic effects of metals.* In: Klassen CD, Amdur MO and Doull J eds. Casarett & Doull's Toxicology: The Basic Science of Poisons. McGraw- Hill, New York. pp 691–736.

Holan, Z.R. & Volesky, B. (1995). *Accumulation of cadmium, lead and nickel by fungal and wood biosorbents.* Appl. Biochem. Biotechnol. 53, 133-146.

Kapoor, A & Viraraghavan, T. (1995). *Fungal biosorption. An alternative treatment option for heavy metal bearing wastewater A review.* Bioresource Technol, 53: 195-206.

Karunasagar, D.; Arunachalam, I.; Rashmi, K.; Naveena Lavanya Latha, I.& Maruti Mohan P. (2003). *Biosorption Of Inorganic And Methyl Mercury By A Biosorbent From Aspergillus niger.* World J. Microbiol. and Biotechnol., 19,291-295

Kuyucak, N. & Volesky, B. *(*1988). *Biosorbents for recovery of metals from industrial solutions* Biotechnol. *Let. 10* (2), *137*-142.

Park, D.; Yun, Y.S.; Jo, J.H. & Park, J.M. (2005). *Mechanism of hexavalent chromium removal by dead fungal biomass of Aspergillus niger.* Water Res 39:533–540.

Pokhrel, D. & Viraraghavan, T. (2006). *Arsenic removal from an aqueous solution by a modified fungal biomass.* Water Res, 40: 549 – 552.

Say, R.; Yilmaz, N. & Denizli, A. 2003. *Biosorption of cadmium, lead, mercury, and arsenic ions by the fungus Penicillium purpurogenum.* Sep. Sci. Technol. 38, 2039–2053.

Schlieper C (Ed), 1972. *Research Methods in Marine Biology.* Sidgwick and Jackson Ltd., London.

USEPA, 2001. U.S.EPA *Workshop on managing arsenic risks to the environment: Characterization of waste, chemistry, and treatment and disposal.* May 2001, Denver,Co, pp107. Available from: http://www.epa.gov/ord/NRMRL/pubs/625r03010/625r03010 total.pdf

Vala, A.K. & Upadhyay, R.V. (2008). *On the tolerance and accumulation of arsenic by facultative marine Aspergillus sp.* Res J Biotechnol, (special issue Proceedings of International Society Biotechnology – 2008 conference) 366-368.

Vala, A.K. (2010). *Tolerance and removal of arsenic by a facultative marine fungus Aspergillus candidus.* Bioresour. Technol. 101, 2565–2567.

Volesky, B. 1994. *Advances in biosorption of metals: Selection of biomass types.* FEMS Microbiol. Rev. 14, 291-302.

In: Bioremediation
Editor: Alexander C. Mason

ISBN: 978-1-61122-730-7

Chapter 19

REMEDIATION OF HEAVY METAL CONTAMINATED SOILS BY PHOSPHATE-BEARING BIOTECHNOLOGICAL PRODUCTS

M. Vassileva[1], A. Medina[4], A. Reyes[1], V. Martos[2,3], S. Shilev[5] and N. Vassilev[*1,2]

[1]Department of Chemical Engineering, Faculty of Sciences, University of Granada, Spain
[2]Department of Plant Physiology, Faculty of Sciences, University of Granada, Spain
[3]Institute of Biotechnology, University of Granada, Spain
[4]Department of Microbial Ecology, Netherlands Institute of Ecology (NIOO-KNAW), The Netherlands
[5]Agricultural University, Plovdiv, Bulgaria

ABSTRACT

Elevated concentrations of heavy metals in soil from anthropogenic sources (fertilizers, amendments) or mining activities pose long-term risk to environmental and sustainable production. The deteriorated physical and biological characteristics of contaminated soils need to be improved to establish a vegetation cover. Heavy metal contaminated soil can be treated by operations based on chemical, physical or biological techniques. An interesting approach applied in in-situ remediation of heavy metal contaminated soil includes a variety of phosphate amendments such as soluble phosphate and P-bearing insoluble sources, mainly rock phosphate (RP). In this article, we would like to present our results on remediation of heavy metal contaminated soils by microbially-treated RP obtained as a result of fermentation processes using agro-industrial wastes as substrates. This biotechnological product exerts its activities by three components: organic matter, partially solubilized RP, and microbial biomass. All experiments based on microbially-treated agro-industrial wastes and RP in soil-plant systems contaminated with heavy metals showed higher efficacy in the presence of arbuscular mycorrhizal (AM) fungi. The overall effect is enhanced plant growth, improved soil

* Corresponding Author

structural characteristics, higher soil enzyme activity, and enhanced diversity of soil microorganisms. Results of these experiments showed that the combination of plant growth enhancement and reduced metal translocation caused by our biotechnological products combined with additional microbial inoculations could be regarded as a promising strategy for remediation of heavy metal contaminated soils. Further strategies and biotechnological schemes are discussed.

INTRODUCTION

During the last decades, heavy metals originating from anthropogenic activities such as manufacturing industry, commercial fertilizers introduced into soils, sewage sludge disposal as well as mining and ore processing caused contamination of the environment. Heavy metals pollution of air, ground and surface water, and soil had gradually become a global problem worldwide. In particular, the accumulation of heavy metals in soil-plant ecosystems may deteriorate the quality and safety of agricultural products. Entering the food chain, these toxic elements which tend to bio-accumulate, affect animal and human health (Guo et al., 2006). There are several techniques aimed at remediation of metal-contaminated soils. Excavation, removal and replacing the topsoil contaminated layer, although effective, is considered too expensive (Rayan et al., 2001). Phytoremediation, solidification, vitrification and chemical remediation form another group of in situ remediation methods which can replace the excavation technique. Chemical remediation consists of amendment of contaminated soils with organic matter or chemical and mineralogical materials thus reducing metal solubility and bioavailability. Amendments such as beringite, red mud, ochre leonardite or lime contain various precursors that reduce the mobility and bioavailability of the toxic elements through different mechanisms (complexation, (co)precipitation or entrapment) (Hamon et al., 2002; Adriano et al., 2004; Kumpiene et al., 2008).The introduction of agricultural wastes and industrial by-products as amendments into heavy metal contaminated soils is accepted as an environmentally friendly and cost-effective approach. Various by-products have been tested as in situ additives to remediate metal contaminated soil. Particularly, P-bearing compounds such as rock phosphate have shown to provoke a high metal mobility reduction (a large number of references are included in the text).

During the last decade, we have developed an efficient biotechnological method for solubilization of insoluble phosphates cultivating acid-producing fungi on media containing agricultural wastes mainly sugar beet wastes (SB) and dry olive wastes (DOW) (Vassilev and Vassileva, 2003). Further introduction of the resulting bio-products which included three components (various types of mineralized organic matter, soluble phosphates and fungal biomass) into soil-plant systems resulted in higher plant growth and P uptake (Vassilev et al., 1997; Vassielva et al., 1998). Recent studies proved another important application of our product as a remediation tool for amelioration of the negative effects of heavy metals on plant-soil systems. The aim of this paper is to describe our studies and explain the advantages of the biotechnological approach to remediation of contaminated soils.

REMEDIATION PROPERTIES OF PHOSPHATE-BEARING SOURCES

Phosphate-bearing materials are widely used for immobilizing heavy metals in contaminated soil. The mechanism of metal immobilization by these materials is based on a formation of geochemically stable metal-phosphates. Particularly natural P-bearing materials such as hydroxyapatite [Ca5(PO4)3OH] and rock phosphate [e.g. Ca5(PO4)3F] are extensively studied for their ability to immobilize metals (Ma et al., 1993; Prasad et al., 2000, 2002; Mavropoulos et al., 2002; Cao et al., 2003, 2004; Martin et al., 2008). As suggested by Ma et al. (1993) immobilization of metals based on reactions with hydroxyapatite occurs in two reaction sequences:

a) Dissolution of original metal species and P-bearing material in soil and
b) Precipitation of dissolved metal and P.

It is accepted that apatite [Ca10–a–bNaa Mgb(PO4)6–x(CO3)xF2 + 0.4x] is the ideal P-bearing material for metal remediation. In this case, a preliminary step is the metal incorporation into the apatite structure through a process of ion exchange with Ca2+. This reaction can facilitate apatite dissolution and phosphate is released simultaneously. Further, a new metal-phosphate solid phase (such as Ca10–xPbx(PO4)6(OH)2) would form provoked by the lower solubility between metal ions and phosphate (Shrestha et al., 2003). Based on these reactions, apatite materials can be used effectively to immobilize a large number of metals such as Pb, Ba, Cd, Co, Cu, Mg Mn, Th, Zn and U in metal contaminated sites (Peng et al., 2009).

In general, the above described series of reaction can be enhanced in the acid condition with a soil pH value of 4 and below (Zhang and Ryan, 1999). Therefore, to dissolve P materials such as rock phosphate, soil acidification may be needed. The acidification, however, may increase the leaching of other trace elements and increase the overall soil toxicity. In order to avoid the negative effects of soil acidification, various strategies are proposed which include addition of more soluble P amendments, liming, etc. (Moon et al., 2000; McGowen et al., 2001).

It is important to note that phosphate-based salts and phosphoric acid are more soluble compared to phosphate rock and therefore more efficient. On the other hand, these easily soluble P-bearing materials caused pH decrease in soil which leads to increase in the metal concentration and facilitates the formation of metal-phosphate minerals (Chen et al., 2007; Melamed et al., 2003).

REMEDIATION PROPERTIES OF ORGANIC MATTER

On the other hand, it is well known that the-bioavailability of heavy metals in soil is affected by the presence of organic matter. Organic matter retains effectively heavy metals forming complexes and chelates of varying stability (Leita et al. 1999). Such heavy metal-organic associations can be found both in soil solution and at the surfaces of either soil constituents or biosolids. In a heavy metal-polluted soil, Kiikila et al. (2002) studied the effect of biosolids as organic immobilizing agents and observed that the exchangeable metal concentration decreased. In addition, the solubility of organic matter determines the mobility of heavy metals. Generally, the complexation of metal ions with insoluble organic matter can lower their mobility. On the contrary, dissolved organic compounds facilitate the formation of soluble metal complexes which in turn increases their mobility. It is accepted that the reactions of complex formation between heavy metals and organic complexants is one of the most important processes in soil which determines the mobility of heavy metals and their bioavailability.

Pascual et al. (1997) demonstrated that the addition of various forms of organic wastes to the soil positively changed some important soil characteristics such as biomass carbon, basal respiration, biomass C/total organic C ratio, and activate the metabolism of soil microorganisms. It is important to note that heavy metals presented in organic matter (sewage sludge or compost) negatively affected the microbial development and metabolic activity which, in turn, decrease the organic matter microbial decomposition thus causing accumulation of organic matter. This phenomenon seems of great importance as microorganisms are deeply involved in geochemical processes. In particular, various microorganisms can mobilize metals through reactions such as autotrophic and heterotrophic leaching, chelation, and methylation. These processes have been well described in a recent review paper (Gadd 2004). In general, they lead to dissolution of insoluble metal-bearing compounds and minerals, and desorption of metals from exchange sites on clay minerals or organic matter in the soil. Heterotrophic metabolism can also lead to leaching when microbial cells release organic acids and siderophores. For example, acidification through organic acid production can supply both protons and metal complexing anions thus leading to metal release (Gadd and Sayer 2000).

In general, mineralization of lignocellulosic agro-industrial wastes by microbial processes and simultaneous solubilization of inorganic insoluble phosphates provides for plants an organic amendment rich in polysaccharide compounds and available P (when RP is applied in the fermentation process). In addition, this mineralized organic matter can be used as C and energy source for activities of soil microorganisms. This effect could be further extended to enhancement of the soil enzyme activities which are considered as major factors contributing to overall soil microbial activity, and soil quality (Garcia et al. 1997). The increase of enzymatic activities in soils is involved in an increase in the availability of nutrients to plants, which in turn have a positive influence on soil fertility thus completing the beneficial set of effects of the biotechnological products developed by our group.

Remediation Properties of Our 3-Component Biotechnological System

Heavy metal polluted soils are generally characterized by poor soil structure, low water-holding capacity, lack of organic matter and nutrient deficiency. Thus, in order to establish the plant cover successfully in a contaminated soil, physical and biological soil characteristics need to be improved. Bearing in mind the above short description of the remediation effects of phosphate-bearing materials and organic matter, our biotechnological scheme, which was successfully proved as a plant growth enhancer and bio-protector (Vassilev et al., 2003, 2006), was employed for the first time in heavy metal contaminated soil rehabilitation activities (Medina et al. 2005; 2006; 2010). In a microcosm experiment, 50 g/kg of the mixture produced after solid-state fermentation processes of SB treatment by *A. niger* in presence or absence of rock phosphate, was introduced into a heavy metal (Zn or Cd) contaminated soil-plant system. The mixture of microbially-mineralized agro-waste, simultaneously (partially) solubilized RP, and the fungal mycelium were the most effective amendment in improving growth and mineral nutrition of the white clover used as a test plant. In combination with the mycorrhizal fungus *Glomus mosseae*, this treatment increased plant growth about 28 times more than in non-mycorrhized control plants. It is important to note that *Trifolium* plants failed to establish in a Cd-contaminated soil without the inclusion of the biotechnological product (Medina et al. 2005). To illustrate the advantage of our bio-product, it should be noted that the effect of *G. mosseae* on plant biomass shifted from +86% (in the treatment without amendment) to +1192% in SB+RP+*A. niger* amended Zn-contaminated soil (Medina et al. 2006). Improved nodule formation was observed in soil-plant systems inoculated with *Rhizobium trifoli* and amended with microbially-treated SB and RP while this process was highly depressed in Zn-contaminated control soil and was nearly zero in control Cd-contaminated soil. As consequence of an enhanced plant biomass, Zn phytoextraction increased by 1832 % over untreated ones. Similar results were registered in Cd-contaminated soil. In a separate phytoremediation experimental scheme using a natural heavy metal multi-contaminated soil, our biotechnological product was successfully tested in combination with autochthonous mycorrhizal fungi and bacteria such as *Bacillus cereus* (Azcon et al. 2009a, 2009b). The bacterium was previously proved as a heavy-metal tolerant microorganism. The SB/RP/*A. niger* amendment increased the content of most of the analyzed elements (B, Cu, Mo, Cd, Cr, and Mn) except for Al. Mycorrhizal and bacterial addition improved this effect for whatever metal present in the multi-contaminated soil. Plant biomass *(Trifolium repens)* highly increased compared to non-amended control which is an important result particularly in phytoremediation experiments.

Each component of the biotechnological product including plant-growth stimulating microorganisms exerts additional effects within the overall beneficial activity. For example, even in heavy metal contaminated soil, the effect of microbially-treated SB on shoot biomass of *T. repens* ranged from 434% (without mycorrhiza) to 549% (with mycorrhiza) increase while in unamended soil, the mycorrhizal effect on shoot biomass was 232% (Medina et al. 2006). However, treated SB/RP increased shoot growth over control by 1006% (in absence of mycorrhiza) and by 1192% (in presence of the AM fungus *Glomus mosseae).* Highly increased root development and the number of nodules formed were registered under the same conditions in mycorrhized plants.

The benefits of organic amendments are due also to the improvement of the physical characteristics of the soil, which in turn favors the establishment and viability of a stable plant cover. In this regard, microbially-treated agro-wastes improved the structural stability of rhizosphere soil of plants to a statistically significant level. This restoration effect of soil structure may depend on the amount and nature of the organic matter added. In a number of publications we have demonstrated that microbially-treated agro-industrial wastes significantly improved soil properties. Aggregate stability, water-soluble C, and water-soluble carbohydrate C were increased in the rhizosphere when semiarid soil was supplemented with *A. niger*- treated DOW and SB alone or in combination with soil microorganisms *Yarrowia lipolytica,* AM fungi and *Rhizobium* using *Dorycnium* plants (Medina et al. 2004a). Root AM colonization was increased considerably by the amendments and by *Y. lipolytica* inoculation.

The low nodulation found in this degraded soil was also increased by *A. niger*-treated DOW/SB/RP. The application of organic amendments to the soil increased enzyme (beta-glucosidase and phosphatase) activities as well as the biodiversity of AM populations. These first results were further confirmed in other works which showed an increase in aggregate stability by the addition of the amendments (microbially-treated SB- or SB+RP) to the soil, while no effect of AM inoculation was observed in heavy metal (Zn) contaminated soil (Medina et al., 2006). It was suggested that microbial population developed after the addition of the SSF-processed wastes (that can be used as C and energy sources for activities of soil microorganisms) could be responsible for soil aggregate formation and stabilization.

Similar beneficial effect of our biotechnological product related to stabilization of soil aggregates was found in acidic semiarid mine tailings (Carrasco et al. 2009). Particularly after soil drying aggregate stability was 66% higher than the control soil. The fermentation mixture of mineralized organic matter and solubilized RP raised the soil pH and increased soil water-soluble carbon under either watering or drying conditions. It was concluded that the addition of microbially-treated SB was more effective than inoculation with an autochthonous bacterium, for stabilizing the soil structure of acidic mine tailings during desiccation, although the overall effectiveness was dependent on the degree of acidity of the tailing. Soil surface structure stabilization of metal mine tailings could reduce erosion, protect soil against degradation and limit the spread of metal contamination particularly when dealing with fragile, semiarid soils which are exposed to a high risk of water erosion. Soil structure has a prevailing role in soil infiltration and biogeochemical processes. Therefore, improved soil structure means increased water retention, nutrient uptake, drainage, aeration and root growth.

An important feature of the proposed biotechnological remediation approach is related to improvement of soil biochemical characteristics. Soil enzymatic activities were improved in heavy metal contaminated soil amended with microbially-treated SB and RP when dehydrogenase, phosphatase and beta-glucosidase activities reached their highest values (Medina et al. 2005; 2006). The increase of enzymatic activities in soils is involved in an increase in the availability of nutrients to plants, which in turn have a positive influence on soil fertility (Garcia et al. 1997). Values of beta-glucosidase activity, that indicates carbohydrates transformation, showed that *A. niger*-treated SB combined with AM inoculation increased this hydrolytic activity which is important as energy source provider for rhizospheric microorganisms. Similarly, dehydrogenase and phosphatase activities (indexes of microbial activity and phosphorus mineralization, respectively) were registered at maximal level when *A. niger*-treated SB/RP and AM fungus were applied, which is an indication of their mutual effect on nutrient cycling and energy flow. In addition, the biotechnological product was found to increase indole-acetic acid (IAA) production in rhizosphere Cd-contaminated soil more than AM-colonization.

It is also interesting to note the enhanced antioxidant enzyme (ascorbate peroxidase, APX; glutathione reductase, GR; superoxide dismutase, SOD; and catalase, CAT) activities in heavy metal multi-contaminated soil-plant, experiments employing our biotechnological product combined with autochthonous microorganisms (arbuscular mycorrhizal fungi and bacteria) (Azcon et al. 2009b). As expected, *A. niger*-treated SB/RP mixture significantly increased micro- and macronutrients and reduced metal concentration, particularly Al and Ni, in plant shoot tissue. Moreover, enhanced antioxidant enzyme activity was observed in amended soil-plant systems. Degenerative reactions related to heavy metal stress are mediated by reactive oxygen species such as free radicals (superoxide and hydroxyl radicals) but also

H_2O_2 and singlet oxygen. Hydroxyl radicals and singlet oxygen are so reactive that their production must be minimized. On the other hand, Cd and H_2O_2 are synthesized at very high rates even under optimal conditions and may trigger toxic reactions leading to lipid peroxidation, protein denaturation and DNA mutations (Bowler et al. 1992). The ability of plants to increase antioxidative protection in order to combat the negative consequences of heavy metal stress appears to be limited as many studies have shown that exposure to elevated concentrations of redox reactive metals reduces rather than increases antioxidative enzyme activities (Schutzendubel and Polle 2002). However, results reported in this study, proved that our biotechnological product alone or in combination with the autochthonous microorganisms make a major contribution to plant antioxidant activities under the conditions of heavy metal stress. It is interesting to note, that the effect of solid-state microbially-treated agro-industrial wastes/simultaneously solubilized RP on antioxidant enzyme activities differed between the antioxidant enzymes: SOD and CAT activities were reduced by amendments while the opposite effect was observed for APX and GR activities.

Different levels of enzyme activity were observed depending on the waste material (DOW, SB, or both) used in the fermentation process and additional microbial inoculants (AM, *Y. lipolytica)* introduced into soil-plant systems (Medina et al. 2004b). For example, dehydrogenase activity increased in treated agrowaste-amended soil, particularly when microbially-treated SB was added, the lowest value was obtained when AM fungi was inoculated but in DOW-amended soil, non-significant responses to microbial treatments or dehydrogenase activity were found. Urease activity was highly increased in the rhizosphere of DOW-or SB-amended plants. The highest value was in the rhizosphere of mycorrhizal plants growing in DOW treatments. The application of treated agro-wastes enhanced protease activity and *Y. lipolytica* significantly increased such activity in DOW-amended soil.

Conclusions

In summary, although further research and field experiments over an extended time course must be carried out before their use within remediation strategies of contaminated areas could be considered a sustainable choice, our biotechnological approach is a feasible technology to reduce the negative effect of toxic anions and cations in contaminated soils. One of the most important lines of further research should be related to the type and chemical characteristics of the phosphate-bearing materials. The material of choice should be as free as possible of heavy metals as natural phosphate-bearing sources are considered by some authors as heavy-metal-contaminators (Sabiha-Javied et al., 2009; Mortvedt and Beaton, 1995). For example, sedimentary rocks contain high concentration of heavy elements (rare earth elements) (Hurst, 1989) etc. along with environment polluting elements: Cd, As, Sb, V, Cr, Zn, Cu, Ni, etc. The amounts of these hazardous elements vary widely not only among various phosphate rock sources but also even in the same deposit (Stocia et al., 1997). Therefore, it is extremely important to study the short- and long-term effect of whichever soil remediation method based on phosphate-bearing rocks monitoring the fate of target (contaminant) and introduced heavy metals in order to prevent additional release of toxic elements to the soil and soil-plant environment.

ACKNOWLEDGMENTS

This work was supported by Projects CTM2005-06955 and CTM2008-03524 (Ministerio de Ciencia e Innovación, España), a Project from the Junta de Andalucía (Proyecto de Exelencia-P09-RNM-5196), and EU COST FA0905

REFERENCES

Adriano DC, Wenzel WW, Vangronsveld J, Bolan NS (2004) Role of assisted natural remediation in environmental cleanup. *Geoderma* 122, 121–142.

Azcon R, Medina A, Roldan A, Biro B, Vivas A (2009a) Significance of treated agro- waste residue and autochthonous inoculates (arbuscular mycorrhizal fungi and *Bacillus cereus)* on bacterial community structure and phytoextraction to remediate soils contaminated with heavy metals. *Chemosphere* 75, 327-334

Azcon R, Peralvarez M, Biro B, Roldan A, Ruiz-Lonsano JM (2009b) Antioxidant activities and metal acquisition in mycorrhizal plants growing in a heavy metal multi-contaminates soil amended with treated lignocellulosic agrowaste. *Appl Soil Ecol* 41, 168-177

Cao R, Ma L, Chen M, Singh S and Harris W (2003) Phosphate-induced metal immobilization in a contaminated site. *Environ Poll,* 122 19-28.

Cao, X.D., Ma, L.Q., Rhue, D.R., Appel, C.S. (2004) Mechanisms of lead, copper, and zinc retention by phosphate rock. *Environ Pollut*, 131, 435–444.

Carrasco L, Caravaca F, Azcon R, Roldan A (2009) Soil acidity determines the effectiveness of an organic amendment and a native bacterium for increasing soil stabilization in semiarid mine tailings. *Chemosphere* 74, 239-244

Chen S, Xu M, Ma Y, Yang J (2007) Evaluation of different phosphate amendments on availability of metals in contaminates soil. *Ecotoxic Environ Safety* 67, 278-285

Gadd GM (2004) Microbial influence on metal mobility and application for bioremediation. *Geoderma* 122, 109-119

Gadd GM, Sayer JA (2000) Fungal transformations of metals and metalloids. In: Lovley DR (Ed.), Environmental-Metal Interactions. *Am Soc Microbiol*, Washington, pp. 237-256

Guo GL, Zhou QX, Ma LQ (2006) Availability and assessment of fixing additives for the in situ remediation of heavy metal contaminated soils: A review. *Environ Monit Assess,* 116, 513–528

Garcia C, Hernandez MT, Costa F (1997) Potential use of dehydrogenase activity as an index of microbial activity in degraded soils. *Commun Soil Sci Plant Anal*, 28, 123–134

Hamon RE, McLaughlin MJ, Cozens G (2002) Mechanisms of attenuation of metal availability in situ remediation treatments.*Environ Sci Technol*, 36, 3991–3996.

Hurst FJ (1989) *The recovery of uranium from phosphoric acid*, IAEA, Vienna, IAEATECDOC. 553.

Kumpiene J, Lagerkvist A, Maurice Ch (2008) Stabilization of As, Cr, Cu, Pb and Zn in soil using amendments – A review. *Waste Manag*, 28, 215-225.

Leita L, Nobili MD, Mondini C (1999) Influence of inorganic and organic fertilization on soil microbial biomass, metabolic quotient and heavy metal bioavailability. *Biol Fertil Soils* 28, 371-376

Ma QY, Traina SJ, Logan TJ (1993) In situ lead immobilization by apatite. *Environ Sci Tech,* 27, 1803–1810.

Martin WA, Larson SL, Felt DR,Wright J, Griggs CS, Thompson M, Conca JL, Nestler CC (2008) The effect of organics on lead sorption onto Apatite. *Appl Geochem*, 23, 34–43.

Mavropoulos, E., Rossi, A.M., Costa, A.M., Perez, C.A.C., Moreira, J.C., Saldanha, M. (2002) Studies on the mechanisms of lead immobilization by hydroxyapatite. *Environ Sci Technol*, 36, 1625–1629.

McGowen SL, Basta NT, Brown GO (2001) Use of di-ammonium phosphate to reduce heavy metal solubility and transport in smelter contaminated soil. *J Environ Qual*, 30, 493–500.

Medina A, Vassilev N, Aguacil MM, Roldan A, Azcon R (2004a) Increased plant growth, nutrient uptake, and soil enzymatic activities in a desertified Mediterranean soil amended with treated residues and inoculated with native mycorrhizal fungi and a plant growth-promoting yeast. *Soil Sci* 169, 260-270

Medina A, Vassileva M, Caravaca F, Roldan A, Azcon R (2004b) Improvement of soil characteristics and growth of *Dorycnium penthaphyllum* by amendment with agro wastes and inoculation with AM fungi and/or the yeast *Yarrowia lipolytica. Chemosphere* 56, 449-456

Medina A, Vassilev N, Barea JM, Azcon R (2005) Application of *Aspergillus-treated* agrowaste residue and *Glomus mosseae* for improving growth and nutrition of *Trifolium repens* in a Cd-contaminated soil. *J Biotechnol* 116, 369-378

Medina A, Vassileva M, Barea JM, Azcon R (2006) The growth-enhancement of clover by *Aspergiilus*-treated sugar beet waste and *Glomus mosseae* inoculation in Zn contaminated soil. *Appl Soil Ecol* 33, 87-98

Medina A, Vassilev N, Azcon R (2010) The interactive effect of an AM fungus and an organic amendment with regard to improving inoculum potential and the growth and nutrition of *Trifolium repens* in Cd-contaminated soils. *Appl Soil Ecol*, 44, 181-189.

Melamed R, Cao X, Chen M, Ma LQ (2003) Field assessment of lead immobilization in a contaminated soil after phosphate application. *Sci Total Environ*, 305, 1–3, 117–127.

Moon DH, Dermantas D, Menounou N (2000) Arsenic immobilization by calcium–arsenic precipitates in lime treated soil. *Sci Total Environ*, 330, 171–185.

Mortvedt JJ, Beaton JD (1995) Heavy metals and radionuclide contaminants in phosphate fertilizers, in: H. Tiessen (Ed.), *Phosphorus in the Global environment: Transfer cycles and Management,* John Wiley and Sons, New York, pp. 93–106.

Pascual JA, Garcia C, Hernandez T, Ayuso, M (1997) Changes in the microbial activity of an arid soil amended with urban organic wastes. *Biol Fertil Soils* 24, 429-434

Peng Jian-feng, Song Yong-hui, Yuan Peng, Cui Xiao-yu, Qiu Guang-lei (2009) The remediation of heavy metals contaminated sediment. *J Hazard Mat*, 161, 633–640.

Prasad, M., Amritphale, S.S., Saxena, S., Chandra, N., Rao, T.C. (2000) Attenuation of heavy metal ions by lean grade phosphorite. *Miner Eng*, 13, 1301–1305.

Prasad, M., Saxena, S., Amritphale, S.S. (2002) Adsorption models for sorption of lead and zinc on francolite mineral. *Ind Eng Chem Res*, 41, 105–111.

Ryan JA, Zhang P, Chou J and Sayers DE (2001) Formation of chloropyromorphite in a lead-contaminated soil amended with hydroxyapatite. *Environ Sci Technol*, 35, 3798-3803.

Sabiha-Javied, T. Mehmood, M.M. Chaudhry, M. Tufail, N. Irfan (2009) Heavy metal pollution from phosphate rock used for the production of fertilizer in Pakistan. *Microchem J*, 91, 94-99.

Schutzendubel A, Polle A (2002) Plant responses to abiotic stresses: heavy metal-induced oxidative stress and protection by mycorrhization. *J Exp Bot* 53: 1351-1365

Shrestha R., Fischer R., Rahner D. (2003) Behavior of cadmium, lead and zinc at the sediment–water interface by electrochemically initiated processes. Colloids Surf. *A: Physicochem. Eng. Aspects,* 222, 261–271.

Stocia L, Georgescu IL, Filip D, Bunus F (1997) Determination of valuable elements in natural phosphate. *J Radioanal Nucl Chem*, 216, 161–163.

Vassilev, N. and Vassileva, M. (2003) Biotechnological solubilization of rock phosphate on media containing agro-industrial wastes. *Appl Microbiol Biotechnol*, 61, 435-440.

Vassilev N, Vassileva M, Nikolaeva I (2006) Simultaneous P-solubilizing and biocontrol activity of microorganisms: Potentials and future trends. *Appl Microbiol Biotechnol* 71: 137-144

Vassilev, N., Fenice, M. and Federici, F. (1997) Olive mill waste water treatment by immobilized cells of *Aspergillus niger* and its enrichment with soluble phosphate. *Proc Biochem,* 32, 617-620.

Vassileva, M. and Vassilev, N. (1998) Rock phosphate solubilization by *Aspergillus niger* grown on olive cake and its further introduction into soil-plant system. *W J Microbiol Biotechnol*, 14, 281-285.

Zhang PC, Ryan JA (1999) Transformation of Pb(II) from cerrusite to chloropyromorphite in the presence of hydroxyapatite under varying conditions of pH. *Environ Sci Technol*, 33, 625–630.

Index

2

A

B

C

D

E

F

G

H

I

J

K

L

M

P

Q

R

S

T

U

V

W

X

Z